SCIENCE

A CLOSER LOOK

Macmillan/McGraw-Hill

Program Authors

Dr. Jay K. Hackett
Professor Emeritus of Earth Sciences
University of Northern Colorado
Greeley, CO

Dr. Richard H. Moyer
Professor of Science Education and
  Natural Sciences
University of Michigan–Dearborn
Dearborn, MI

Dr. JoAnne Vasquez
Elementary Science Education Consultant
NSTA Past President
Member, National Science Board
  and NASA Education Board

Mulugheta Teferi, M.A.
Principal, Gateway Middle School
Center of Math, Science, and Technology
St. Louis Public Schools
St. Louis, MO

Dinah Zike, M.Ed.
Dinah Might Adventures LP
San Antonio, TX

Kathryn LeRoy, M.S.
Chief Officer
Curriculum Services
Duval County Schools, FL

Dr. Dorothy J. T. Terman
Science Curriculum Development Consultant
Former K–12 Science and Mathematics Coordinator
Irvine Unified School District, CA
Irvine, CA

Dr. Gerald F. Wheeler
Executive Director
National Science Teachers Association

Bank Street College of Education
New York, NY

Contributing Authors

Dr. Sally Ride
Sally Ride Science
San Diego, CA

Lucille Villegas Barrera, M.Ed.
Elementary Science Supervisor
Houston Independent School District
Houston, TX

American Museum of Natural History
New York, NY

Contributing Writer

Ellen C. Grace, M.S.
Consultant
Albuquerque, NM

The McGraw·Hill Companies

Macmillan/McGraw-Hill

2014 impression

Copyright © 2011 by The McGraw-Hill Companies, Inc. All rights reserved. Except as permitted under the
United States Copyright Act, no part of this publication may be reproduced or distributed in any form or by
any means, or stored in a database or retrieval system, without prior permission of the publisher.

Send all inquiries to:
Macmillan/McGraw-Hill
8787 Orion Place
Columbus, OH 43240-4027

FOLDABLES is a registered trademark of The McGraw-Hill Companies, Inc.

ISBN: 978-0-02-288008-8
MHID: 0-02-288008-9

Printed in the United States of America.

11 12 13 DOR 17 16 15

Be a Scientist

◄ Making a model can help you understand how something works.

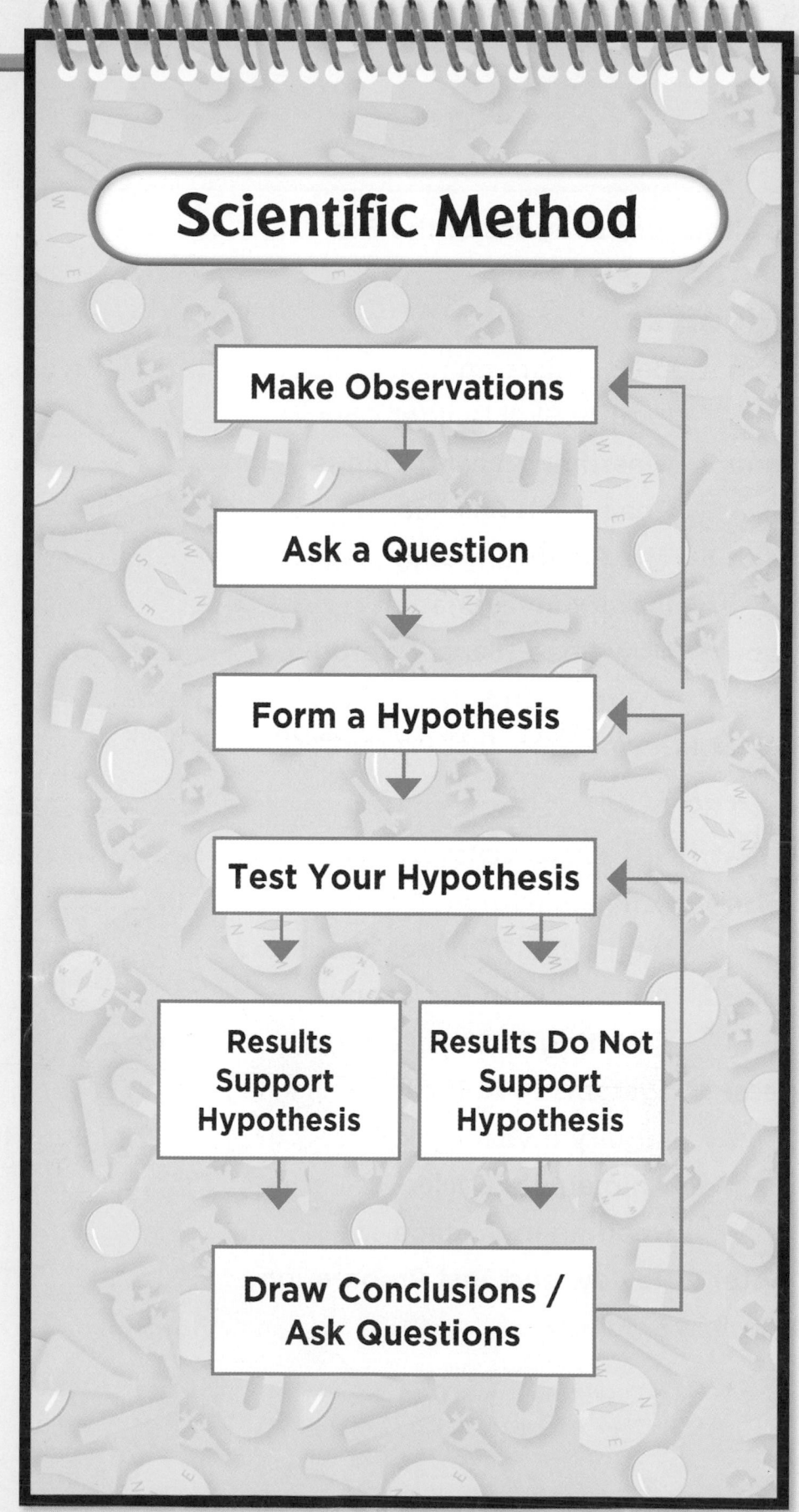

Scientific Method

Make Observations

↓

Ask a Question

↓

Form a Hypothesis

↓

Test Your Hypothesis

↓   ↓

Results Support Hypothesis   **Results Do Not Support Hypothesis**

↓   ↓

Draw Conclusions / Ask Questions

Life Science

UNIT A Living Things

UNIT B Ecosystems

Earth and Space Science

UNIT C Earth and Its Resources

UNIT D Weather and Space

Physical Science

UNIT E Matter

UNIT F Forces and Energy

Online Resources

Animations

 **Science in Motion** Animations online at **www.macmillanmh.com**

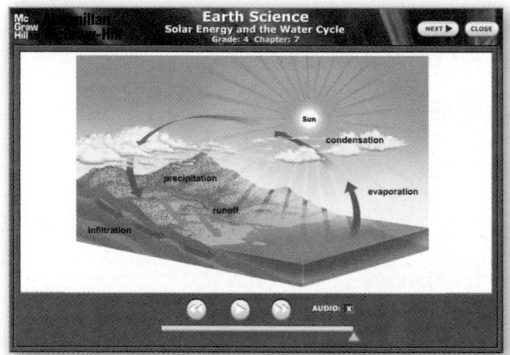

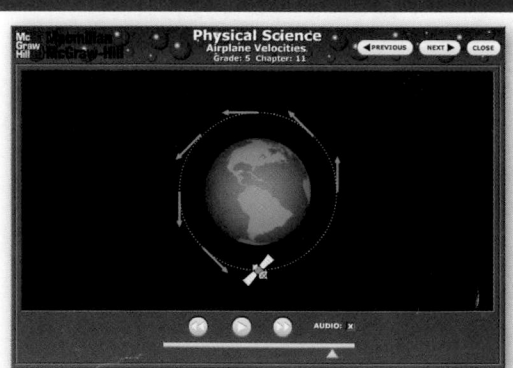

Additional Student Resources

Visit **www.macmillanmh.com** for additional student resources.

OSE Online Student Edition

See the book online.

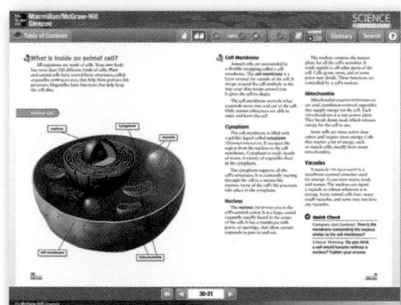

Vocabulary Games

Test your vocabulary knowledge.

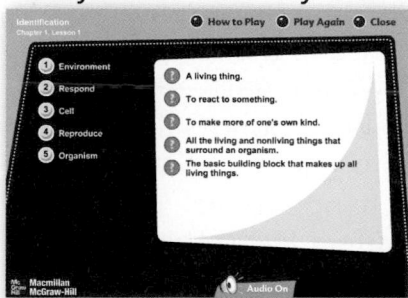

e-Career

Learn about science careers.

e-Journal

Discover and write about science.

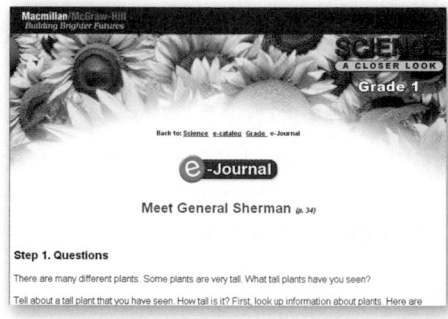

e-Glossary

Hear terms and review definitions.

e-Review

Watch an animated summary and take a quiz for each lesson.

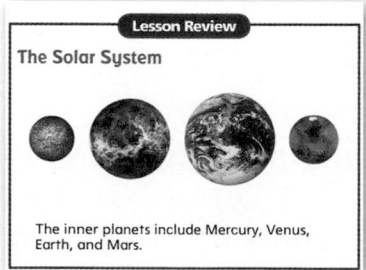

Other Online Resources

View images from NASA or see real-world connections at the American Museum of Natural History online.

Activities and Investigations

Life Science

xiv

Earth and Space Science

Activities and Investigations

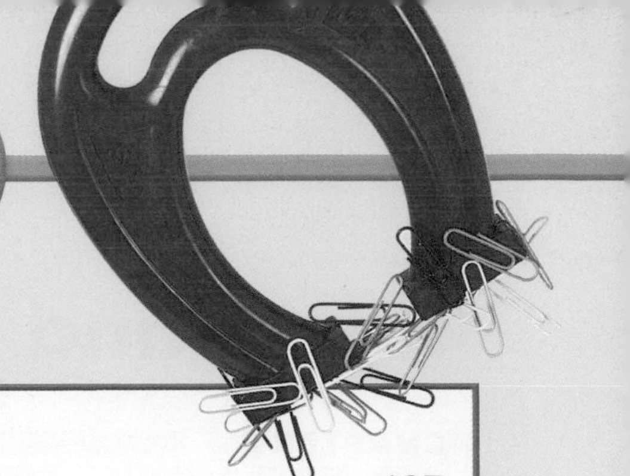

Physical Science

Be a Scientist

Mount Etna is the largest active volcano in Europe.

Mount Etna, Italy

The Scientific Method

Look and Wonder

The islands of Indonesia have many active volcanoes. What happens inside Earth that sends these clouds of ash and gas into the sky?

Francesca studies volcanoes in their natural settings.

Jim studies volcanoes in the laboratory.

Explore

What do you know about volcanoes?

- Why are some mountains volcanoes?
- What happens when a volcano erupts?
- Why do some volcanoes explode more violently than others?

How do scientists find answers to these questions?

Jim Webster and Francesca Sintoni are geologists (jee•OL•uh•jists). They work at the American Museum of Natural History in New York City. Geologists are scientists who study what goes on inside and outside Earth. Jim and Francesca are curious about volcanoes. They want to understand more about why volcanoes erupt.

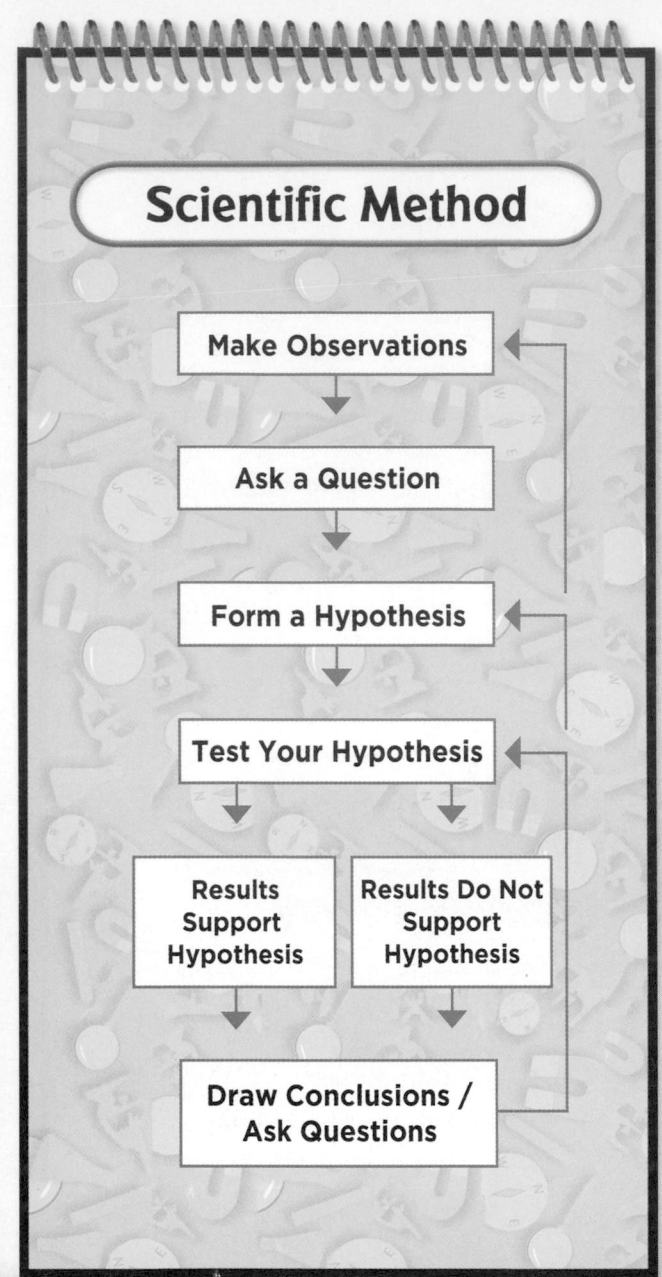

Scientific Method

Make Observations

↓

Ask a Question

↓

Form a Hypothesis

↓

Test Your Hypothesis

↓

| Results Support Hypothesis | Results Do Not Support Hypothesis |

↓

Draw Conclusions / Ask Questions

What do scientists do?

More than one million people live in the city of Naples, Italy. This city lies in the shadow of an active volcano named Mount Vesuvius (vuh•SEW•vee•uhs). It has erupted explosively many times during the past 2,000 years. "It's very dangerous," says Francesca, who lives in Italy. She studies Mount Vesuvius.

The Scientific Method

Francesca and Jim want to know what causes volcanoes like Mount Vesuvius to erupt. To find out, they use the scientific (si•uhn•TIF•ik) method. The **scientific method** is a process that scientists use to answer questions. This method helps them explain the natural world. The steps in the scientific method guide their investigations. Not every step needs to be followed in order every time.

From Naples you can see Mount Vesuvius. It last erupted in 1944.

Asking Questions

Volcanoes are filled with melted rock called *magma*. Magma is found deep inside Earth. Sometimes a gas is present in the magma. The gas may have water vapor, chlorine, or other substances in it.

When magma erupts from a volcano, lavas (LAH•vuhz) form. Many lavas are filled with small holes. These holes were once bubbles of gas in the hot magma.

Jim and Francesca ask why some volcanic eruptions are more explosive than others. They already know that water vapor affects how volcanoes erupt. Based on what they know, Jim and Francesca make a prediction. They predict that other substances will also affect volcanic eruptions. One variable (VAYR•ee•uh•buhl) they want to test is a substance called chlorine. A **variable** is something that changes, or varies.

Form a Hypothesis

1. Ask many "why" questions.

2. Look for connections between important variables.

3. Suggest possible explanations for those connections.

▶ **Make sure the explanations can be tested.**

Forming a Hypothesis

Jim and Francesca form a hypothesis (hi•PAH•thuh•sis). A **hypothesis** is a statement that can be tested to answer a question. Their hypothesis states that if magma has chlorine, then a volcano will have a larger explosion.

Francesca and Jim want to know why volcanoes erupt the way they do.

How do scientists test their hypotheses?

Can Jim and Francesca do research in an active volcano? No! Instead they use a laboratory, or lab for short. An instrument in Jim's lab models the heat and pressure deep inside a volcano. "We're trying to imitate the temperature and pressure inside Earth's crust," Jim explains.

Selecting a Strategy

To test their hypothesis, Jim and Francesca need to collect evidence. They decide to perform a set of experiments. An **experiment** is a scientific test that can be used to support or disprove a hypothesis. The pair design a set of experiments to test the effects of chlorine.

Planning a Procedure

Jim and Francesca write the steps of their procedure clearly. That way, they and others can repeat their experiments. Why? Good experiments are done again and again. If the results are similar, the evidence is stronger.

The plan is to add known amounts of chlorine to volcanic rock samples. Chlorine is the only variable they will change. The variable that changes in an experiment is the *independent variable*. Most experiments test only one independent variable at a time.

A good experiment also has *controlled variables* that are kept the same. Here, the scientists plan to control the mass, pressure, and temperature of each sample. How will they know if chlorine has any effect? They will count the number of holes in each rock. These holes are their dependent variable.

The holes in volcanic rock were once gas bubbles that formed inside magma.

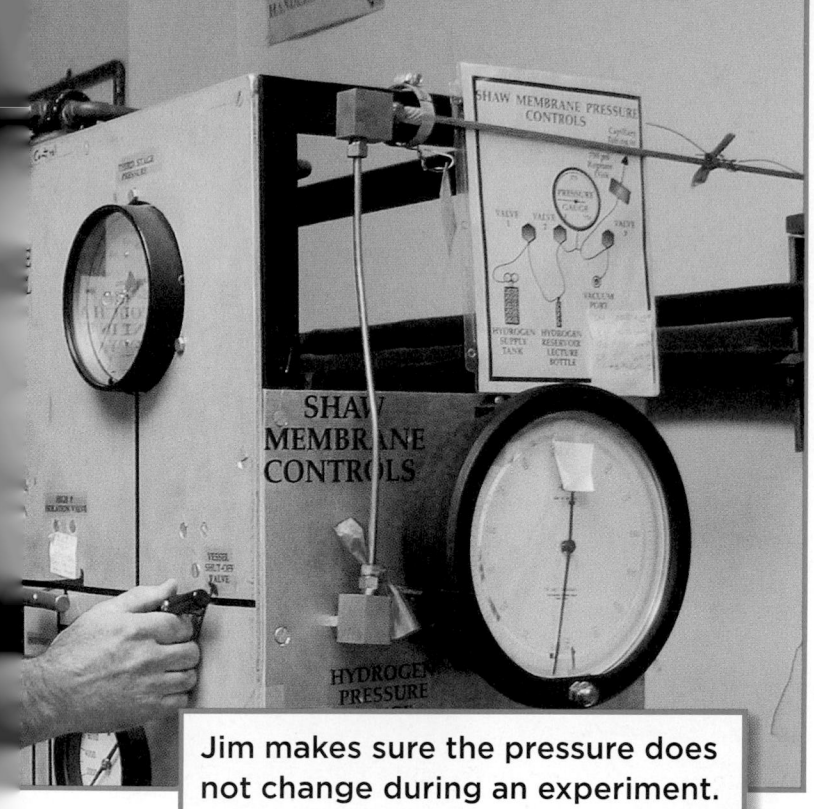

Jim makes sure the pressure does not change during an experiment.

Collecting Data

Jim and Francesca follow their plan. They pour crushed rock and water into tiny metal capsules. They add different amounts of chlorine. One capsule has no chlorine.

Francesca puts the sealed capsules inside a strong steel cylinder. Then Jim increases the pressure inside the cylinder. He also increases the temperature to about ten times hotter than a pizza oven!

After one week, it is time to cool the cylinder and open it. Jim and Francesca open the capsules. They observe the cooled rocks under a microscope. They count and record the number of holes. Later, they repeat the experiment exactly. They make sure the data are dependable.

Test Your Hypothesis

1. Think about the different kinds of evidence needed to test the hypothesis.

2. Choose the best strategy to collect this data.

   • **perform an experiment** (in the lab)
   • **observe the natural world** (in the field)
   • **make and use a model** (on a computer)

3. Plan a procedure and gather data.

▶ **Make sure the procedure can be repeated.**

The rock is crushed into small particles.

7

How do scientists analyze data?

When Jim and Francesca collect data, they keep careful records of their observations. They record how much chlorine went into each capsule. They carefully describe each tiny piece of cooled rock. They record the number of holes. Then they organize all this data in a way that makes sense.

A lab assistant looks at each sample with an electron microscope. ▼

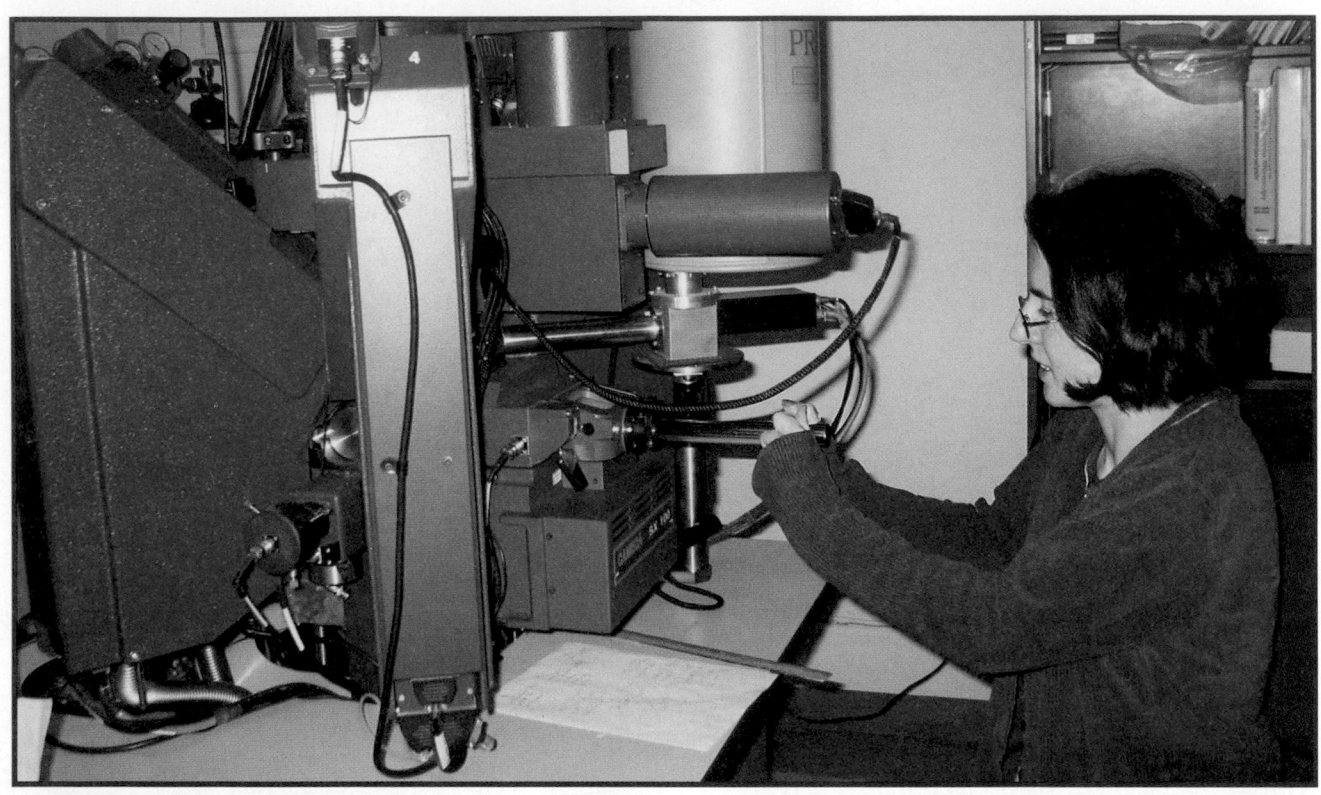

Comparing Samples

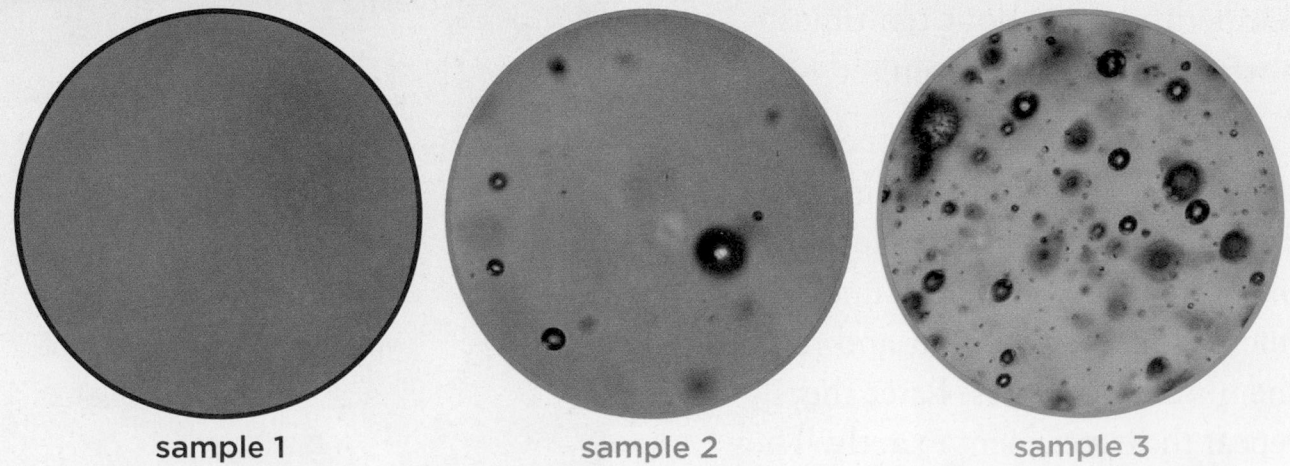

sample 1   sample 2   sample 3

Data Chart

Run	Temperature	Pressure	Chlorine	Bubble
1	920°C	**200 MPa**	0%	none
2	920°C	**200 MPa**	0.8%	some
3	920°C	**200 MPa**	0.9%	many

Looking for Patterns

The table above has some of the results from Jim and Francesca's study. In total, they ran about 50 experiments. Each one took about one week to complete. That means it took almost a year to collect their data!

After Jim and Francesca organize all their data, they look for patterns. What do their data show? When a sample has more chlorine, the cooled rock has more holes. The control sample, without chlorine, has no holes at all.

Checking for Errors

As they go along, Jim and Francesca review their procedures. They check that the experiments were run correctly. If they find any errors, they cannot use the data. Errors mean they must try again.

Analyzing the Data

1. Organize the data as a table, graph, diagram, map, or group of pictures.

2. Look for patterns in the data that show connections between important variables in the hypothesis being tested.

▶ Make sure to check the data by comparing it to data from other sources.

How do scientists draw conclusions?

Now Jim and Francesca must decide whether their data support their hypothesis. They compare their results with lavas from Mount Vesuvius and other explosive volcanoes. This comparison allows them to draw their conclusion.

Does more chlorine in the magma cause a bigger explosion? "Yes, it does!" Francesca exclaims.

The results of an experiment do not always support the tested hypothesis. This can be a useful outcome. When a hypothesis is not supported, scientists ask why. They may decide to test the hypothesis with new experiments using different methods.

Sometimes scientists conclude that a hypothesis is incorrect. When this happens, they often form a new hypothesis. Then they follow the steps of the scientific method once again.

Pumice is rock from an explosive volcano.

Mount Saint Augustine volcano, Alaska

Communicating

Jim and Francesca report their conclusions. This way, other scientists can do the same experiment and compare their results. Many scientists share their results so people can learn from their work.

Asking New Questions

A scientist's results may lead to new questions. Jim wants to know if chlorine affects eruptions at other volcanoes too. What other gases affect the size of eruptions? What else happens when a volcano erupts?

Today Jim studies Mount Saint Augustine volcano in Alaska. Like Mount Vesuvius, Mount Saint Augustine is an active volcano. It makes up its own island in Alaska's Cook Inlet.

Drawing Conclusions

1. Decide whether the data clearly support or do not support the hypothesis.

2. If the results are not clear, rethink the procedure.

3. Write the results to share with others.

▶ Make sure to ask new questions.

Think, Talk, and Write

1. Why is the scientific method useful to scientists?

2. What other questions about volcanoes can you think of? Choose one. Form a hypothesis that could be tested.

3. What could scientists do if their data disproved their hypothesis?

Jim visits Mount Saint Augustine with other scientists. Together, they make new observations.

Focus on Skills

Scientists use many skills as they apply the scientific method. Inquiry (IN•kwuh•ree) skills help you gather information and answer questions about the world around you. Here are some important inquiry skills that all scientists use:

▲ What observations can you make about the squirrel in this photograph?

Observe Use your senses to learn about an object or event.

Form a Hypothesis Make a statement that can be tested to answer a question.

Communicate Share information with others.

Classify Place things with similar properties into groups.

Use Numbers Order, count, add, subtract, multiply, or divide.

Make a Model Assemble something that represents an object, a system, or a process.

Scientists form a hypothesis before they begin an experiment.

A data table is a good way to organize information. ▶

Carton A	
Prediction	
Day	Observations
1	
2	
7	
10	

Use Variables Identify things that can control or change the outcome of an experiment.

Interpret Data Use information that has been gathered to answer questions, solve problems, or compare results.

Measure Find the size, distance, time, volume, area, mass, weight, or temperature of something.

Predict State a likely result of an event or experiment based on facts or observations.

Infer Form an idea or opinion from facts or observations.

Experiment Perform a test to support or disprove a hypothesis.

Inquiry Skill Builder

In each chapter of this book, you will find an Inquiry Skill Builder. These features will help you practice the skills that scientists use every day.

It is important for a scientist to use variables during an experiment. ▶

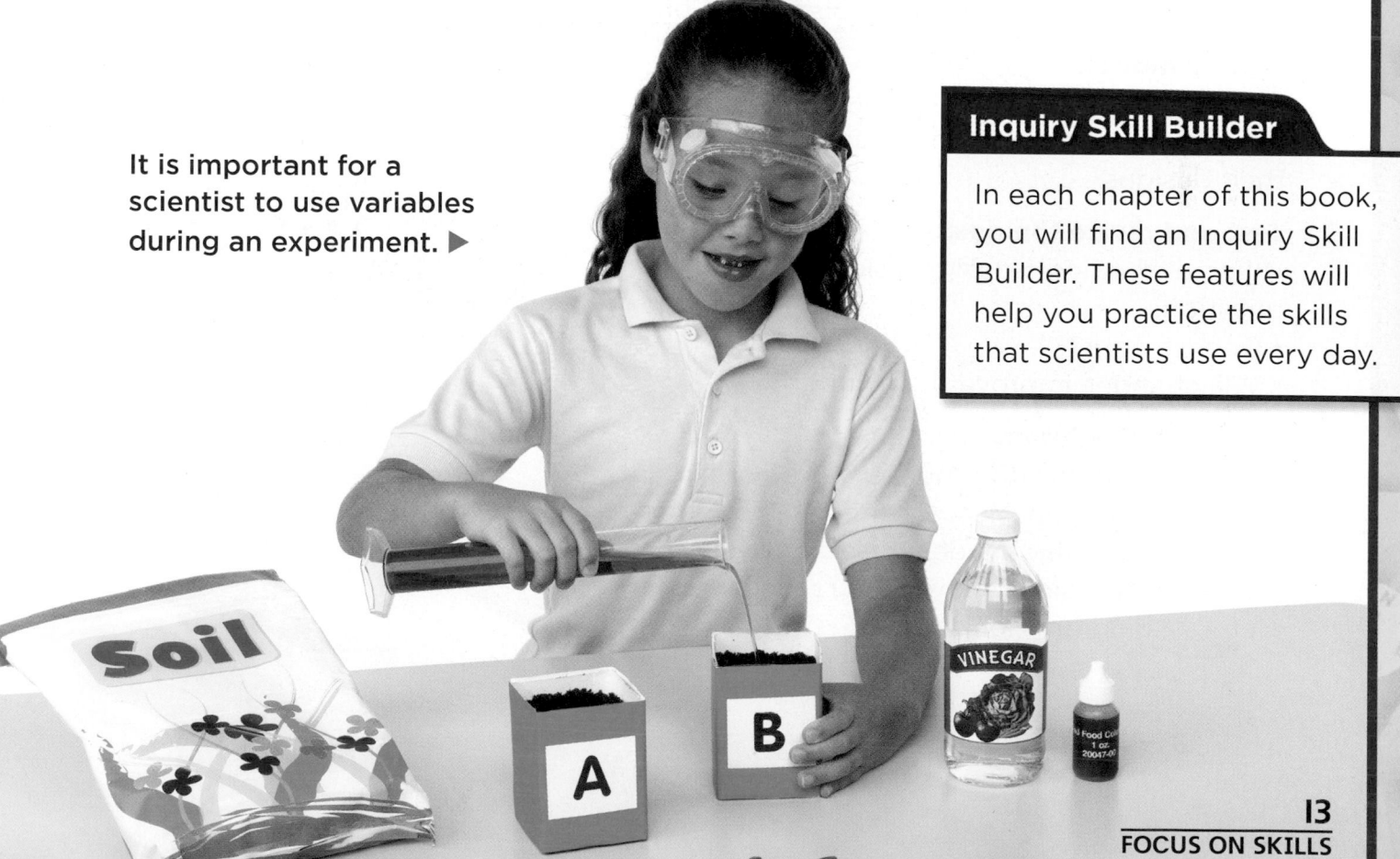

Soil

VINEGAR

A   B

Science and Technology:
The Design Process

When scientists have a problem, they must design a solution. Sometimes the scientist must invent a brand new solution. Other times they might change an existing design.

▶ Learn It

How could you design a bridge? Use the **design process** to help you design a solution.

1 **Identify and describe the problem.**
In order to solve the problem, you must understand it. How much distance must the bridge span? How much weight must it hold?

2 **Propose a solution.**
Your solution should include the information needed to solve the problem. Consider the materials needed and how much time you have to solve the problem.

3 **Build a model.**
A model is a small-scale or full-size replica of an object. Architects and engineers use models to test their designs.

4 **Test and revise the design.**
When evaluating a design, ask the following questions.
• Does it work?
• Will changes improve the solution?

5 **Explain the solution.**
Finally communicate how you solved or did not solve the problem. Most designs are not perfect when they are first made. Present your design in a group discussion or a written report. Include pictures, diagrams, or photos.

▶ Try It

Materials masking tape, soda straws, paper clips, construction paper, rubber bands, pennies, plastic cup

1. Use the **design process** to construct a bridge from common classroom materials. Build the bridge between two level desks or stacks of books. Place the books or desks 0.5 meters (about 1.5 feet) apart. The bridge must support a plastic cup containing 20 pennies.

2. Illustrate what the bridge will look like before beginning. Label your illustration and list the materials.

3. Build your design.

4. Test your design. *Does it hold the cup with 20 pennies?*

5. If the bridge fails the 20-penny test, change your design. Then retest the design.

6. Explain your solution to the class.

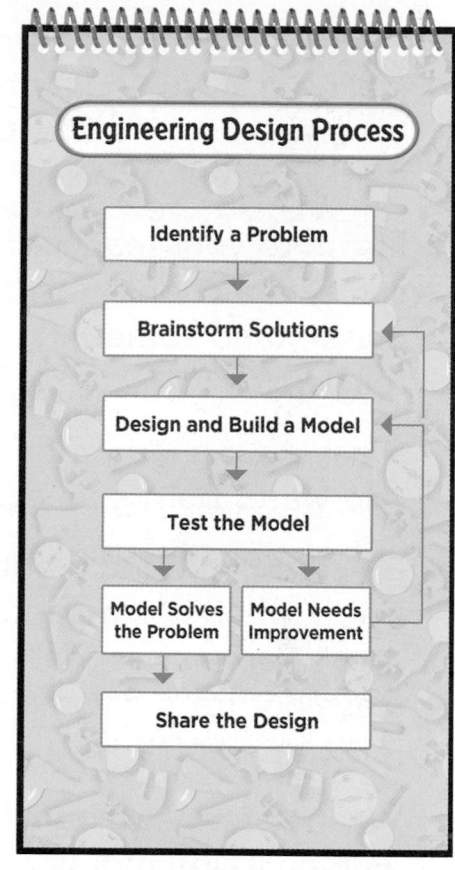

Communication is the last step in the design process.

▶ Apply It

1. How could you improve the design of your bridge to support a cup with 40 pennies?

2. Compare any problems you and others in your class might have faced. Were there any common problems?

3. If you were designing a real bridge, why would it be important to build a model?

4. How could you use the design process to solve a real-life problem?

5. Research bridges built long ago. How has bridge design changed? How do bridges improve human lives?

Safety Tips

In the Classroom

- Read all of the directions. Make sure you understand them. When you see "⚠ **Be Careful,**" follow the safety rules.

- Listen to your teacher for special safety directions. If you do not understand something, ask for help.

- Wash your hands with soap and water before an activity.

- Be careful around a hot plate. Know when it is on and when it is off. Remember that the plate stays hot for a few minutes after it is turned off.

- Wear a safety apron if you work with anything messy or anything that might spill.

- Clean up a spill right away, or ask your teacher for help.

- Dispose of things the way your teacher tells you to.

- Tell your teacher if something breaks. If glass breaks, do not clean it up yourself.

- Wear safety goggles when your teacher tells you to wear them. Wear them when working with anything that can fly into your eyes or when working with liquids.

- Keep your hair and clothes away from open flames. Tie back long hair, and roll up long sleeves.

- Keep your hands dry around electrical equipment.

- Do not eat or drink anything during an experiment.

- Put equipment back the way your teacher tells you to.

- Clean up your work area after an activity, and wash your hands with soap and water.

In the Field

- Go with a trusted adult—such as your teacher or a parent or guardian.

- Do not touch animals or plants without an adult's approval. The animal might bite. The plant might be poison ivy or another dangerous plant.

Responsibility
Treat living things, the environment, and one another with respect.

Living Things

Bees and flowers are
both made of cells.

Dragons of the Sea

Leafy seadragons look super cool! Let's call the one in the photo Lennie. He looks pretty awesome because he has a headdress with a matching outfit. Don't you think he's the best-dressed fish around? Seadragons can be in the ocean off southern Australia. Luckily, predators have a hard time finding seadragons because they look so much like something else in the ocean – seaweed! When they are ready to eat, they suck up tasty baby shrimp that happen to be drifting by. They do this by extending their long, tube-like snouts. Just one more thing that makes seadragons unique!

leafy seadragon

 **Write About It**

Response to Literature The name "leafy seadragon" sounds almost like a plant. Is a leafy seadragon a plant or an animal? How can you tell? Write an essay to compare and contrast plants and animals.

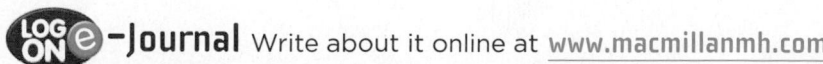

CHAPTER 1

Kingdoms of Life

 **The Big Idea** **What are living things and how are they classified?**

coral reef in Indo-Pacific Ocean, Indonesia

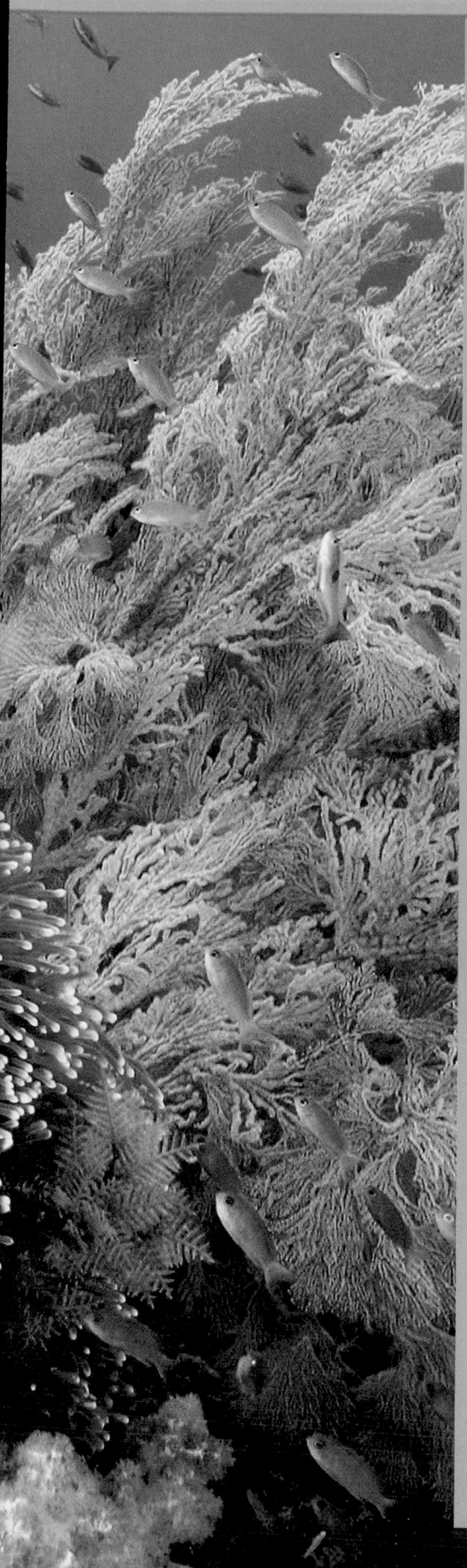

Big Idea Vocabulary

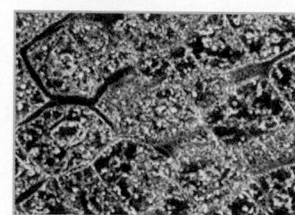

cell the smallest unit of living matter (p. 22)

organism a living thing that carries out five basic life functions (p. 22)

trait a characteristic of a living thing (p. 34)

kingdom the largest group into which organisms can be classified (p. 35)

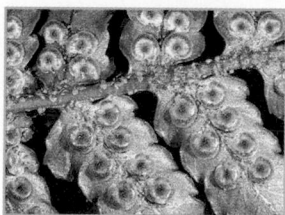

spore one of the cells in a seedless plant that grows into a new organism (p. 52)

seed an undeveloped plant with stored food sealed in a protective covering (p. 60)

 LOG ON Visit www.macmillanmh.com for online resources.

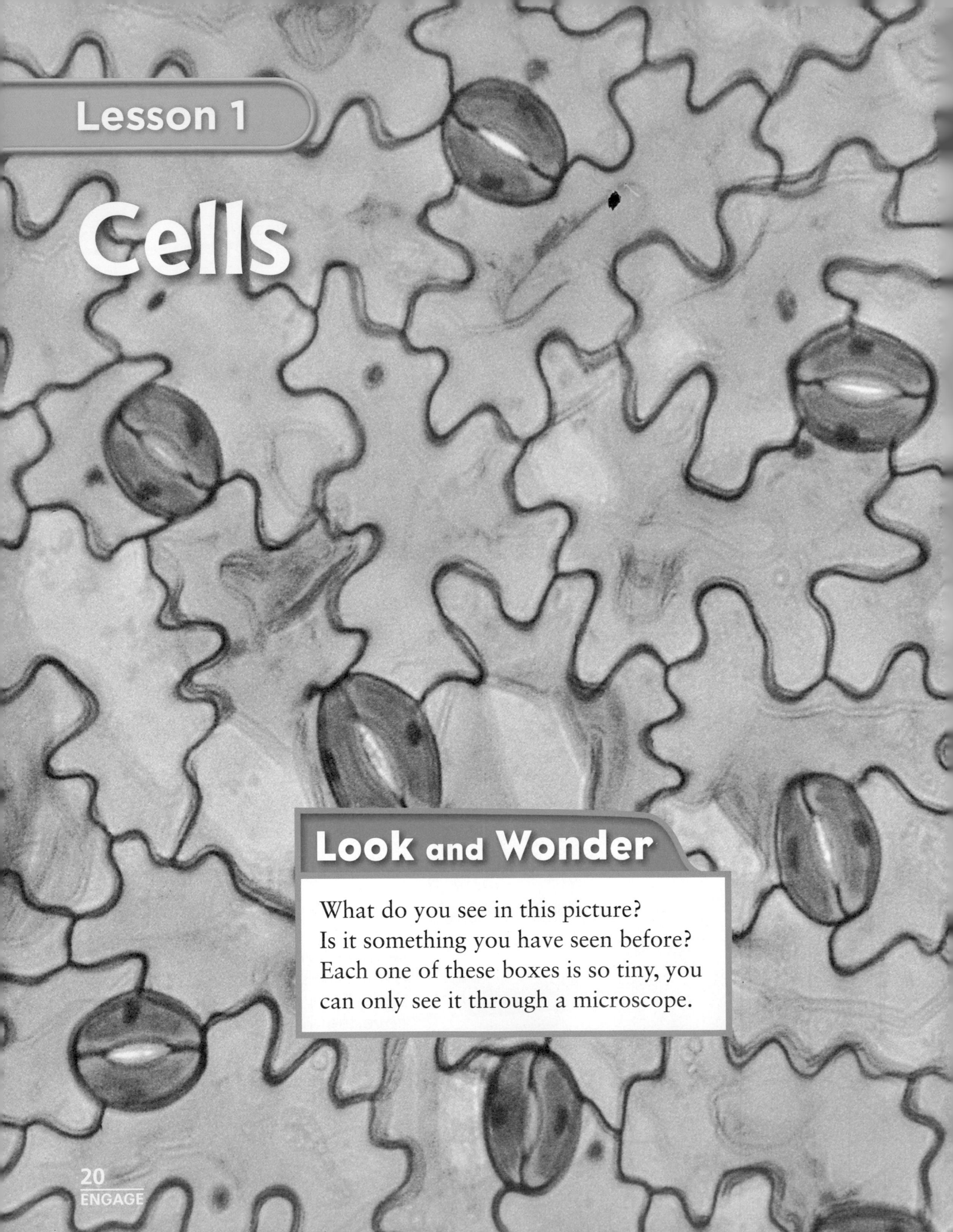

Cells

Look and Wonder

What do you see in this picture?
Is it something you have seen before?
Each one of these boxes is so tiny, you
can only see it through a microscope.

What are living things made of?

Purpose

Use different ways of observing to explore the parts of plants.

Procedure

1. **Infer** Draw an onion plant. Label its parts. How might each part help the plant live?

2. Ask your teacher to cut the plant lengthwise. Draw and label in a data table what you see.

3. **Observe** Look at the onion skin and the leaf with a hand lens. Draw and label in a data table what you see.

4. Ask your teacher for the prepared slides. One has an onion skin. The other has a leaf. Look at the slides under a microscope. Draw them in your data table. Use high and low power.

- onion
- leaf
- hand lens
- prepared slides of onion skin and leaf
- microscope

Draw Conclusions

5. **Communicate** How did your observations change as you looked more closely?

6. **Interpret Data** What do both the onion skin and the leaf seem to be made of?

Explore More

What might you see if you looked at the onion's roots? Make a plan to test your idea. Then try it!

Step 3

▶ Essential Question

How are living things organized?

▶ Vocabulary

cell, p. 22

oxygen, p. 22

organism, p. 22

tissue, p. 27

organ, p. 27

organ system, p. 27

▶ Reading Skill

Compare and Contrast

Different   Alike   Different

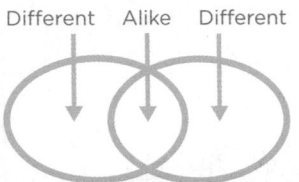

▶ Technology

e-Glossary and e-Review online at www.macmillanmh.com

What are living things?

You know that plants and animals are living things. How do you know? For one thing, plants and animals have cells (SELZ). A **cell** is the smallest unit of living matter. Ants and onion plants are made of cells. You are too!

Living Things Have Needs

A living thing may have millions of cells. It may have only one cell. In any case, all living things need water, food, and a place to live. Most living things also need **oxygen** (AHK•sih•jun)—a gas in air and water.

Living Things Reproduce

Scientists use the word *organism* in place of *living things*. An **organism** is a living thing that carries out five basic *life functions,* or jobs.

One life function is for organisms to make more of their own kind. The birds below are albatross. The chick is their offspring. *Offspring* is a term we use for the young of living things. To make more of one's own kind is to *reproduce.*

Living things reproduce.

Living things grow.

Other Life Functions

As a snake grows bigger, it sheds its skin. Not all organisms shed their skin. But they all grow and develop.

How do organisms get energy for growing? They use food! Woodchucks eat flowers. Plants make their own food. After they eat, organisms must get rid of wastes. Owl pellets show what food an owl ate.

Lastly, all organisms respond to changes in their environment. Why are all the sunflowers in the photo below facing the same way? Like all plants, they grow toward the light.

Is It a Living Thing?			
Life Function	Lizard	Rock	Car
Does it grow?	✔	✘	✘
Does it use fuel to get energy?	✔	✘	✔
Does it get rid of wastes?	✔	✘	✔
Does it reproduce?	✔	✘	✘
Does it respond to changes in its environment?	✔	✘	✘

Read a Table

How can you tell if a car is a living thing?

Clue: See if it performs all five life functions.

Quick Check

Compare and Contrast How are plants different from computers?

Critical Thinking What makes you a living thing?

Living things get rid of wastes.

Living things use food for energy.

Living things react to changes.

How do plant and animal cells compare?

All cells have smaller parts that help them stay alive. But all cells are not the same. Plants and animals share some of the same cell parts. Plant cells also have some things that animal cells do not have.

Plant Cells Have Chlorophyll

Most plant cells have green parts called *chloroplasts* (KLOR•uh•plasts). They are filled with a green substance called *chlorophyll* (KLOR•uh•fil). This substance helps plants make food using the Sun's energy. Animal cells do not have chloroplasts or chlorophyll.

Plant Cells Have Cell Walls

Plant cells also have sturdy *cell walls*. Cell walls give the cell a shape like a box. Animal cells have a cell membrane but not a cell wall. Animal cells tend to have a round shape.

① **cell wall**
This stiff structure protects and supports the plant cell.

② **mitochondrion**
(mi•tuh•KAHN•dree•un)
Food is burned here to give the cell energy.

③ **chloroplast**
The plant cell's food factory has chlorophyll.

④ **nucleus** (NEW•klee•us)
This controls all cell activities.

⑤ **chromosome**
(KROH•muh•sohm)
These control how the cell develops.

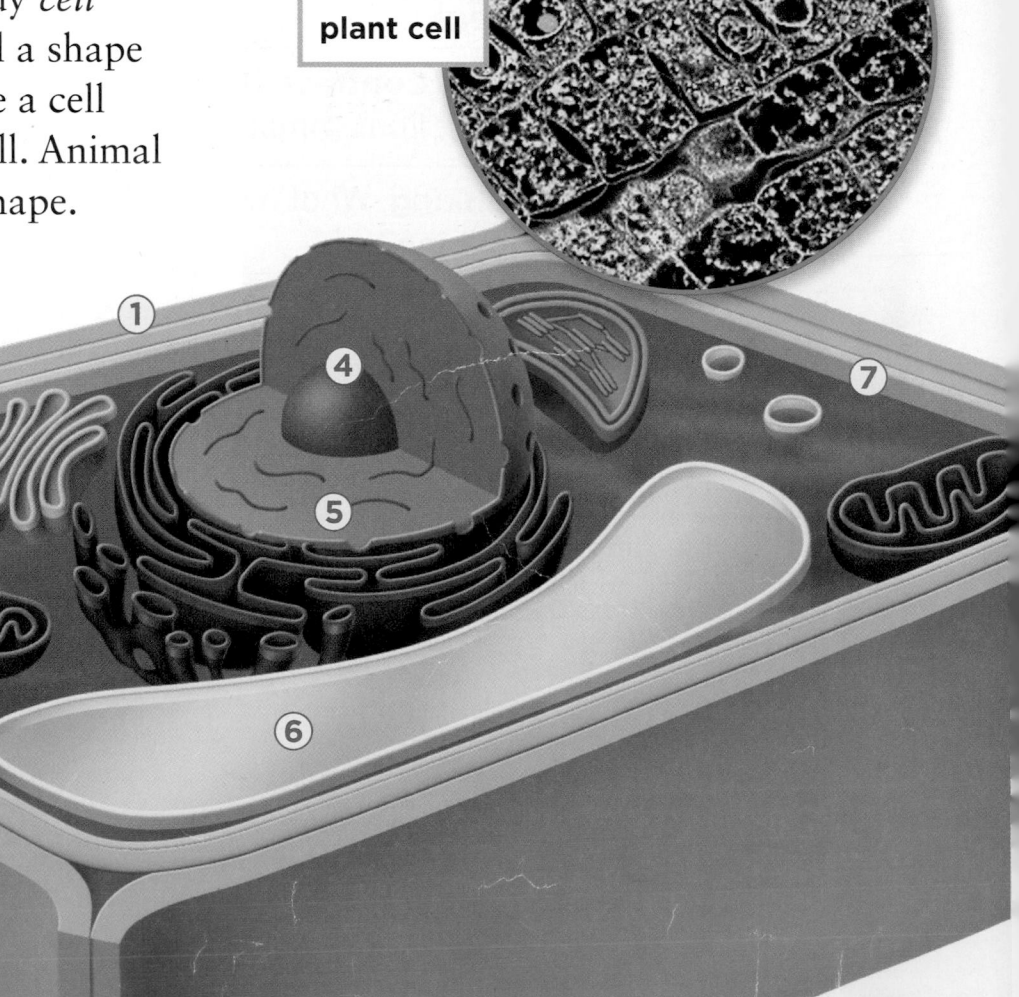

plant cell

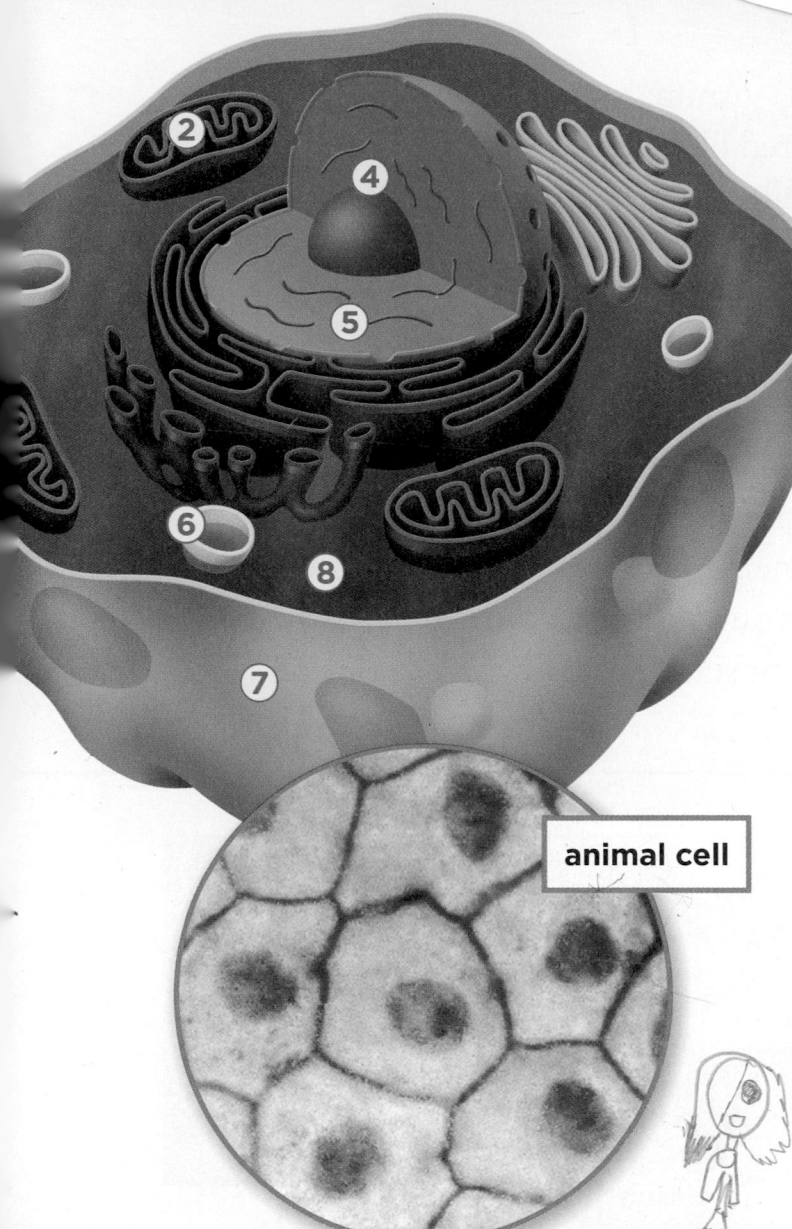

animal cell

	Plant Cells	Animal Cells

cell wall	✔	✘
mitochondria	✔	✔
chloroplasts	✔	✘
nucleus	✔	✔
chromosomes	✔	✔
vacuole	large	small
cell membrane	✔	✔
cytoplasm	✔	✔

Read a Table

How are a plant cell and an animal cell alike? How are they different?

Clue: Read down the list of cell parts. Compare the plant cell side by side with the animal cell.

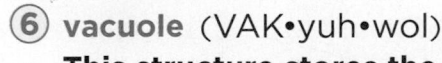

⑥ **vacuole** (VAK•yuh•wol)
This structure stores the cell's food, water, and wastes. Plant cells have one or two vacuoles. Animal cells have many.

⑦ **cell membrane**
This thin covering is found outside the cell. In plants, it is inside the cell wall.

⑧ **cytoplasm** (SI•tuh•pla•zum)
Filling the cell is a substance that is like jelly. It is mostly water. It also has important chemicals.

 Quick Check

Compare and Contrast How is a cell wall different from a cell membrane?

Critical Thinking Which cell part functions <u>most</u> like your brain? Explain your answer.

How are cells grouped?

What makes your heart different from your skin? The cells are different! When an organism has many cells, its cells tend to do different jobs.

For instance, many plants have root cells. Their job is to take in water and nutrients. Root cells do not make food, so they have no chloroplasts. Other cells in the plant make food.

Animals take in substances from red blood cells. Red blood cells look like soccer balls without any air inside. They have the important job of carrying oxygen and other matter through the body.

Nerve cells carry messages from one part of an animal's body to another. When you want to walk, nerve cells carry the message from your brain to your leg. Then your muscle cells help move your leg.

Levels of Organization

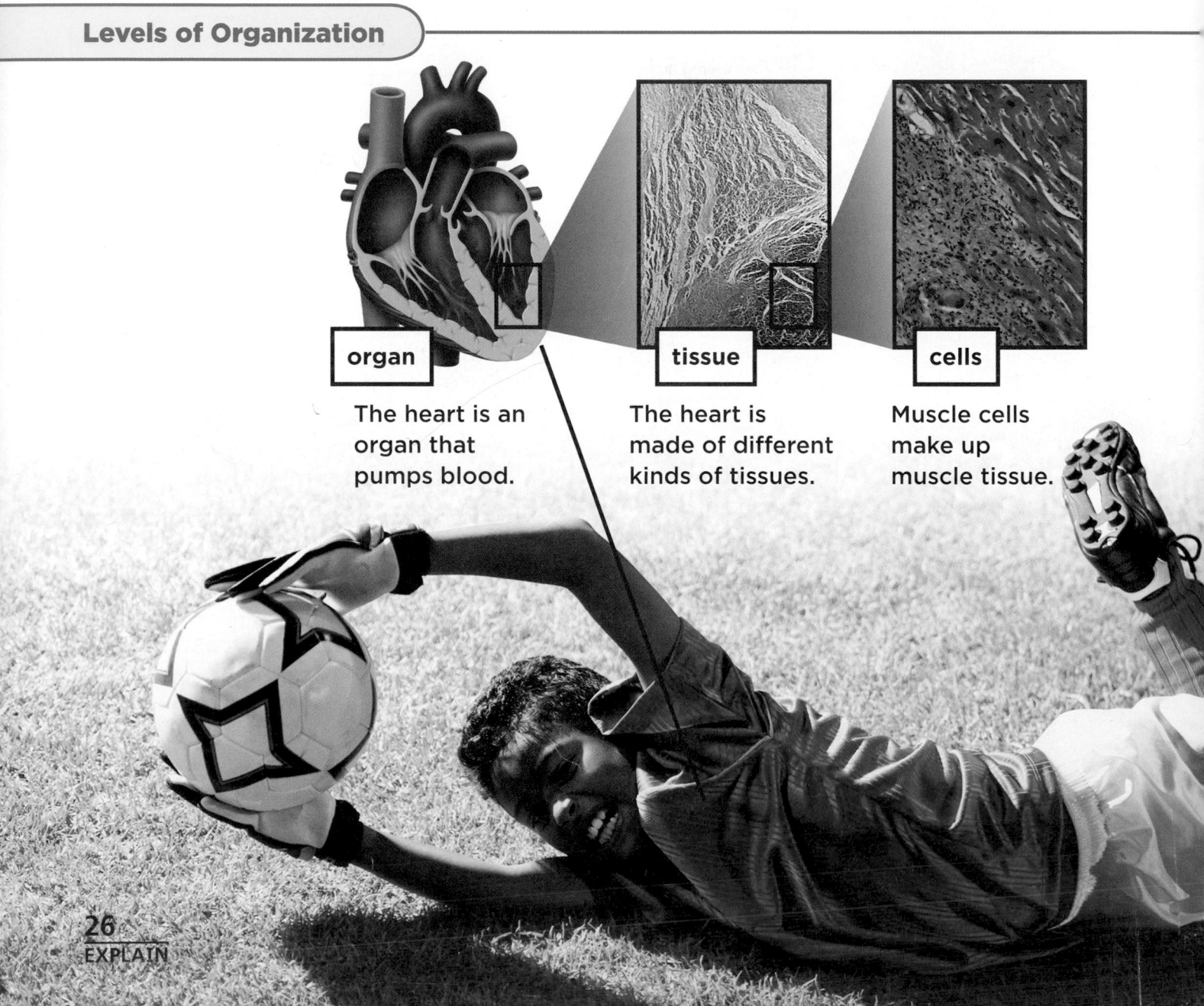

organ

The heart is an organ that pumps blood.

tissue

The heart is made of different kinds of tissues.

cells

Muscle cells make up muscle tissue.

Cells Make Up Tissues

In organisms with many cells, the ones that do the same job are grouped together. These cell groups form tissues (TIH•shewz). A **tissue** is a group of similar cells that work together to perform a job.

Tissues Make Up Organs

Tissues can also group together. When they do, they form an **organ**. The tissues in an organ work together to carry out a job. For example, your heart pumps blood.

Organs Make Up Organ Systems

Organs work together in an **organ system** to perform a life function. Your heart is part of the circulatory system. It moves blood throughout your body.

Quick Lab

Cells, Tissues, and Organs

1. One by one, call out a cell name—brain, nerve, or muscle—in that order. Write your cell name on a card.

2. **Make a Model** Pair up with a student who is the same cell type. For example, two nerve cells should pair up. This represents nerve tissue.

3. Model an organ by forming a larger group. For example, all "nerve" students form a group. "Brain" students should join this group.

4. Find a way to model an organ system.

✔ Quick Check

Compare and Contrast How is an organ different from a tissue?

Critical Thinking Why do different living things need different organs?

How can you see cells?

Some things are too small to see with your eyes alone. Most cells are that tiny. Bacteria (bak•TIR•ee•uh) are the smallest cells of all!

Microscopes

To see most cells, you need to use a microscope (MI•kruh•skohp). A microscope works like a hand lens. It makes small things look much bigger.

The microscopes scientists use are a lot more powerful than the ones you use. Some can make a cell look hundreds of thousands of times larger!

Microscopes are also used to study viruses. *Viruses* are even smaller than cells. Viruses cannot reproduce on their own. Instead, they force living cells to make new copies of the virus.

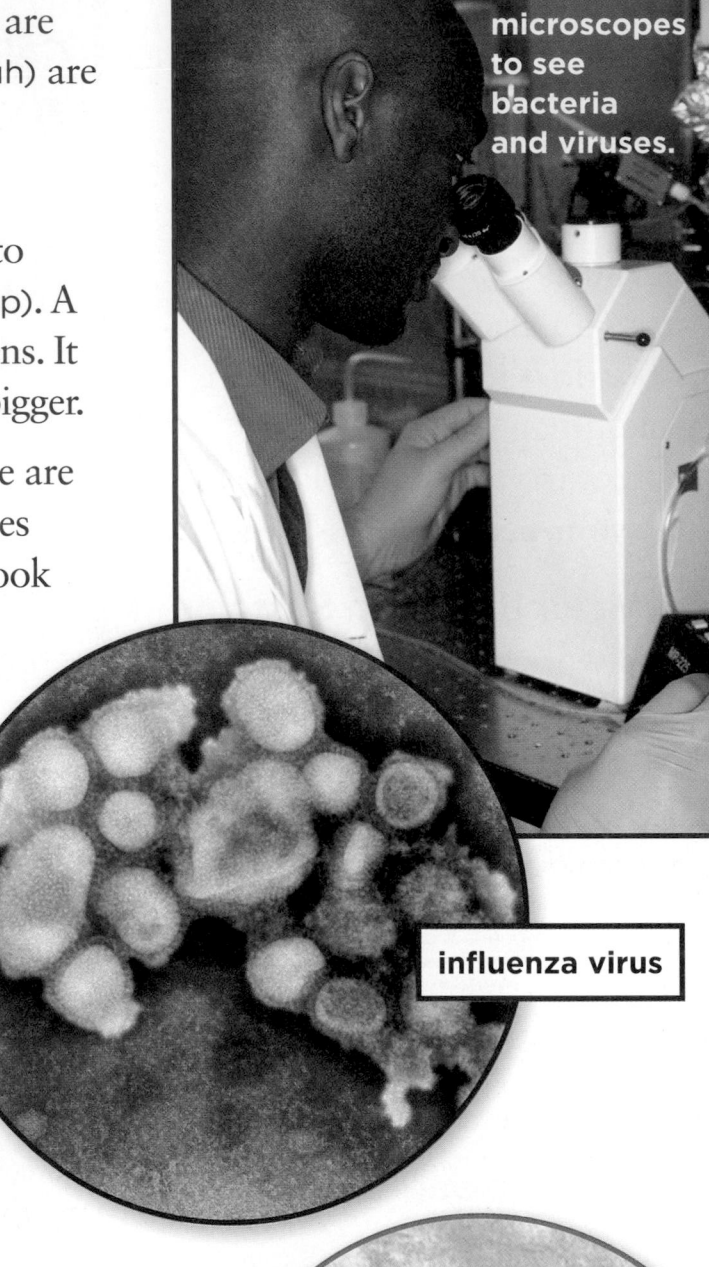

Scientists use microscopes to see bacteria and viruses.

influenza virus

 Quick Check

Compare and Contrast How is a hand lens like a microscope? How is it different?

Critical Thinking Is a virus an organism? Explain.

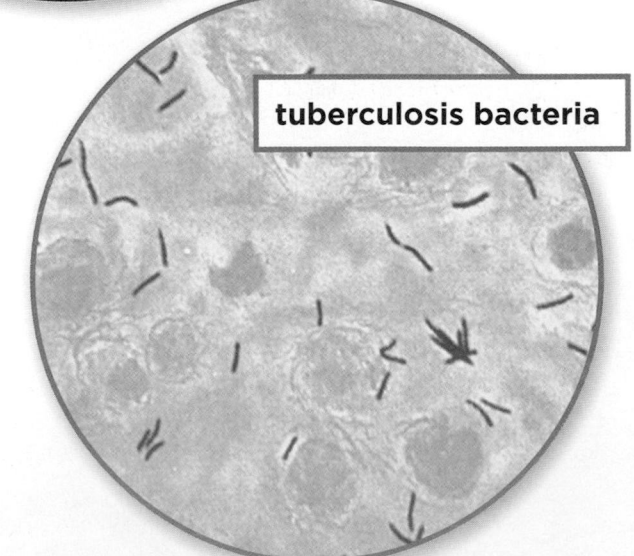

tuberculosis bacteria

Lesson Review

Visual Summary

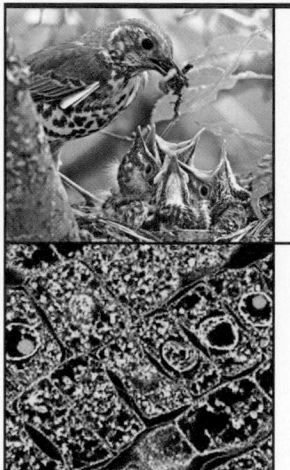

Living things are made of cells. Cells help organisms perform five basic life functions.

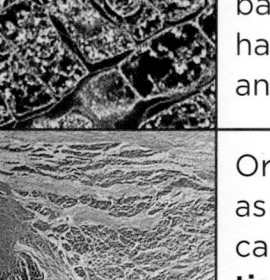

Plant cells and animal cells share several basic parts. Plant cells have some parts that animal cells do not.

Organisms can exist as single cells. Cells can be organized into **tissues, organs, and organ systems.**

Make a **FOLDABLES®** Study Guide

Make a trifold book. Use it to summarize what you read about living things.

Living things...	Plant cells and animal cells...	Tissues, organs, and organ systems...

Think, Talk, and Write

1 Vocabulary The _____ controls the activities of the cell.

2 Compare and Contrast How are plant and animal cells alike? How are they different?

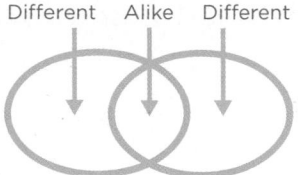

Different   Alike   Different

3 Critical Thinking Can one cell be a living thing? Explain why or why not.

4 Test Prep Which of these parts is only in plant cells?

A mitochondrion

B chloroplast

C cell membrane

D chromosome

5 Test Prep Most plant cells

A are shaped like boxes.

B have tiny vacuoles.

C do the same job.

D are round.

6 Essential Question How are living things organized?

Writing Link

Write a Story
Write a story that starts by looking at an organism from far away. Describe it as you move closer and closer, until the cells are visible. What would you see each time you get closer?

Math Link

Estimate
Professor Bubica sees 38 cells in her microscope. If she uses a less powerful microscope, she will see five times as many. How many cells will she see in the second microscope?

Focus on Skills

Inquiry Skill: Observe

You have read about organ systems. An organ system performs a job. Plants have a system of cells and tubes to move water from the ground into the plant's cells. How do scientists know this? They **observe** plants.

▶ Learn It

When you **observe**, you use one or more of your senses to learn about the world around you. Even though scientists know a lot about plants, they continue to observe them. Scientists are always learning new things about plants. They record their observations so they can share information with others. They use their observations to try to understand things in our world. You can too!

▶ Try It

In this activity you will **observe** how water moves through a plant. Remember to record your observations.

Materials **water, jar, blue food coloring, spoon, celery stalk, scissors**

1 Pour 100 milliliters of water into a jar. Add a few drops of blue food coloring into the jar. Stir the contents with a spoon.

2 Use scissors to cut about 3 centimeters off the bottom of a fresh celery stalk. Put the stalk in the jar of water. Record the time when you do this.

3 Observe the celery for 30 minutes. Record your observations. Use your observations to describe how water moves through a plant.

▶ **Apply It**

Now **observe** how water travels through other plants. Repeat your inquiry using a white flower, such as a carnation. Record your observations. Then share them with your classmates.

What I Did	What I Observed

Classifying Living Things

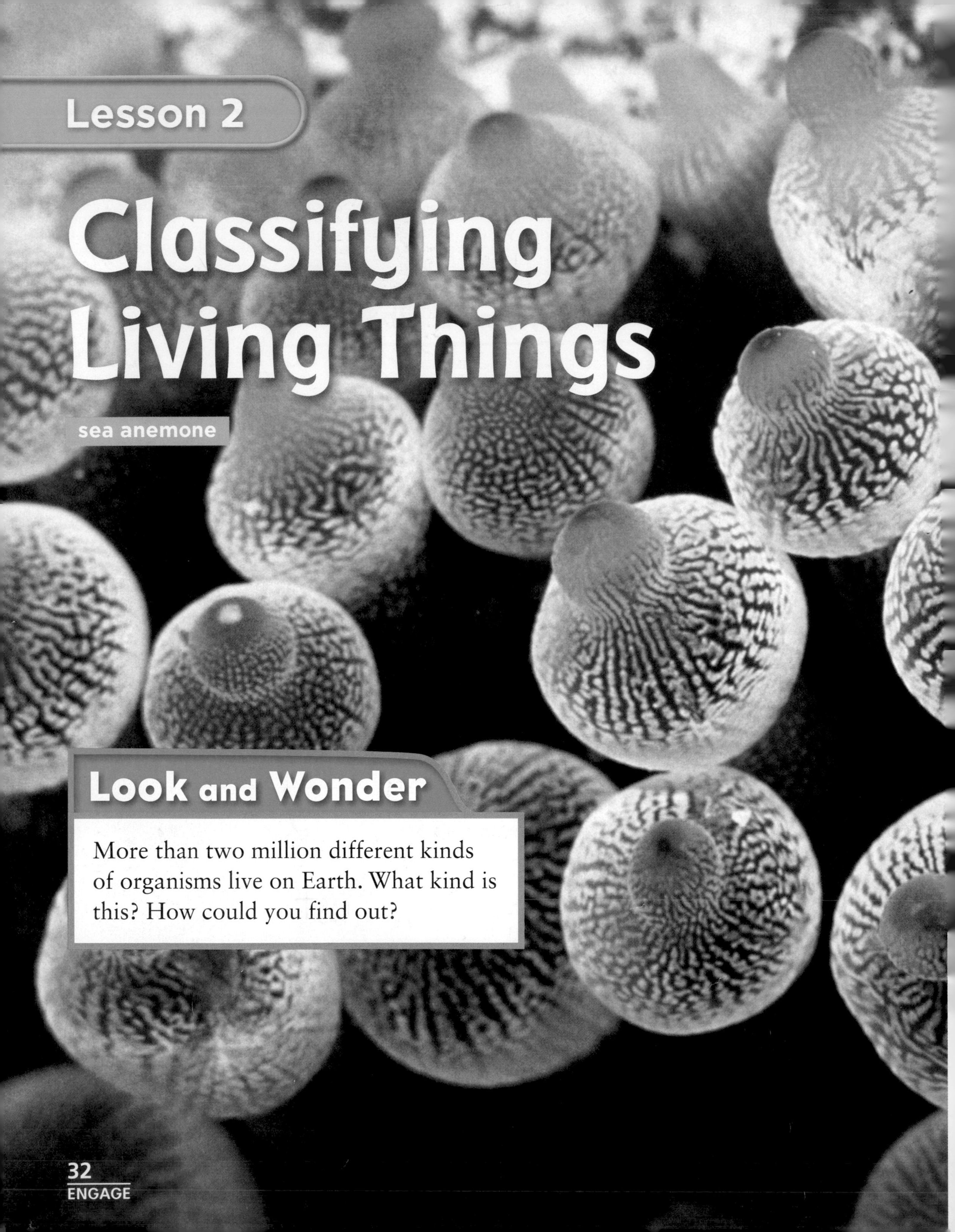

sea anemone

Look and Wonder

More than two million different kinds of organisms live on Earth. What kind is this? How could you find out?

How are organisms classified?

Purpose

Explore how to sort animals and plants into groups using different characteristics.

Procedure

1 Choose ten different animals and plants. You may pick organisms that you see in your neighborhood. You may also use some of the organisms shown here. Make a card for each organism you choose.

2 **Observe** How are the organisms alike? How are they different? Do the animals have wings, beaks, or tails? Do the plants have seeds or flowers? Make a data table. Record each organism's characteristics.

3 **Classify** Sort your cards into groups that have similar characteristics. This is one way scientists classify animals and plants.

Draw Conclusions

4 **Observe** What are the characteristics of the organisms in each group? Make a list.

5 **Predict** Will your classification work for other organisms? Think of other animals or plants that could be placed into each group.

Explore More

Find out how other students sorted their organisms. Are their groups the same as yours? Which characteristics did other students use? Compare them to the characteristics of your organisms.

Materials

- paper
- scissors
- colored markers

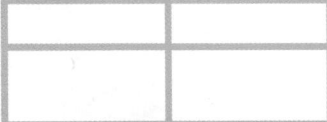
Mushrooms are fungi, not plants. Fungi do not make their own food.

How are living things classified?

Have you ever tried to sort laundry? How do you do it? You might not have known it, but sorting laundry is a way to classify items. When you *classify,* you place things that share properties together in groups.

To classify, you have to decide which characteristics to use to define the groups. For example, you can sort laundry by color. Living things can be organized into groups too.

Traits

To classify organisms into groups, scientists study many traits. A **trait** is a characteristic of a living thing.

Scientists look at body form and how an organism gets food. They observe how, or if, it moves. They also study the number of cells, whether the cell has a nucleus, and its cell parts.

For many years, scientists could not agree on a way to classify living things. People often used different names to describe the same organism. Over time, a system of classification came about.

Classifying Organisms

Kingdom	archaea	bacteria	protists	fungi	plants	animals
Number of Cells	one	one	one or many	one or many	many	many
Nucleus	no	no	yes	yes	yes	yes
Food	make their own or get food from other organisms	make their own or get food from other organisms	make their own or get food from other organisms	get food from other organisms	make their own food	get food from other organisms
Move from Place to Place	yes	yes	yes	no	no	yes

Six Kingdoms

Scientists divide living things into six kingdoms. A **kingdom** is the largest group into which organisms can be classified. All the members of a kingdom share the same basic traits.

Plants have their own kingdom. So do animals. There are two kingdoms of organisms with one cell and no nucleus. These organisms have many other traits that are different too. There is also a kingdom for *protists* (PROH•tists) and one for *fungi* (FUN•ji).

Read a Chart

How are archaea and bacteria different from the other four kingdoms?

Clue: Find the columns for both kingdoms. Compare the data side by side with other columns.

 **Quick Check**

Classify Into which kingdom would you classify an organism that has many cells, does not make its own food, and moves?

Critical Thinking Some bacteria make their own food. Why are they not classified as plants?

How are organisms grouped within a kingdom?

Squirrels and lizards belong to the animal kingdom even though they are very different. To further classify animals, scientists divide them into smaller groups.

The next group down is a *phylum* (FI•lum). Members of a phylum have at least one major trait in common, such as having a backbone.

A phylum is broken down into smaller groups called *classes*. Each class has even smaller groups of *orders*. Orders have *families*.

The chart shows these groups from largest to smallest. Each grouping has fewer and fewer members. The smaller the group, the more similar the organisms in it are to each other. The smallest are *genus* (JEE•nus) and *species* (SPEE•seez).

The eastern red squirrel is a member of the animal kingdom.

Kingdom

Members of the animal kingdom move and eat food.

Phylum

Members of this phylum share at least one major characteristic, such as having a backbone.

Class

Members of this class produce milk for their young.

Order

Members of this order have long and sharp front teeth.

Family

Members of this family have a bushy tail.

Genus

Members of this genus climb trees.

Species

A species is made up of only one type of organism.

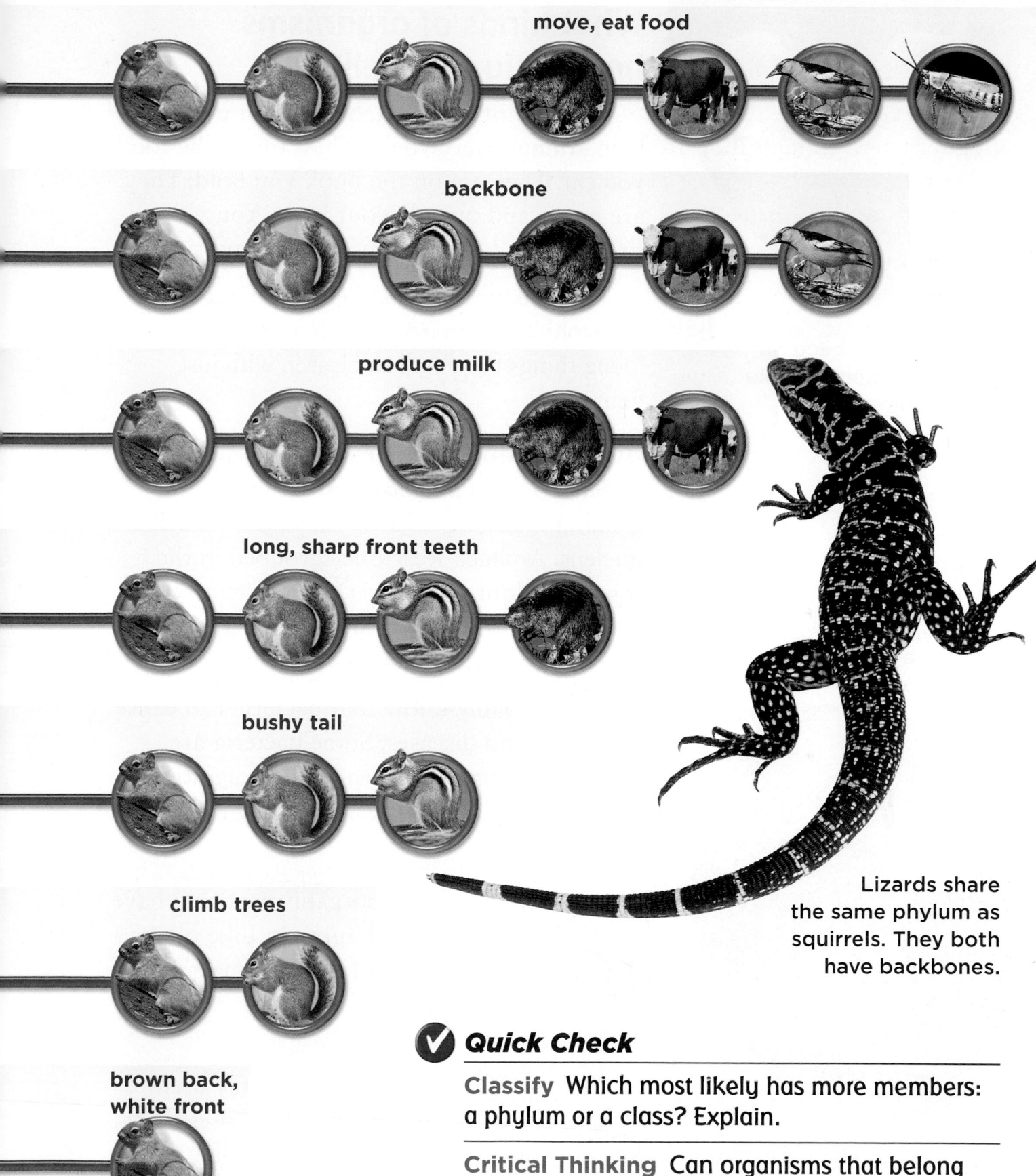

move, eat food

backbone

produce milk

long, sharp front teeth

bushy tail

climb trees

brown back,
white front

Lizards share
the same phylum as
squirrels. They both
have backbones.

✔ Quick Check

Classify Which most likely has more members:
a phylum or a class? Explain.

Critical Thinking Can organisms that belong
to different kingdoms be in the same phylum?
Explain why or why not.

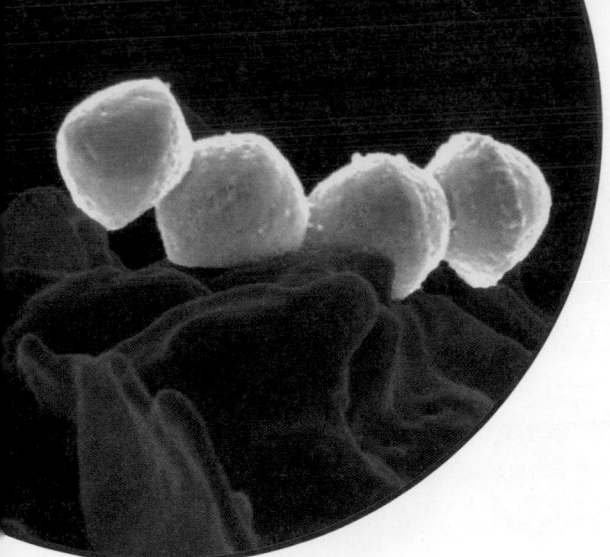

Streptococcus bacteria can cause infections like "strep" throat.

What kinds of organisms have only one cell?

You cannot see them, but there are tiny living things everywhere. They are on the food you eat. They are on the book you hold. They are inside and outside your body. You will find them in lakes, oceans, ponds, and rivers.

Tiny organisms are called microorganisms (MI•kroh•OR•guh•ni•zumz). *Microorganisms* are living things too small to be seen with just your eyes.

Bacteria and Archaea

Bacteria and *archaea* (ar•KEE•uh) are the smallest microorganisms. They have no cell nucleus. Archaea were once grouped in the same kingdom as bacteria. Some bacteria break down dead plant and animal matter for food. Other bacteria make their own food.

You probably know that bacteria can cause infections and diseases. Some bacteria are helpful. You have bacteria in your digestive system. They help break down your food.

Fungi

Some *fungi* are microorganisms. Fungi have traits of both plants and animals. Like plants, their cells have a cell wall. Like animals, fungi cells do not have chloroplasts. They cannot make their own food.

Yeast is a fungus that is commonly used to make bread. It makes the dough rise. Yeast has only one cell. Some fungi have many cells. Fungi cells have a cell nucleus.

Yeast is a kind of fungus.

FACT Not all bacteria cause disease.

Protists

Members of the protist kingdom also have a cell nucleus. Protist cells have different parts that perform different jobs. A *paramecium* (per•uh•MEE•see•um) has a structure that pumps out extra water from inside the cell.

Some protists, such as algae (AL•jee), make their own food. Others get food by eating other organisms.

Like bacteria and fungi, most protists are harmless. Many are even helpful. Protists are a food source for other organisms. However, some protists can cause serious diseases, such as malaria.

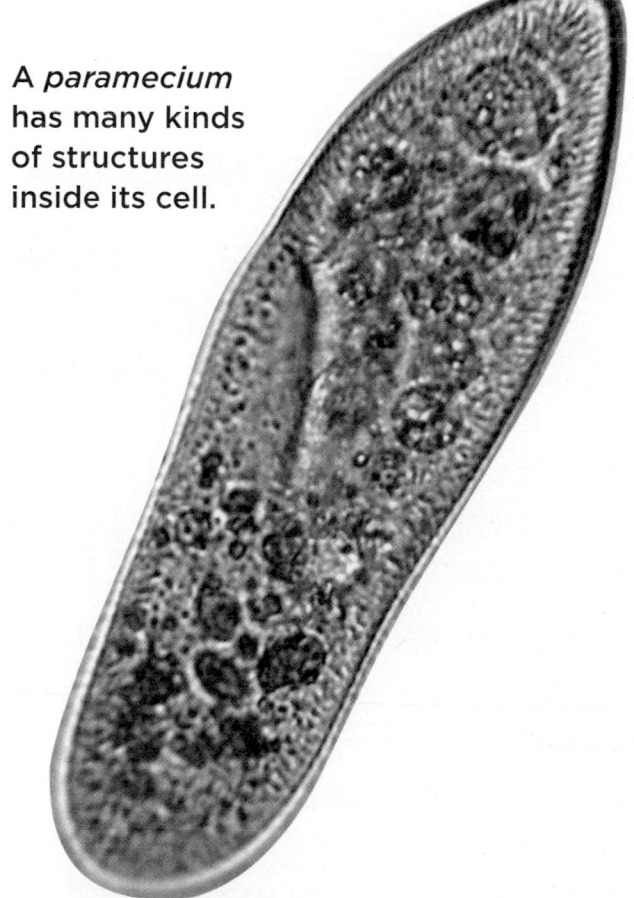

A *paramecium* has many kinds of structures inside its cell.

≡ Quick Lab

Observe a One-Celled Organism

1. **Observe** Using a microscope, look carefully at the organism on your slide.

2. **Classify** Is your organism made of one cell or many cells? How can you tell?

3. The microscope you are using is not strong enough to view individual bacteria. What kind of organisms are you observing?

✔ Quick Check

Classify How can you tell the difference between protists and bacteria?

Critical Thinking How can observing cells under a microscope be useful when identifying organisms?

How are organisms named?

Scientists use a naming system to classify living things. Each kind of organism has its own name. The first part of the organism's name is its genus. The second part is its species. By using these names, scientists can identify and study specific organisms.

Scientists have named about 1.7 million species on Earth. Countless more have yet to be named!

Genus and Species

Wolves and coyotes belong in the genus *Canis* (KAY•nus). Dogs belong in this genus too. Members of the genus *Canis* look similar. They all eat meat. However, the species in this genus have different traits. One trait is color. Red wolves are *Canis rufus*. Gray wolves are *Canis lupus*. Coyotes are *Canis latrans*.

 Quick Check

Classify How do scientists use names to classify organisms?

Critical Thinking How would a scientist name an organism that has just been discovered?

Naming Organisms

Genus *Canis*

gray wolf
(Canis lupus)

coyote
(Canis latrans)

Read a Diagram

Does the term *lupus* refer to a genus or a species?

Clue: A species is the smallest classification group.

Lesson Review

Visual Summary

Organisms **can be grouped** by kingdom, phylum, class, order, family, genus, and species.

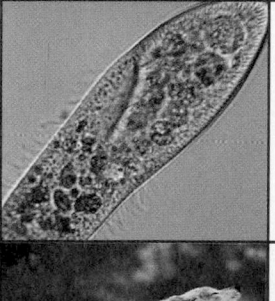

Bacteria, protists, and fungi belong to kingdoms that include **single-celled organisms.**

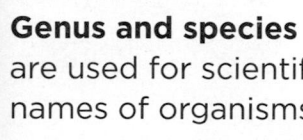

Genus and species are used for scientific names of organisms.

Make a FOLDABLES Study Guide

Make a four-tab book. Use it to summarize what you learned about classifying living things.

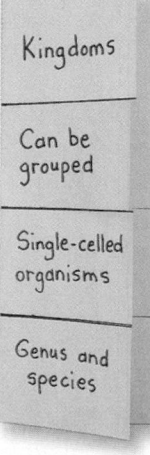

Kingdoms

Can be grouped

Single-celled organisms

Genus and species

Think, Talk, and Write

1 **Vocabulary** Plants, animals, fungi, and _____ are the four kingdoms that have organisms with many cells.

2 **Classify** Many birds eat the seeds of the rose plant, *Rosa rugosa.* What is this plant's genus and species?

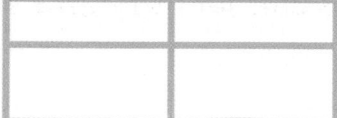

3 **Critical Thinking** How can classification of a poisonous organism help save someone's life?

4 **Test Prep** Which statement about the number of species is true?

   A Kingdoms contain the most.
   B A phylum contains the most.
   C Orders have the fewest.
   D Kingdoms have the fewest.

5 **Test Prep** All the organisms in this kingdom make their own food.

   A fungi
   B protists
   C bacteria
   D plants

6 **Essential Question** How can living things be grouped?

 **Writing Link**

Write an Essay
Think about the common traits of cats. Write an essay that tells how a cat is different from a dog.

 **Math Link**

Solve a Problem
A family of plants contains four different *genera* (the plural of genus). Each genus has three species. How many plant species are in this family?

Red Tide
A Bad Bloom at the Beach

You're ready for some fun in the sun. But when you get to the beach, it's closed. Then you notice that the water is a strange color. You can put your swimsuit away. Your beach is a victim of red tide!

Red tide isn't actually a tide. It is ocean water that is blooming with a harmful kind of algae. These one-celled organisms are poisonous to the sea creatures that eat them. The water isn't always red, either. Sometimes it's orange, brown, or green.

a bloom of red tide

Connect to

AMERICAN
MUSEUM ö
NATURAL
HISTORY

at www.macmillanmh.com

An outbreak of red tide can do a lot of damage. On the coast of Florida, one killed tens of thousands of fish, crabs, birds, and other small animals within a few months. It also killed large animals like manatees, dolphins, and sea turtles. Red tides can also make people sick if they eat infected shellfish.

Scientists are working to predict where and when red tides occur. They measure the amount of algae along coastlines. They use data collected from satellites to study wind speed and direction. This information helps scientists predict where blooms may develop. With their predictions, scientists help warn local agencies about future red tides.

Infer

▶ Use information that you already know.

▶ List the details in the text that support your inference.

Write About It

Infer

1. What could you infer about a closed beach with reddish-colored water?

2. How could the prediction of red tides be helpful to people?

 **e-Journal** Research and write about it online at www.macmillanmh.com

The Plant Kingdom

Look and Wonder

Have you ever wondered where food comes from? You might say, "The supermarket!" But where does food really come from? The story begins with the Sun and leaves. What do leaves have to do with making food?

How are leaves different from each other?

Make a Prediction

How do leaves from different plants differ from each other? Make a prediction.

Test Your Prediction

1 Observe Use the hand lens to observe both leaves carefully. What do you notice?

2 Communicate Record your observations in a chart like the one shown. How are the leaves different?

Leaf Trait	Leaf A	Leaf B
Texture		
Color		
Size		
Shape		

Draw Conclusions

3 Infer Tell what each leaf trait on the chart is for. For example, you might infer that fuzzy leaves are for catching rain. Colored leaves might be for attracting insects. Record your ideas.

Explore More

Which leaf traits do both leaves have in common? Tell what each shared trait is for. Make a plan to test your idea.

Materials

- leaves from two plants
- hand lens

Step 1

Read and Learn

▶ **Essential Question**

What are plants?

▶ **Vocabulary**

root, p. 49

root hair, p. 49

stem, p. 49

photosynthesis, p. 50

stomata, p. 51

transpiration, p. 51

respiration, p. 51

spore, p. 52

▶ **Reading Skill** ✓

Infer

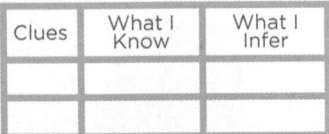

Clues	What I Know	What I Infer

▶ **Technology** 🔵LOG ON🔺

e-Glossary, e-Review, and animations online at www.macmillanmh.com

How do we classify plants?

Plants come in all sizes, shapes, and colors. Some are so small you can barely see them. Others can be as tall as skyscrapers. In all, there are about 400,000 different kinds of plants.

Classifying by Structures

One way to classify plants is by their parts, or structures. Scientists look at the shapes of leaves, the kinds of stems, and the shapes of roots. Some plants do not have these structures. We can use this fact to sort plants into two groups. One group has plants with roots, stems, and leaves. The other group has no roots, stems, or leaves.

All plants need to move water from the ground into their cells. Plants with roots and stems have a system of tubes for this. Mosses and other plants that lack such structures grow close to the ground. They do not need a tube system. They can take in water directly from the soil.

Wort plants do not have roots, stems, or leaves.

Classifying by Seeds

What do you find when you bite into an apple? At the apple's core are seeds. If you plant these seeds, they grow into apple trees. Then those trees make more apples and more seeds.

We can classify plants by whether they have seeds. Most plants that you are familiar with have seeds. In fact, most of the plants that have roots, stems, and leaves also have seeds and fruit.

Horsetail plants grow about one meter tall. They have stems that look like horses' tails. They also have roots and leaves. But horsetails have no seeds or fruit. How do they reproduce? Horsetails produce offspring from spores. Other plants with roots, stems, and leaves use spores as well.

The violet plant has roots, stems, and leaves. Its offspring grow from seeds.

 Quick Check

Infer You find a plant without roots, stems, or leaves growing close to the ground. Do you think it has seeds? Explain.

Critical Thinking Think of your favorite plant. How would you classify it based on what you have read?

The horsetail plant has roots, stems, and leaves, but no seeds.

How do plants get what they need?

About 400 years ago, a Dutch scientist named Jan van Helmont wanted to know how plants meet their needs. He planted a seedling in a pot of soil. He watered it regularly. After five years, the seedling became a small tree. Only a tiny amount of soil was missing from the pot. Where did the plant get the material to grow?

Van Helmont concluded that most of the material came from the water. He was partly correct. Almost 100 years later, scientists found that most of the material comes from *carbon dioxide*, a common gas in the air.

We now know that trees and other green plants use the energy from sunlight to make their own food. The key ingredients are water and carbon dioxide.

Van Helmont's Experiment

In his experiment, Jan van Helmont discovered that growing plants use only a small amount of soil.

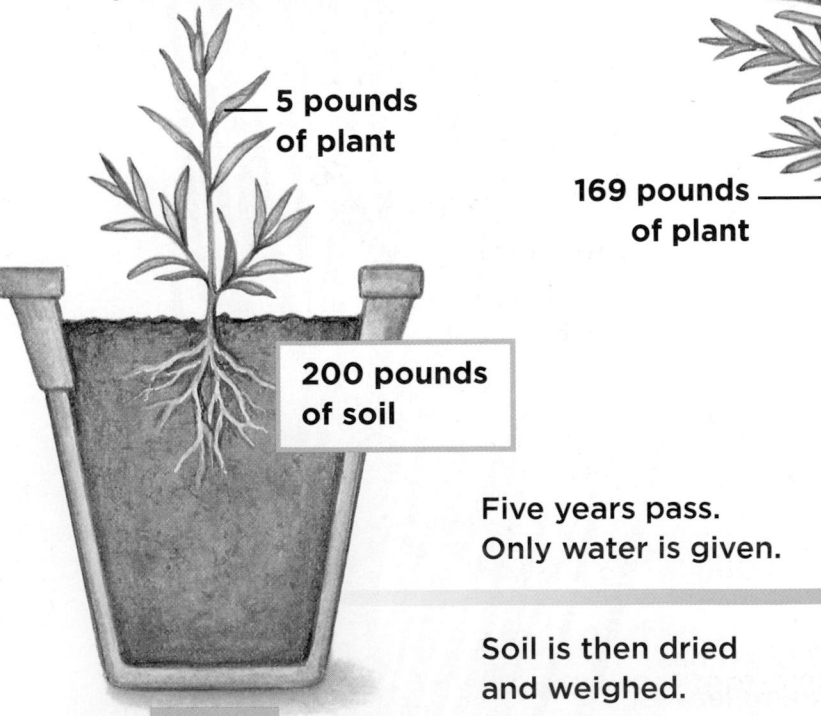

5 pounds of plant

200 pounds of soil

169 pounds of plant

199 pounds and 14 ounces of soil

Five years pass. Only water is given.

Soil is then dried and weighed.

Start

End

The Role of Roots

Roots take up water and nutrients from the ground. They also keep plants firmly in the soil. Some roots even store food.

Roots are covered with **root hairs**. These are thin cells that look like thread. Root hairs take in the water and nutrients that plants need.

All roots do the same jobs, but different plants have different kinds of roots. Carrots and dandelions have one large root called a *taproot*. Grasses have *fibrous* (FI•brus) roots that spread out into the soil.

taproot

fibrous root

The Role of Stems

A plant's stem grows above the ground. The **stem** moves food, water, and nutrients throughout the plant. Stems also hold the plant upright so it does not fall over.

There are two kinds of stems. Most trees and shrubs have woody stems. Woody stems protect the plant and give it extra support. Smaller plants have stems that are soft, green, and bendable. They rely on the pressure of watery sap for support.

Woody stems are strong. They cannot bend.

Nonwoody stems are soft and bendable.

 Quick Check

Infer Why do most trees have woody stems instead of nonwoody stems?

Critical Thinking Why is it important that a plant's roots allow water to move in only one direction?

Why are leaves important?

Like all living things, plants need energy. Animals eat food to get energy. Plants make their own food. Most plants use leaves to collect light from the Sun.

Photosynthesis

Plants use the energy in sunlight to make food from water and carbon dioxide. This is the process of **photosynthesis** (foh•toh•SIN•thuh•sus). Most photosynthesis takes place in the leaves of plants.

Photosynthesis begins when sunlight hits the leaf. The light energy goes into the plant cells. Inside the chloroplasts, chlorophyll collects the light energy.

When the chloroplasts have enough energy, a change takes place. Water and carbon dioxide combine to form plant sugars, or food. The cells let oxygen out as a waste product. Animals use this oxygen to breathe.

Photosynthesis

Leaves take in sunlight.

Plants give off oxygen.

Read a Diagram

Which gas is made by the plant?

Clue: Follow the arrows pointing away from the plant. Read all labels.

 Science in Motion Watch how photosynthesis occurs at www.macmillanmh.com

Leaves take in carbon dioxide from the air.

Roots take in water and nutrients from the soil.

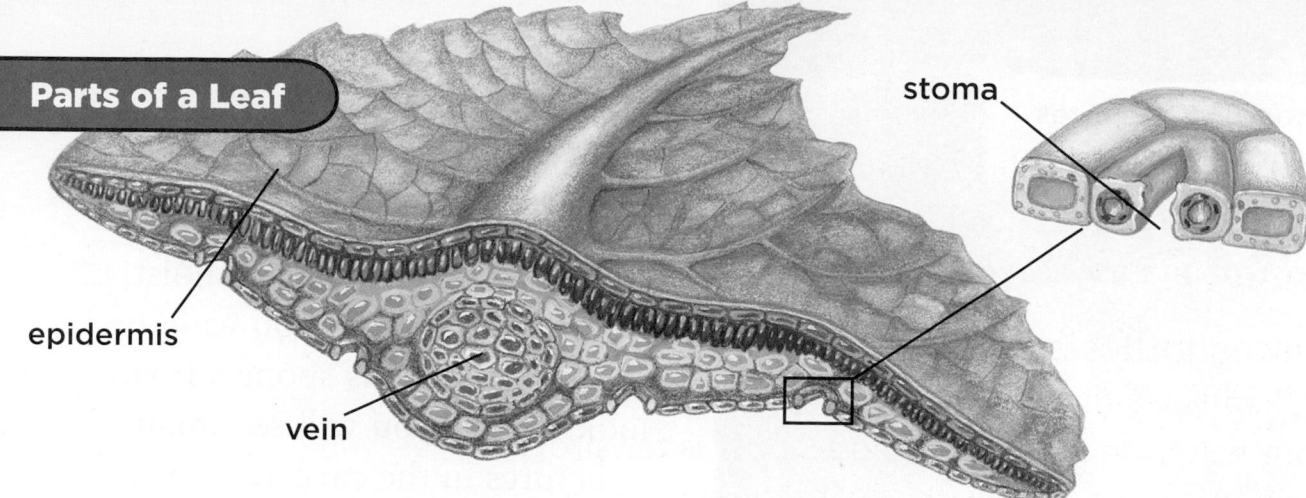

stoma

epidermis

vein

Collecting Carbon Dioxide

Where do leaves get carbon dioxide gas for photosynthesis? Carbon dioxide enters through tiny holes on the bottom of leaves. The holes are called **stomata** (STOH•muh•tuh). A single one is a *stoma* (STOH•muh).

Collecting Water

The roots of a plant take up water from soil. Small tubes called *veins* (VAYNZ) carry the water from the roots to the stem. Veins also move the water into each leaf.

Leaves have a thin covering to keep water in. This layer is called the *epidermis*. It protects leaves the way skin protects your body.

Transpiration

If a plant has enough water, its stomata stay open. Water escapes. This process is called **transpiration**. At other times, the plant closes its stomata. The water stays inside the leaf.

Respiration

Like all living things, plants need energy. They get it from the sugars made during photosynthesis. The sugars provide plants with food.

Veins carry sugars from the leaves to the rest of the plant. Inside each cell, mitochondria break down the sugars. The energy stored inside the sugars is released. This process is called **respiration**. During respiration, cells take in oxygen and give off carbon dioxide and water.

Respiration takes place in both plant and animal cells. It also takes place in protists, fungi, and most bacteria. Some cells can also break down sugars without using oxygen.

 **Quick Check**

Infer Why does a plant need both photosynthesis and respiration to survive?

Critical Thinking Desert plants often keep their stomata closed during the day. Why?

Mosses use spores to make new plants.

What are mosses and ferns?

In the cool forests of North America, the ground is a moist, green carpet. When you walk on this carpet, it feels spongy. If you look closely, you will see small structures in the carpet. They are mosses. Look around the forest and you will find the delicate leaves of ferns.

Spores

Ferns and mosses are seedless plants. Instead of using seeds to make new plants, they make spores. A **spore** is a cell in a seedless plant that can grow into a new plant.

How Spores Grow

Spores grow inside tough spore cases. The cases protect the spores from too much heat or too little water. When the spore cases open, the spores are released. They drift through the air and then settle.

Spores that land on damp ground can grow into new moss or fern plants. Like adult plants, spores need light, nutrients, and water to grow. If their needs are met, adult plants can make spores.

This fern has spore cases on the underside of its leaves.

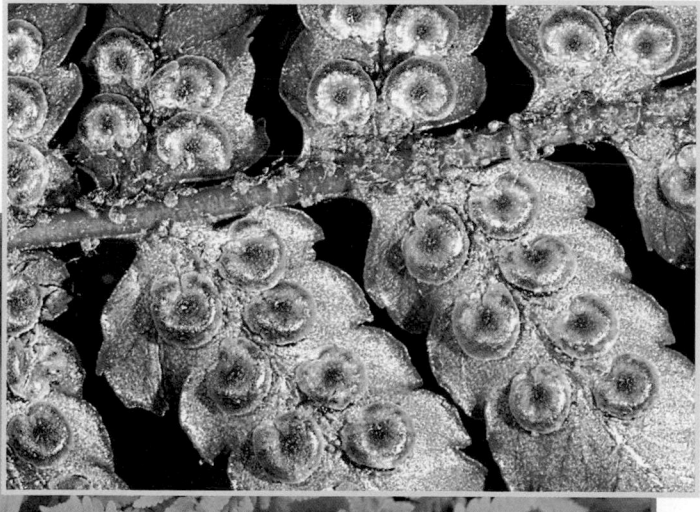

≡ Quick Lab

How Do Mosses Get Water?

1. Cut a sponge into strips of different lengths.

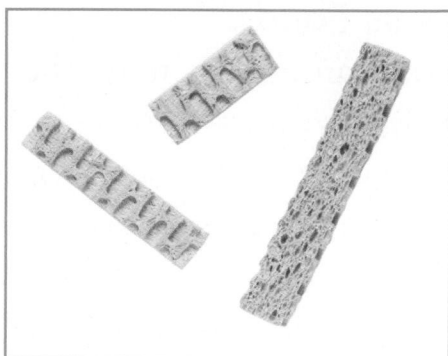

2. Pour water into several paper cups. Place the short end of each sponge into a cup of water. Some sponges will stand taller than others.

3. **Observe** What happens to the water?

4. Which sponge got completely wet first? Which took longest to become wet?

5. **Infer** Why do you think mosses are so small?

✔ Quick Check

Infer How are mosses and ferns alike? How are they different?

Critical Thinking Can a spore grow into a new plant if it is buried below the soil? Explain.

How do we use plants?

Plants do much more than just add beauty to our world. Plants give us the food we eat and are useful in many other ways.

Plants as Food

Lettuce and spinach are the leaves of plants. If you eat carrots or beets, you are eating roots. Celery and asparagus are stems. Broccoli and cauliflower are flowers. Rice and beans are seeds. You may think all fruits taste sweet, but tomatoes and cucumbers are fruits too.

Medicines and More

People have used certain plants as herbs or medicines for a long time. Today, we are finding more medicines that come from plants. We use trees to build things like furniture and toys. Plants can be burned as fuel for heating or cooking. Plants are used for clothing too. Flannel and denim come from the cotton plant.

✔ Quick Check

Infer Why are cucumbers fruits and not vegetables?

Critical Thinking Are there plant parts that people do not use? Explain.

Plants That People Eat

Read a Photo

Which plants are fruits? Which are roots?

Clue: Classify each as having seeds or growing in soil.

FACT Vegetables do not have seeds.

Lesson Review

Visual Summary

Plants are classified by the presence of **roots, stems, and leaves** or by whether they have seeds.

Photosynthesis is the process of making food in the presence of sunlight. Respiration turns food into energy.

Mosses and ferns have spores instead of seeds. People **use plant parts** for food, medicine, and clothing.

Make a **FOLDABLES**® Study Guide

Make a layered-look book. Use it to summarize what you learned about the plant kingdom.

The Plant Kingdom

Plants are classified by...
Roots, stems, and leaves
Photosynthesis
Mosses and ferns
Uses of plants parts

Think, Talk, and Write

1. **Vocabulary** Mosses and ferns use _____ to make new plants.

2. **Infer** A scientist compared transpiration in a water lily and a desert cactus. Which plant would you expect to have a higher rate of transpiration? Why?

3. **Critical Thinking** Why do mosses grow so close to the ground?

Clues	What I Know	What I Infer

4. **Test Prep** In which part of a plant does <u>most</u> photosynthesis take place?

   A woody stems

   B leaves

   C roots

   D root hairs

5. **Test Prep** Photosynthesis requires all of the following <u>except</u>

   A light.

   B carbon dioxide.

   C water.

   D oxygen.

6. **Essential Question** What are plants?

 **Writing Link**

Plants for Dinner

Plan a dinner that includes at least four different parts of a plant. Describe each dish. Write a recipe for one of the dishes.

 **Art Link**

Photosynthesis Art

Make a poster showing the steps in photosynthesis and respiration. Include how water, carbon dioxide, oxygen, and sunlight are involved.

Be a Scientist

Materials

microscope and prepared slide of root

marker

water

2 plastic jars

plastic wrap

2 rubber bands

scissors

onion plants with roots

How do root hairs affect the amount of water a plant can absorb?

Form a Hypothesis

Root hairs are found on the roots of most plants. They help plants take in water and nutrients. Does the number of root hairs affect the amount of water a plant can absorb? Write your answer in the form "If a plant has more root hairs, then..."

Test Your Hypothesis

1 **Observe** Look at the root slide with a microscope. Draw and describe the root and root hairs.

2 **Measure** Pour 100 mL of water into two jars. Mark the water level with a marker. Cover each jar with plastic wrap. Secure the plastic wrap with a rubber band.

3 Poke a small hole in the plastic wrap of one jar. Push one onion plant bulb through the hole. The bulb and roots should be under the water. Cut 3 cm off the ends of the leaves.

4 Take another onion plant bulb. Rub the root hairs off the roots using your fingernails. Repeat step 3 with the second jar.

5 **Observe** Check the jars every hour. Record your observations. Note any changes to the plants or the water level in the jars.

Step 3

Draw Conclusions

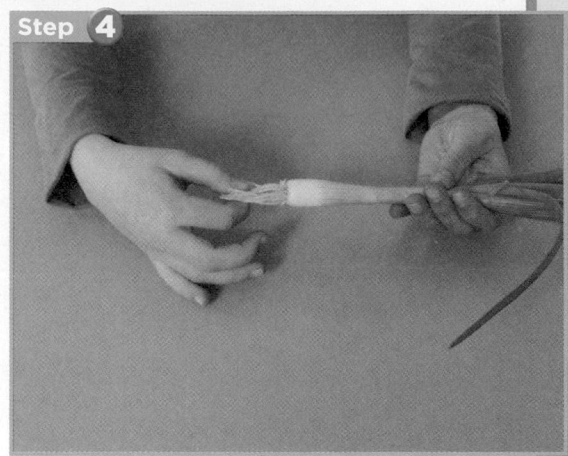

Step **4**

6 **Communicate** Did your results support your hypothesis? Why or why not?

7 What did you observe when you looked at the slide of the root?

8 **Interpret Data** What happened to the water level in each jar during the activity? Why do you think this happened?

9 **Infer** What would happen to a plant if the root hairs were damaged by insects?

Guided Inquiry

How do taproots and fibrous roots differ?

Form a Hypothesis

How does the type of root affect the amount of water a plant can absorb? Do fibrous roots absorb more water than taproots? Write a hypothesis.

Test Your Hypothesis

Design a model to illustrate how the type of root affects the amount of water a plant can absorb. Use classroom materials in your model. Write out the materials you need and the steps you will follow. Record your results and observations.

Draw Conclusions

Did your results support your hypothesis? Why or why not?

Open Inquiry

What else would you like to learn about roots? Write a hypothesis. Then design an experiment to answer your question.

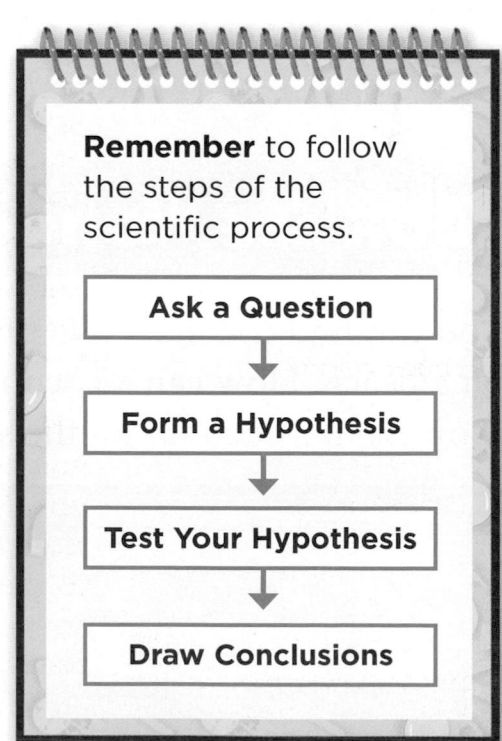

Remember to follow the steps of the scientific process.

Ask a Question

↓

Form a Hypothesis

↓

Test Your Hypothesis

↓

Draw Conclusions

How Seed Plants Reproduce

Look and Wonder

You can find coconuts like this one on a beach. How can a coconut tree grow from a coconut on the sand?

Does a seed need water to grow?

Make a Prediction

What happens to a seed that does not get any water? Make a prediction.

Test Your Prediction

① Line each cup with a folded paper towel. Crumple two more paper towels. Place one in each cup to hold the lining in place. Label one cup *water* and the other *no water*.

② Place one bean about two centimeters from the bottom of each cup. Place it between the paper towel lining and the side of the cup.

③ **Use Variables** Fill the graduated cylinder with water. Drip the water into the cup labeled *water*. Add just enough water to wet the paper towels. Do not wet the paper towels in the other cup.

④ **Observe** Place both cups in a sunny spot. Keep the lining of the cup labeled *water* moist. Observe the beans every day for ten days. Record your observations in a table.

Draw Conclusions

⑤ Compare your results with those of your classmates. How are they alike? Different?

⑥ **Infer** Was your prediction correct? Does a seed need water to grow?

Explore More

Does a seed need sunlight to grow? Write a hypothesis. Plan an experiment to test it.

Materials

- **2 plastic cups**
- **paper towels**
- **marker**
- **lima beans**
- **small graduated cylinder**
- **water**

Step **3**

Step **4**

My Prediction	What Happens

Read and Learn

▶ **Essential Question**
How do seed plants grow and reproduce?

▶ **Vocabulary**

seed, p. 60

reproduction, p. 62

ovary, p. 62

pollination, p. 63

fertilization, p. 63

germination, p. 64

life cycle, p. 65

▶ **Reading Skill** ✔
Predict

My Prediction	What Happens

▶ **Technology** LOG ON
e-Glossary and e-Review online at www.macmillanmh.com

How do we classify seed plants?

What are some of the plants you see every day? Chances are, they are seed plants.

Grasses, trees, and other seed plants grow from seeds. A **seed** is a plant that is not fully formed. The coating of a seed protects the plant inside.

There are about 250,000 species of seed plants. How could you classify all of them? What would you look for?

Comparing Seeds and Structures

Seeds come in all shapes and sizes. You can use those differences to classify seed plants. A watermelon has many flat, slippery seeds. A peach has just one seed, or pit, that is hard and round. A cherry seed is smaller and smoother.

Plants that make seeds also have roots, stems, and leaves. These parts can be very different among plants. They can also help you classify seed plants. Coconuts are giant seeds that come from certain palm trees. The trees are tall and sturdy, with woody stems. Watermelons and grapes grow on vines.

Each seed in a watermelon can make a new plant.

FACT ▶ In nature, all fruits have seeds.

Flowers and Cones

We can also classify seed plants by where they store seeds. Most seed plants have flowers that bear fruit. The fleshy tissue of the fruit protects the seeds inside.

Have you ever seen a pinecone? Pinecones come from conifers (KAH•nuh•furz). A *conifer* is a seed plant that has no flowers or fruit.

If they have no fruit, where do conifers store their seeds? Pine trees and other conifers bear seeds on the surface of cones. Most conifers have male and female cones. The male cones are smaller than the females.

The male cones produce a yellow powder called *pollen*. A gust of wind can blow pollen grains from a male cone to a female. If the pollen lands on a female cone, the seeds develop there.

Conifer Seeds

Read a Photo

Which pinecone is the female? Which is the male?

Clue: Smaller male pinecones produce pollen.

✔ *Quick Check*

Predict A friend gives you a seed from a plant that does not make fruits or flowers. What kind of plant is it likely to grow?

Critical Thinking Could conifer trees reproduce without wind? Explain your answer.

Conifers grow well in cold climates.

How do seeds form?

Think about a delicate red rose or a colorful tulip. Flowers may look pretty or smell nice, but they do not make food for a plant. What is the job of a flower?

Reproduction

You know that all living things can make more of their kind. **Reproduction** (ree•pruh•DUK•shun) is how living things make offspring.

Flowering plants use flowers to reproduce. Most flowers have male and female parts. The male part is the *stamen,* which includes the anther. The *anther* makes pollen that has male sex cells. The female part of a flower is the pistil, which includes the ovary (OH•vuh•ree). The *pistil* makes female sex cells, or eggs. The **ovary** stores the eggs. To reproduce, flowering plants combine male and female sex cells. How do these cells join together?

The bright colors and sweet smells of flowers attract bees.

Pollination

A gust of wind can blow pollen from an anther to a pistil. More often, plants rely on birds, insects, or other pollinators. *Pollinators* are animals that carry pollen from one flower to another. For instance, bees suck up *nectar*—a sugary liquid inside a flower. If the bee touches an anther, pollen sticks to it. At the next flower, the pollen falls from the bee onto the tip of the pistil. Moving pollen onto the pistil is called **pollination**.

Fertilization

The next step is to move the male sex cells to the ovary. The pollen grows a long tube. Male sex cells travel inside the tube to the ovary. Inside the ovary, a male sex cell combines with an egg. Fertilization occurs when the male sex cell joins with the egg. Fertilization is the process that forms a seed.

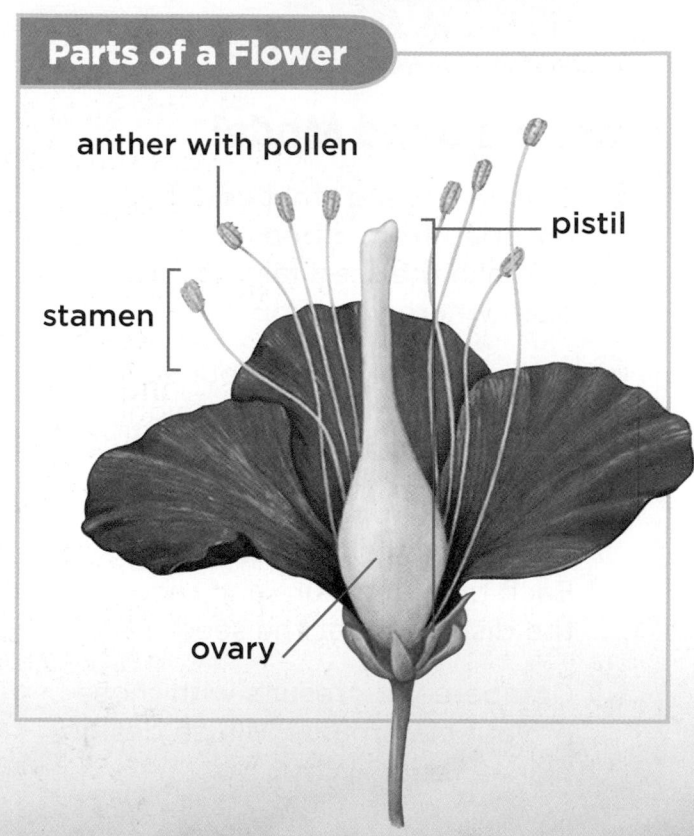

Parts of a Flower

anther with pollen

pistil

stamen

ovary

✔ *Quick Check*

Predict What would happen to a plant if there were fewer pollinators in its environment?

Critical Thinking How does nectar help both the plant and the pollinator?

Make a Seed Model

1 Design a "seed" that could be dropped from a tree. Your goal is to make the seed fall the farthest from where you drop it.

2 **Make a Model** Build your seed. Make sure it includes a weight, such as a paper clip.

3 **Measure** Drop your model from the same height several times. Each time, measure and record the distance that the seed travels.

4 Compare your results with those of your classmates. Which design worked best? Why?

How do seeds grow?

Inside each seed are an unformed plant and a food source. A layer called the *seed coat* surrounds and protects them. In addition to food, seeds need water and warm temperatures in order to grow.

Under the right conditions, the seed will sprout. First, the seed coat splits open. A root pushes through the crack. Then, a tiny stem grows upward. One or two leaves appear on the stem. The process is called **germination** (jur•muh•NAY•shun).

A *seedling* is the young plant that grows from a seed. A seedling needs water, light, and nutrients to grow. If its needs are met, the seedling will grow into an adult plant. The adult reaches *maturity*.

3 **A flower forms.**

4 **The adult makes new seeds.**

2 **The seedling has a stem, roots, and leaves.**

1 **The seed sprouts.**

Life Cycle of a Seed Plant

germination

adult plant

pollination

animal eats fruit

fruit with seeds

fertilization

Life Cycles

For some plants, it takes only days for the seed to reach maturity. For others, it takes years. No matter how long it takes, all seed plants go through the same stages.

Look at the diagram above. It shows the life cycle of a berry plant. A **life cycle** includes the stages of germination, maturity, and reproduction. *Death* marks the end of a plant's life. Some plants grow, reproduce, and die in a single season. Others, such as some conifers, can live more than 5,000 years.

When a plant dies, its life ends, but the life cycle does not. Seed plants make new seeds. The life cycle continues with new seedlings.

Seeds on the Move

Seeds need space to grow, but they cannot move on their own. Some, like dandelion seeds, are easily carried by wind. Others have prickly surfaces that can be caught on animal fur.

Many seed plants bear fruits. When an animal eats the fruit, the seeds pass through the animal's body. Then, the seeds can germinate on the ground.

 Quick Check

Predict What might happen to seed plants if there were no animals?

Critical Thinking Squirrels often bury seeds in the ground to store them for winter. How does this help make new plants?

How are plants alike and different from their parents?

Seeds grow into plants that look similar to their parent plants. For example, all foxglove plants have tall stems and flowers shaped like trumpets. However, offspring plants are not exact copies of their parents.

Color, size, and shape are inherited (in•HER•uh•tid) traits. *Inherited traits* are characteristics passed from parent to offspring. When sex cells from two parents combine, some offspring might end up with different traits than their parents. Others might end up with similar traits. For example, offspring might have differently colored flowers than their parents. They might grow taller or shorter than their parents.

foxglove

The flowers of a foxglove plant can have different colors.

The offspring of these trees look similar to their parents.

ginkgo

oak

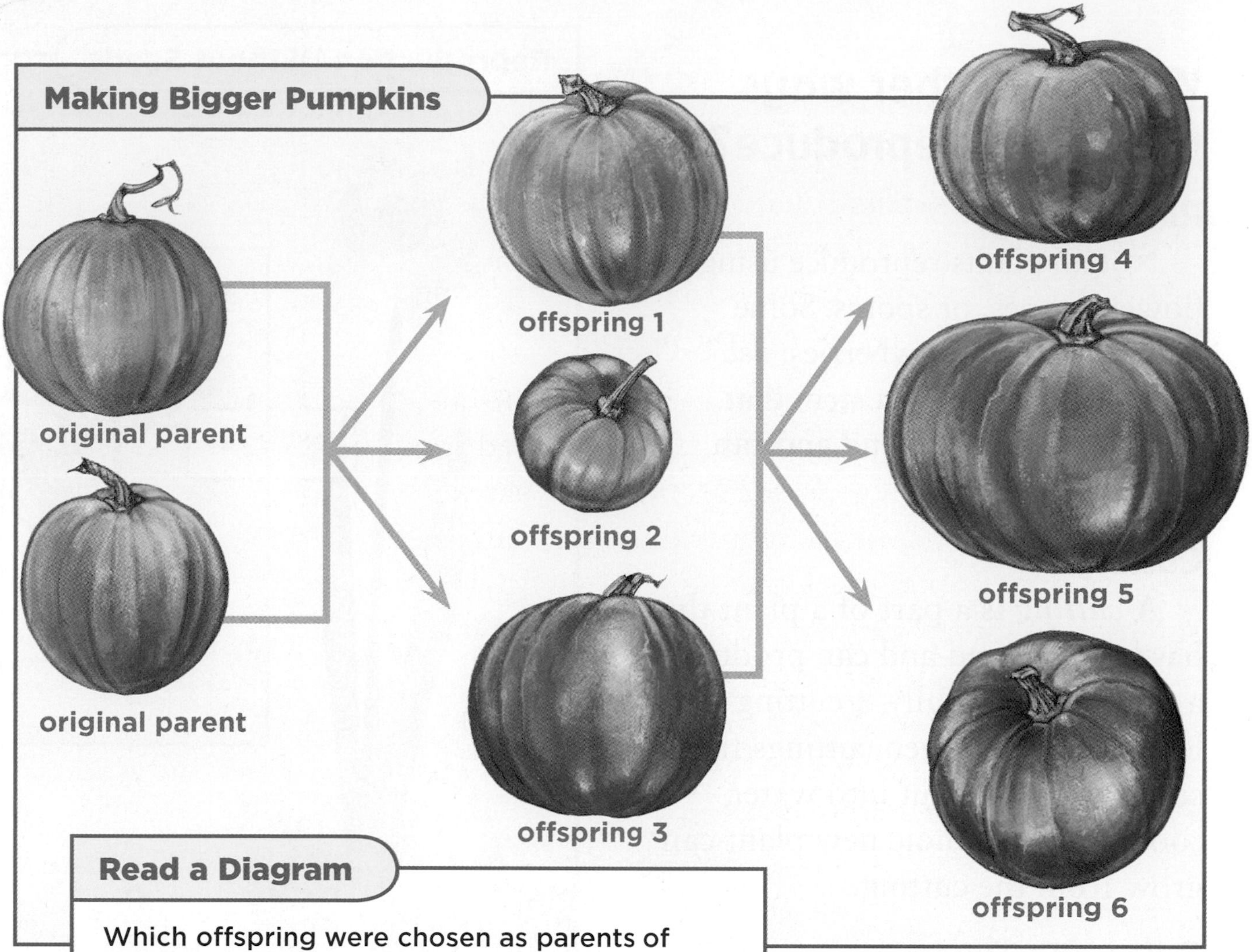

original parent

original parent

offspring 1

offspring 2

offspring 3

offspring 4

offspring 5

offspring 6

Which offspring were chosen as parents of offspring 4, 5, and 6?

Clue: Trace the direction of the lines and arrows.

Choosing Traits

Farmers can put inherited traits to good use. Suppose you wanted to grow a giant pumpkin. You might start by choosing two adult plants that produced the largest pumpkins in the crop. Then you could be a pollinator and move pollen from one plant to the other. What happens if you choose two offspring with the biggest pumpkins and pollinate again? Over time, you get bigger and bigger pumpkins.

Some fruit and vegetable growers pollinate their plants. They grow plants with traits that people prefer. Larger fruits are one good example. Can you think of others?

 **Quick Check**

Predict What kind of offspring might result from a tall parent and a short parent?

Critical Thinking If you were a farmer, how might you make a watermelon with fewer seeds?

What are other ways plants can reproduce?

Runners

Not all plants reproduce using flowers, cones, or spores. Some plants, such as strawberries, use runners. A *runner* is a stem that grows along the ground and can make new plants.

Cuttings

A *cutting* is a part of a plant that has been clipped and can produce a new plant. Usually, a cutting is a leaf or a stem. When cuttings from some plants are put into water, roots grow. A whole new plant can grow from the cutting.

Bulbs and Tubers

A *bulb* is a stem that grows under the ground. Tulips and onions grow from bulbs. A *tuber* is a storage part of a plant. Potatoes are tubers. If you plant a potato by itself, more potatoes will grow.

Reproduction Without Seeds

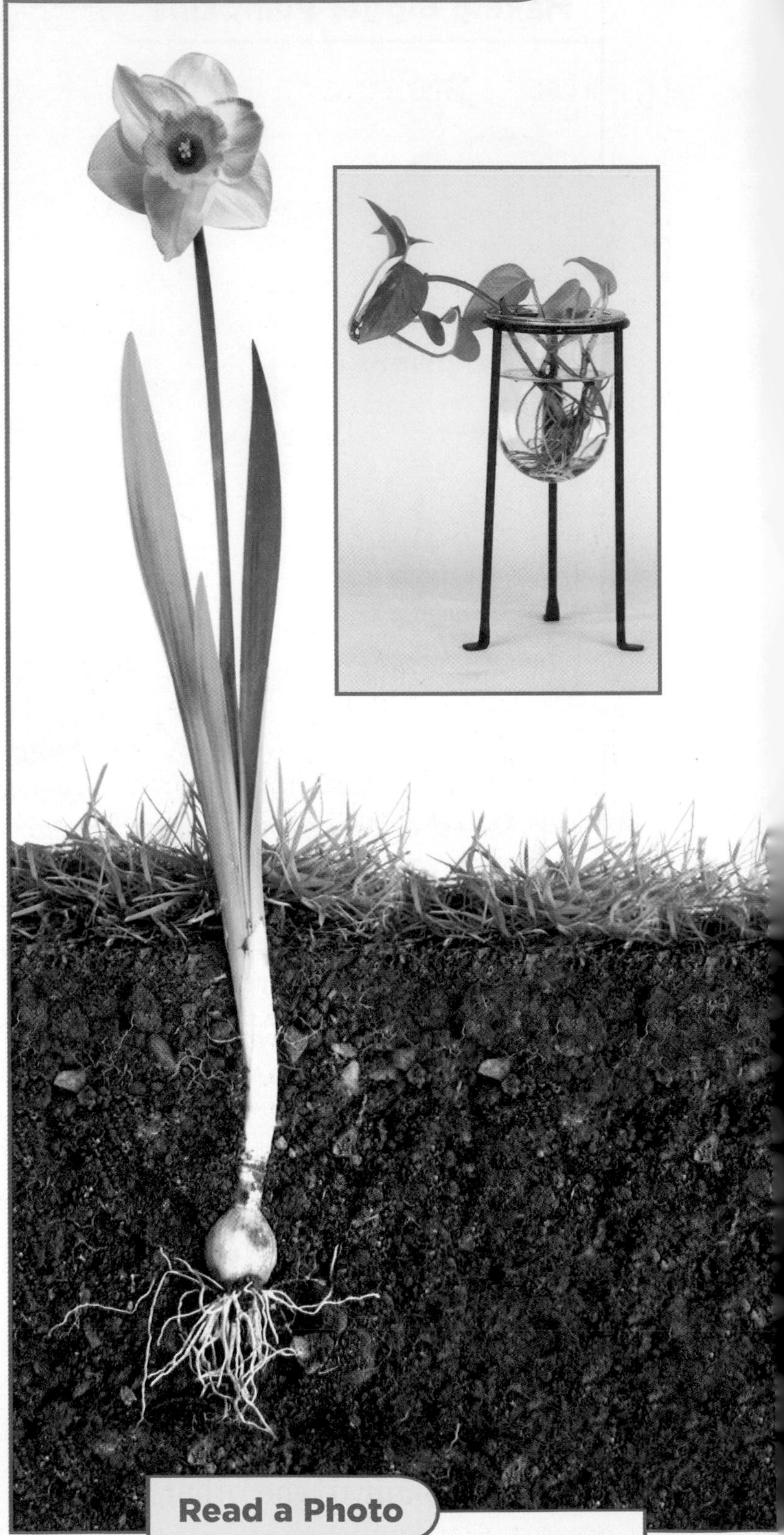

Read a Photo

Which kinds of plant reproduction are shown here?

Clue: Look closely at how the plants are growing.

✔ *Quick Check*

Predict If you bury a potato in soil, what might happen in a few weeks?

Critical Thinking Suppose you want to start a strawberry patch. You do not have any seeds. How could you grow strawberries?

Lesson Review

Visual Summary

There are different ways to **classify seed plants.** The two main groups are flowering plants and conifers.

The **life cycle of a seed plant** includes pollination, fertilization, and germination.

The offspring of seed plants **inherit traits** from their parents. Not all plants have **flowers, cones, or spores.**

Make a FOLDABLES Study Guide

Make a folded book. Use it to summarize what you learned about reproduction in seed plants.

Classify seed plants
Seed plant life cycle
Inherited traits
Flowers, cones
Spores

Think, Talk, and Write

1. **Vocabulary** The movement of pollen from the anther to the pistil is called _____.

2. **Predict** What might happen to flowering plants if there were no more animal pollinators?

My Prediction	What Happens

3. **Critical Thinking** Conifers do not produce fruit. How might this affect the way conifers grow in an area?

4. **Test Prep** Where does fertilization take place in a plant?

   A in the ovary

   B in the anther

   C in the stamen

   D in the pistil

5. **Test Prep** Where is pollen made?

   A in the ovary

   B in the anther

   C in the pistil

   D in the seed

6. **Essential Question** How do seed plants grow and reproduce?

Writing Link

Describe Plants
Choose two different kinds of seed plants. Find out more about them. Write a paragraph about how you can tell them apart.

Math Link

Solve a Problem
A tree produces 3,000 seeds. Squirrels bury half of the seeds. In winter, the squirrels find and eat half of the seeds they buried. How many seeds are left in the ground?

Dandelions and Me

When I was eight, Dad and I drove away from the city to look for a new house. We waited outside the real estate agent's office. I began to pick the dandelions growing along the edge of the steps. I was so sad about leaving our home, I started crying.

"Bet I can make you laugh," said Dad. He was holding a puffball—a dandelion turned to seed. "What did the wolf do in *The Three Little Pigs*?"

"Well, he huffed and he puffed ... ." As I said that, Dad took a deep breath and blew it out. Pieces of the puffball floated off like little white umbrellas. Some got carried away by the wind. "Look at that!" I laughed. "They're going to a new place too."

"And just like us, they'll be fine," Dad added. So I took a deep breath, blew, and sent more dandelion seeds off to their new homes.

Personal Narrative

A personal narrative

▶ tells a story about a personal experience;

▶ uses the first-person point of view ("I") to tell the writer's feelings;

▶ tells the events in a sequence that makes sense.

Write About It

Personal Narrative Think about a time you saw seeds being carried from place to place. Write a personal narrative about the event. Tell how it made you feel.

 **-Journal** Research and write about it online at **www.macmillanmh.com**

Parts of a Whole

Not every seed becomes a new plant. Of all the seeds a plant makes, only a fraction of them will grow. A fraction is a part of a whole.

The number of seeds that do grow may differ for different plants. Suppose a sunflower plant and a thistle plant each make 100 seeds. One fourth ($\frac{1}{4}$) of the sunflower's seeds grow into new plants. Two fifths ($\frac{2}{5}$) of the thistle's seeds grow into new plants. Which plant makes the most new plants from 100 seeds? Use fraction strips to find out.

Fractions

Fraction strips can show how fractions are related.

▲ This fraction strip shows that $\frac{2}{5}$ is greater than $\frac{1}{4}$.

Solve It

A tomato and a pepper each have 100 seeds. Two thirds ($\frac{2}{3}$) of the tomato seeds grow into new plants. Two fifths ($\frac{2}{5}$) of the pepper seeds grow into new plants. Which grows more plants?

Visual Summary

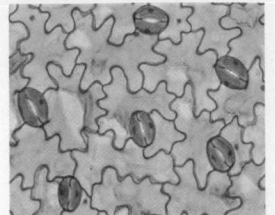

Lesson 1 Cells are the basic building blocks of all living things.

Lesson 2 Living things are classified by kingdom, phylum, class, order, family, genus, and species.

Lesson 3 The roots, stems, and leaves of a plant help it get water and nutrients, support itself, and make food.

Lesson 4 Seed plants use flowers, fruits, and cones to make seeds that will produce new plants.

Make a **FOLDABLES®** Study Guide

Tape your lesson study guides to a piece of paper as shown. Use your study guide to review what you have learned in this chapter.

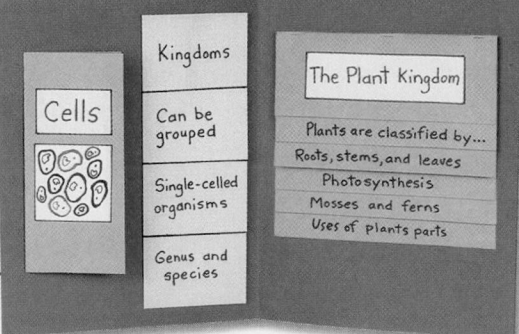

Vocabulary
DOK 1

Fill each blank with the best term from the list.

cell, p. 22   **reproduction,** p. 62

kingdom, p. 35   **spore,** p. 52

life cycle, p. 65   **stem,** p. 49

organism, p. 22   **trait,** p. 34

photosynthesis, p. 50   **tissue,** p. 27

1. The smallest unit of living matter is a _____.

2. All living things make offspring through _____.

3. A plant's _____ moves food, water, and nutrients throughout the plant.

4. The largest group into which an organism can be classified is a _____.

5. The stages of growth a plant goes through are part of its _____.

6. A group of similar cells that do a job together is called _____.

7. The ability to make their own food is a _____ that all plants share.

8. A plant makes food from sunlight, carbon dioxide, and water during _____.

9. A seedless plant can make a new plant by producing a(n) _____.

10. Anything that can carry out the five basic life functions is a(n) _____.

LOG ON **e-Glossary** Words and definitions online at www.macmillanmh.com

Answer each of the following.

11. Classify What are some ways that plants can be classified?

12. Observe Find a plant in or around your school or home. Describe the plant. Include details about its appearance. List the functions of each plant part that you observe.

13. Critical Thinking What could you infer if you looked into a microscope and saw a cell with cell walls? Explain your answer.

14. Personal Narrative Share an experience you have had with an animal in the genus *Canis*. Explain why the experience was meaningful.

15. Look at the picture of a carrot shown below.

Which part of this plant do people eat?

A stem	**C** fruit
B cone	**D** root

16. True or False *A seed is a living thing.* Is this statement true or false? Explain.

17. True or False *The phylum is larger than the class.* Is this statement true or false? Explain.

18. True or False *The nucleus of a cell burns food and releases energy.* Is this statement true or false? Explain.

19. True or False *Mosses and ferns reproduce using seeds.* Is this statement true or false? Explain.

The Big Idea

20. What are living things and how are they classified?

Performance Assessment
DOK 2

Make a Cell Model

1. Make a model of a plant leaf cell using things you can eat. Use different food items for each cell part.

2. Make sure that your model has the traits of a plant leaf cell—that it is shaped like a box and green. Label each part of the cell.

3. Write a short paragraph explaining the job of each cell part.

Test Preparation

1 Look at the apple shown below.

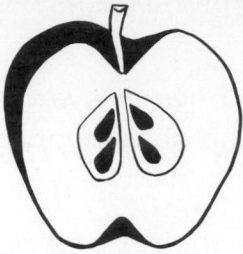

How does fruit help
a plant reproduce?

A Pollen is protected by the fruit.

B Pollen is moved by the fruit.

C Seeds are protected by the fruit.

D Spores are protected by
the fruit.
DOK I

2 In some ways mushrooms are
similar to plants.

What makes mushrooms different
from plants?

A Mushrooms cannot make their
own food.

B Mushrooms cannot move from
place to place.

C Mushroom cells do not have
cell walls.

D Mushroom cells do not have
a nucleus.
DOK 2

3 Which plant uses flowers
to reproduce?

A fern

B moss

C pine tree

D sycamore tree
DOK 2

4 What is true of <u>all</u> living things?

A They have tissues.

B They can move.

C They use energy.

D They change shape.
DOK I

5 Which of these do your
cells contain?

A cell wall

B chlorophyll

C chloroplast

D cytoplasm
DOK 2

6 Which trait will a tulip tree
<u>most likely</u> pass on to its
offspring?

A scar on its bark

B yellow-green flower

C diseased leaf

D broken branch
DOK 2

7 Which of these <u>most likely</u> shows a model of a plant cell?

A

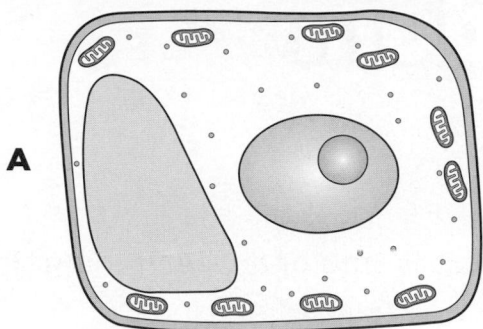

B

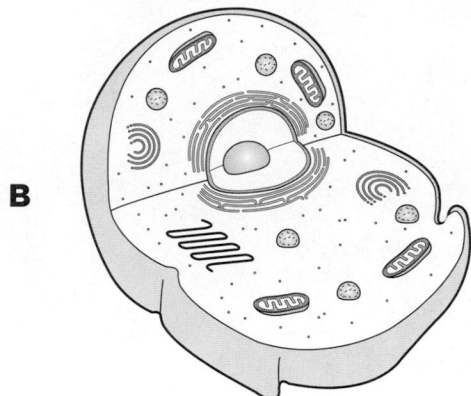

C

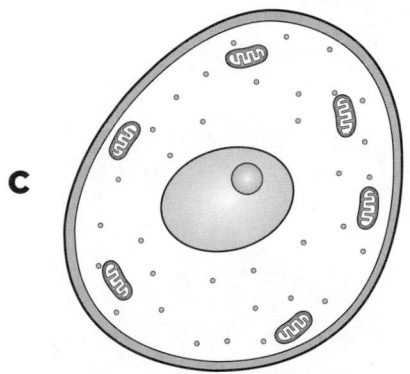

D

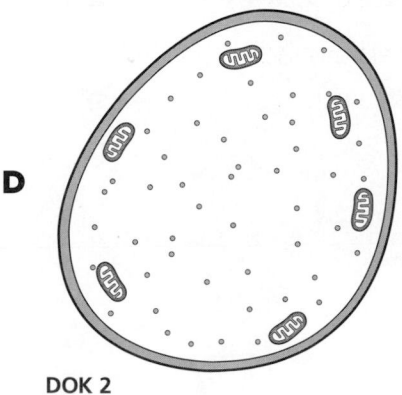

DOK 2

8 The diagram below shows the life cycle of a berry plant.

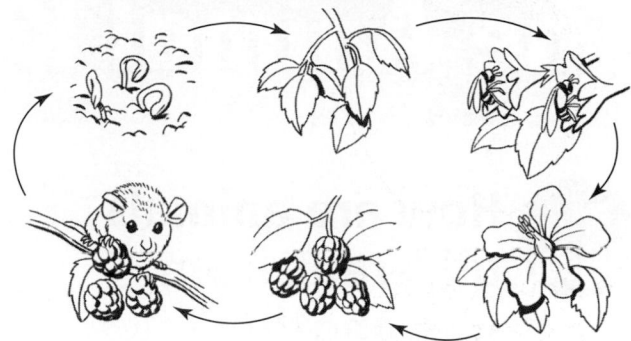

Part A

Bees and mice play different roles in this berry plant's life cycle.

Describe two ways these animals help berry plants complete their life cycle.

DOK 2

Part B

Suppose that either bees or mice were removed from the environment.

How would berry crops be affected? Explain why you think so.

DOK 2

Check Your Understanding

Question	Review	Question	Review
1	pp. 60–61	5	pp. 24–25
2	pp. 34–35, 38	6	pp. 66–67
3	pp. 52, 60–65	7	pp. 24–25
4	pp. 22–23	8	pp. 62–65

CHAPTER 2

The Animal Kingdom

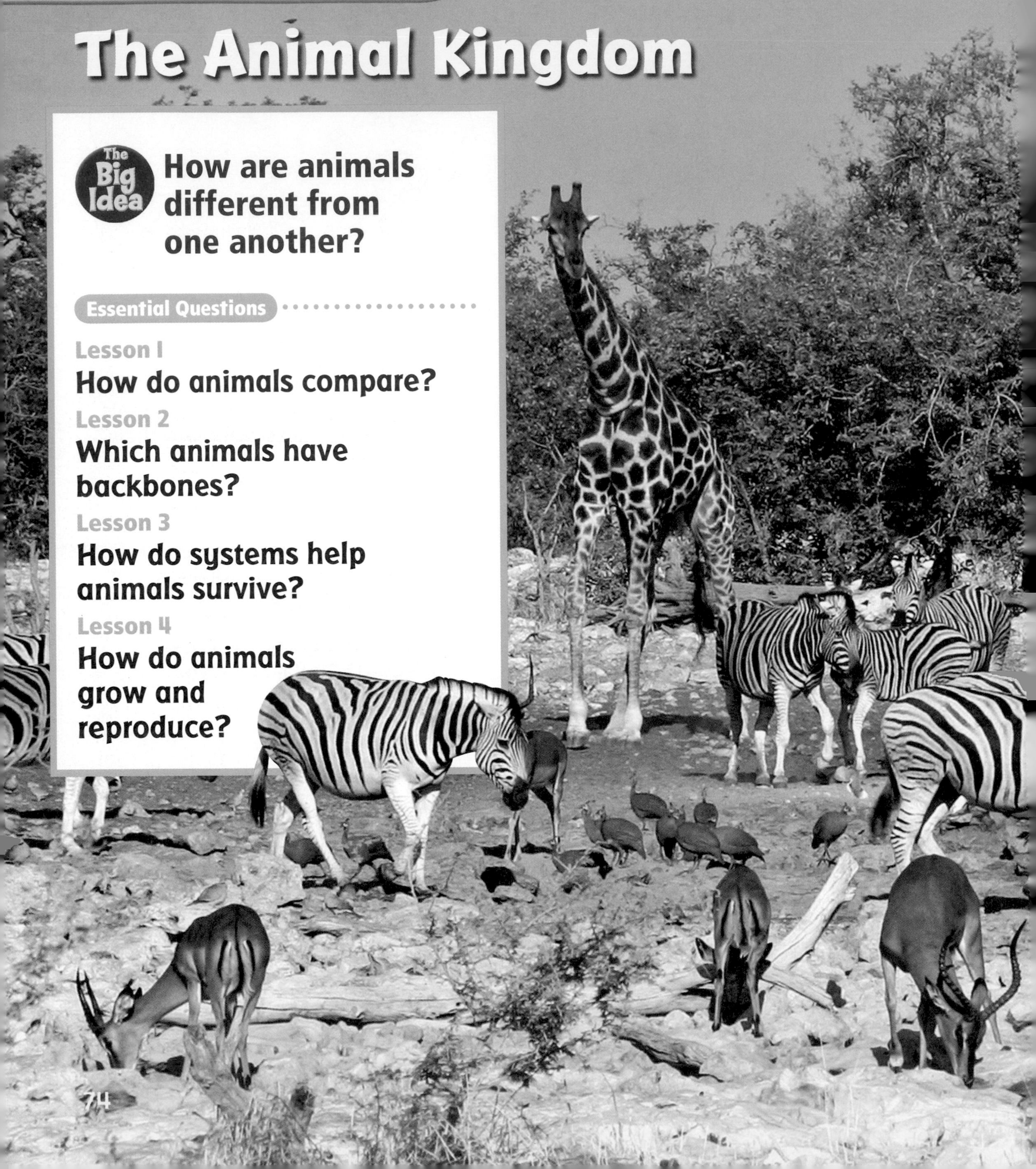

The Big Idea How are animals different from one another?

Essential Questions

Lesson 1
How do animals compare?

Lesson 2
Which animals have backbones?

Lesson 3
How do systems help animals survive?

Lesson 4
How do animals grow and reproduce?

 # Big Idea Vocabulary

invertebrate an animal without a backbone (p. 79)

vertebrate an animal with a backbone (p. 90)

muscular system the organ system made up of muscles that move bones (p. 100)

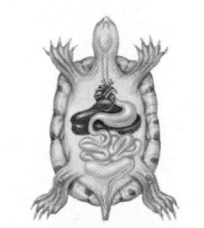

digestive system the organ system that breaks down food for fuel (p. 104)

metamorphosis a life cycle including four stages of growth: egg, larva, pupa, and adult (p. 112)

heredity the passing of traits from parent to offspring (p. 114)

 Visit www.macmillanmh.com for online resources.

Animals without Backbones

Look and Wonder

Here is a riddle. What has thousands of teeth, one foot, and no backbone? A snail, that's what! Which other animals lack a backbone?

What makes an earthworm an animal?

Make a Prediction

What characteristics does an earthworm have that make it an animal? Make a prediction.

Test Your Prediction

1 Take the earthworm out of the terrarium. Place it on a damp paper towel. What does the earthworm do? Watch how it moves. Record your observations.

2 **Observe** Gently touch the worm with your finger. Watch what happens. Record your observations.

3 **Observe** After a few days, observe the terrarium. Do you notice any changes in the earthworm's environment?

Draw Conclusions

4 **Communicate** How did the earthworm respond when you touched it?

5 **Infer** Does the earthworm have a skeleton? How can you tell?

6 What characteristics does the earthworm have that make it an animal?

Explore **More**

Observe other kinds of animals. Do they have the same characteristics as the earthworm?

Materials

- **living earthworm**
- **soil**
- **leaves**
- **damp paper towel**

Step **2**

Earthworm Observations	
How it moves	
What happens when touched	
How the environment changes	

Read and Learn

▶ **Essential Question**
How do animals compare?

▶ **Vocabulary**

invertebrate, p. 79

sponge, p. 80

cnidarian, p. 80

mollusk, p. 81

echinoderm, p. 81

endoskeleton, p. 81

arthropod, p. 82

exoskeleton, p. 82

▶ **Reading Skill** ✔
Main Idea and Details

Main Idea	Details

▶ **Technology** LOG ON
e-Glossary and e-Review online
at www.macmillanmh.com

What are invertebrates?

How would you describe animals? Think about your pets, your friends' pets, or animals in a zoo. Don't forget that you are an animal too.

One way to describe animals is by their similarities and differences. How are animals similar? All animals are made of many cells. Most animals can move on their own. Like all living things, animals grow, change, reproduce, and respond to their environment. They get the energy they need from eating food.

Symmetry

One difference between animals is how their bodies are shaped. Most animals have symmetry (SIH•muh•tree). *Symmetry* means that parts of the animal's body match up with other parts around a midpoint or line. Some kinds of animals have no symmetry.

▲ A sea urchin has symmetry.

◀ A sponge is the simplest kind of animal. It does not have symmetry.

Invertebrate Groups

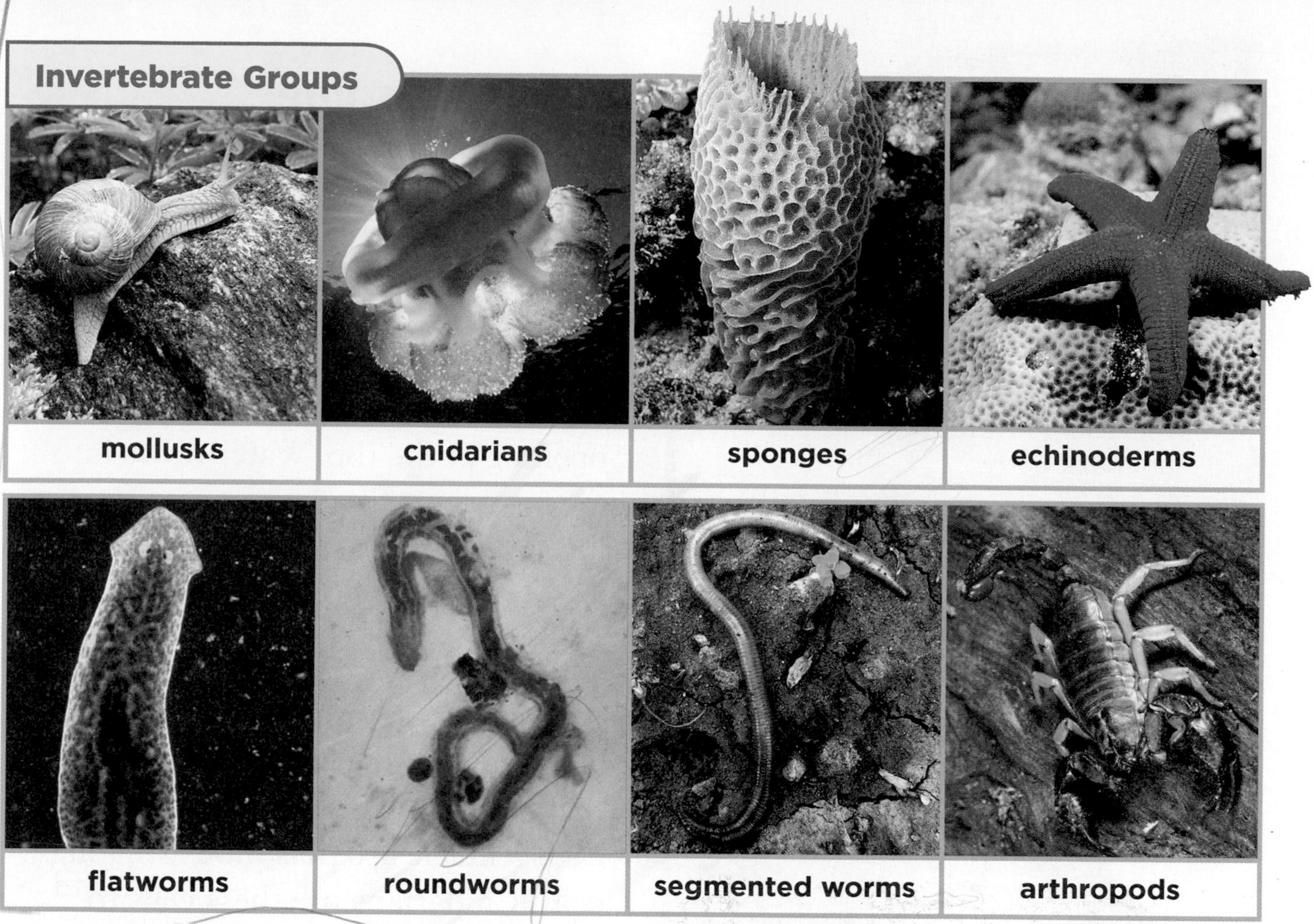

mollusks	cnidarians	sponges	echinoderms
flatworms	roundworms	segmented worms	arthropods

Backbone or No Backbone

Another way that animals differ is whether they have a backbone. A *vertebrate* (VUR•tuh•brayt) is an animal that has a backbone. An animal without a backbone is an **invertebrate** (in•VUR•tuh•brayt).

More than 95 out of every 100 animals are invertebrates. They come in all shapes and sizes. Some, like ants and earthworms, can fit in your palm. Others can grow much larger. Some have a hard, outer covering. Others have a support structure inside their bodies. The chart shows the eight main groups of invertebrates.

Read a Photo

Which animal group does not have symmetry?

Clue: Look at each animal. See whether the parts on one side match up with the parts on the other side.

✓ **Quick Check**

Main Idea and Details What are two ways in which animals differ?

Critical Thinking Could an animal with a backbone have no symmetry? Explain.

How Jellyfish Move

1 Make a Model Blow up a balloon. Hold the end of the balloon tight so the air cannot escape. The balloon models the hollow shape of a jellyfish. The air inside represents water that fills the animal's body.

2 What do you think will happen when you let go of the balloon?

3 Observe Let go of the balloon. What happens? How does this model the way a jellyfish moves?

What are some invertebrates?

Sponges

A **sponge** is the simplest kind of invertebrate. Sponges do not have symmetry. Sponges live underwater. Most are shaped like a sack, with an opening at the top. Water flows into the sponge. It filters the water for food. The adults stay in one place. Their offspring float or swim.

Cnidarians

A **cnidarian** (ni•DAYR•ee•un) is an animal with armlike parts called tentacles (TEN•tih•kulz). At the tip of each tentacle are poisonous stinging cells. Cnidarians use these cells to stun prey. Jellyfish and corals are cnidarians.

Corals are cnidarians.

Mollusks

Do you collect seashells? Those shells come from invertebrates with soft bodies. These invertebrates are **mollusks**. Most mollusks have shells and live in water. Snails and slugs are the only mollusks that live on land.

Some adult mollusks, such as clams and oysters, stay attached to one place. Others, such as squid and octopuses, swim freely.

The fileclam is a mollusk that lives among coral reefs.

Echinoderms

Sea urchins are echinoderms (ih•KI•nuh•durmz). An **echinoderm** has spiny skin. It also has an internal support structure. This structure is called an **endoskeleton** (EN•doh•SKE•luh•tun).

 Quick Check

Main Idea and Details What do sponges, cnidarians, mollusks, and echinoderms have in common?

Critical Thinking Why do you think all sponges live underwater?

Before and After

Read a Photo

What happens to an octopus when it is threatened?

Clue: Look at the color and shape of the octopus in both pictures.

What are arthropods?

The largest invertebrate group is arthropods (AR•thruh•podz). **Arthropods** have jointed legs and a body that is divided into sections. Some, like crabs and shrimps, breathe with gills. Others, like insects and some spiders, breathe through open body tubes.

Every arthropod has an **exoskeleton** (EK•soh•SKE•luh•tun). This is a hard covering that protects the body. It also keeps in moisture so the animal doesn't dry out.

Lobsters, bees, and scorpions are arthropods. Insects are by far the largest arthropod group with almost one million species. Centipedes and millipedes are relatives of the insects.

These animals are all insects.

bee

moth

✔ Quick Check

Main Idea and Details What are the features that all arthropods share?

Critical Thinking Are all insects arthropods? Are all arthropods insects? Explain.

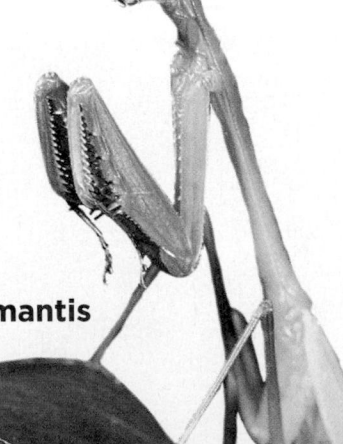
praying mantis

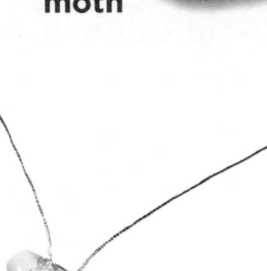
beetles

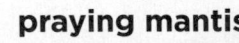

FACT Many arthropods shed their exoskeletons so they can grow bigger.

Insects

ant

Insects have one pair of antennae, three pairs of legs, one or two pairs of wings, and three body sections.

Arachnids

Arachnids (uh•RAK•nudz) include spiders, ticks, and scorpions. They have four pairs of legs, two body sections, and fangs.

spider

Crustaceans

crab

Crustaceans (krus•TAY•shunz), such as crabs and shrimps, have two pairs of antennae and two to three body sections. They can chew.

Centipedes and Millipedes

centipede

Centipedes have one pair of legs on each body section. A millipede has two pairs of legs on each body section.

How are worms classified?

You may think all worms look alike, but there are several groups. Three of these are *flatworms*, *roundworms*, and *segmented worms*.

Flatworms

True to their name, flatworms have flat bodies. They have a head, simple eyes, and a tail. Flatworms are the simplest worms. Most are harmless. Some live inside the bodies of other animals.

Roundworms

Roundworms have thin bodies with pointed ends. They are not as thin as flatworms. Roundworms have a one-way digestive system. Food comes into one opening. Wastes leave through another. Most roundworms live inside the bodies of other animals.

▲ An earthworm is a segmented worm.

Segmented Worms

Earthworms, sandworms, and leeches are segmented worms. Their bodies are divided into segments. The segments are identical, except for the head and tail ends. Each end has an opening for the digestive system.

Most segmented worms live on land. Unlike flatworms and roundworms, there are few segmented worms that live inside another animal's body.

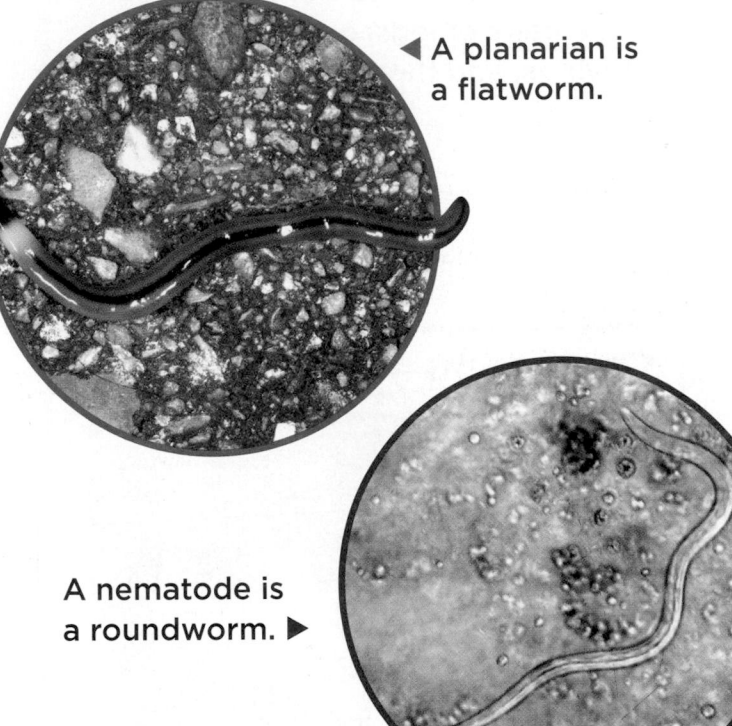

◄ A planarian is a flatworm.

A nematode is a roundworm. ▶

✔ Quick Check

Main Idea and Details Describe the three groups of worms.

Critical Thinking Where do worms that live inside the bodies of other animals get their food?

Lesson Review

Visual Summary

 Invertebrates—such as sponges, cnidarians, echinoderms, and mollusks—are animals without backbones.

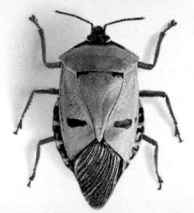 The largest invertebrate group is **arthropods.** They have jointed legs and a body that is divided into sections.

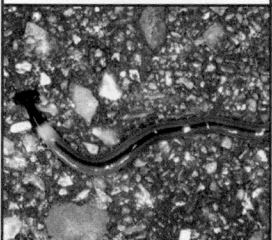 The three main groups of **worms** are flatworms, roundworms, and segmented worms.

Make a FOLDABLES Study Guide

Make a three-tab book. Use it to summarize what you learned about animals without backbones.

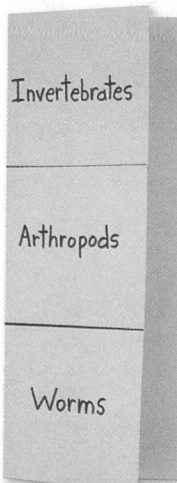

Invertebrates

Arthropods

Worms

Think, Talk, and Write

1 **Vocabulary** When one side of an animal matches up exactly with the other side, that animal has _____.

2 **Main Idea and Details** What are the advantages of having an exoskeleton? The disadvantages?

Main Idea	Details

3 **Critical Thinking** Why do soft animals like sponges and cnidarians not live on land?

4 **Test Prep** Which is an invertebrate?
  A eagle
  B fish
  C snake
  D shrimp

5 **Test Prep** Which characteristic do mollusks and arthropods share?
  A They have backbones.
  B They do not have backbones.
  C They have exoskeletons.
  D They do not move.

6 **Essential Question** How do animals compare?

 **Writing Link**

Write a Story
Choose an invertebrate from this lesson. Write a story from its point of view. Tell what it is like to live the life of this animal.

 **Art Link**

Make a Poster
Make a poster that shows the major invertebrate groups. Use drawings or photographs. Label each group.

Focus on Skills

Inquiry Skill: Classify

You know that some animals are vertebrates and others are invertebrates. This grouping depends on whether an animal has a backbone. Scientists group, or **classify,** living things based on the traits they share.

One way to classify animals is by the presence of a backbone. You can also classify animals by other traits, such as symmetry. Symmetry describes how the body parts are arranged.

jellyfish

▶ Learn It

When you **classify,** you place things that share properties into groups. Classifying is a good way to organize data. You can probably remember the properties of a few groups. It is harder to remember those properties when you have thousands of groups!

It is important to keep good notes when you classify. Your notes can help you see why things belong in the same group. They also help you classify things in the future.

red fox

▶ Try It

Classify animals by their symmetry. Most animals, like butterflies, have *bilateral* (bi•LA•tuh•rul) symmetry. This means that their two sides are alike. Others, like sea stars, have *radial* (RAY•dee•ul) symmetry. That means their body parts stretch out from a central point. A few animals have no symmetry at all.

1 Look at each animal pictured on these pages.

2 Write each animal's name on a chart like the one shown.

3 Mark an *X* on the chart to show the kind of symmetry each animal has.

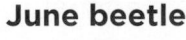

June beetle

My Observations

Animal	Bilateral Symmetry	Radial Symmetry	No Symmetry
June beetle			
red fox			
barrel sponge			

bat

desert tortoise

▶ **Apply It**

1. Study the data on your chart. How many animals have radial symmetry? Bilateral symmetry? How many animals have no symmetry?

2. Look at magazines or books for photos of animals. Add them to your chart or make a new chart to compare them.

3. **Classify** the symmetry of each new animal. Try to find at least one animal that shows each type of symmetry.

4. Now classify all the animals in a new way. Use size, color, or any other property you choose. Share your findings with the class.

barrel sponge

Animals with Backbones

Look and Wonder

Elephants are the largest land animal. The males can weigh as much as 6,800 kilograms (15,000 pounds)! Nearly all the large animals in the world have something in common that helps support their weight. Do you know what it is?

What does a backbone do?

Make a Prediction

Which can support more weight—an animal with a backbone or one without? Make a prediction.

- modeling clay
- pencil

Test Your Prediction

1. **Make a Model** Using clay, make a model of an animal with four legs but no backbone.

2. Now make an identical clay model, this time with a backbone. Make your model the same size and shape as the first. Assemble the "backbone" by molding clay around a pencil.

3. **Observe** Use balls of clay to add weight to your models. How much weight can each model hold before it breaks?

Step **2**

Draw Conclusions

4. Which model supported the weight better—the one with a backbone or without?

5. What advantage does a backbone give to a land animal?

6. **Infer** What advantage would a backbone give to an animal that lives underwater?

Explore More

Make a third model that uses pencils for the legs and the backbone. How does your new model compare to the others? What do the pencils in the legs represent?

Step **2**

Essential Question

Which animals
have backbones?

Vocabulary

vertebrate, p. 90

warm-blooded, p. 90

cold-blooded, p. 90

amphibian, p. 92

reptile, p. 92

bird, p. 93

mammal, p. 94

Reading Skill ✔

Compare and Contrast

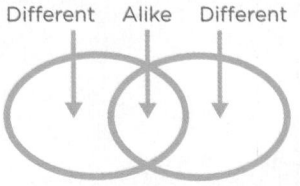

Different   Alike   Different

Technology

e-Glossary and e-Review online
at www.macmillanmh.com

What are vertebrates?

Did you know that you are part of the same large group as fish, frogs, snakes, birds, and mice? Each has a backbone, and so do you!

A backbone is what makes **vertebrates** different from invertebrates. The backbone is an important part of a vertebrate's endoskeleton. It supports the animal's body. A backbone allows larger and heavier animals to move about.

Some vertebrates, such as birds and mammals, are **warm-blooded**. Their body temperature does not change much. These animals get heat from breaking down food. This helps them keep a constant body temperature.

Fish, amphibians, and reptiles are **cold-blooded**. They cannot keep constant body temperatures. Instead, their body temperatures change with their surroundings. Cold-blooded animals get heat from outside their bodies.

Vertebrates are grouped into seven classes. These are jawless fish, cartilaginous (kar•tuh•LA•juh•nus) fish, bony fish, amphibians, reptiles, birds, and mammals.

Can you spot the backbone of this fish?

reptiles

jawless fish

amphibians

bony fish

warm-blooded

birds

Fish

The three classes of fish are jawless, cartilaginous, and bony fish. The first two have skeletons made of *cartilage* (KAR•tuh•lij). Cartilage is the same rubbery material that is in your outer ears or the tip of your nose.

Jawless fish have boneless mouths. The mouth acts like a suction cup. Lampreys are one example. Cartilaginous fish include sharks and rays.

The largest class of vertebrates is the bony fish. Their skeletons are made of bone, and they are covered in scales. Tuna and goldfish are bony fish.

✔ Quick Check

Compare and Contrast How are the three classes of fish alike? How are they different?

Critical Thinking Why might warm-blooded vertebrates eat more often than cold-blooded ⸏rates?

e groups are warm-blooded?
⸏looded?

e colors of the labels.

What are some other vertebrate groups?

Amphibians

Frogs, toads, and salamanders are amphibians (am•FIH•bee•unz). **Amphibians** are cold-blooded vertebrates that spend part of their lives in water and part on land.

Like all amphibians, a frog begins its life in water. Frog eggs hatch into tadpoles. Tadpoles have gills that allow them to live in water—but not on land. As they grow, these parts change. The tadpoles grow lungs. This allows the adults to live on land.

An amphibian's skin needs to stay moist. Even though it has lungs, the adult also breathes through its skin. If the skin dries out, the animal will not survive. That is why most amphibians live near water.

Reptiles

Lizards, snakes, turtles, and alligators are reptiles (REP•tilez). **Reptiles** are cold-blooded vertebrates that live on land.

Unlike amphibians, reptiles have dry skin. Their skin is covered with plate-like scales. This strong, waterproof covering helps reptiles live on land.

Reptiles cannot breathe through their skin. They use lungs. When they reproduce, a tough cover keeps their eggs from drying out.

Amphibians and Reptiles

lizard

frog

Read a Photo

How are the frog and lizard different?
Clue: Observe the surroundings of each animal.

Birds are the only animals that have feathers.

Birds

Birds are warm-blooded vertebrates with feathers. Feathers are light, yet they keep birds warm and dry. Birds also have beaks and two legs with clawed feet. Their feet have scales, like reptile skin.

Even though all birds have feathers, not all birds can fly. For birds that do fly, other traits come in handy. Light, hollow bones help. So do powerful lungs. The shape of their wings and strong flight muscles help them lift off the ground.

Birds lay eggs with strong shells. Most birds keep their eggs warm by sitting on them until they hatch.

FACT Reptile skin is dry, not slimy.

How Birds Fly

1 **Measure** Cut a strip of paper about 5 centimeters wide and 20 cm long.

2 **Make a Model** Put the top 2 cm of the strip between the cover and the first page of an open book. Close the book. The paper models a wing.

3 Hold the book near your mouth with the long side horizontal. Gently blow across the top of the paper.

4 What happens to the paper when you blow across it?

5 **Infer** The shape of a bird's wing is like an airplane wing. How might this shape help birds fly?

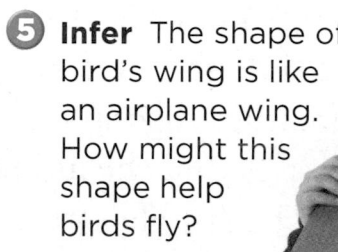

✔ *Quick Check*

Compare and Contrast How do the body coverings of amphibians, reptiles, and birds compare?

Critical Thinking Would a lizard be able to live in a very cold environment? Why or why not?

What are mammals?

Did you know that you are a mammal? A **mammal** is a warm-blooded vertebrate with fur or hair. Mammals can live in trees, water, and most other places on Earth.

Mammals care for their offspring. The three main groups of mammals are classified by how the young are born. Most give birth to live young. But some lay eggs. Females produce milk to feed their young.

The loris is a mammal with a good sense of sight.

 Quick Check

Compare and Contrast How do mammals differ from each other? How are they alike?

Critical Thinking A scientist discovers a new animal. She thinks it may be a mammal. How can she be sure?

Groups of Mammals

Mammals That Lay Eggs The only mammals that lay eggs are the duck-billed platypus and the spiny anteater.

Mammals with Pouches Kangaroos, koalas, and opossums carry their young in pouches until the young are grown.

Mammals That Develop Inside Sheep, bats, apes, and all other mammals develop inside the mother's body.

Lesson Review

Visual Summary

Vertebrates are animals with backbones. They can be warm-blooded or cold-blooded. There are seven classes.

Fish, amphibians, and **reptiles** are cold-blooded. **Birds** are warm-blooded and have feathers.

Mammals are warm-blooded and have hair or fur. The young develop in three different ways.

Make a FOLDABLES Study Guide

Make a layered-look book. Use it to summarize what you learned about animals with backbones.

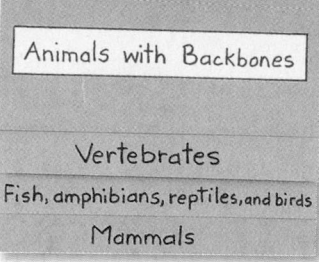

Animals with Backbones

Vertebrates

Fish, amphibians, reptiles, and birds

Mammals

Think, Talk, and Write

1. **Vocabulary** A _____-blooded animal uses heat from its surroundings to stay warm.

2. **Compare and Contrast** How are the seven classes of vertebrates alike? How are they different?

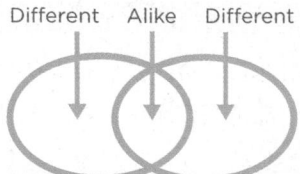

Different   Alike   Different

3. **Critical Thinking** A newt looks like a lizard, but it is really an amphibian. What traits must a newt have?

4. **Test Prep** Birds and mammals <u>most likely</u>

   A have backbones and give milk.

   B have backbones and lay eggs.

   C have backbones and look after their young.

   D lay eggs and are cold-blooded.

5. **Essential Question** Which animals have backbones?

Writing Link

Write a Story
What vertebrates live in your neighborhood? Choose one. Write a paragraph that explains what kind of vertebrate it is. Include its characteristics.

Math Link

Solve a Problem
A blue whale can weigh 100 tons. One ton equals 2,000 pounds. How many pounds does the blue whale weigh? How many times heavier than you is the blue whale?

Gentle Giants

Manatees are very large mammals that live in the water. These gentle giants live in Florida. Manatees are *endangered*. This means there are very few left. Scientists are worried that they might die out. How did this happen?

Manatees live along the coast, where people often go boating. These animals are curious, but they move very slowly. Sometimes they come close to people in boats and bump into them. Many have been hurt or killed by boat propellers.

People build things along the coast. This affects the air, the water, and the land. Many plants that manatees eat have died. Manatees prefer warm water, but water temperatures along the coast may be cooling.

Explanatory Writing

Good explanatory writing

▶ explains how to do a task or how something happened;

▶ gives clear details that are easy to follow.

Write About It

Explanatory Writing
Find out more about another endangered animal. Write a short explanation of why it is endangered.

 **LOG ON** **e-Journal** Research and write about it online at **www.macmillanmh.com**

Protecting Animals

Scientists are working to protect manatees and other endangered animals. The United States Fish and Wildlife Service, or USFWS, keeps a list of species that are endangered. The Karner blue butterfly and San Francisco garter snake are two examples. The table below shows the number of endangered animal species throughout our nation.

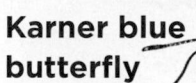

Karner blue butterfly

Making a Hundredths Box

▶ A hundredths box can help you write percents. The word *percent* means "out of 100." So 25 percent means 25 out of 100. To show 25 percent, shade in 25 of 100 boxes.

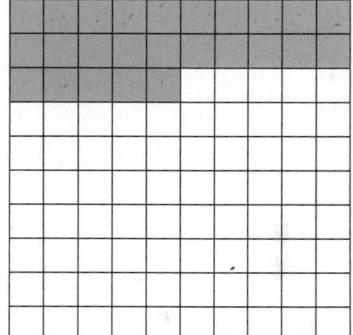

▶ A percent can also be written as a fraction. 25 percent = $\frac{25}{100}$ = $\frac{1}{4}$

Endangered Animal Species in the United States		
Animal Group	Number of Endangered Species	Percentage of Total
mammals	68	17
birds	76	19
reptiles	14	3
amphibians	13	3
fishes	75	18
insects and spiders	59	14
other invertebrates	105	26
Total	410	100

Source: USFWS Threatened and Endangered Species System, Sept. 2006

Solve It

Use the table to solve these math problems.

1. What percentage of endangered species are insects and spiders?

2. Which vertebrate group has the highest percentage of endangered species?

3. Vertebrates make up what percentage of endangered species?

Systems in Animals

ostrich

Look and Wonder

Did you know that birds can run? Ostriches run the fastest—nearly 64 kilometers per hour (40 miles per hour)! They use their powerful leg muscles to escape danger quickly. What other body systems help animals survive?

How does an earthworm respond to light?

Form a Hypothesis

How will an earthworm react to light? Write a hypothesis.

Test Your Hypothesis

1 Gently place an earthworm on a moist paper towel.

2 **Observe** Use the hand lens to watch the earthworm for several minutes. What does it do? Does the worm stay in one place or does it move around? Record your observations.

3 **Experiment** Shine a flashlight on the earthworm for about one minute. Watch how it reacts. Record your observations in a chart.

4 Repeat step 3 three more times. Record your observations.

Draw Conclusions

5 **Interpret Data** Did your results support your hypothesis? What happens when an earthworm is exposed to light?

6 How might an earthworm sense light?

Explore More

Can an earthworm sense light when under the ground? Form a hypothesis. Design an experiment to answer the question.

Materials

- **earthworm**
- **moist paper towel**
- **hand lens**
- **flashlight**

Step 2

▶ **Reading Skill** ✔

Cause and Effect

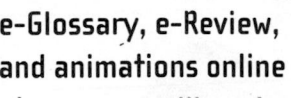

Cause → Effect	
	→
	→
	→
	→

▶ **Technology** 🔵LOG ON

e-Glossary, e-Review, and animations online at www.macmillanmh.com

How do animals move and sense changes?

Animals have different organ systems. A system is a group of parts that work together.

The Skeletal and Muscular Systems

Bone is living tissue. A vertebrate's bones make up its **skeletal system** (SKE•luh•tul SIS•tum). This is the frame that supports an animal's body. It also protects the organs inside.

The skeletal system works with another system to allow vertebrates to move. This is the **muscular** (MUS•kyuh•lur) **system**. It is made of strong tissues called *muscles* (MUH•sulz). To move, the muscles shorten and pull on the bones.

How do invertebrates move? Most have some kind of muscular system. Earthworms wriggle by shortening and stretching their muscles.

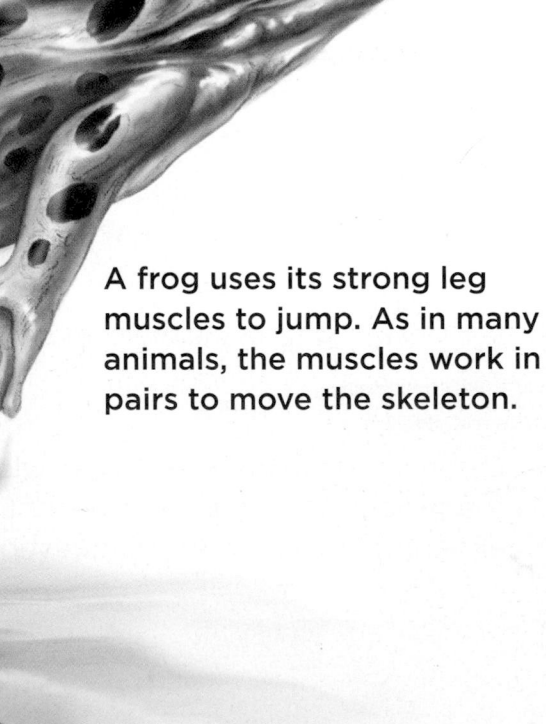

A frog uses its strong leg muscles to jump. As in many animals, the muscles work in pairs to move the skeleton.

The Nervous System

The master control system of the body is the nervous (NUR•vus) system. The **nervous system** is made of nerve cells.

Invertebrates have simple nervous systems. A sponge, for example, has only a few scattered nerve cells. Vertebrates have complicated nervous systems. In vertebrates, millions of nerve cells work together as *nerves*.

The nervous systems of vertebrates consist of a brain, spinal cord, nerves, and sense organs. These help animals use senses—such as sight, hearing, taste, touch, and smell—to detect changes in their surroundings.

Owls have a keen sense of sight. Large eyes help them see at night.

✓ Quick Check

Cause and Effect How do the skeletal and muscular systems work together?

Critical Thinking How is the nervous system important to the other body systems?

The dolphin's brain sends a message to jump. The message travels through the dolphin's nerves to its muscles. Then the dolphin leaps into action!

Make a Model Lung

1. Your teacher will cut the bottoms off of a plastic bottle and a balloon. Stretch the balloon over the bottom of the bottle. Tape it in place.

2. Put one end of a straw into the neck of a whole balloon. Wrap a rubber band around the neck and straw. Make a tight seal.

3. Insert the straw and balloon into the top of the bottle. Fix them in place with modeling clay. The balloon should hang inside the bottle.

4. **Make a Model** Push and pull on the balloon that is stretched over the bottom of the bottle. What happens?

5. **Infer** The diaphragm (DI•uh•fram) is a muscle that inflates the lungs. Which part of your model represents the diaphragm? How does the model show how lungs work?

How do air and blood travel in the body?

The Respiratory System

All animal cells need oxygen. To get oxygen from the air into their cells, most animals use a respiratory (RES•pruh•tor•ee) system. The **respiratory system** brings oxygen to the blood and removes wastes, like carbon dioxide gas, from the blood.

Some small invertebrates, like worms, do not need such a system. Gases move easily into and out of their tissues. Larger animals need a respiratory system. They use organs such as gills or lungs to exchange gases with the water or the air.

Adult salamanders have lungs. Like all amphibians, they also breathe through their skin.

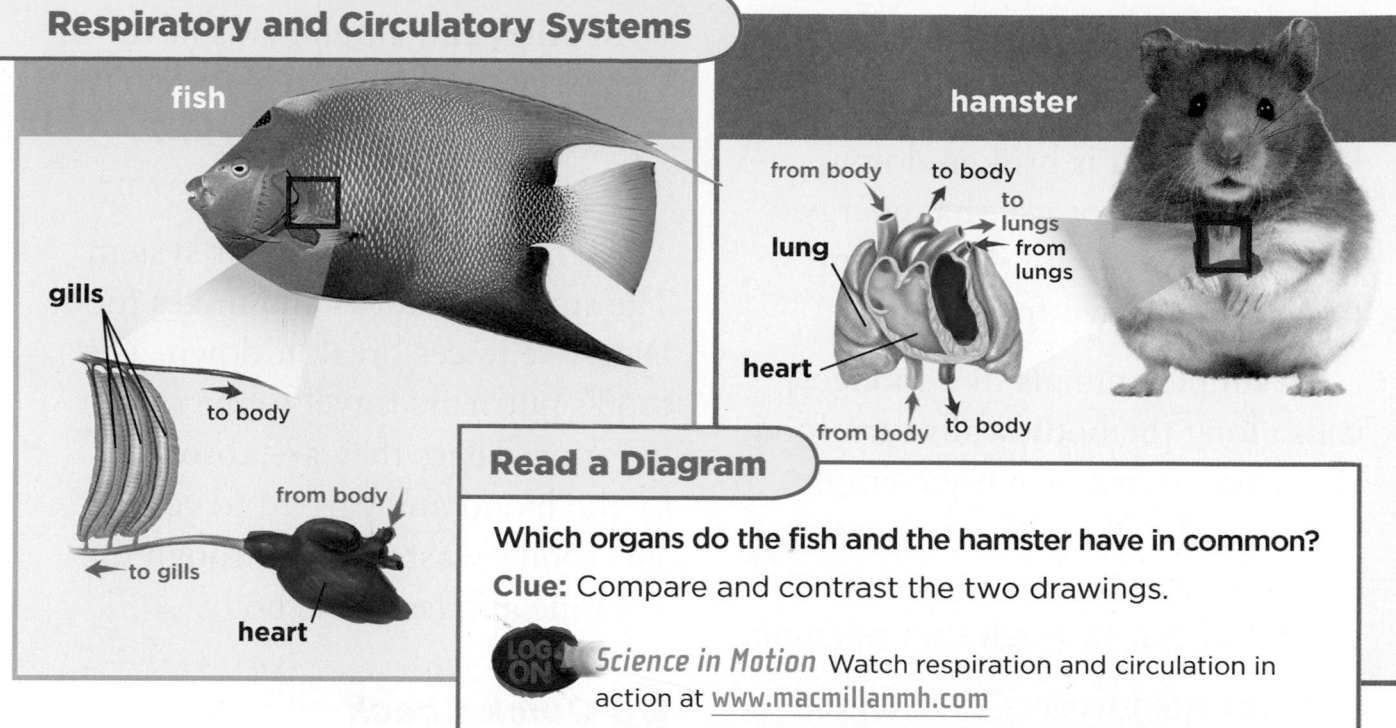

Respiratory and Circulatory Systems

fish

gills

to body

to gills

from body

heart

hamster

from body

to body

to lungs

from lungs

lung

heart

from body

to body

Read a Diagram

Which organs do the fish and the hamster have in common?

Clue: Compare and contrast the two drawings.

Science in Motion Watch respiration and circulation in action at www.macmillanmh.com

The Circulatory System

The heart, blood, and blood vessels make up the circulatory (SUR•kyuh•luh•tor•ee) system. The job of the **circulatory system** is to move blood through the body. The blood carries oxygen, food, and water to the body's cells. It also removes the cells' wastes.

The heart is the main organ of the circulatory system. It has strong muscle tissue to pump lots of blood.

The hearts of most fish have two parts, or chambers. An amphibian heart has three chambers. Mammals and birds have hearts with four chambers. Sponges and cnidarians do not have hearts. In fact, they have no circulatory system at all!

The Excretory System

When cells break down food and other chemicals, they produce wastes. The **excretory** (EK•skruh•tor•ee) **system** removes these wastes.

The liver, kidneys, bladder, skin, and lungs are excretory organs. The liver and kidneys filter wastes from the blood. The bladder stores liquid wastes. The skin sweats to remove excess minerals. Lungs remove waste gases from cells. So do gills.

✓ Quick Check

Cause and Effect What would happen if blood did not pick up oxygen in the lungs?

Critical Thinking What do the respiratory and circulatory systems have in common?

FACT Blood is actually a liquid tissue.

How is food broken down?

Animals take in food for energy. Until that food is broken down, body cells cannot use that energy. The **digestive** (di•JES•tiv) **system** helps break down food.

In simple animals like sponges, cells along the body walls turn food into small particles. Other simple invertebrates have digestive systems with one opening. Food and wastes enter and exit through that opening.

Segmented worms have digestive systems with two openings. Food enters through the mouth. Wastes exit through the tail end.

Reptiles and amphibians have a more complex digestive system. Study the diagram below. How many digestive organs are shown?

Mammals have a similar system. The stomach churns and mixes food. Digestive juices break it down. The food's nutrients travel to the small intestine. There they are absorbed by the blood and carried to cells. The food's wastes pass through the intestine and leave the body.

✔ Quick Check

Cause and Effect What happens to the food that a dog eats?

Critical Thinking Predict what will happen to an animal that has a damaged digestive system.

The Digestive System

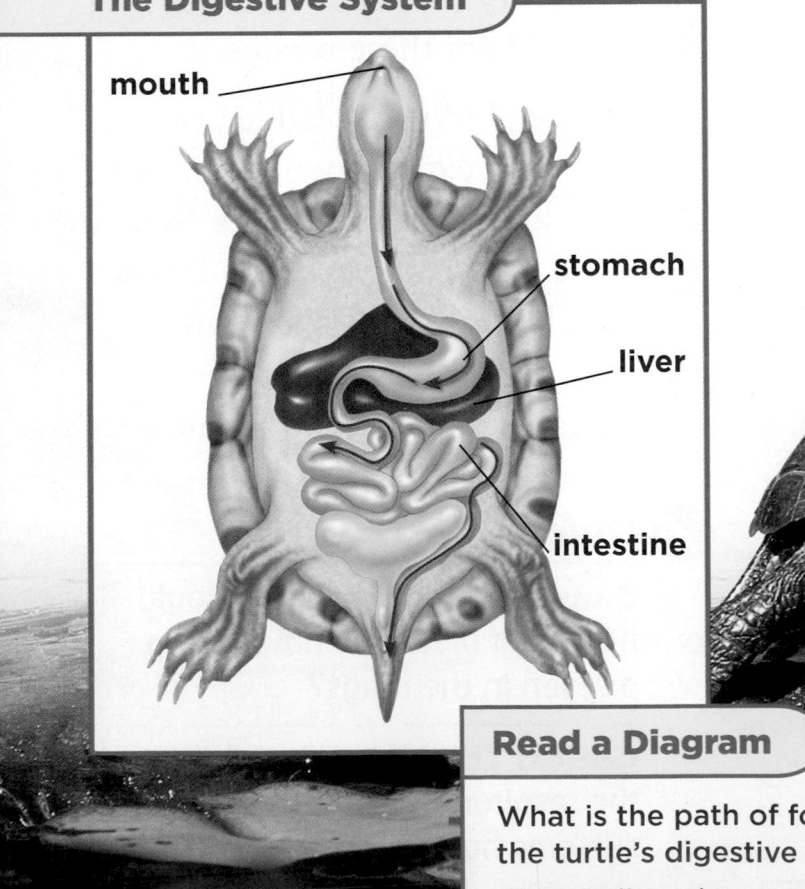

mouth

stomach

liver

intestine

Read a Diagram

What is the path of food in the turtle's digestive system?

Clue: Follow the arrows.

Lesson Review

Visual Summary

 Animals use **nervous, skeletal,** and **muscular systems** to respond and move.

 The **circulatory** and **respiratory systems** transport blood and oxygen.

 The **digestive system** breaks down food for energy. The **excretory system** removes the food waste.

Make a **FOLDABLES**® Study Guide

Make a three-tab book. Use it to summarize what you learned about animal systems.

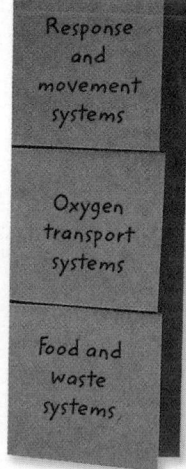

Response and movement systems

Oxygen transport systems

Food and waste systems

Think, Talk, and Write

1 **Vocabulary** The _____ system takes in oxygen from air or water.

2 **Cause and Effect** How does the nervous system cause the muscular and skeletal systems to move your arm?

Cause → Effect
→
→
→
→

3 **Critical Thinking** You climb a flight of stairs and find that your heart starts beating faster and you take deeper breaths. Why does physical activity affect you this way?

4 **Test Prep** The stomach is part of which body system?

- **A** digestive
- **B** nervous
- **C** skeletal
- **D** circulatory

5 **Test Prep** The excretory system

- **A** takes in oxygen.
- **B** supports the muscular system.
- **C** breaks down food.
- **D** gets rid of wastes.

6 **Essential Question** How do systems help animals survive?

 **Writing Link**

Write an Essay

Which of the five senses is most important to you? Write an essay about how it helps you understand changes in your environment.

 **Art Link**

Draw a Diagram

Draw a diagram of an entire animal. Show the main organs of each body system. Attach labels that name the organs and explain what they do.

Be a Scientist

craft sticks

glue

contact paper

scissors

basin of water

Structured Inquiry

How do feet help birds move in water?

Form a Hypothesis

Birds travel in the air, on the land, and in the water. How does the shape of a bird's feet help it to swim? Write your answer in the form "If a bird has feet that are … , then it will move better in the water."

Test Your Hypothesis

1. **Make a Model** Spread out three craft sticks in a fan shape. Glue the sticks in place. This is the frame for your bird foot.

2. Follow step 1 to make a second bird foot.

3. Cover the top and bottom of the first bird foot with contact paper. Cut the paper to the correct size around the outside of the foot. Leave the second foot uncovered.

4. **Observe** Drag each foot through a basin of water several times slowly. Observe the amount of water that gets pushed aside each time. Record your observations.

Step 1

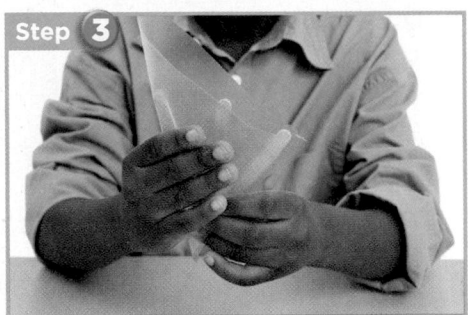

Step 3

Step 4

Draw Conclusions

5 **Interpret Data** Which foot moved more water? Which foot is better suited for swimming?

6 **Infer** What kind of feet do your models represent in real life?

Pekin duck

Guided Inquiry

How do teeth help animals eat?

Form a Hypothesis

Many animals have front teeth that are shaped differently from their back teeth. How does the shape of their teeth help animals eat different foods? Write a hypothesis.

Test Your Hypothesis

Make a plan to test how different shapes of teeth are used for eating different kinds of foods. Choose foods that animals might eat, such as carrots, corn, meat, or seeds. Write the steps you will follow. Then conduct your experiment. Record your results and observations.

Draw Conclusions

What can you conclude about the different shapes of teeth? Which ones are better for eating which kinds of foods? Why?

Open Inquiry

What other questions do you have about animal structures? Design an investigation to answer one of your questions. Write the steps so that another group can do the experiment by following your procedure.

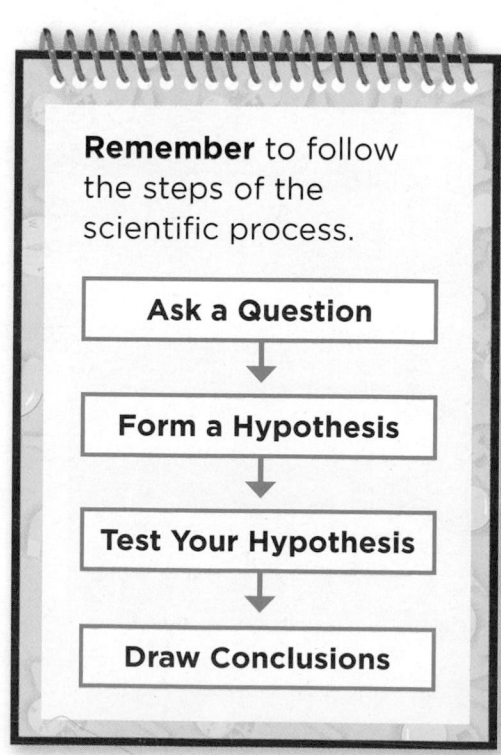

Remember to follow the steps of the scientific process.

Ask a Question

↓

Form a Hypothesis

↓

Test Your Hypothesis

↓

Draw Conclusions

Animal Life Cycles

caterpillar

Look and Wonder

When this caterpillar is ready, it will spin a cocoon. There it will live for a short time. The next time it appears, it will have wings! How does a caterpillar change to a butterfly?

How does a caterpillar change as it grows?

Purpose

Explore how a caterpillar changes into a butterfly.

Procedure

1. Your teacher will give you a caterpillar. Place it gently inside the butterfly kit. Put the food and water in the proper place inside the kit.

2. **Observe** Look carefully at your caterpillar each day. Record any changes in a table like the one shown.

Draw Conclusions

3. **Interpret Data** How many different forms did your caterpillar take? Describe each form.

4. How does the final stage of a caterpillar's life compare with the first stage?

5. **Predict** Do all caterpillars go through these same life stages? Design an investigation that you could test.

Explore More

Do other animals have different stages of development? How could you find out?

Materials

- butterfly kit

Step 2

Observations		
Day	Body Changes	Behavior Changes
1		
2		
3		
4		

What are the stages of an animal's life?

The caterpillar goes through many changes as it grows into a butterfly. These stages of growth and change make up an organism's life cycle. All living things follow a pattern of birth, growth, reproduction, and death.

Each different organism has its own kind of life cycle. A penguin, for example, changes slowly as it grows. The chick hatches from its egg after many weeks. It depends on its parents for warmth, shelter, and food. Soon the chick grows a thick coat of downy feathers. This helps it stay warm and dry by itself.

As the penguin gets older, waterproof feathers grow in place of the down. They keep the adult dry while it swims in search of food.

A moth begins life as an egg.

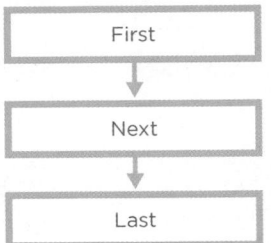

penguin egg

newly hatched chick

young penguin

Life Spans

An organism's **life span** is how long it can usually live in the wild. A moth has a life span of about one week. The oldest recorded age for a human is 122 years! Scientists do not know why some animals have longer life spans than others.

A boa constrictor can live as long as 20 years.

✔ Quick Check

Sequence What are the main stages of an animal's life?

Critical Thinking The average life span of a mouse is three years. Why don't all mice live for three years?

adult penguin

penguin

The life span of a skunk is about three years.

Koi fish can live to be 100 years old!

What is metamorphosis?

Most young animals look like smaller copies of their parents. Kittens look like small cats. Chicks look like small birds. Other young animals don't look like their parents at all. Take insects, for example.

Butterflies and most other insects go through metamorphosis (me•tuh•MOR•fuh•sis). The process of **metamorphosis** has several separate growth stages.

Incomplete Metamorphosis

Damselflies are insects that go through *incomplete metamorphosis*. In this kind of metamorphosis, the difference in growth stages is hard to see.

As the insect grows and changes, it *molts*. Molting is when an animal gets too large for its exoskeleton. It sheds the exoskeleton and grows a new one. Other insects that go through incomplete metamorphosis are grasshoppers and termites.

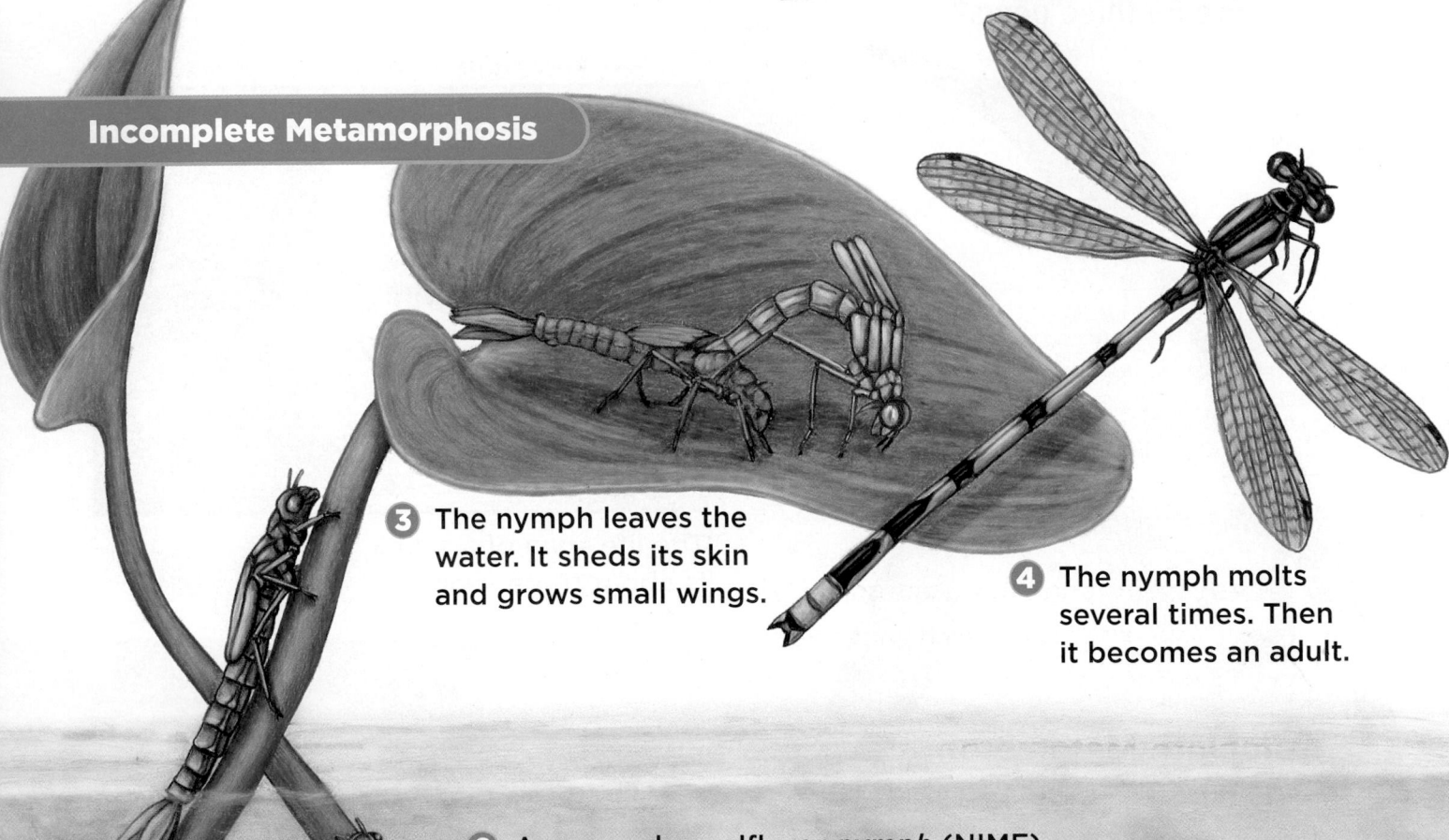

Incomplete Metamorphosis

3 The nymph leaves the water. It sheds its skin and grows small wings.

4 The nymph molts several times. Then it becomes an adult.

2 A young damselfly, or *nymph* (NIMF), hatches from an egg.

1 A female damselfly lays eggs on the stem of a water plant.

Egg

① A female butterfly lays eggs on a leaf.

Larva

② A wormlike *larva* hatches from the egg. It begins to eat the leaf.

Pupa

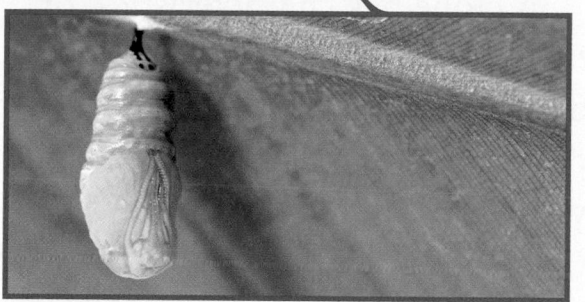

③ The larva becomes a *pupa* (PYEW•puh) and develops adult tissues and organs.

Adult

④ The adult animal is a butterfly. It has six legs, two pairs of wings, and a pair of long antennae. Butterflies can reproduce.

Read a Photo

After which stage does a caterpillar become a butterfly?

Clue: Follow the numbered captions.

Complete Metamorphosis

Butterflies go through *complete metamorphosis*. Look at the sequence of photos. Each growth stage looks different. Beetles, flies, and mealworms also go through complete metamorphosis.

✓ Quick Check

Sequence What are the stages in the complete metamorphosis of a butterfly?

Critical Thinking What stages of life do humans seem to go through? List some of them.

How do animals reproduce?

The life cycle of every animal includes reproduction. This is when parent animals make offspring.

One Parent

Cnidarians and other simple invertebrates reproduce by *budding*. A bud forms on the adult's body. The bud slowly develops into a new animal. After some time, the bud breaks off. It grows into an adult.

Sea stars and other echinoderms can reproduce by *regeneration* (rih•je•nuh•RAY•shun). This is when a whole animal develops from just a part of the original animal.

Both budding and regeneration produce clones. A **clone** is an exact copy of its parent.

Birds need two parents to produce eggs. One parent may guard the eggs while the other parent gets food.

Inheriting Traits

A clone has characteristics, or *traits*, that are identical to those of its parent. When traits are passed from parent to offspring, we say those traits are *inherited* (in•HER•uh•tid). All animals inherit traits from their parent or parents. The passing of these traits is called **heredity** (huh•REH•duh•tee).

Like all mammals, zebras reproduce with two parents.

Kinds of Reproduction

	Method	Number of Parents
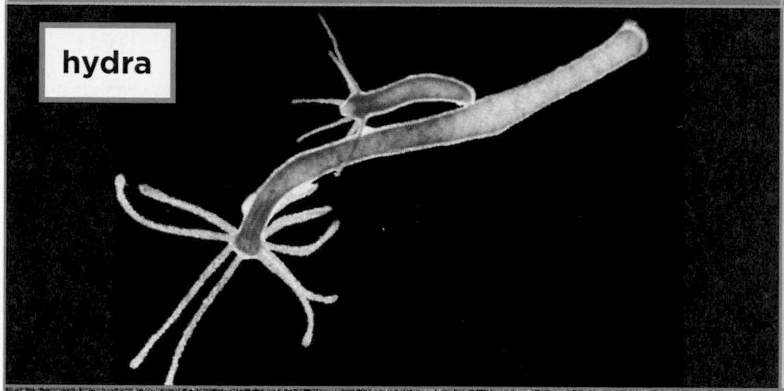 hydra	budding	1
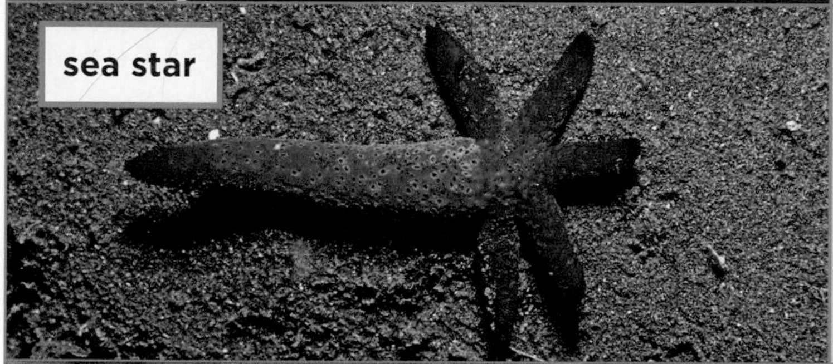 sea star	regeneration	1
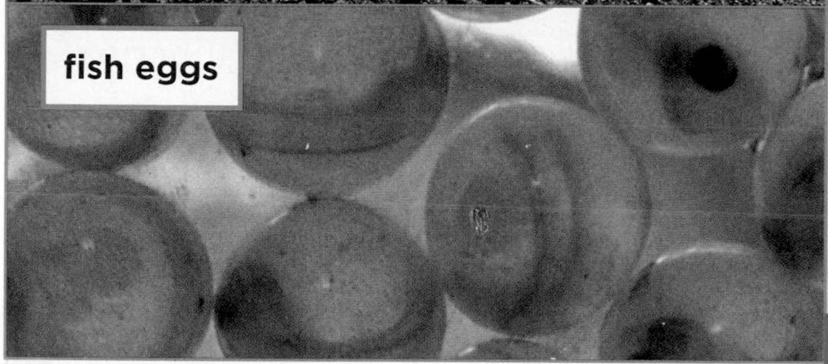 fish eggs	fertilization	2

Read a Table

Which animal reproduces with two parents?

Clue: Look at the data in the last column.

Two Parents

Another kind of reproduction requires cells from two parents. The female cell is called an *egg*. The male cell is a *sperm*. When an egg and a sperm join, offspring are produced. This joining is called *fertilization* (fur•tuh•luh•ZAY•shun).

As the fertilized egg grows, it is called an *embryo* (EM•bree•oh). An embryo has traits from both its parents. It is not identical to either parent.

Quick Check

Sequence Describe the steps that take place in the formation of an embryo.

Critical Thinking Can offspring from two parents be clones? Explain.

Animal Cards

1. Make two sets of inherited traits. Use yellow cards for female parents and orange cards for male parents.

2. Write an example of a trait on each yellow card. On each orange card, write different examples of the same traits.

3. **Use Numbers** Choose one card from each set. How many combinations of different traits can you make?

4. **Infer** How do the cards model reproduction with two parents?

Traits   Traits

What is inherited?

Look at the pigs in the photograph below. Some of their traits, such as color, are determined by heredity. But how are the pigs acting, or *behaving*? Is behavior an inherited trait? Perhaps.

An **inherited behavior** is a set of actions that parents pass on to their offspring. The simplest kind is a reflex, like blinking. A less simple example is instinct. **Instinct** is a way of acting that an animal does not have to learn. Birds build nests, and spiders spin webs, by instinct.

Not all behaviors are inherited. Some are learned. Learning can happen when an animal interacts with its environment or with others. A **learned behavior** occurs when an animal changes its behavior through experience. Do you ride a bicycle? Bicycling is a learned behavior.

✔ Quick Check

Sequence Describe how you could teach a learned behavior.

Critical Thinking What are other examples of learned behavior?

Lesson Review

Visual Summary

The **life cycle** of an animal includes birth, growth, reproduction, and death.

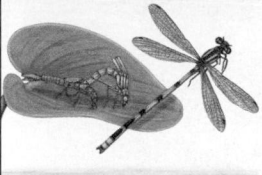

There are two kinds of **metamorphosis**—complete and incomplete.

Animals **reproduce** when parents make offspring. Offspring with two parents **inherit traits** from both.

Make a FOLDABLES® Study Guide

Make a trifold book. Use it to summarize what you learned about animal life cycles.

Main Idea	What I learned...	Sketches
Life cycle		
Metamorphosis		
Traits		

Think, Talk, and Write

1. **Vocabulary** _____ takes place when a sperm and an egg combine.

2. **Sequence** Describe what happens during the incomplete metamorphosis of a damselfly.

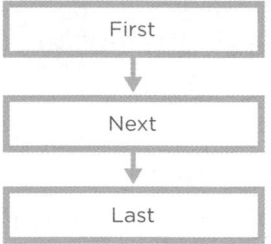

First
↓
Next
↓
Last

3. **Critical Thinking** Why are some behaviors, such as eating, inherited instead of learned?

4. **Test Prep** Which animal's metamorphosis has a pupa stage?
   - **A** frog
   - **B** damselfly
   - **C** butterfly
   - **D** grasshopper

5. **Essential Question** How do animals grow and reproduce?

Writing Link

Fictional Narrative
Choose an animal that goes through complete metamorphosis. Write a story from the point of view of that animal. Include all the stages of the animal's metamorphosis.

Math Link

Solve a Problem
One minute after fertilization, an egg splits. It becomes two cells. The cells continue to split every minute. How long will it take until there are 64 cells?

LOG ON e-Review Summaries and quizzes online at www.macmillanmh.com

Meet Christopher Raxworthy

The island of Madagascar lies off the southeast coast of Africa. This island has plants and animals found nowhere else in the world. Chris Raxworthy is a scientist at the American Museum of Natural History. He has spent many years gathering information about these animals and their habitats.

On the island, you can find unusual creatures like the Mantella poison frog. This tiny frog's vivid colors warn predators to stay away. Female frogs lay 20–30 eggs on land in a clump. The eggs are hidden in moist areas near streams. When it rains, tadpoles hatch. They wriggle their way into the stream. In about a year, the tadpoles turn into adults.

Adult Mantella poison frogs are only about one inch long.

Chris studies amphibians and reptiles in their habitats.

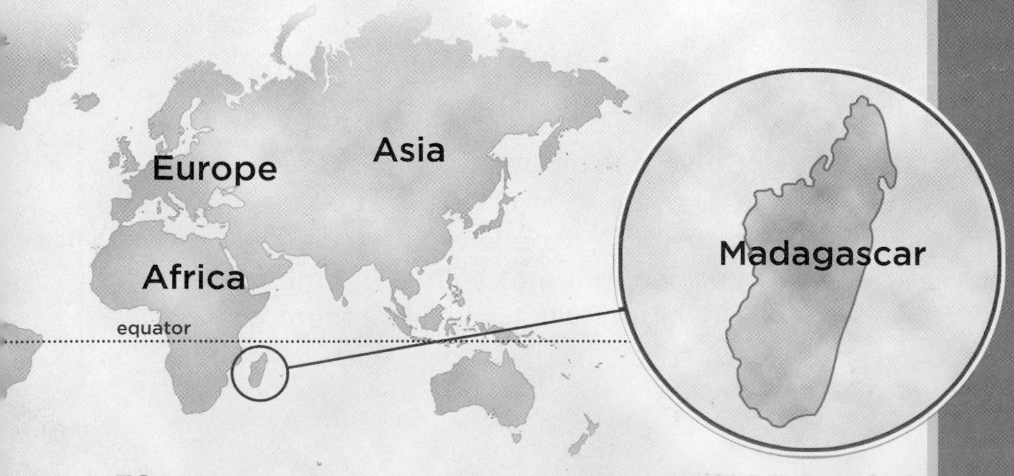

Europe

Asia

Africa

equator

Madagascar

Dwarf dead leaf chameleons grow to about $3\frac{1}{2}$ inches long.

Waterfalls are common in the rain forests of Madagascar.

The dwarf dead leaf chameleon also makes its home on Madagascar. This unusual animal gets its name because it looks just like a dead leaf. During the day, the chameleons hide out among the dead leaves of the rain forest floor. At night, they climb up into the low branches to sleep. Female chameleons lay 2 or 3 large eggs in the leaves on the forest floor. Baby chameleons hatch about ten weeks later. In 9 to 12 months, they are all grown up.

The tropical climate of Madagascar is home to lots of living things. But their habitats are in trouble. Forests are being cut down for farmland and cattle grazing. As the forests disappear, so do the animals that live there. Chris and other scientists want to make sure that creatures like the Mantella poison frog and dwarf dead leaf chameleon don't disappear forever.

Write About It

Compare and Contrast How does the life cycle of the Mantella poison frog compare to the life cycle of the dwarf dead leaf chameleon?

 **e-Journal** Research and write about it online at www.macmillanmh.com

Compare and Contrast

▶ To compare, use words such as *like* and *both*.

▶ To contrast, use words such as *unlike* and *but*.

Connect to

 AMERICAN MUSEUM OF NATURAL HISTORY

at www.macmillanmh.com

Visual Summary

Lesson 1 Invertebrates are animals without backbones.

Lesson 2 Vertebrates are animals with backbones.

Lesson 3 Animals have body systems that help them carry out basic life functions.

Lesson 4 Animals go through stages of growth and change during their life cycles.

Make a FOLDABLES Study Guide

Tape your lesson study guides to a piece of paper as shown. Use your study guide to review what you have learned in this chapter.

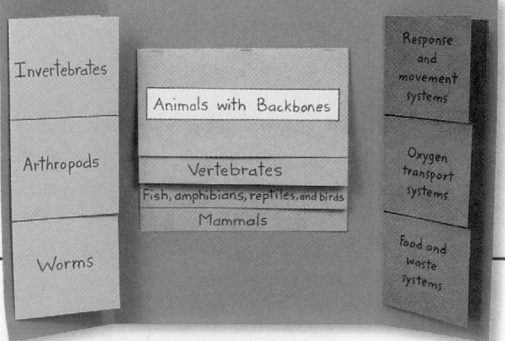

Fill each blank with the best term from the list.

bird, p. 93

digestive system, p. 104

exoskeleton, p. 82

heredity, p. 114

instinct, p. 116

invertebrates, p. 79

metamorphosis, p. 112

nervous system, p. 101

reptile, p. 92

vertebrate, p. 90

1. More than 95 out of every 100 animals are _____.

2. Food is broken down by the _____.

3. A(n) _____ is any animal that has a backbone.

4. Arthropods have a hard _____ that protects their bodies.

5. A(n) _____ is warm-blooded and has feathers.

6. Butterflies go through a process of change called _____.

7. The brain and sense organs are part of an animal's _____.

8. A lizard is a cold-blooded vertebrate belonging to the _____ class.

9. Traits are passed from parent to offspring through _____.

10. A(n) _____ is a behavior that an animal does not have to learn.

LOG ON **e-Glossary** Words and definitions online at www.macmillanmh.com

Answer each of the following.

11. **Main Idea and Details** What is the purpose of the circulatory system? Provide details to support your answer.

12. **Classify** Choose an animal discussed in the chapter. Classify it using what you have learned—as vertebrate or invertebrate, warm-blooded or cold-blooded, and so on. Explain each answer.

13. **Critical Thinking** How might a fish control its body temperature?

14. **Explanatory Writing** How are echinoderms and arthropods alike? How are they different? Provide examples of each.

15. **Classify** Which class of vertebrates has cold-blooded organisms that lay eggs on land?

16. **Classify** Which classes of vertebrates have organisms that <u>most likely</u> care for their young?

17. **True or False** *Damselflies go through complete metamorphosis.* Is this statement true or false? Explain.

18. **True or False** *All fish have bones.* Is this statement true or false? Explain.

19. Which body system carries messages to the other body systems?

   A excretory system

   B nervous system

   C respiratory system

   D muscular system

The Big Idea

20. How are animals different from one another?

Performance Assessment
DOK 2

Make an Invertebrate Reference Book

Make a picture book showing the invertebrates in this chapter.

1. Write the name of each invertebrate you studied.

2. List the invertebrates in alphabetical order in your book. Draw a picture of the animal next to its name.

3. Write down all the information you have learned about the animal.

4. Write a paragraph that explains how two animals from your book are similar and different.

Test Preparation

1 Which animal is classified as an invertebrate?

A octopus

B chicken

C deer

D eagle

DOK I

2 The picture below shows a hydra reproducing.

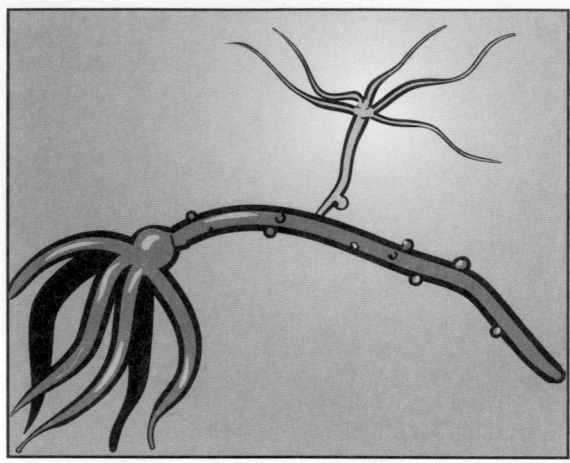

The hydra's offspring will <u>most likely</u> have

A none of the parent's traits.

B some of the parent's traits.

C half of the parent's traits.

D all of the parent's traits.
DOK 2

3 In animals, which system is responsible for communication within the body?

A respiratory system

B digestive system

C skeletal system

D nervous system
DOK 2

4 Which animal <u>most likely</u> cares for its young?

A bird   **C** insect

B frog   **D** snake
DOK I

121A

5 Which shows complete metamorphosis?

A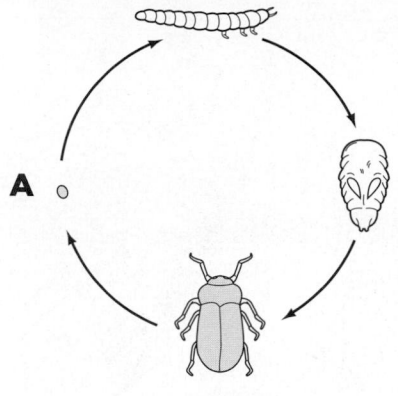

B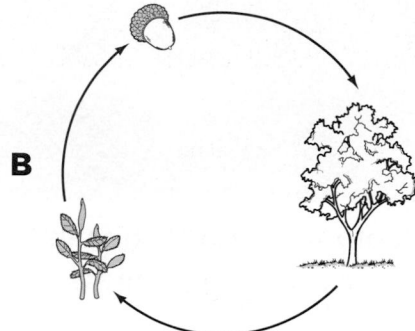

C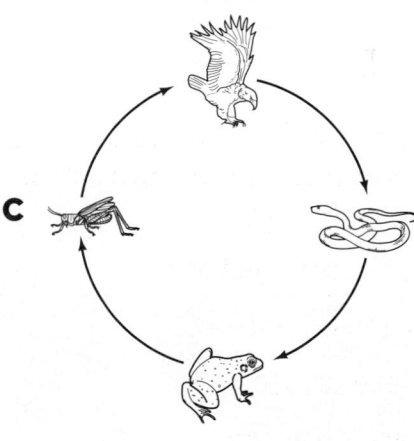

D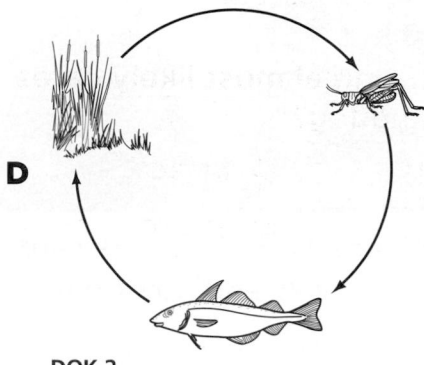

DOK 2

6 The pictures below show a fish and a rabbit.

fish

rabbit

Name two body structures the fish has that the rabbit does <u>not</u> have.
DOK 1

Tell how each structure helps the fish survive in its environment.
DOK 2

Check Your Understanding

Question	Review	Question	Review
1	pp. 78–81	4	pp. 92–93, 110–114
2	pp. 114–115	5	pp. 112–113
3	pp. 100–104, R20–R21	6	pp. 90–94, 102–103

Zoologist

Do you love animals? If so, you might like to be a zoologist (zoh•AH•luh•jist). Zoologists are animal scientists. They study all kinds of animals. Some zoologists work in laboratories. Others study animals in their natural habitats. Still others work in museums or zoos.

All zoologists have college degrees. Most have more than one degree. If this career sounds appealing, start exploring it! Ask your parent or guardian to take you to a local zoo, aquarium, or animal shelter. You might even meet a zoologist there.

▲ Zoologists study animals in their natural habitats.

Bird Handler

What do eagles, hawks, and owls have in common? They are all birds of prey. Birds of prey have good vision for catching prey. They also have very strong claws!

People who care for these birds in wildlife centers are trained bird handlers. Some bird handlers show the animals to visitors. Others collect birds from the wild.

To be a bird handler, you need training. You have to catch the birds without harming them. You need to feed them and keep them healthy. Bird handlers know how to be safe around these majestic birds!

▲ Trained bird handlers care for hawks and other birds.

LOG ON ⊜-Careers at www.macmillanmh.com

Ecosystems

Literature

Magazine Article

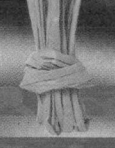

Sea Otters

Key to the Kelp Forest

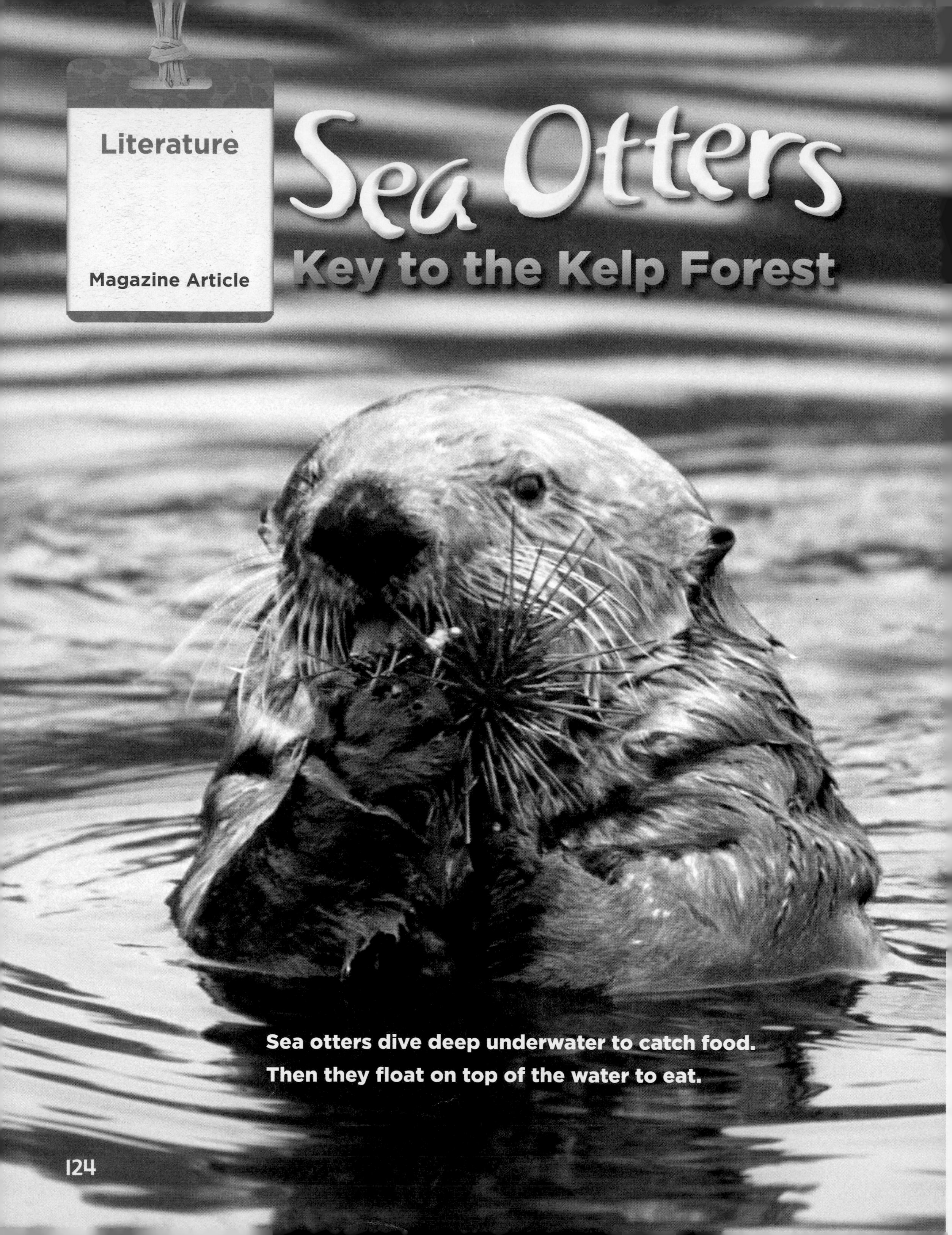

Sea otters dive deep underwater to catch food. Then they float on top of the water to eat.

Sea otters are incredible sea mammals.

They love to be in the water? Unlike whales and seals, sea otters do not have a layer of blubber to keep warm. This can make them cold in the water! Luckily, the do have very thick fur that helps to keep them warm. Sea otters have more hair per square inch than we even have on our entire head – 700,000. That's a lot!

Sea otters eat a bunch of food for energy. This also helps them to stay warm. Sea otters like to eat clams and mussels. Their favorite food is the purple sea urchin. They have to hunt for their food in underwater forests of giant seaweed known as kelp.

Sea otters are very important to the kelp forest. Sea urchins eat loose bits of kelp, and the sea otters dive down to catch and eat them! If there aren't enough sea otters around, the number of sea urchins becomes out of balance, and then they start to eat at the kelp plants. This makes the plants break off and float away! If the sea forest is destroyed, all of the animals lose their home.

Oil spills and diseases caused by pollution are some of the problems that sea otters face. Luckily, people are working to keep the sea otters safe. Not only is this good for them, but it's also good for the many other animals that depend on them!

 **Write About It**

Response to Literature Research another place where plants and animals depend on each other. Write a report describing how the plants and animals interact.

 **e-Journal** Write about it online at www.macmillanmh.com

CHAPTER 3

Exploring Ecosystems

 **The Big Idea** Where do plants and animals live and how do they depend on each other?

Essential Questions • • • • • • • • • • • • • • • • • •

Lesson I

How do the parts of an ecosystem interact?

Lesson 2

How do ecosystems compare?

Lesson 3

How do organisms get energy?

ecosystem the living and nonliving things in an environment and all their interactions (p. 131)

habitat the home of an organism (p. 131)

biome one of Earth's large ecosystems with its own kind of climate, soil, plants, and animals (p. 138)

deciduous forest a forest biome with many kinds of trees that lose their leaves each autumn (p. 140)

producer an organism, such as a plant, that makes food (p. 150)

energy pyramid a diagram that shows the amount of energy at each level of a food chain (p. 156)

 Visit www.macmillanmh.com for online resources.

Introduction to Ecosystems

Look and Wonder

Environments contain many things, both living and nonliving. What are the living things and nonliving things in this environment?

What can you find in an environment?

Make a Prediction

What living and nonliving things might you find in your environment? Make a prediction.

Materials

- tape measure
- 4 clothespins
- ball of yarn
- hand lens

Test Your Prediction

1 **Measure** Mark off an area of ground that is about 1 meter square. Push a clothespin in the ground at each corner. Wrap yarn around the tops of the four clothespins as a border.

2 **Observe** Using your hand lens, look at the living and nonliving things inside the square.

3 Make a data table to record what you see. Label each object as living or nonliving.

4 **Communicate** Share your findings with a classmate. Compare what was in the environments each of you observed.

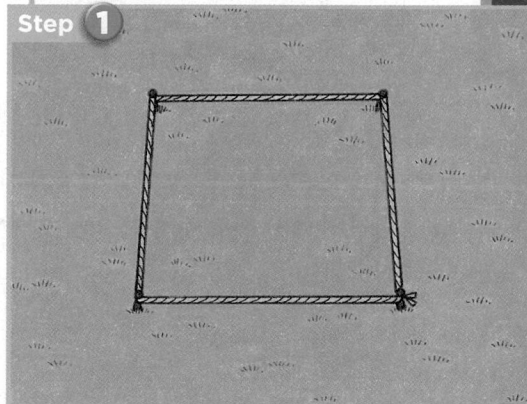

Step **1**

Draw Conclusions

5 **Classify** How many different kinds of living and nonliving things were in your environment? Which did you see more of?

6 Does your data match your prediction? How?

7 How did your data differ from your classmate's? How were the data similar?

Explore More

Would you get the same results if you looked at a different square meter in the same environment? Try it. Compare and contrast your results. Include the results of your classmates.

▶ Essential Question

How do the parts of an ecosystem interact?

▶ Vocabulary

biotic factor, p. 130

abiotic factor, p. 130

ecosystem, p. 131

habitat, p. 131

population, p. 132

community, p. 132

▶ Reading Skill ✔

Fact and Opinion

Fact	Opinion

▶ Vocabulary

e-Glossary and e-Review online at www.macmillanmh.com

What is an ecosystem?

Who or what is around you right now? You probably see classmates or your teacher. Perhaps there are books. You might be at a desk. Your classroom environment contains both living and nonliving things.

Scientists call the living things in an environment **biotic** (bi•AH•tik) **factors**. Plants, animals, and bacteria are all biotic factors. You are a biotic factor too!

Your desk, pen, pencil, and textbook are abiotic (ay•bi•AH•tik) factors. **Abiotic factors** are the nonliving things in an environment. Abiotic factors include water, soil, and light. Climate (KLI•mut) is another abiotic factor. *Climate* refers to the typical weather pattern in an environment. The study of how biotic and abiotic factors interact is called *ecology* (e•KAH•luh•jee).

A Pond Ecosystem

An environment's biotic and abiotic factors interact in an **ecosystem** (EE•koh•sis•tum). An ecosystem can be small, like a single log, or very large, like a forest.

All the living things in an ecosystem depend on the nonliving things to survive. For example, to lay its eggs, a frog depends on the water in a pond. The living things also depend on one another. The frog rests on a lily pad. It eats flies.

The place in an ecosystem where an organism lives is its **habitat**. Different ecosystems have different types of habitats. A forest ecosystem has fallen logs and trees. The logs provide a habitat for spiders and mushrooms. The trees provide a habitat for birds and squirrels.

Read a Diagram

What biotic and abiotic factors might you find in a pond ecosystem?

Clue: Classify the living and nonliving things in the picture.

≡ Quick Lab

Sun and Shade

① Obtain two plants of the same type that are of similar size. Label one pot *shady* and the other *sunny*.

② **Use Variables** Put the plant labeled *sunny* in bright sunlight. Place the other plant in a dark or shady location.

③ Keep the soil in both plants moist using the same amounts of water.

④ Observe each plant every day for two weeks. Record your observations in a data table.

⑤ **Interpret Data** Which variable did you investigate? What can you conclude about how well each plant grew? Explain.

✔ Quick Check

Fact and Opinion *Forest ecosystems are more interesting than pond ecosystems.* Is this statement a fact or an opinion? Explain.

Critical Thinking What kinds of biotic and abiotic factors do you depend on?

Read a Photo

What populations are visible in these two ecosystems?

Clue: Try to name the plants and animals in the photographs.

What are populations and communities?

Like all habitats, a pond is home to many different organisms. Each kind of organism is a member of a different species. A **population** is all the members of a species that live in an ecosystem. For example, the bullfrogs in a pond make up one population. Water lilies are another.

All of the populations in an ecosystem make up a **community**. A pond community may have populations of bullfrogs, fish, water lilies, and dragonflies. The size of a community depends on factors such as food, shelter, and light. Communities in warm, moist ecosystems tend to outnumber those in cold or dry places.

Studying Ecosystems

When scientists want to know about an ecosystem, they look at its populations and communities. A change in a population can affect the community. The reverse is also true.

 Quick Check

Fact and Opinion The algae population is more important than the beetle population in a pond. Is this a fact or an opinion? Explain.

Critical Thinking How can a change in one population affect the entire community in an ecosystem?

Lesson Review

Visual Summary

 An ecosystem includes both **biotic and abiotic factors.**

 Each biotic factor in an ecosystem has its own **habitat.**

 All of the populations in an ecosystem form a **community.**

Make a FOLDABLES Study Guide

Make a three-tab book. Use it to summarize what you have read about ecosystems.

Biotic and abiotic factors

Habitat

Community

Think, Talk, and Write

1. **Vocabulary** How are populations different from communities?

2. **Fact and Opinion** *An entire ecosystem can exist under a single rock.* Is this a fact or an opinion? Explain.

Fact	Opinion

3. **Critical Thinking** How do light and temperature affect the ability of an ecosystem to support life?

4. **Test Prep** Which are abiotic factors?
   A snow, wind, rain
   B air, bacteria, temperature
   C plants, birds, insects
   D rain, wind, flowers

5. **Test Prep** All the populations in an ecosystem make up a(n)
   A habitat.
   B abiotic factor.
   C community.
   D pond.

6. **Essential Question** How do the parts of an ecosystem interact?

 **Math Link**

Solve an Equation
An elephant eats about 155 pounds of food every day. How many pounds of food would a population of nine elephants eat in one day?

 **Art Link**

Picture an Ecosystem
Draw a picture of your favorite ecosystem. Include all the biotic and abiotic factors that you can think of. Label the different factors.

LOG ON e-Review Summaries and quizzes online at www.macmillanmh.com

Inquiry Skill: Predict

Scientists use what they know about a subject to plan their experiments. You know that plants depend on air, soil, light, and water. Knowing this information, you can investigate plants and their needs. You can **predict** what might happen in an experiment.

▶ Learn It

When you **predict**, you state the possible results of an event or experiment. You base your statement on what you already know. First, you tell what you think will happen. Then, you conduct your experiment. Finally, you analyze your results to determine whether your prediction was correct.

▶ Try It

How do you **predict** that a seed will grow in polluted soil? Use what you have learned about plants and ecosystems to make your prediction. Write your prediction. Then, experiment to see whether your prediction was correct.

Materials **2 milk cartons, measuring cup, soil, 10 bean seeds, water, safety goggles, graduated cylinder, vinegar, red food coloring**

1 Label one milk carton *A* and the other one *B*. Place one cup of soil in each carton. Push 5 bean seeds just below the surface of the soil in each carton. Water the soil just until it is moist.

2 ⚠ **Be Careful.** Wear safety goggles. Measure 80 milliliters of vinegar into the measuring cup. Place 5 drops of red food coloring in the vinegar. Carefully pour the liquid into carton B.

3 Place the cartons near a sunny window. Add equal amounts of water to each carton every 2-3 days. Observe both cartons after 2 days, 7 days, and 10 days. Write your observations in a chart like the one shown.

Carton A	
Prediction	
Day	Observations
1	
2	
7	
10	

Carton B	
Prediction	
Day	Observations
1	
2	
7	
10	

4 In which carton did the seeds grow better? Compare your results to your prediction. Was your prediction correct?

5 Carton B models polluted soil. Use a spoon to dig up the soil in carton B. Can you still see the food coloring? What does this tell you about pollution?

▶ **Apply It**

Now that you have learned to think like a scientist, make another prediction. How do you **predict** different amounts of water will affect a plant's growth? Plan an experiment to find out whether your prediction is correct.

Biomes

Acadia National Park, Maine

Look and Wonder

Acadia National Park has 11 species of amphibians, 40 mammal species, and more than 270 different birds. Do they all live in forests like this one? Why do different organisms live in different places?

How much sunlight reaches a forest floor?

Purpose
Model how sunlight affects the kinds of plants that can grow on a forest floor.

Procedure

1. Use the materials to build paper trees with leaves. Remember that trees need many leaves to make their own food.

2. **Make a Model** With your classmates, tape all of the trees upright to a piece of cardboard. This is a model of a forest.

3. Shine a lamp at an angle above the forest. Mark the places on the cardboard that get little or no light.

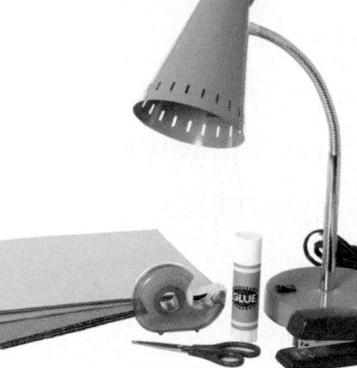

Materials

- construction paper
- scissors
- large piece of cardboard
- tape
- glue
- stapler
- lamp

Draw Conclusions

4. **Interpret Data** How would you describe the light that reaches a forest floor?

5. **Infer** What kinds of plants grow on the floor of a forest? Can you think of any examples?

Explore More

What other factors affect the kinds of plants that can grow on the forest floor? Do some research. Report on your findings.

Step 1

Read and Learn

Essential Question
How do ecosystems compare?

Vocabulary

biome, p. 138
grassland, p. 140
deciduous forest, p. 140
tropical rain forest, p. 141
desert, p. 142
taiga, p. 142
tundra, p. 143

Reading Skill
Summarize

Technology
e-Glossary and e-Review online at macmillanmh.com

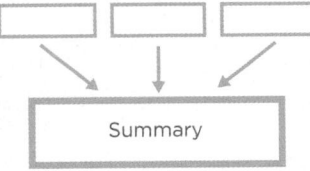

What is a biome?

Suppose you are in an unknown place somewhere in the world. You look around for clues about where you are. The ecosystem stretches out all around you. All you see is sand.

Where are you? You are probably in a desert biome (BI•ohm). A **biome** is a large ecosystem that has its own kinds of plants, animals, and soil. Biomes also have their own patterns of temperature and rainfall. Some biomes are so large, they stretch across an entire continent!

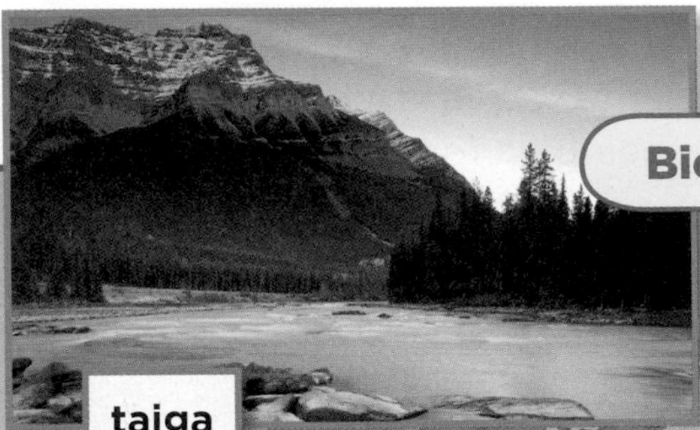

taiga

Biomes of the World

Legend	
tundra	grassland
taiga	desert
deciduous forest	mountain
tropical rain forest	polar ice

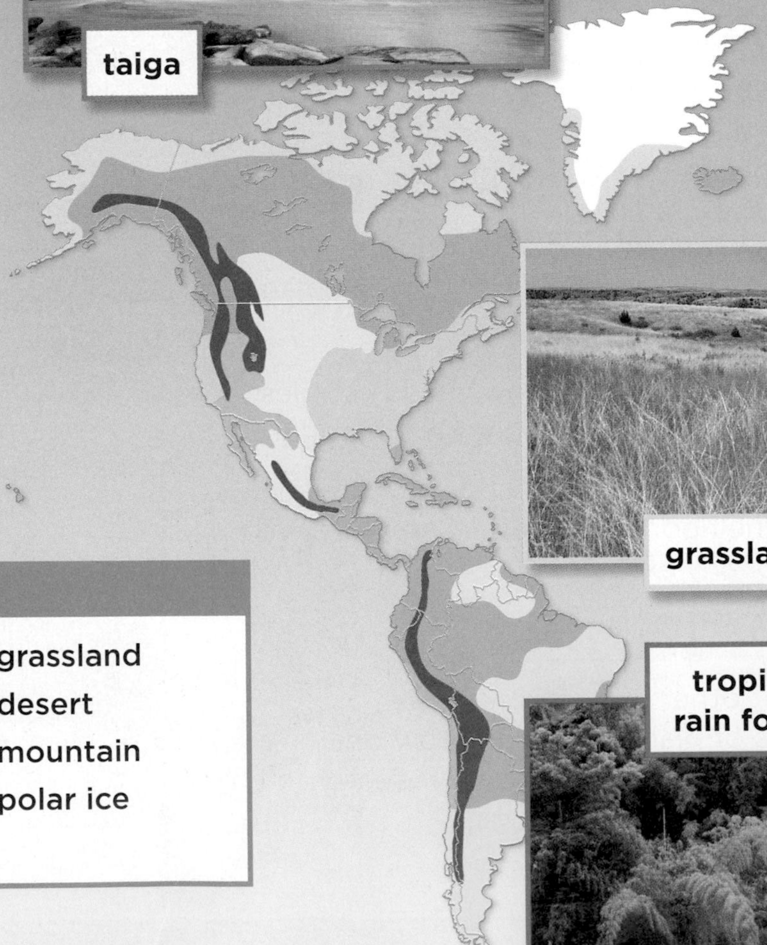

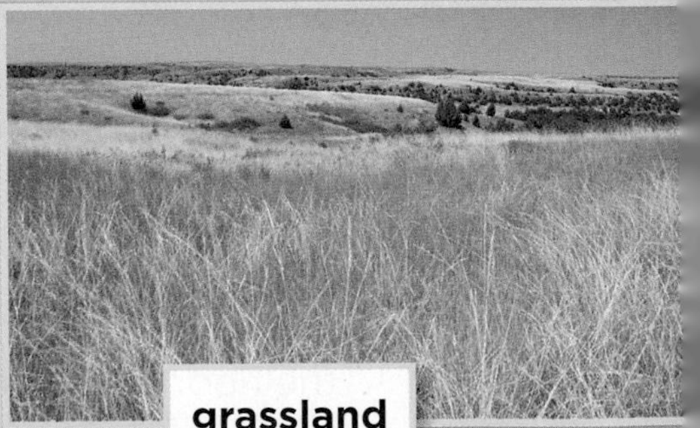

grassland

tropical rain forest

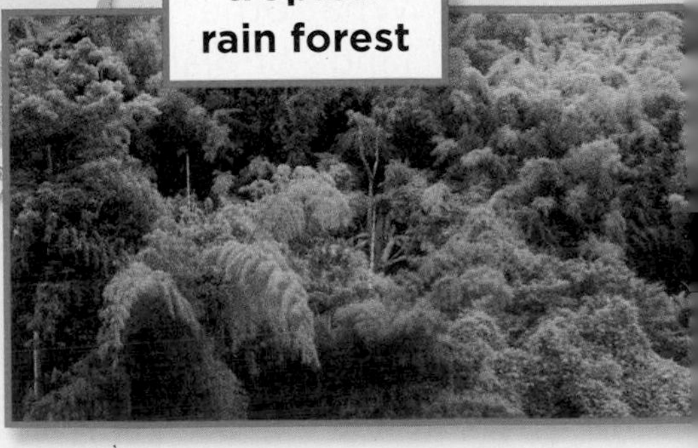

The Major Biomes

Look at the map. Do you notice the different colors? You are looking at the world's major biomes.

Near the equator are tropical rain forests rich with plants and animals. Closer to the poles are taiga and tundra biomes. Fewer organisms live there. Earth also has dry grasslands and sandy deserts. You may be most familiar with the deciduous (dih•SIH•juh•wus) forest biome.

Quick Check

Summarize What is a biome? Describe each of the six biomes shown in the photographs.

Critical Thinking Which biome do you live in? Use the map to help you answer the question.

Read a Map

Which biomes are found in the United States?

Clue: Compare the colors on the map to the legend.

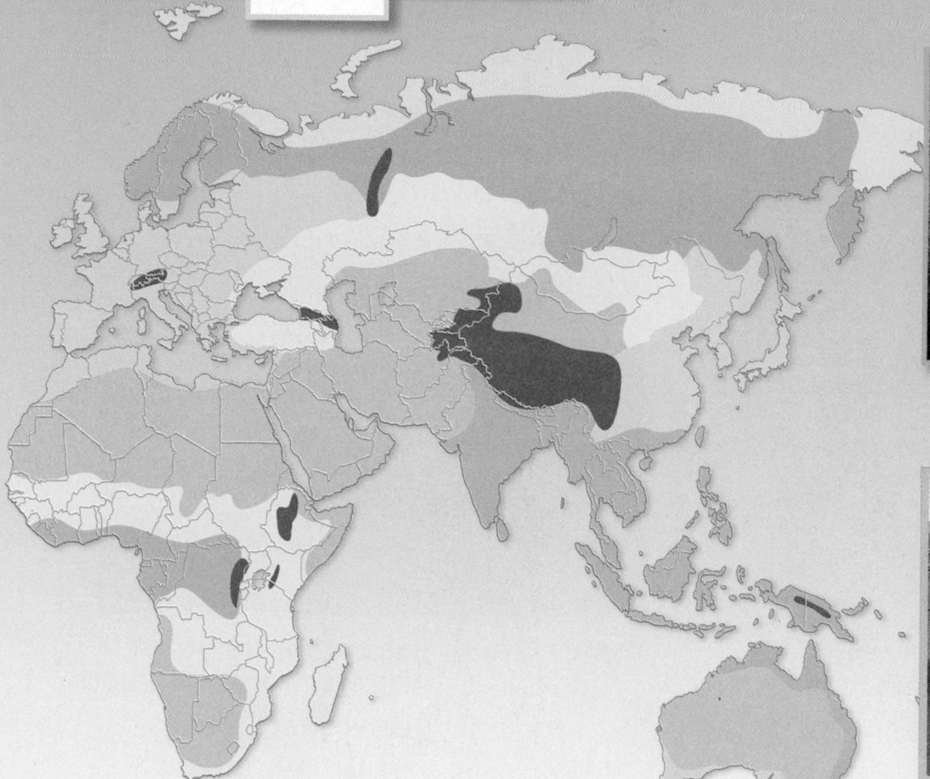

tundra

deciduous forest

desert

What are grasslands and forests?

Grasslands

As its name tells you, a **grassland** is a biome where the main kind of plant is grasses. Grasslands get some, but not much, rainfall.

A *prairie* is a kind of grassland that has a mild climate. During hot, dry summers, the grasses often burn. This burning gives prairies a rich soil that is good for farming. One example of a prairie is in the Great Plains region of the United States.

A *savanna* is a grassland with shrubs and few trees. It is warm all year round. Savannas, like the ones in Africa, get more rain than prairies get. They also have less fertile soils.

Deciduous Forests

Have you ever seen the leaves on trees change colors in the fall? If you have, then you have probably seen a deciduous forest biome.

Earth has many different kinds of forests with all kinds of trees. **Deciduous forests** have trees that lose their leaves each year. Common trees include oaks and maples.

Deciduous forests often have cold winters and warm summers. The trees have broad leaves that keep much of the Sun's light from reaching the ground. When the leaves fall to the ground, they quickly decay. This makes the soil very rich. The ground may be covered with flowers, ferns, and small shrubs.

Grassland

bison

Temperature	cool winters, warm to hot summers
Rainfall	moderate
Soils	fertile

Deciduous Forest

raccoon

Temperature	cold to moderate winters, warm summers
Rainfall	year-round
Soils	fertile

Tropical Rain Forests

Along and near Earth's equator are **tropical rain forests**. These biomes are hot and damp with plenty of rainfall. Rain forest trees are so tall and leafy that almost no light reaches the ground.

Parts of a Forest

Most forests have three distinct parts. The *canopy* (KA•nuh•pee) spreads over the forest like a huge umbrella. The top leaves of the tallest trees receive the most light.

Beneath the canopy is the *understory*. Here, the dense layer of leaves makes it very damp. Plants such as orchids live on tree trunks.

The lowest part of a forest is its *floor*. It is so dark on the forest floor, few kinds of plants can grow.

Tree Orchids

Read a Photo

How are the orchid's roots different from the roots of ordinary plants?

Clue: Where is the orchid growing?

Tropical Rain Forest

chameleon

Temperature	warm year-round
Rainfall	wet year-round
Soils	poor in nutrients

✓ Quick Check

Summarize Name and describe two types of forest biomes.

Critical Thinking How is soil an important abiotic factor in grasslands and forests?

Desert

sidewinding adder

Temperature	extremely variable
Rainfall	very little
Soils	dry and thin

Taiga

moose

Temperature	cold winters, mild summers
Rainfall	moderate
Soils	poor in nutrients

What are deserts, taiga, and tundra?

Deserts

Every continent has at least one desert. A **desert** is a barren biome with very little rainfall. The world's largest desert is the Sahara in Africa.

Desert biomes can be hot or cold. In the daytime, hot deserts can pass 100°F (38°C). At night, the air cools to around 25°F (-4°C).

Few organisms can live in such an extreme environment. Desert plants, like the cactus and yucca, are able to survive long periods with no rain. Desert animals can go for long periods with little water.

Taiga

The largest biome in the world is the taiga (TI•guh). The **taiga** is a cool forest biome in the upper regions of the north. In fact, *taiga* is the Russian word for "forest."

The taiga also contains lakes and ponds. They formed thousands of years ago. Then, glaciers covered the land in Alaska, Canada, and parts of Europe.

Conifers such as fir and spruce trees grow almost everywhere in the taiga. These trees provide most of the world's lumber.

Taiga winters are long and cold. Black bears and other mammals have thick fur to keep them warm. In summer, melting snow provides the animals with plenty of water.

FACT Not all deserts are hot. The average temperature in the Gobi Desert is below freezing!

Tundra

caribou

Temperature	cold and long winters, short and cool summers
Rainfall	some in summer
Soils	frozen

Tundra

The **tundra** is a cold, dry biome without trees. The ground is frozen all year. Winters are long and icy cold. Summers are short and cool. Very little rain falls in the tundra.

Plants in the tundra grow close to the ground. They have shallow roots that help them get water from melting snow. Tundra plants include wildflowers, mosses, and grasses. Different kinds of lichen (LI•kun) grow on rocks.

Tundra animals can be large. Mammals such as caribou and musk oxen have an extra layer of fat. The fat keeps them warm. Many animals *hibernate,* or go into a deep sleep, for the winter. Some birds fly south to find warmer temperatures.

Quick Lab

Biome Soils

1 **Observe** Examine each soil sample carefully. Record your observations in a table.

Soil Sample	Time	Amount of Water
Soil 1		
Soil 2		
Soil 3		

2 Label three pots *1, 2,* and *3.* Place an equal amount of each soil type in its labeled pot.

3 Have your partner hold a pot over a pan. Pour in 120 mL of water.

4 **Measure** Record the time it takes for the water to drain. Measure and record the amount of drained water. Repeat with the other pots.

5 **Infer** Which soil holds water best? How might this affect how plants grow in different biomes?

✔ Quick Check

Summarize Describe the taiga, tundra, and desert biomes.

Critical Thinking How are the desert and tundra biomes similar?

Are there water biomes?

On land, biomes are grouped by plant life, climate, and other factors. Water ecosystems are grouped by the kind of water and organisms that can be found there. They are not always grouped as biomes.

Freshwater Ecosystems

In a freshwater ecosystem, the water either flows over land or it stands still. Streams and rivers have flowing water. Fish and other organisms have structures to cope with that flow. Lakes and ponds have still water.

Wetlands and Estuaries

A *wetland* is an ecosystem where it is wet during some or all of the year. Some have fresh water. Two other kinds of wetlands are salt marshes and mangrove swamps. These are both estuaries. An *estuary* is an ecosystem where ocean and freshwater ecosystems meet.

Ocean Ecosystems

You could think of the ocean as one big saltwater ecosystem. Earth has five oceans that are all connected. When scientists study ocean ecosystems, they consider the depth and distance from the shore. Algae are the main producers in the ocean. Algae live near the ocean's surface where there is enough sunlight to grow.

Quick Check

Summarize What are the main water ecosystems?

Critical Thinking Does an estuary contain salt water? Why?

Alligators live in swamps and other wetlands.

Visual Summary

Land ecosystems can be grouped as **biomes,** such as **grasslands, deciduous forests,** and **tropical rain forests.**

Harsher conditions are found in the **desert, taiga,** and **tundra** biomes.

Water ecosystems include rivers, lakes, wetlands, estuaries, and oceans.

Make a **FOLDABLES** Study Guide

Make a four-tab book. Use it to summarize what you read about biomes.

Biome

Grasslands, temperate forests, tropical rain forests

Taiga, tundra, desert

Water ecosystems

Think, Talk, and Write

1 Vocabulary The ground is always frozen in the _____ biome.

2 Summarize Choose a biome. Write a summary of its features. Include descriptions of the abiotic and biotic factors. Use other reference materials if you need them.

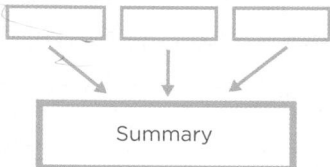

Summary

3 Critical Thinking You have a seed from a rain forest. You plant it in desert soil. The plant does not grow. Why?

4 Test Prep Which biomes have many trees?

A deciduous forests, tropical rain forests, deserts

B tropical rain forests, deciduous forests, tundra, taiga

C taiga, deciduous forests, tropical rain forests

D taiga, tundra, grasslands, deserts

5 Essential Question How do ecosystems compare?

 **Writing Link**

Write a Brochure
Write a travel brochure for a biome of your choice. Include pictures and descriptions that would persuade people to visit it.

 **Social Studies Link**

Humans and Biomes
Choose a biome. Research how people live there. What types of homes do they live in? What kinds of foods are grown there?

MUSEUM MAIL CALL

buckeye butterfly

Scientists at the American Museum of Natural History work to protect endangered habitats everywhere. They collect stories from people around the world to learn about these environments.

American Museum of Natural History   Send   Cancel

Dear Museum Scientists,

My name is Tommy. I live on the coast of Florida, near a mangrove swamp.

My mom is a tour guide who shows people the amazing creatures that live in the mangrove trees and surrounding wetlands. I am writing to you because I am worried about what is happening near my home. The mangroves are home to many animals, including storks, butterflies, snakes, and crabs. Mangrove roots provide shelter for fish and shrimp. The mangroves also protect the coast from wind, waves, and floods.

Lately many new neighborhoods are being built. This construction is replacing many mangroves with stores, homes, marinas, airports, and parking lots. What will happen to the animals that call the mangroves home? There must be a way for us to live together with the mangroves and animals.

Tommy

mangrove land crab

Many different animals live in Florida's mangroves.

purple gallinule

Draw Conclusions

▶ Use your own knowledge about the subject.

▶ Support your conclusions with information in the text.

Write About It

Draw Conclusions What might happen to the plants and animals of Florida's wetlands if people continue to build there?

LOG ON e-Journal Research and write about it online at www.macmillanmh.com

Connect to

AMERICAN
MUSEUM ō
NATURAL
HISTORY

at www.macmillanmh.com

Relationships in Ecosystems

Look and Wonder

The frog is the hunter. The grasshopper is the hunted. Both animals need energy to live and grow. Where does that energy come from?

How much energy do living things use?

Purpose

Model how energy passes from one organism to another in an ecosystem.

Procedure

1. Work in groups of four. Make labels for *Sun*, *plant*, *plant eater*, and *meat eater*.

2. **Measure** Cut a 1-meter strip of butcher paper. This represents energy that living things can use. Make a mark every 10 centimeters along the strip.

3. **Make a Model** Each student takes a label. *Sun* begins by passing the energy strip to *plant*.

4. *Plant* cuts off 10 cm from the strip. *Plant* holds the larger section and passes the smaller section to *plant eater*.

5. *Plant eater* cuts off 1 cm and passes the smaller section to *meat eater*.

Draw Conclusions

6. **Infer** Why do you think the energy strip gets cut before it gets passed on?

7. **Use Numbers** How much energy is available to the meat eater compared to the plant? Compared to the plant eater?

Explore More

What might happen if the plant could not make its own food energy? Design a test to find out.

Materials

- markers
- label paper
- butcher paper
- meter stick
- scissors

Step 4

Read and Learn

▶ **Essential Question**

How do organisms
get energy?

▶ **Vocabulary**

producer, p. 150

consumer, p. 151

decomposer, p. 151

food chain, p. 152

food web, p. 154

competition, p. 155

energy pyramid, p. 156

▶ **Reading Skill** ✔

Draw Conclusions

Text Clues	Conclusions

▶ **Technology**

e-Glossary, e-Review,
and animation online
at www.macmillanmh.com

How do organisms depend on one another?

To understand an ecosystem, scientists look at the relationships and roles of organisms within a community.

Producers

Every organism in an ecosystem relies on producers. **Producers** are organisms that make their own food using the energy in sunlight. Producers on land include green plants, such as grasses and trees. In lakes and oceans, the main producers are algae. Many other protists are producers too.

Roles in an Ecosystem

Producers make food using sunlight.

Consumers eat producers.

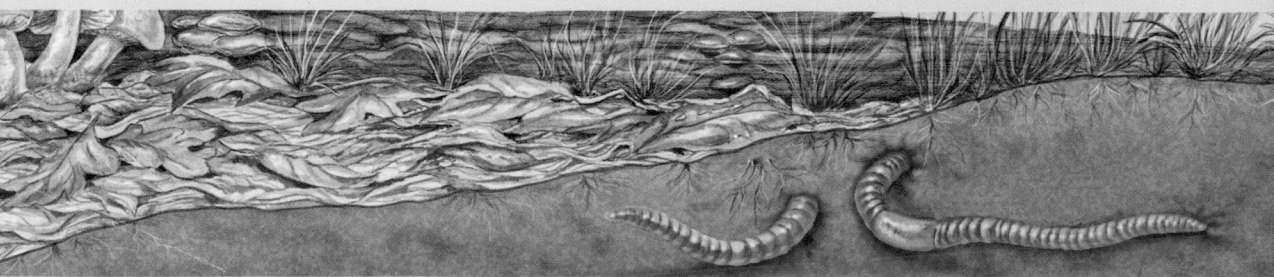

Decomposers break down dead and decaying organisms.

Consumers

Organisms who cannot make their own food are called **consumers**. Birds, mammals, and other consumers get energy from eating other organisms.

We can classify consumers by the kinds of food they eat. *Herbivores* (HUR•buh•vorz) eat only producers. Porcupines and most other rodents are herbivores. So are rabbits and deer.

Some animals eat producers and consumers. These are *omnivores* (AHM•nih•vorz). Opossums, raccoons, and bears are all omnivores. *Carnivores* (KAR•nuh•vorz) are animals that eat herbivores and omnivores. Ospreys and other birds of prey are carnivores. So are cats, tigers, and lions. Sharks are carnivores too.

Decomposers

Some organisms break down dead producers, consumers, and wastes. They turn dead and decaying matter into simpler substances. These organisms are **decomposers**. Worms, bacteria, fungi, and other decomposers get energy this way. They return substances to the ecosystem as nutrients. In other words, decomposers are recyclers.

✔ Quick Check

Draw Conclusions What would happen if producers disappeared?

Critical Thinking Do consumers get energy from the Sun? Explain.

≡ Quick Lab

Observe a Decomposer

1 Moisten four food samples. Place each one in a plastic bag.

2 Seal the bags. Put them in a warm, dark place.

⚠ **Be Careful!** Do not open the bags after you seal them.

3 **Observe** Check the bags each day. Record your observations.

4 **Communicate** How did the foods change? Why did this happen?

herbivore

omnivore

carnivore

blue heron

sunfish

mayfly

algae

How does food travel in a pond?

Clue: The arrows in a food chain point to the next consumer.

 Science in Motion Watch decomposers in action at www.macmillanmh.com

What is a food chain?

Every organism needs energy to live and grow. The energy in an ecosystem comes from the Sun.

Look at the animals shown on these pages. None of them can use the Sun's energy directly. The energy from sunlight is stored in food. That energy passes from one organism to another in a **food chain**. The energy in a food chain moves from producers to consumers.

A Pond Food Chain

Algae and green plants are first in the pond food chain. Algae capture the Sun's energy during photosynthesis. They store it in their cells as sugars.

What happens when a plant eater, such as a mayfly, eats the algae? The insect uses oxygen to release the energy stored in the algae it ate. It uses some of that energy to move, grow, eat, and reproduce. It stores some of the energy in its tissues.

A meat eater like the sunfish might snap up the mayfly. A blue heron may then eat the sunfish. As you can see, even the heron gets some of the Sun's energy that was passed along the chain.

All the plants and animals in the pond become food for decomposers after they die. Bacteria and other decomposers break down the dead tissues into simple nutrients that other living things can use.

A Land Food Chain

Food chains on land usually start with grasses, trees, or other green plants. In the example on the right, the thistle plant is the producer. The caterpillar is the herbivore that munches on its leaves. The mantis, skink, and owl are the other consumers, in that order.

Food chains can be shorter than the ones shown here. For example, a skink might eat a caterpillar instead of a mantis. The skink might die before it gets consumed by the owl.

Where are the decomposers in both food chains? Decomposers are often left out of food chain diagrams. This is because they take part in every step of the food chain.

Notice the arrows shown in the diagrams. The direction of the arrows is important. Each arrow in a food chain points away from the organism that is eaten. The arrow shows the direction that energy flows.

✓ Quick Check

Draw Conclusions Why is the term *food chain* a good description of the relationships shown on these pages?

Critical Thinking What is the longest food chain you can think of? Draw and label a diagram showing your food chain.

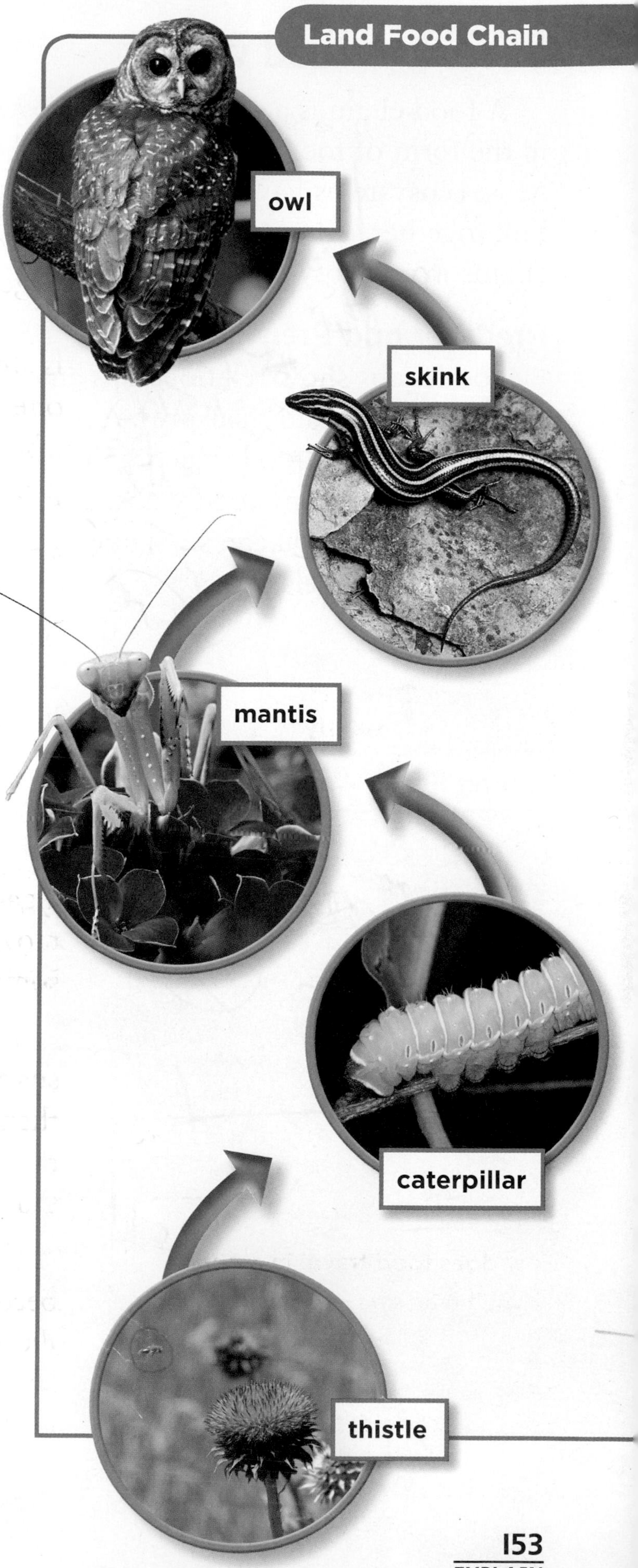

owl

skink

mantis

caterpillar

thistle

What is a food web?

A food chain is a good model of how energy travels in the form of food. However, it shows only one path. Most ecosystems have many different food chains that link together. A **food web** shows how all the food chains in an ecosystem are connected.

Predator and Prey

Food webs show relationships among predators (PRE•duh•torz) and their prey. A *predator* is a carnivore that hunts for its food. The organisms that it hunts are called *prey*. In many food webs, organisms are predators as well as prey. You can see some of these relationships in the diagram below.

Ocean Food Web

Read a Diagram

Which predator in this food web has the greatest number of prey organisms?

Clue: Remember, the arrows point away from the prey and toward the predators.

Competition

A food web shows that a single organism can take part in more than one food chain. When this happens, competition can result. **Competition** is the struggle between organisms for food, water, and other needs.

Look at the land food web. It has different herbivores, such as deer, small birds, and mice. What if they all ate the same plants? The three populations would compete for the food. One population might win out. The other populations would die out unless they found a different food or moved to a different place.

Competition is not limited to animals. In the forest, small plants and flowers compete with tall trees for nutrients. Plants may compete for sunlight too.

Individuals in a population also compete with each other. You may have watched squirrels in a park compete for nuts. All living things on Earth can be considered part of one giant food web.

Quick Check

Draw Conclusions In the ocean food web, which animals compete with the killer whale for fish?

Critical Thinking List four different food chains in the land food web at the right.

What is an energy pyramid?

An **energy pyramid** shows the amount of energy in an ecosystem. Producers are always at the bottom, or base, of the pyramid. Producers use only a small amount of the energy in sunlight to make food. The plant's cells burn some of the food that they make. The plant stores the rest in its stems, leaves, and roots.

The next level of the pyramid shows herbivores. A herbivore must eat a lot to stay alive. This is because only about 10 percent of a plant's energy passes to the herbivore.

Where does rest of the energy go? Some is lost as heat when the animal breaks down the plant tissue. Some plant tissue cannot be broken down, and is passed as waste.

Omnivores and carnivores make up the next levels of the pyramid. At each level consumers lose about 90 percent of the remaining energy. The animals at the top get just a tiny fraction of the original energy.

✔ Quick Check

Draw Conclusions Why do ecosystems have more producers than consumers?

Critical Thinking An ocean food chain can have more carnivores than a land food chain. Why?

Energy Pyramid

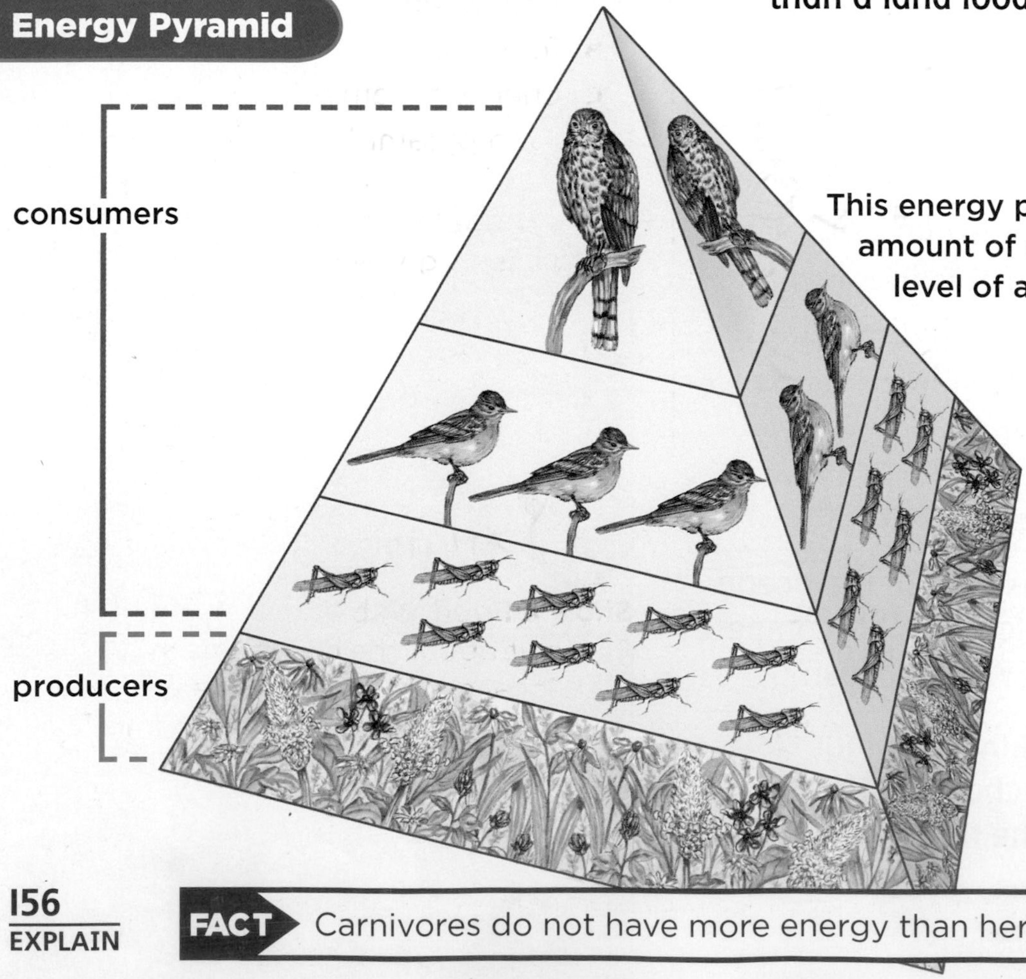

consumers

producers

This energy pyramid shows the amount of energy at each level of a food chain.

FACT Carnivores do not have more energy than herbivores.

Lesson Review

Visual Summary

In an ecosystem, food is made by **producers,** eaten by **consumers,** and broken down by **decomposers.**

Food chains and **food webs** show the relationships between organisms in an ecosystem.

Energy pyramids show the amount of energy in an ecosystem.

Make a FOLDABLES Study Guide

Make a four-tab book. Use it to summarize what you read about relationships in ecosystems.

Producers, consumers, decomposers

Food chains

Food webs

Energy pyramid

Think, Talk, and Write

1. **Vocabulary** What is an omnivore? Give three examples.

2. **Draw Conclusions** Scientists are doing a survey of an ecosystem. So far, they have counted more carnivores than herbivores. Is the survey complete? Why or why not?

Text Clues	Conclusions

3. **Critical Thinking** Why do carnivores usually have sharper teeth than herbivores?

4. **Test Prep** Two food chains can combine to form a(n)
   - A   ecosystem.
   - B   food web.
   - C   energy pyramid.
   - D   food pyramid.

5. **Essential Question** How do organisms get energy?

Math Link

Figure the Number of Carnivores
Ecosystems have 10 times as many herbivores as carnivores. How many carnivores would you expect to find if there were 4,250 herbivores?

Art Link

Show a Food Web
Find out about the organisms in your local environment. Make a poster showing each organism and the food web that connects them.

The Moth That Needed the Tree

The yucca moth of the Mojave Desert spreads the pollen of yucca trees. It also does something very unusual. When the moth visits a flower on the yucca tree, it pokes a hole in the flower's ovary. Then it places its own eggs in the flower! The moth leaves pollen on the flower as well. This helps the plant reproduce.

The moth's eggs and the tree's seeds grow at the same time. The seeds become food for the moth's offspring. All of this happens inside the flower! The young moths get the food they need. They also stay safe from predators. The yucca moth and yucca tree depend on each other.

yucca flower

Expository Writing

Good expository writing

▶ supports the main idea with facts and details;

▶ organizes facts and details to show causes and effects;

▶ draws a conclusion based on the information presented.

 **Write About It**

Expository Writing Research another example of how insects and plants depend on each other. Write a report with facts and details from your research.

 Research and write about it online at **www.macmillanmh.com**

How Many Monarchs?

Each winter, about 180 to 280 million Monarch butterflies travel from the north toward Mexico. There the climate is warmer, and the butterflies can survive.

The milkweed plant is the Monarch's main source of food. Today, people are building in places where milkweed grows. Monarchs are having trouble finding enough food for their journey south. This change in the food chain means that fewer butterflies make the trip each year. Their numbers have been reduced by many millions.

Place Value								
A place-value chart can help you understand the values of large whole numbers.								

hundred millions	ten millions	millions	hundred thousands	ten thousands	thousands	hundreds	tens	ones
1	0	5,	8	3	7,	5	0	9

▲ Read this number as one hundred five million, eight hundred thirty-seven thousand, five hundred nine.

 **Solve It**

An average population of Monarch butterflies is one million, nine hundred fifty-eight thousand, thirty-three. Write this number in a place-value chart.

Visual Summary

Lesson 1 The living things in an ecosystem interact with each other. They depend on the nonliving things.

Lesson 2 Earth's land can be divided into six major biomes. Each has its own characteristics.

Lesson 3 Energy is passed from one organism to another in an ecosystem.

Make a FOLDABLES Study Guide

Tape your lesson study guides to a piece of paper as shown. Use your study guide to review what you have learned in this chapter.

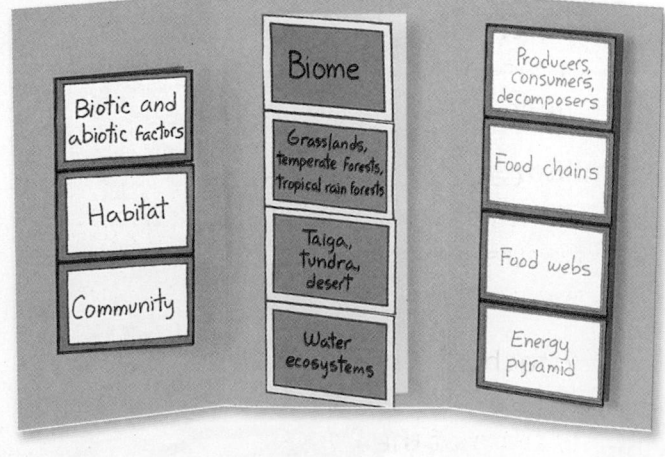

Fill each blank with the best term from the list.

biome, p. 138	**food web,** p. 154
consumer, p. 151	**habitat,** p. 131
decomposers, p. 151	**population,** p. 132
desert, p. 142	**producer,** p. 150
ecosystem, p. 131	**taiga,** p. 142

1. Two or more food chains that share links make a(n) _____.

2. The biome that gets the least amount of rainfall is the _____.

3. An organism that cannot make its own food is called a(n) _____.

4. A large ecosystem with its own kind of plants and animals is called a(n) _____.

5. The place where an organism lives is its _____.

6. An organism that uses energy in sunlight to make food is a(n) _____.

7. Most of the plants found in the _____ stay green all year.

8. The biotic and abiotic factors of an environment make up an entire _____.

9. All the individuals of a species that live in an ecosystem form a(n) _____.

10. Organisms that get their energy from dead and decaying matter are called _____.

LOG ON e-Glossary Words and definitions online at www.macmillanmh.com

Answer each of the following.

11. **Fact and Opinion** *A desert biome is a bad place for organisms to live.* Is this a fact or an opinion? Explain.

12. **Expository Writing** Explain how abiotic factors are important to an ecosystem. Use details to support your explanation.

13. **Predict** Suppose you are camping in the northeastern United States. What kinds of plants and animals would you expect to see there?

14. **Critical Thinking** Suppose a company started building houses in a grassland. What might happen to the food chain in this area?

15. **Main Idea and Details** Explain the difference between a food chain and a food web.

16. **Interpret Data** Which organisms in the energy pyramid are consumers? Which are producers?

17. A porcupine eats roots and leaves. The porcupine is a(n)

   A herbivore.   **C** predator.

   B omnivore.   **D** producer.

18. **True or False** *An energy pyramid shows all the food chains in an ecosystem.* Is this statement true or false? Explain.

19. **True or False** *All deserts are hot.* Is this statement true or false? Explain.

20. Where do plants and animals live and how do they depend on each other?

Performance Assessment

DOK 2

Biome Mobile

Make a mobile showing six biomes. Use a coat hanger, string, crayons or markers, and construction paper.

1. Cut one sheet of construction paper into six pieces. Write the name of a different biome on each piece.

2. For each biome, list at least four plants and animals that live there. Illustrate them on both sides of the paper.

3. Attach each piece of paper to a string. Tie each string onto the bottom of the hanger.

1 Look at the picture of a lake habitat.

Which of these is an abiotic factor?

A turtle   **C** rock

B grass   **D** bird

DOK 1

2 Look at the map shown below.

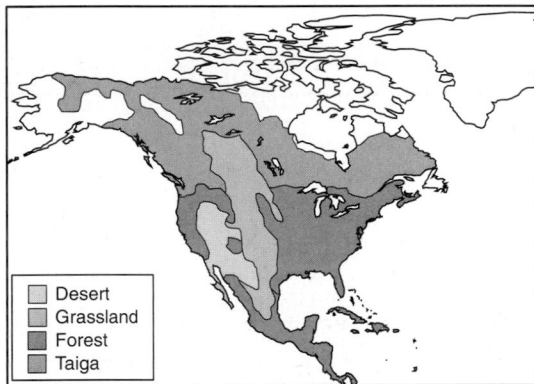

☐ Desert
☐ Grassland
☐ Forest
☐ Taiga

How can this map <u>most likely</u> be used?

A to predict what the weather will be like in an area

B to predict what the ecosystems will be like in an area

C to learn what the oceans and lakes will be like in an area

D to learn what the consumers will be like in an area

DOK 2

3 In which way is the Sahara similar to a tundra?

A There is little sunlight.

B There is much rainfall.

C The temperatures there change greatly from day to night.

D The animals and plants there must survive with little water.

DOK 2

4 Which animal will <u>most likely</u> compete with the coyote in this food web?

A lizard   **C** rattlesnake

B mouse   **D** tortoise

DOK 2

5 How do herbivores get energy?

A by consuming plants

B by consuming animals

C by making it themselves

D by consuming predators

DOK 1

6 Look at the diagram below.

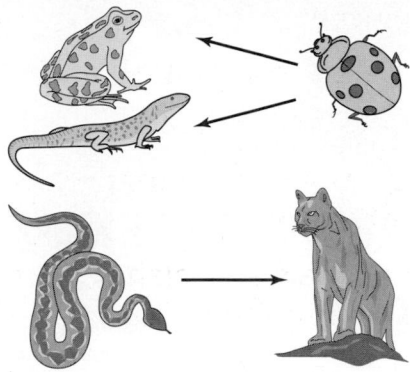

Which describes the flow of energy?

A from the beetle to the frog

B from the frog to the lizard

C from the cougar to the snake

D from the cougar to the frog
DOK 1

7 Look at the concept map.

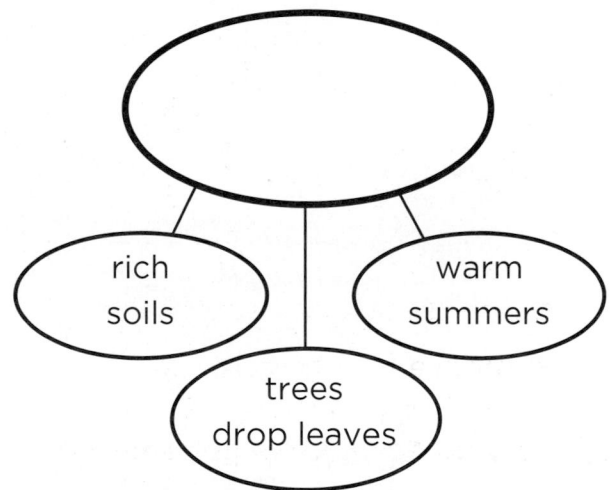

Which biome belongs in the empty oval?

A grassland   **C** deciduous forest

B tundra   **D** taiga
DOK 1

8 Look at the energy pyramid below.

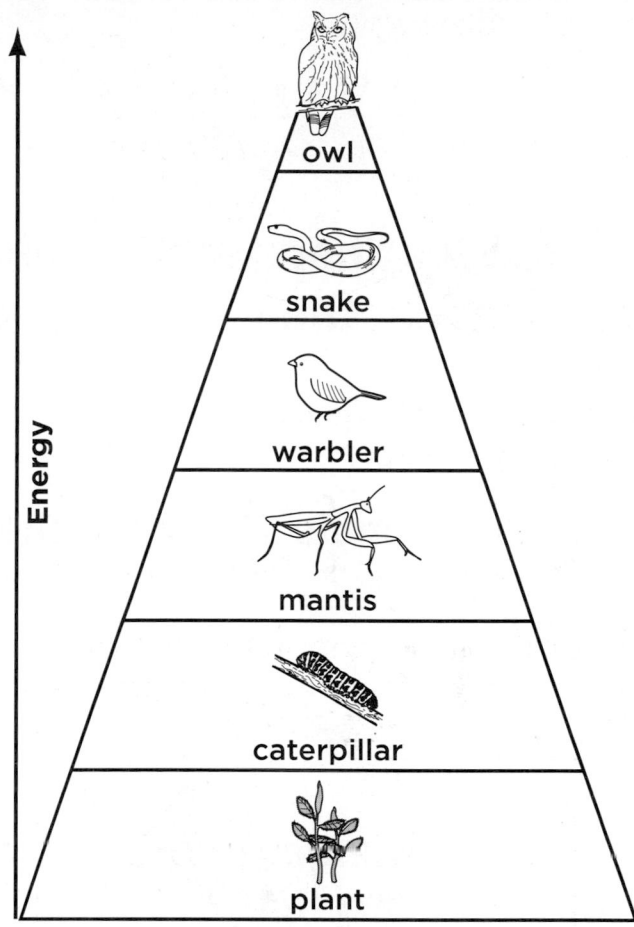

Suppose warblers were removed from this ecosystem. Predict what would happen to owls and mantises. Explain each prediction.
DOK 3

Check Your Understanding

Question	Review	Question	Review
1	pp. 130–131	5	pp. 150–154
2	pp. 138–144	6	pp. 152–153
3	pp. 142–143	7	pp. 138–144
4	pp. 154–155	8	pp. 130–131, 150–156

CHAPTER 4

Surviving in Ecosystems

 **The Big Idea** What happens to living things when their environments change?

Essential Questions ·

Lesson 1

How do animals survive in their environments?

Lesson 2

How do plants survive in their environments?

Lesson 3

How can changes in an environment affect the organisms that live there?

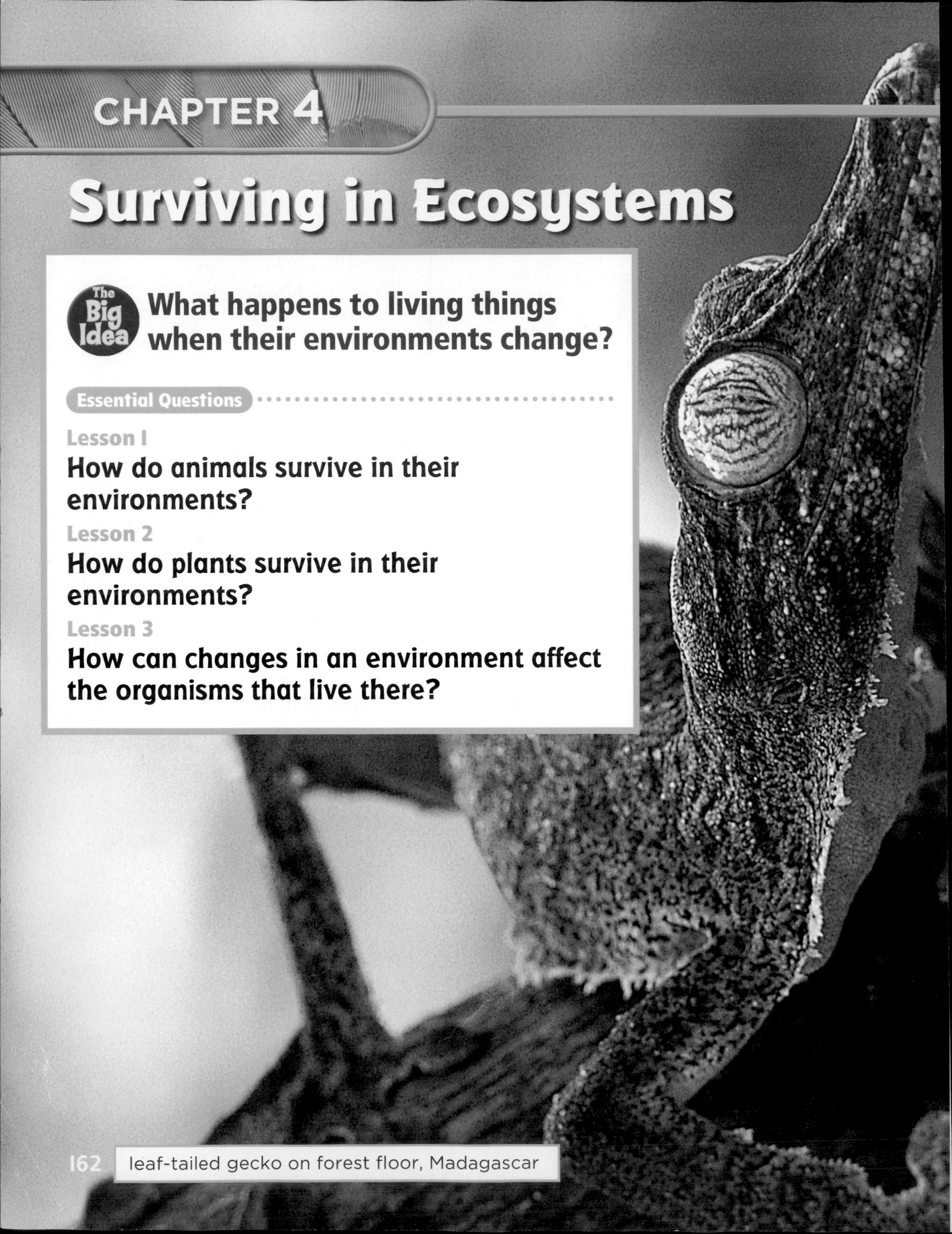

leaf-tailed gecko on forest floor, Madagascar

 # Big Idea Vocabulary

adaptation a trait or behavior that helps a living thing survive in its environment (p. 166)

camouflage an adaptation in which an animal can hide by blending in with its surroundings (p. 168)

mimicry when one kind of living thing has similar traits to another (p. 169)

tropism the response of a plant to something in its environment (p. 177)

accommodation an individual organism's response to changes in its ecosystem (p. 188)

extinct when the last of a species dies (p. 189)

 Visit www.macmillanmh.com for online resources.

Animal Adaptations

Look and Wonder

The snowcap hummingbird feeds on flowers in the rain forests of Costa Rica. Could this bird survive without its long and narrow beak?

Does the shape of a bird's beak affect what it eats?

Purpose

Explore how the shape of a bird's beak affects the kind of food the bird eats.

Procedure

1 **Predict** The first four materials represent different types of bird beaks. Which beak works best for picking up rice grains? Pieces of foam? Water? Record your predictions.

2 **Experiment** Try picking up each food with each type of beak. Record your results in a chart like the one shown.

My Results				
Beak	**Rice**	**Foam**	**Water**	**My Prediction**
spoon				
fork				
straw				
chopsticks				

Draw Conclusions

3 **Infer** Were your predictions correct? Which beaks are best for picking up small, hard things? Large things? Soft things? Explain.

Explore More

How does the shape of an animal's claw affect what it eats? Make a prediction. How could you test your ideas? Work with a partner to try it.

Materials

- spoon
- fork
- straw
- chopsticks
- rice grains
- foam packing material
- cup of water

Step **2**

Read and Learn

▶ **Essential Question**
How do animals survive in their environments?

▶ **Vocabulary**
adaptation, p. 166
hibernate, p. 168
camouflage, p. 168
mimicry, p. 169

▶ **Reading Skill** ✔
Predict

My Prediction	What Happens

▶ **Technology** LOG ON
e-Glossary and e-Review online
at www.macmillanmh.com

What are adaptations?

Every ecosystem has its challenges. Organisms can meet those challenges with the right set of traits. These traits help them meet their needs.

Traits can be physical characteristics. For example, a bird's beak is a physical characteristic. Different beak types are suited to different kinds of food. A hummingbird's beak is long and narrow. It gets its food from thin flowers. An owl's beak is curved and pointed. How do you think the owl gets its food?

Certain kinds of traits are adaptations (a•dap•TAY•shunz). **Adaptations** are physical characteristics or behaviors that help a living thing survive in its environment. To *survive* is to continue to live. A polar bear's fur helps it keep warm. The trunk of an elephant allows it to grasp things and feed itself.

A camel's hump stores fat for times when food is scarce.

arctic fox

fennec fox

Desert Adaptations

You have learned that deserts are dry biomes. Desert animals have adaptations that save water.

A sandgrouse is a desert bird with feathers that soak up water. This allows it to carry water to its young in the nest. A kangaroo rat is a mammal that never needs to drink. It gets water from food.

Many animals have adaptations for staying cool in hot deserts. The fennec fox has large ears that give off heat. Its fur is thinner than the fur of its relatives in cooler climates.

Camels have all kinds of adaptations for desert life. They can close their nostrils to keep out sand. They store fat in humps. The fat gives them energy when there is not much food available. Wide hooves help camels walk on sand.

The kangaroo rat gets water from the seeds it eats. ▶

✔ Quick Check

Predict What might happen to a desert animal if you moved it to the tundra?

Critical Thinking How are an eagle's claws and a giraffe's neck similar?

What are some other adaptations of animals?

Animals that live in hot climates need to stay cool. Animals in cool climates need to stay warm. You'll find different adaptations depending on the environment you are in.

Behaviors

Some adaptations are behaviors. Northern black bears avoid the cold by hibernating (HI•bur•nayt•ing). When an animal **hibernates**, it lives off its body fat and uses very little energy.

Some animals survive by leaving when the temperature changes. Many birds migrate (MI•grayt) from cooler to warmer places. To *migrate* is to change location periodically.

The dormouse hibernates in its nest.

Camouflage

Some animals blend in with their environment. This adaptation is called **camouflage** (KA•muh•flahj). It helps animals hide. The arctic fox and the arctic hare change color with the seasons. In winter their fur matches the white snow. In summer their fur turns brown and matches the soil.

Camouflage

Read a Photo

What adaptations help this snow leopard survive in its environment?

Clue: Compare the animal to its surroundings.

The hover fly on the left mimics the honeybee on the right.

Mimicry

Look closely at the two insects above. On the right is a honeybee. Honeybees defend themselves with stingers. The other insect looks like a honeybee, but it is a hoverfly (HUH•vur•fli). Hoverflies do not have stingers.

By looking like a honeybee, the hover fly avoids predators. A predator might eat a regular insect but not eat a bee. When one kind of living thing looks like another kind, it is called **mimicry** (MIH•mih•kree).

Body Structures

Animals often have body parts that are adaptations. Some snakes and lizards have poison glands in their jaws. Their bite can hurt or kill a predator. Hedgehogs are covered with hard spines. If a predator comes near, they curl into a ball. A predator would not want to eat a spiny ball!

Quick Lab

Mimicry Model

1 **Observe** The animal on the right is a nonpoisonous scarlet king snake. At the bottom is a poisonous coral snake. Compare the two snakes.

2 **Make a Model** Use markers, modeling clay, or other materials to model each snake. How are they alike? Different?

3 How could this saying save someone's life? "Red touches yellow, kill a fellow. Red touches black, good for Jack."

✓ Quick Check

Predict Would you find an animal that hibernates in a tropical rain forest? Why or why not?

Critical Thinking Why do you think most poisonous animals lack camouflage?

A hedgehog rolls into a ball when a predator is near.

These reef fish eat the algae that grow on the shells of sea turtles.

How else do animals survive?

Some adaptations involve interactions among organisms. Living things interact in harmful and helpful ways.

Harmful Interactions

Fleas make their homes in the fur of mammals. The mammal gives the flea food and shelter. The flea eats the mammal's blood. Fleas also lay eggs deep in the fur.

What does the flea give to the animal? The mammal gets red, itchy skin. Scratching can cause skin rashes and hair loss. In this kind of interaction, only one animal is helped. The other animal is harmed.

Helpful Interactions

The shells of sea turtles can become covered with algae. Some fish will swim with sea turtles and eat the algae off the shells. The fish get food. In turn, the turtle gets its shell cleaned.

The Egyptian plover is a bird that lives with Nile crocodiles. When the crocodile opens its mouth, the bird picks off any leeches. The bird gets a meal. The crocodile gets clean gums!

 Quick Check

Predict Which of the animals discussed above would survive without the other?

Critical Thinking What other examples of harmful interactions can you think of?

FACT Some bacteria that live inside people are helpful.

Lesson Review

Visual Summary

Adaptations are traits that help living things survive in their environments.

Animals have many adaptations. **Examples include** hibernating, migrating, camouflage, and mimicry.

Organisms interact with each other in ways that can be harmful or helpful to their survival.

Make a FOLDABLES Study Guide

Make a layered-look book. Use it to summarize what you learned about animal adaptations.

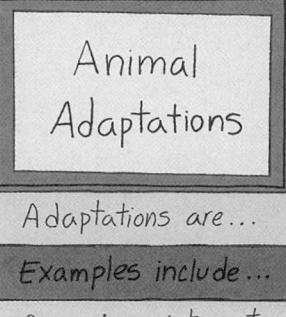

Animal Adaptations

Adaptations are...

Examples include...

Organisms interact...

Think, Talk, and Write

① **Vocabulary** An adaptation in which one living thing looks similar to another living thing is called _____.

② **Predict** Suppose someone moved a polar bear to a rain forest. Predict how the bear might survive there.

My Prediction	What Happens

③ **Critical Thinking** Choose a familiar animal. Where does it live? Describe its adaptations to that environment.

④ **Test Prep** **When an animal blends in with its surroundings, it is using**

  A hibernation.

  B camouflage.

  C migration.

  D mimicry.

⑤ **Test Prep** **During cold winters, some mammals survive by**

  A saving water.

  B hibernating.

  C shedding their fur.

  D scratching at fleas.

⑥ **Essential Question** How do animals survive in their environments?

Writing Link

Write a Story

Write a story in which your main character has adaptations that help it survive in its environment.

Social Studies Link

Look-Alike Products

How do companies use mimicry to sell products? Why do you think they do this? Give some examples.

Focus on Skills

Inquiry Skill: Form a Hypothesis

You have learned how adaptations such as camouflage help animals survive in their environments. Consider the skin of the salamander on the right. It lives in the forests of the southern United States. Its colors and patterns give the salamander camouflage in a forest. Scientists study camouflage to learn more about adaptations. They use what they learn to **form a hypothesis** that they can test.

▶ Learn It

When you **form a hypothesis**, you make a statement about what you think is true. A good hypothesis is based on observations or collected data. It is important that your hypothesis can be tested.

▶ Try It

Form a hypothesis about how camouflage affects the ability of a predator to find prey. Will it take more time or less time? Write your answer in the form "If an animal is camouflaged, then the amount of time to find it…"

Materials tracing paper, colored pencils or crayons, scissors, stopwatch

1. Choose a small animal from this lesson. Trace its outline on two sheets of tracing paper. Carefully cut out both shapes.
   ⚠ **Be Careful.** Scissors are sharp!

2. Color one copy of your animal to blend in with some visible part of your classroom. Record its traits in a chart like the one shown. Leave the other copy alone.

3. Predict how long it will take someone to spot the animal that is camouflaged. You cannot hide the animal under or behind anything.

Traits	
Color	
Shape	
Size	
Pattern	

4. Make a data table like the one shown below. Then choose four classmates as "predators." They will wear blindfolds or leave the classroom while you place your prey.

5. Test one predator at a time. Begin timing when your classmate takes off the blindfold or returns to the room. Record how long it takes each predator to find the prey.

6. Repeat steps 4–5 with the animal that is not camouflaged.

7. Compare both sets of data. How well do they match your prediction? Do you notice any patterns?

Time To Spot Prey	
Predator 1	
Predator 2	
Predator 3	
Predator 4	

▶ **Apply It**

Decide whether your results support or disprove your hypothesis. Explain your answer.

1. If your findings support your hypothesis, then choose a different variable to test. Does an animal's shape or size affect how fast a predator finds it? **Form a hypothesis** that you can test. Then test it. Record and interpret your results.

2. If your findings do not support your hypothesis, go back and review your procedure. Was your animal well camouflaged? If not, try changing its colors or location. Review your original hypothesis. Does it still make sense? Form a new hypothesis if you need to. Then repeat the investigation using a different design or method.

Plants and Their Surroundings

Look and Wonder

Have you ever seen tree roots growing in air? Mangrove trees grow where no other trees can. They flourish in salt water, where the soil has little oxygen. Why would the mangrove tree need roots like these?

Explore

How do plants respond to their environments?

Make a Prediction

Plants need sunlight to live. If the light is blocked, how will a plant respond? Make a prediction.

Test Your Prediction

1. ⚠ **Be Careful.** Handle scissors carefully. Cut an opening in one side of one end of a shoe box.

2. **Measure** Cut two dividers from the cardboard. Make them as tall as the shoe box but 3 centimeters narrower.

3. Place the dividers upright across the inside of the box. Tape the first divider to the same side as the opening you cut in step 1. Tape the other divider a few inches away on the opposite side, as shown. Put a plant in the end of the box opposite the opening. Put the lid on the box. Turn the opening toward bright sunlight.

4. **Observe** Every 3–4 days for several weeks, remove the lid to water your plant. Observe and measure its growth. Record your observations in a data table.

Step 3

Materials

- shoe box
- scissors
- cardboard
- ruler
- tape
- potted plant

Draw Conclusions

5. **Interpret Data** What happened to the plant? Why?

6. **Infer** How did the plant get sunlight? How does this model plants that live on the forest floor?

Explore More

Would a seed germinate in the box you made? Design an investigation to find out. Use several lima bean seeds placed in a damp paper towel.

▶ Essential Question
How do plants survive in their environments?

▶ Vocabulary
stimulus, p. 176

tropism, p. 177

▶ Reading Skill ✔
Problem and Solution

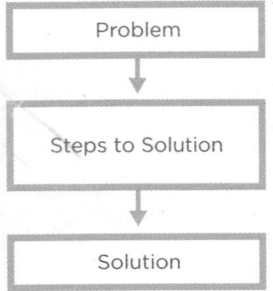

Problem

↓

Steps to Solution

↓

Solution

▶ Technology
e-Glossary, e-Review, and animation online at www.macmillanmh.com

How do plants respond to their environments?

Plants cannot move around the way most animals can. Yet plants can react to changes in their environment. A **stimulus** (STIM•yuh•lus) is something in an environment that causes a living thing to respond.

Stimulus and Response

How does a plant react, or *respond*, to a stimulus? It changes its direction or pattern of growth. Light, water, and gravity can each be a plant stimulus.

Plants respond to light by growing toward the source of the light. Plants respond to water by growing their roots toward the water's source. The roots of most plants grow downward—the same direction as the pull of gravity. The stems of most plants grow upward, away from gravity.

Tropism Experiment

Read a Diagram

Which variable was tested in this experiment?

Clue: The diagram shows the results. Look at the differences in the two plants.

Like most plants, the stems of water lilies grow toward the sunlight.

Tropism

A **tropism** (TROH•pih•zum) is the response of a plant to something in its environment. The responses of plants to light, water, and gravity are tropisms. Plants also show tropisms to chemicals and heat.

What causes a tropism? The British scientist Charles Darwin did an experiment to find out. He took two growing plant shoots. He covered the tip of one shoot with a cap made of tinfoil. He let the other shoot grow normally.

The results were clear. The shoot covered in foil did not bend toward the light. Darwin concluded that there was something in the tip that caused the shoot to bend. Later experiments showed that this "something" was a chemical that all plants have. Plants use this chemical to grow.

Quick Lab

Drying Time

1. Wet two paper towels. Roll one into the shape of a tube. Lay the other towel flat on a tray.

2. **Measure** Place both towels in bright sunlight or under a lamp. Record the time they take to dry.

3. Look at the maple leaf and the branch of pine needles. Which towel is like the maple leaf? Which is like a pine needle?

maple leaf

pine needles

4. **Infer** Which can get more sunlight—a maple leaf or a pine needle? Which holds more moisture? How is each kind of leaf adapted to its environment?

✔ Quick Check

Problem and Solution How could you test a plant's response to a chemical such as vinegar?

Critical Thinking Some people think that plants respond to music. How could you test this hypothesis?

Read a Photo

What adaptations help the prickly pear cactus live in the desert?

Clue: Compare the plant to its surroundings.

LOG ON *Science in Motion* See adaptations of other desert plants at **www.macmillanmh.com**

What are some plant adaptations?

Like animals, plants have adaptations for various environments. Just as camels and other desert animals need to conserve water, so do desert plants. A cactus is a good example. It has soft tissue that holds water just like a sponge. It also has a thick, waxy cover to keep the water inside.

Plants in temperate forests have different adaptations. Cold winter air can damage leaves. There is less liquid water in the environment during winter. Most trees here lose their leaves in winter. This protects them from drying out. Without leaves, a tree cannot make food. Instead, the tree uses stored food. In spring the tree grows new leaves and begins storing food for the next winter.

 Quick Check

Problem and Solution How is it possible for plants to live in many different environments?

Critical Thinking What might happen if you brought a desert plant into a humid greenhouse?

The bright red color of the Peruvian lily is an adaptation that attracts pollinators.

Lesson Review

Visual Summary

 A plant responds to a **stimulus** by changing the way it grows.

 Plants have **tropisms** to light, water, gravity, heat, and chemicals.

Plants have **adaptations** that help them live in different environments.

Make a FOLDABLES® Study Guide

Make a trifold book. Use it to summarize what you learned about how plants respond to their surroundings.

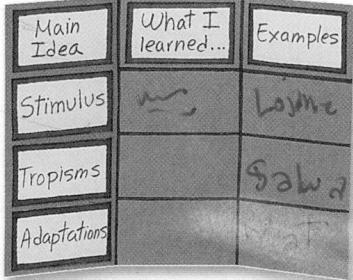

Think, Talk, and Write

1. **Vocabulary** What is a stimulus?

2. **Problem and Solution** How could you show that plants respond to changing temperatures?

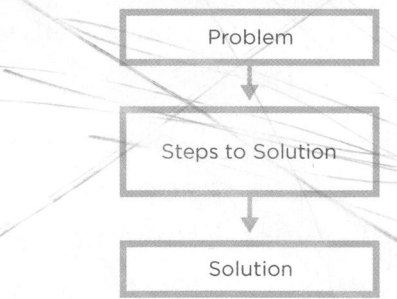

Problem

↓

Steps to Solution

↓

Solution

3. **Critical Thinking** How are the adaptations of a desert plant different from those of a rain forest plant?

4. **Test Prep** **Which word describes a plant's response to its environment?**

   A tropical

   B tropism

   C gravity

   D stimulus

5. **Essential Question** How do plants survive in their environments?

Math Link

Solve a Problem
A plant is 6 cm tall. It grows 0.5 cm each day. How tall will the plant be after one week? After three weeks?

Art Link

Make a Poster
Use reference books to learn more about plant adaptations. Make a poster of your findings.

A Field of Sun

There it was—a field of gold. There were hundreds, or maybe thousands, of tall sunflowers in full bloom. They seemed to stand in rows like soldiers, with their faces all pointed to the east.

A sunflower in full bloom is a beautiful sight. The head is made up of more than 1,000 tiny flowers. These flowers later produce seeds. Golden petals circle the head. They look a bit like a tutu on a dancer or a mane on a lion.

The sunflower's stem is tall and straight. It holds the head to the east. This protects the seeds from the hot rays of the Sun. Budding sunflowers follow the movement of the Sun. In the morning, they point east. In the late afternoon, they point west. At night, they point east again.

Descriptive Writing

Good descriptive writing

▶ includes details that tell how something looks, sounds, smells, tastes, or feels;

▶ uses words that describe.

Write About It

Descriptive Writing

Do some research about another plant. Write a description of how this plant reacts to its environment.

e-Journal Research and write about it online at **www.macmillanmh.com**

Sunflowers in full bloom face the east.

Comparing Plant Angles

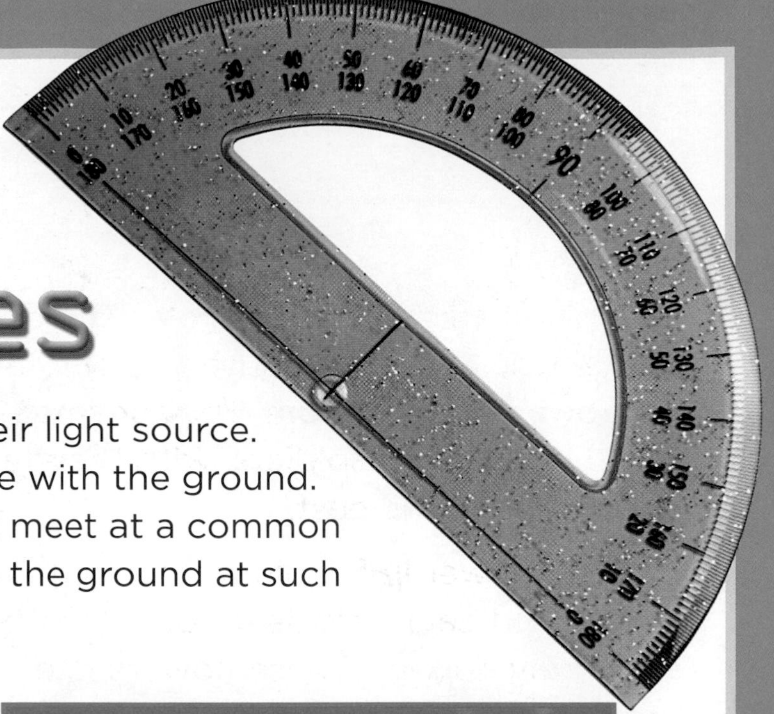

Plants grow in the direction of their light source. This makes their stems form an angle with the ground. An angle is formed by two lines that meet at a common endpoint. The stem of a plant meets the ground at such an endpoint.

Solve It

1. Which diagram shows an acute angle? Which shows an obtuse angle?

2. Are there any right angles shown in the diagrams? How do you know?

Classifying Angles

▶ A right angle has a square corner where the lines meet.

▶ An obtuse angle has a wider opening than a right angle.

▶ An acute angle has a smaller opening than a right angle.

Changes in Ecosystems

Look and Wonder

These seedlings are growing in dry, cracked mud. Was the soil always this dry?

How can a change to an ecosystem affect living things?

Make a Prediction

How can a period of little to no rain affect living things? Make a prediction.

Test Your Prediction

1 Write *hawk* on a yellow card, *lizard* on a green card, and *fox* on a red card. These are the predators. Write *prey* on the rest of the cards.

2 Each player takes one predator card. Mix ten prey cards of each color and stack them on the table. Put the other prey cards aside.

3 **Make a Model** Take turns drawing a prey card. Keep only the ones that match the color of your predator card. Return the others to the bottom of the pile. After every three turns, add a new prey card to the deck. This models the growth of the community. Play for 12 rounds. Count the cards left in the pile.

4 A long period of no rain kills half of the prey. Remove three prey cards of each color. Play again. After every six turns, add a prey card to the deck. Play for 12 rounds. Count again.

Draw Conclusions

5 **Use Numbers** How many cards were left at the end of each game?

6 **Infer** What did the model in step 3 represent? Did your results match your prediction?

Explore More

Would your results change if there were fewer predators? Make a prediction and test it.

Materials

- **18 green index cards**
- **18 yellow index cards**
- **18 red index cards**

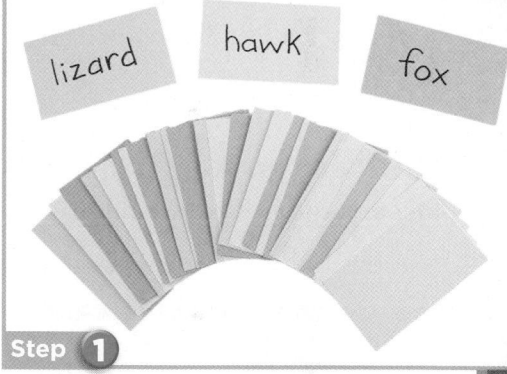

lizard hawk fox

Step **1**

Step **3**

▶ **Essential Question**
How can changes in an environment affect the organisms that live there?

▶ **Vocabulary**
accommodation, p. 188

endangered, p. 189

extinct, p. 189

▶ **Reading Skill** ✔
Cause and Effect

Cause → Effect
→
→
→
→

▶ **Technology**
e-Glossary and e-Review online at www.macmillanmh.com

What causes an ecosystem to change?

It may be hard to notice, but ecosystems are always changing. Some changes make it difficult for living things to survive.

Natural Events

In most ecosystems, change is part of a natural pattern. Volcanoes can fill a valley with ash. Hurricanes can destroy coastal wetlands. A lot of rain can cause landslides, turning hills into rivers of mud. Too little rain can cause a *drought* (DROWT). During droughts, the soil can dry up.

It can take a long time for an ecosystem to recover from such changes. Mount Saint Helens is a volcano in the state of Washington. In 1980, it erupted. Ash and lava killed nearby plants. The ecosystem needed many years to recover.

Read a Photo

How do the two photos show a cause and effect?

Clue: Read the dates in the captions.

Natural Change in Ecosystems

Mount Saint Helens in 1980

Mount Saint Helens in 1995

Living Things

Ecosystems can be changed by living things, such as locusts. A locust is a kind of grasshopper. In small numbers, locusts pose little danger. But in some places, giant swarms of locusts can gather in search of food. A swarm can have 50 million locusts in it! The locusts eat any plants along their path. They can leave a whole community without food.

Some living things can have a helpful effect on an ecosystem. Have you ever seen a "gator hole" in a wetland? An alligator uses its feet, tail, and snout to churn up the muddy water. These movements create a hole. Slowly, the hole fills with water.

Gator holes help alligators survive during droughts. The effect does not stop there. Birds and other animals move to gator holes when their own habitats get too dry. There they find food, water, and shelter.

✔ Quick Check

Cause and Effect What might happen to a wetland that is hit by a hurricane?

Critical Thinking How might an alligator benefit from a drought?

In large numbers, locusts can cause a lot damage farmers' crops.

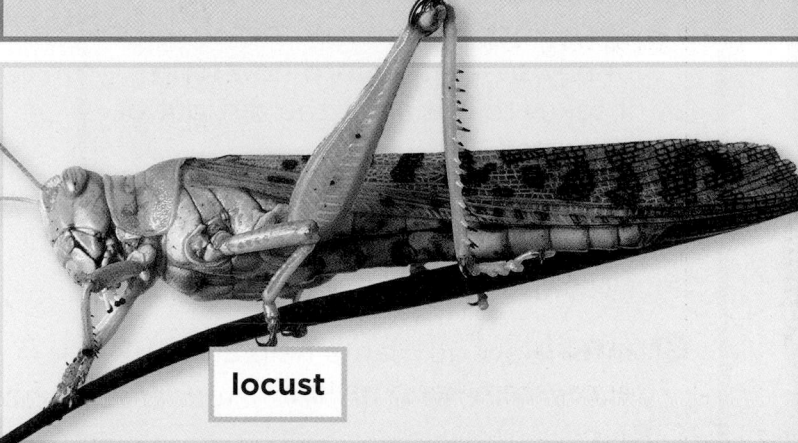

locust

alligator

Gator holes help many animals survive periods of drought.

Holding Soil Together

1. Take a pot that has a seedling growing in it. Fill an empty pot of the same size with soil.

2. **Measure** Shake the dirt out of each pot onto a large sheet of newspaper. Record how long it takes to get all of the dirt out of each pot.

3. Which pot took longer to get the soil out? What caused it to take longer?

4. **Infer** How do plants help an ecosystem keep its soil?

How do people change ecosystems?

Like other living things, people change their surroundings. Some changes are helpful. Other changes can harm an ecosystem.

Deforestation

Often trees are removed to build houses and other buildings. When people cut down forests, it is called *deforestation*. This destroys many forest habitats. Living things lose their homes and sources of food.

Overpopulation

People need places to live and work. The more people there are, the more they use and consume. Water and space become harder to find. When too many individuals live in an area, it is called *overpopulation*. This can happen with any species, not just humans.

Manufactured Ecosystems

a ship being sunk

concrete "reef balls" on the sand

Pollution

Cars, trucks, and power plants give off gases. These gases can harm the air we breathe. Adding harmful things to the air, water, or land is *pollution*. Litter is a form of pollution too. Pollution can kill plants and animals in an ecosystem.

Protection

People may cause problems to ecosystems, but they can also be helpful. People are driving less and using hybrid cars. They treat wastes to remove harmful substances.

You can help too. You can plant new trees. You can recycle paper, glass, and plastic. You can turn off the water when brushing your teeth. Can you think of other ways to help ecosystems?

The green liquid flowing out of this drainage pipe is pollution.

 Quick Check

Cause and Effect What happens to populations of plants and animals when a forest is cleared?

Critical Thinking How are deforestation and overpopulation related to each other?

a subway car being sunk

Read a Photo

In what ways do people help rebuild underwater ecosystems?

Clue: Look at what is being put into the water in each photo.

▲ A city building is not a natural ecosystem for a bird.

What happens when ecosystems change?

A deer lifts his head to sniff the air. The smell of fire fills the forest. Flames rush through the trees. The struggle for survival begins. How do the animals survive?

Accommodating

When an ecosystem changes, some living things change their behaviors and habits. An individual's response to change is called an **accommodation** (uh•kah•muh•DAY•shun).

A fire can destroy the main food supply of an animal in a forest. Some animals, like deer, can change their diet. They may eat tree bark instead of leaves. Others will use new plants or materials as shelter.

Moving Away

Not all animals can accommodate ecosystem changes. Some must find new places to live. Their search for food, water, and shelter may take them far away.

It may sound odd, but natural fires can benefit a forest ecosystem. Natural fires thin forests in some places. Young trees and smaller shrubs replace the trees. This provides new sources of food for forest animals.

The passenger pigeon was hunted into extinction in 1914.

The Tasmanian tiger was declared extinct in 1936. This was the last known individual.

Extinction

Some living things cannot accommodate an ecosystem change. If an organism does not meet its needs after a change, it will die. Sometimes an entire species can slowly disappear.

A living thing that has few of its kind left is **endangered** (in•DAYN•jurd). Some endangered plants and animals can become extinct (ik•STINGT). A species is **extinct** when the last of its kind dies.

Seed ferns became extinct millions of years ago.

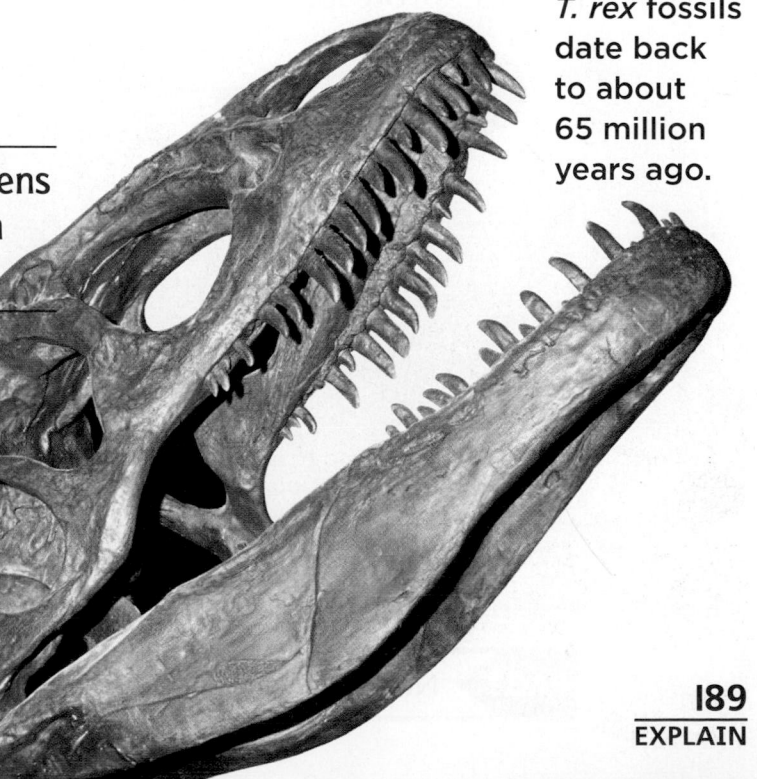

T. rex fossils date back to about 65 million years ago.

✔ Quick Check

Cause and Effect What happens to the plants and animals in an ecosystem after a fire?

Critical Thinking Why are plants usually the first organisms to spread to a new area?

How can people prevent extinction?

Once an animal or plant is extinct, it is gone forever. How does this happen? Sometimes it is because of people. When people move into wild places, they change the land. They build houses and grow crops. They hunt. They bring in new kinds of species. Organisms may not survive all these changes.

Large numbers of giant panda bears once lived in China. Pandas eat bamboo. When people began to cut down the bamboo forests, the pandas could not find enough to eat. Today, pandas are endangered.

Scientists are trying to prevent the extinction of panda bears. Large areas of land in China are preserved for pandas to live. Now, panda cubs are born in these protected places.

The brown pelican faced extinction in 1970. It has recovered along the U.S. Atlantic Coast. Elsewhere, it is still endangered.

 **Quick Check**

Cause and Effect Why do living things become extinct? What happens when they do?

Critical Thinking What would happen to pandas if scientists did not help them have cubs?

The giant panda is an endangered species.

FACT Not all endangered species become extinct.

Lesson Review

Visual Summary

Changes in ecosystems are caused by natural events, living things, and human activities.

When ecosystems change, living things move, accommodate, or become extinct.

People can help protect living things and their ecosystems.

Make a FOLDABLES® Study Guide

Make a three-tab book. Use it to summarize what you learned about changes in ecosystems.

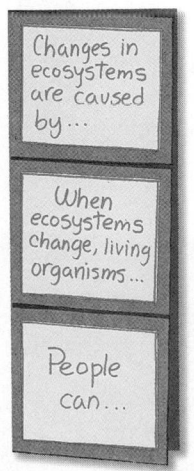

Changes in ecosystems are caused by...

When ecosystems change, living organisms...

People can...

Think, Talk, and Write

1. **Vocabulary** When a species no longer exists, it is _____.

2. **Cause and Effect** What happens when a forest is cut down to build a power plant?

Cause →	Effect
→	
→	
→	
→	

3. **Critical Thinking** Today, people ship goods around the world. Sometimes they move plants and animals by accident. How can such accidents cause changes in an ecosystem?

4. **Test Prep** **Which of these are natural events that change ecosystems?**

   A flood, pollution, deforestation

   B earthquake, overpopulation, fire

   C hurricane, flood, landslide

   D farming, recycling, overpopulation

5. **Essential Question** How can changes in an environment affect the organisms that live there?

 **Writing Link**

Write an Article

Write a magazine article that encourages people to preserve natural spaces. Explain how this will help protect plants and animals.

 **Social Studies Link**

Do Research

Learn about endangered plants and animals in your area. Find out how many are left. What is being done to protect them?

Mail Call

Scientists at the American Museum of Natural History collect stories from people around the world. The stories help scientists learn about local environments.

Dear Museum Scientists,

My name is Clara. I live in a small town in Southern California. The hills around our town are covered with evergreen shrubs. The land is very dry and there are not a lot of trees. This environment is called chaparral.

We did not get a lot of rain here last summer. In August, a lightning storm started a wildfire in the chaparral. When I walked through the area after the fire, all I saw were gray ashes and dead shrubs.

a chaparral environment

Connect to

AMERICAN
MUSEUM Ö
NATURAL
HISTORY

at **www.macmillanmh.com**

It's April now, and I hiked through the burnt chaparral last week. I brought my field guide with me so I could look up the plants and animals I saw.

The chaparral has changed so much! There are fields of wildflowers blooming everywhere. I found a hillside monkey flower and scarlet larkspur. My guidebook told me that these flowers have seeds that can stay dormant for several years. They need fire, heat, or smoke to sprout. The wildflowers have attracted insects like honeybees. The birds and animals are back too! I saw a cactus wren and jackrabbits. My guidebook explained that the low bushes provide shelter for jackrabbits and nesting for cactus wrens. I can't wait to go back to see how the chaparral will change even more!

Your friend,
Clara

Predict

► Think over what you know about a topic.

► Use what you read to tell what might happen.

 **Write About It**

Predict Read the letter again. Predict what the chaparral will be like next year. What might happen to the environment if there is a drought? Write your predictions in the form of a paragraph.

 **e-Journal** Research and write about it online at www.macmillanmh.com

Visual Summary

Lesson 1 Animals have adaptations that help them survive in their environments.

Lesson 2 Plants have adaptations and can respond to their environments.

Lesson 3 When an ecosystem changes, some living things survive and some do not survive.

Make a FOLDABLES Study Guide

Tape your lesson study guides to a piece of paper as shown. Use your study guide to review what you have learned in this chapter.

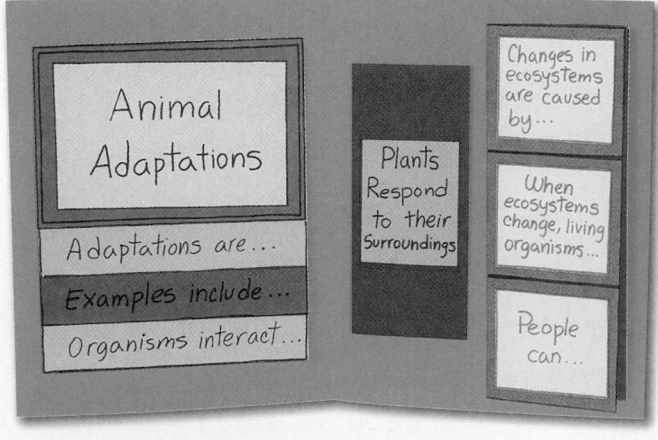

Fill each blank with the best term from the list.

accommodation, p. 188   **hibernate,** p. 168

adaptation, p. 166   **migrate,** p. 168

camouflage, p. 168   **mimicry,** p. 169

endangered, p. 189   **stimulus,** p. 176

extinct, p. 189   **tropism,** p. 177

1. A trait or behavior that helps a living thing survive in its environment is a(n) _____.

2. When all the individuals of a species have died, the species is _____.

3. To survive cold winters, some animals _____.

4. An animal that blends in with its surroundings uses _____.

5. Something in the environment that causes an organism to respond is called a(n) _____.

6. When a species has traits that resemble those of a different species, it uses _____.

7. The response of a plant to light, water, or gravity is a(n) _____.

8. An individual's response to a change in its ecosystem is called a(n) _____.

9. Animals _____ when they change their location periodically.

10. A living thing is _____ when few of its kind are left.

Answer each of the following.

11. Predict Your class plants tulip bulbs inside and outside a greenhouse. Will all the tulips bloom at the same time? Explain why or why not.

12. Form a Hypothesis Species can become endangered when their habitats change. Choose an animal species to research. Form a hypothesis about what might happen to the ecosystem if the species became extinct.

13. Critical Thinking Suppose scientists discovered a new species of animal living in the desert. What adaptations might the animal have?

14. Descriptive Writing Describe three ways in which people change ecosystems.

15. Critical Thinking Compare how camouflage and mimicry protect an animal. Use examples in your answer.

16. True or False *A tropism is an adaptation.* Is this statement true or false? Explain.

17. True or False *Many birds survive winter by hibernating.* Is this statement true or false? Explain.

18. True or False *Changes in an environment always harm living things.* Is this statement true or false? Explain.

19. Only a few individuals of a species are living. This species is

  A camouflaged.   **C** extinct.

  B hibernating.   **D** endangered.

20. What happens to living things when their environments change?

Adaptation Cards

1. Choose five kinds of adaptations you learned about in this chapter. Using available resources, find a different plant or animal with each adaptation.

2. Make a matching card game. Draw your organisms on five separate index cards. Describe each organism's adaptations on five other cards. Exchange your set of cards with a partner. Using your partner's cards, match each organism to its adaptation.

Analyze Your Results

Review the organisms with your partner. Together, write about other adaptations these organisms have. What similarities and differences do you notice?

1 This deciduous tree has adaptations to help it survive in its environment.

Which is an adaptation of the deciduous tree?

A growing new leaves to save energy

B keeping its leaves to save energy

C storing food for use in winter

D storing water for use in summer
DOK I

2 Look at the picture below.

How does this behavior <u>most likely</u> help the dog survive?

A by getting rid of hair

B by getting rid of fleas

C by getting rid of an itch

D by getting rid of extra skin
DOK 2

3 The picture below shows a forest after a fire.

How does the fire affect forest animals?

A It changes their habitat.

B It increases their numbers.

C It makes their habitat larger.

D It increases their food supply.
DOK I

4 The data table shows the populations of four different species of snails.

Snail Population Sizes			
	Year 1990	Year 2000	Year 2010
Species 1	2,000	2,500	2,300
Species 2	2,000	300	1,200
Species 3	2,000	2,700	3,400
Species 4	2,000	700	200

Which snail species will <u>most likely</u> become extinct?

A species 1

B species 2

C species 3

D species 4
DOK 2

5 Which is the <u>best</u> example of an adaptation?

    A the large ears of a desert fox

    B the white fur of a grassland mouse

    C the flat teeth of a predator

    D the sharp teeth of a herbivore
    DOK 2

6 A law is passed to protect endangered species. What can the law be expected to do?

    A make organisms extinct

    B make organisms endangered

    C prevent organisms from becoming extinct

    D prevent organisms from damaging a habitat
    DOK 1

7 Look at the butterfly shown below.

Which feature helps it survive in its environment?

    A bones

    B lungs

    C teeth

    D wings
    DOK 1

8 This bear sleeps through winter.

This seasonal adaptation is known as hibernation. Explain how hibernation helps the bear survive.
DOK 1

Give an example of another way animals are adapted to changing seasons.
DOK 2

9 What is a tropism? Give an example.
DOK 1

10 Explain how the tropism you described in question 9 helps the organism survive in its environment.
DOK 2

Check Your Understanding			
Question	Review	Question	Review
1	pp. 166, 178	6	pp. 189–190
2	pp. 166–170	7	pp. 166–169
3	pp. 184–189	8	pp. 168–169
4	p. 189	9	pp. 176–177
5	pp. 166–167	10	pp. 176–177

Nature Photographer

Picture yourself deep in a forest or below the ocean's surface. There is no one else around. You are ready to capture a special moment on film.

To be a nature photographer, you need to take classes in art and photography. You should also enjoy being outdoors.

A nature photographer needs patience. A single shot might take days or even weeks to get. You may face harsh conditions, like swarms of insects or cold rains. But when you finally capture that perfect photo, it is all worthwhile!

▲ A nature photographer must know about living things and their habitats.

Forester

Do you love the outdoors? You should think about a career as a forester. Foresters manage forests or wilderness areas. Their job is to help protect the land and make the best use of it.

To become a forester, you need a college degree in life science. Your studies will help you decide whether land is safe for hiking, camping, or hunting. You will learn to care for seedlings. You might teach others how to care for the land.

▲ A forester teaches people about forest ecosystems.

LOG ON e-Careers at www.macmillanmh.com

Earth and Its Resources

Flowing water can change
the shape of land.

waterfall at Point Lobos State Reserve, California

Lichen
Life on the Rocks

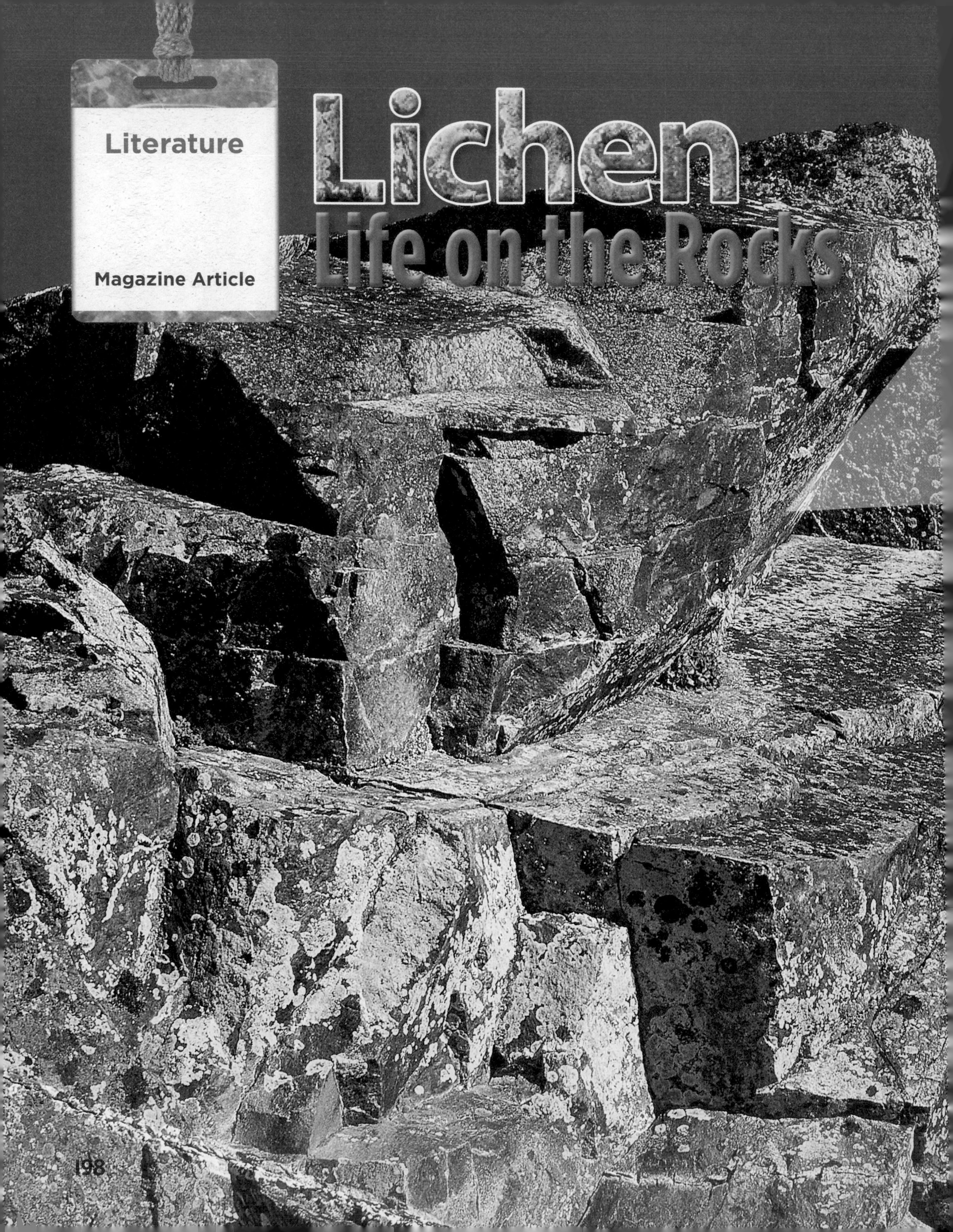

A lichen is actually two things, alga and fungus.
Algae is a tiny green protist that uses sunlight to make its own food. Fungi grow and feed on dead things. They help to decompose dead things and even recycle them.

Lichens can be found on a rock's surface or tree bark. If you find them on a rock's surface, they are crusty and flat. If you find them on tree bark, they are leafy. There's even another kind of lichen that grows in clumps on the ground.

The algae and fungus work together. What a team! They are great partners because the alga makes food for both of them and the fungus protects the alga from drying out due to sunlight. It makes a "house" for the alga. Lichens can live in places were very few other living things are able.

On a bare rock, a lichen can form a crust over it. The lichen grows very slowly, breaking down the rock. Tiny cracks in the rock occur from the acids lichens give off. After that, tiny cracks become filled with water, and when the water freezes, the cracks widen. As time passes, bits of soil that blow around get caught in the lichen, which allows plant seeds to take root. You many even see a rock covered with soil and plants after many years!

Write About It

Response to Literature This article tells you that lichen is not one thing but two. What are the two parts of a lichen? How can a lichen change rocks? Write a summary. Use your own words to explain what this article is about.

LOG ON **e-Journal** Write about it online at **www.macmillanmh.com**

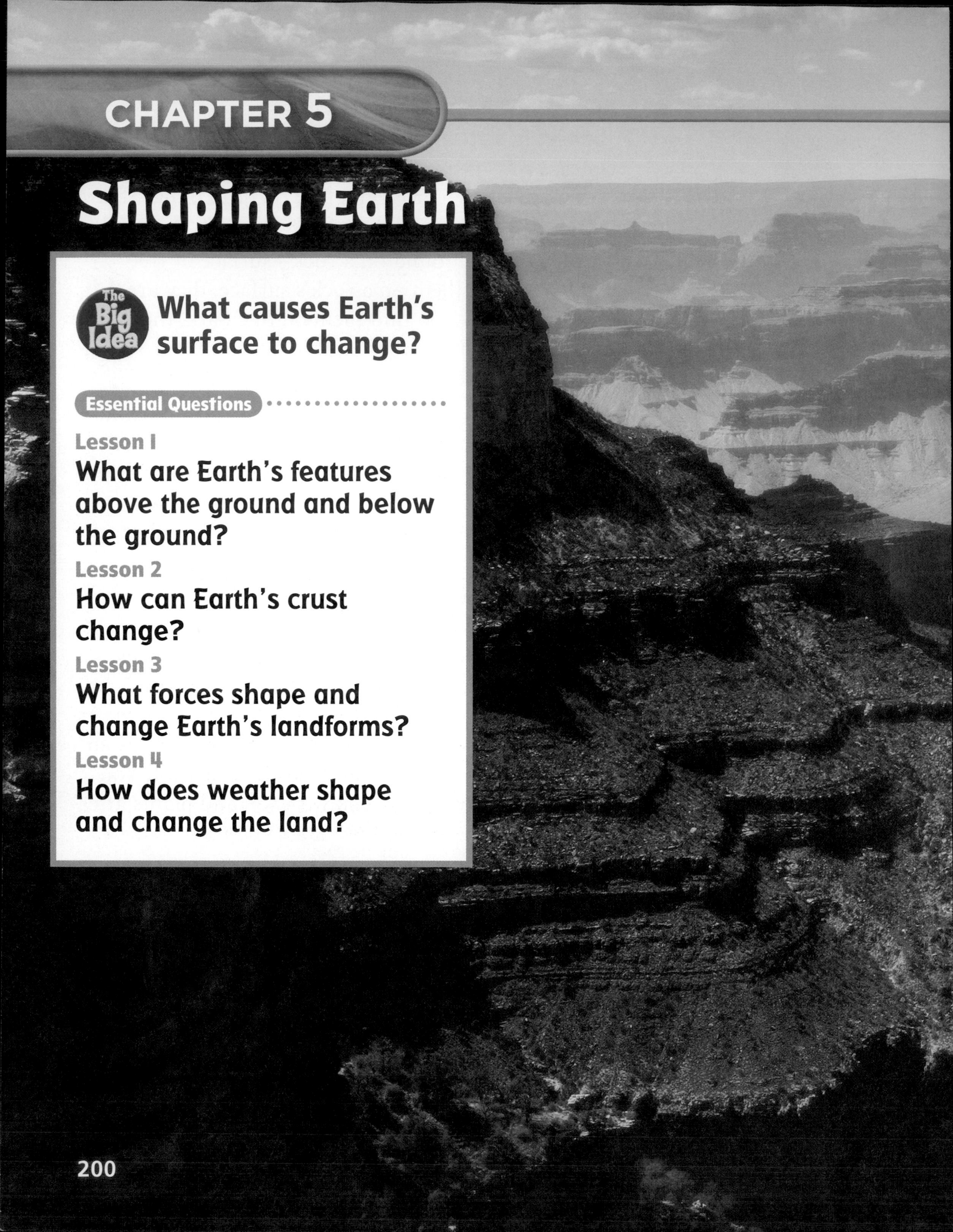

CHAPTER 5

Shaping Earth

 **What causes Earth's surface to change?**

Essential Questions ·

Lesson 1

What are Earth's features above the ground and below the ground?

Lesson 2

How can Earth's crust change?

Lesson 3

What forces shape and change Earth's landforms?

Lesson 4

How does weather shape and change the land?

crust rock that makes up the Moon's and Earth's outermost layers (p. 208)

earthquake a sudden shaking of the rock that makes up Earth's crust (p. 216)

seismograph a tool that graphs seismic waves as wavy lines and helps scientists detect earthquakes (p. 218)

weathering a process that breaks rocks into smaller pieces (p. 226)

erosion the weathering and removing of rock or soil (p. 228)

hurricane a violent, swirling storm with strong winds and heavy rains (p. 240)

 Visit www.macmillanmh.com for online resources.

Earth

Look and Wonder

In the Bisti Badlands of New Mexico, the ground is dry and coarse. Odd features cover the landscape. How would you describe these features?

What shapes can the land take?

Purpose
Explore some of the features on Earth's surface.

Procedure

1 **Observe** Draw some different shapes that you have seen on Earth's surface. Think about places you have visited. Recall places you have seen in magazines, movies, on television, or the Internet. You can also look for pictures in this chapter.

2 **Make a Model** Choose one of these land shapes. Shape some modeling clay to show how the land looks. Add as many details as you can.

3 **Communicate** Discuss your model with a partner. How are your models alike? How are they different?

Draw Conclusions

4 **Classify** Do you recognize any of the land shapes? Name them if you can.

5 Do you think the land always had these shapes? If not, how did they come about?

Explore More

Describe the shape of the land where you live. Are any of the models like that land? If not, make a new model to show what the land looks like near you.

Materials

- modeling clay

Step **2**

Step **2**

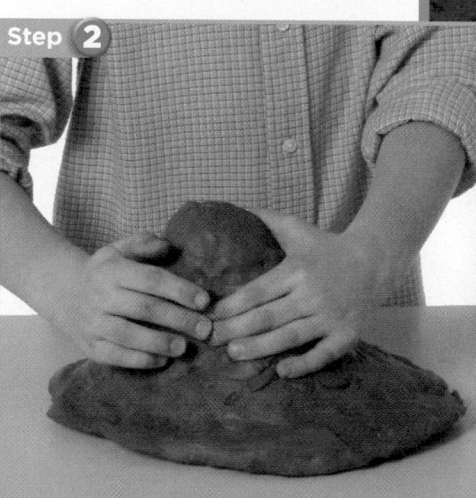

▶ **Essential Question**
What are Earth's features above the ground and below the ground?

▶ **Vocabulary**

crust, p. 208

mantle, p. 208

outer core, p. 208

inner core, p. 208

▶ **Reading Skill** ✓
Draw Conclusions

Text Clues	Conclusions

▶ **Technology**
e-Glossary and e-Review online at www.macmillanmh.com

What does Earth's land look like?

Viewed from space, Earth's land might seem flat. From close up, you can see many natural features on Earth's surface. These features are called *landforms*.

Tallest and Flattest

The tallest and most visible landforms are *mountains*. Most rise steeply to a peak at the top. Others have a gentle slope. Some are *volcanoes* formed by melted rock.

A *plain* is the flattest kind of landform. Plains are vast areas of land without hills or mountains.

Landforms Shaped by Water

Flowing water can shape the land. Streams and rivers can cut small *channels* or larger *gullies* where they flow. Strong flows can create deep *valleys*. In some places, rivers form narrow, steep sided valleys called *canyons*. The Grand Canyon is more than 1 kilometer deep!

Waves wear away land too. Waves can make a beach flat and smooth or sharp and rocky.

Landforms Shaped by Wind

Gusts of wind can pile sand into large mounds in deserts and on beaches. These mounds are called *sand dunes*. Wind can also combine with water to make mountains steeper and valleys deeper.

The Continental United States

① ③

③

②

④

② ④

Read a Map

Describe landforms you might see on a trip across the United States.

Clue: Match the numbers on the map to the numbers on the pictures.

✔ Quick Check

Draw Conclusions What can landforms teach us about Earth's history?

Critical Thinking Compare how wind and water shape the land.

What does it look like where water meets land?

Water always flows downhill. What happens when it gets to the bottom of a landform?

River Deltas

As the land gets flatter, the flow of water in a river begins to slow. If the river empties into an ocean, the water moves even slower. It drops off bits of sand and soil it carried. The bits form a *delta*—a landform that is sometimes shaped like a triangle.

Drainage Basins

Small rivers empty into bigger rivers. A *drainage* (DRAY•nij) *basin* is the area of land drained by flowing water. Much of central North America, for example, is a drainage basin for the Mississippi River.

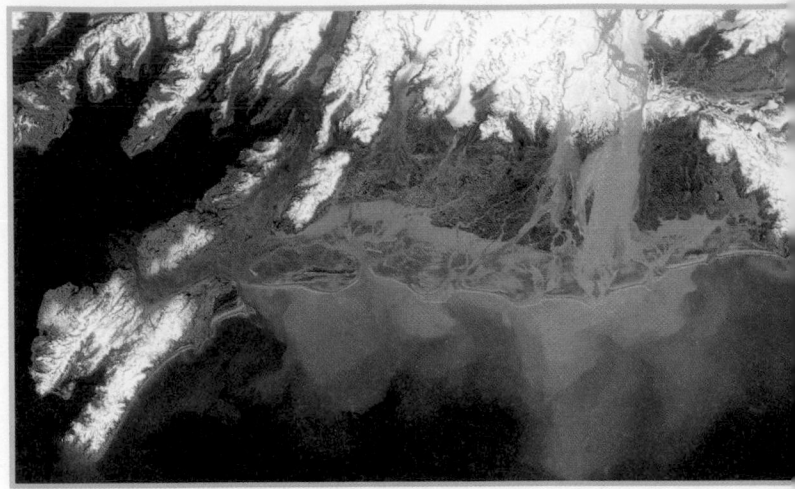

This delta formed where the Copper River meets Prince William Sound in Alaska.

Continental Shelves

It is a sunny day at the beach. You want to swim with your friends. You move farther from shore, but the water is still too shallow. Where are you? You are on a continental shelf. The *continental shelf* is part of the land that is covered by the sea. It can stretch seaward for miles.

From Land to Sea

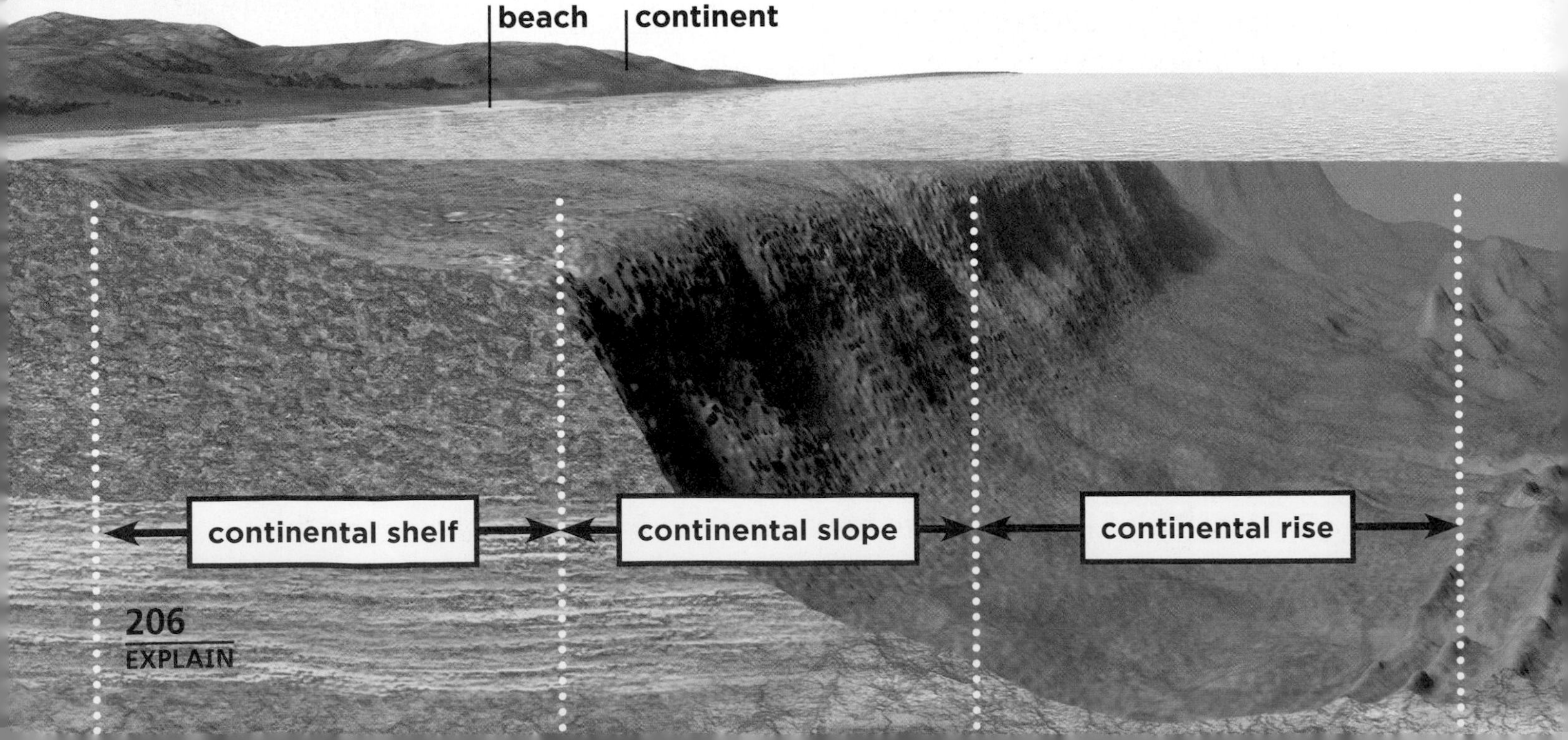

beach | continent

◄─ **continental shelf** ─► ◄─ **continental slope** ─► ◄─ **continental rise** ─►

Beyond the Continent

Beyond the continental shelf is an area of land called the *continental slope*. This land is the steeper part of the continent that slopes down toward the ocean floor. Underwater canyons can form on the slope.

At the base of the slope lies the *continental rise*. The rise connects the continent and ocean floor.

Most of the ocean floor is flat and without features. However, long mountain ranges stretch along the bottom of the oceans. These are *ocean ridges*. The ocean floor is spotted with undersea volcanoes too. The ocean floor also has deep trenches; some are more than 8 kilometers (5 miles) deep.

Read a Diagram

How does Earth's surface change between the continent and ocean floor?

Clue: Trace a line from the beach to the ocean ridge.

ocean ridge

Quick Lab

Drain Away

1. **Make a Model** Place some modeling clay in a plastic or metal container. Shape it into landforms. Include at least one mountain, valley, and plain. Draw a map of your model.

2. Spray water onto the top of the model mountain and along its sides. Repeat until you see water collecting in the container.

3. **Observe** How did the water flow over the model? Where did the water collect?

4. Use your map to describe the drainage basin in your model.

✔ Quick Check

Draw Conclusions Are oceans drainage basins? Explain.

Critical Thinking How do features of the ocean floor compare to those on land?

What is below Earth's surface?

You have read about some of the features on Earth's surface. What is Earth like beneath those features?

Look at the diagram. It shows the interior of Earth. Scientists divide Earth into four main layers.

① **Crust** Rock that makes up Earth's outermost layer. It is brittle and can crack easily. Earth's landforms and underwater features are found on the crust.

② **Mantle** Layer of rock below the crust. It is solid, but some of the rock can move or change shape at high pressures and temperatures.

③ **Outer Core** Liquid layer below the mantle. It is made mostly of melted iron.

④ **Inner Core** Sphere of solid material at Earth's center. It is the hottest part of Earth. It is probably made of iron.

 **Quick Check**

Draw Conclusions Which is the thinnest layer of Earth?

Critical Thinking How can scientists study Earth's core?

Lesson Review

Visual Summary

Earth's landforms include mountains, plains, canyons, and other features formed by water and wind.

Earth's water drains into rivers, channels, drainage basins, and oceans. The ocean has features similar to land.

Earth's layers are the crust, mantle, outer core, and inner core.

Make a FOLDABLES® Study Guide

Make a three-tab book. Use it to summarize what you read about Earth's features.

Earth's landforms

Earth's Water

Earth's layers

Think, Talk, and Write

1 Vocabulary The first layer of Earth beneath the crust is the _____.

2 Draw Conclusions Where would you go to find a delta?

Text Clues	Conclusions

3 Critical Thinking How could you use an apple to model Earth's interior?

4 Test Prep A wide, flat landform is called a

A plain.

B valley.

C canyon.

D trench.

5 Test Prep What is Earth's outermost layer?

A inner core

B outer core

C mantle

D crust

6 Essential Question What are Earth's features above the ground and below the ground?

 **Writing Link**

Write an Essay

Write an essay about the different landforms that you would visit on a trip across the United States. You can also research and write about specific places, such as the Great Lakes or the Grand Canyon.

 **Math Link**

Writing Fractions

Oceans cover about 70 percent of Earth's surface. How can you show 70 percent as a fraction? Express the fraction in simplest form. How could you model this fraction?

Focus on Skills

Inquiry Skill: Experiment

A mountain takes thousands of years to form. Yet it can change in a day. Landslides are one way a mountain changes quickly. Soil and rocks suddenly slide downhill. Houses and trees may be carried away. To study landslides, scientists do **experiments**. Then they draw conclusions from their results.

▶ Learn It

When you **experiment**, you make and follow a procedure to test a hypothesis. A procedure is a set of numbered steps that tell what to do first, next, and last. It is important to record your observations while you follow your procedure. Your observations can help you draw a conclusion from your results. It is always good to run your test several times. That way, you know whether your results are true.

▶ Try It

Do you think there is a way to lessen the damage from landslides? Write your idea in the form of a hypothesis. Then **experiment** to test your hypothesis.

Materials soil, planter box, rocks, twigs, wooden blocks, water, watering can, plastic tablecloth, 2 or 3 books, cookie tray

1 Spread dry soil in the bottom of a long planter box. Push rocks and twigs into the soil. Place small, wooden blocks as houses. Sprinkle the soil with water until it is damp.

2 Cover a table with a plastic tablecloth. Set the box on it. Prop up one end of the box with two or three large books. Place a cookie tray under the other end of the box.

3 Predict what would happen to the soil in a heavy rainstorm. Record your prediction in a chart like the one below.

Step	Predictions	What I Observed
step 3 no wall or barrier		
step 5 wall near top		
step 6 barrier near bottom		

4 Use a watering can to pour water at the high end of the box. Record what happens.

5 Repeat steps 1 and 2. This time, bury a block in the soil near the top of the box to make a wall. Then repeat steps 3 and 4.

6 Repeat steps 1 and 2. This time, make a fence or barrier. Put a piece of wood as wide as the box into the low end of the box. Then repeat steps 3 and 4. If you have time, rerun the entire experiment from step 1 through step 5.

7 Review your results. Do they support your hypothesis? Explain.

8 Can anything be done to prevent or ease the damage caused by landslides? Use your results to explain your answer.

▶ Apply It

1 What do you think would happen if you placed the barrier near the top of the hill? What if one barrier were at the top and one were at the bottom? What if every house had a fence around it? Write a hypothesis for one of these ideas or form your own.

2 Design an **experiment** to test your hypothesis. Remember to record your observations, run several tests, and draw conclusions from your results.

The Moving Crust

Look and Wonder

These mountains in Grand Teton National Park probably began as flat layers of rock. What made them rise and tilt? Did something happen to Earth's crust?

How can Earth's crust change shape?

Make a Prediction

Predict how flat rock layers will react to pressure.

Test Your Prediction

① Flatten three pieces of colored clay into thin layers. Press the layers on top of each other, like a sandwich.

② **Make a Model** ⚠ **Be Careful.** Use the plastic knife to cut the stack of clay into two squares. These are your models of rock layers.

③ Press two wooden blocks flat against opposite sides of one clay square. Move the blocks slowly and firmly toward each other. Record your observations.

④ Cut a slice at an angle from the top to the bottom of the second clay square. Repeat step 3 on this new model. Record your observations.

Draw Conclusions

⑤ How did squeezing change the clay layers? What landforms do the models now resemble?

⑥ **Interpret Data** Did slicing the layers make a difference in how the clay reacted to squeezing? Explain.

⑦ **Infer** Using your model, explain how rock layers can be folded or lifted.

Explore More

Compare how the clay model reacts to slow steady pressure and strong sudden pressure. Make new models and test them.

Materials

- modeling clay (3 colors)
- plastic knife
- 2 wooden blocks

Step ①

Essential Question

How can Earth's crust change?

Vocabulary

fault, p. 215

plateau, p. 215

fold, p. 215

mountain, p. 215

earthquake, p. 216

seismic wave, p. 218

seismograph, p. 218

volcano, p. 220

Reading Skill ✔

Cause and Effect

Cause	→	Effect
	→	
	→	
	→	
	→	

Technology

e-Glossary and e-Review online at www.macmillanmh.com

How does Earth's crust move?

You have learned that Earth is made up of layers. The crust is the thinnest, outer layer. The mantle lies beneath the crust.

Plates

Earth's surface is broken into several huge plates of rock. *Plates* are made of crust and the upper part of Earth's mantle. The crust and mantle are solid but the upper mantle can flow. When the mantle flows, Earth's plates move.

Earth's plates move about as slowly as your fingernails grow. The moving plates get pushed and pulled, changing the crust. You cannot see or feel most of the changes. Others you cannot miss!

Mountains in the Making

Fault-block mountain can form where plates get pushed or pulled.

Faults

As plates get pushed and pulled, cracks called faults form in the crust. A **fault** is a crack in the crust along which movement takes place.

Rocks on either side of a fault can slide up or down compared to the other side. Faults can sometimes form *fault-block mountains*. If the lifting is spread over a wide area, a plateau (pla•TOH) may form. A **plateau** is a high landform.

Folds

Some plates meet at the edges of continents. If the land scrunches up between them, a fold forms. A **fold** is a bend in the rock layers.

If the land keeps scrunching, folds become mountains. A **mountain** is a tall landform. As time passes, wind and rain can break off pieces and change the shape of the mountain.

fault-block mountain

folded mountain

✔ *Quick Check*

Cause and Effect What are two ways that mountains can form?

Critical Thinking Why do mountains form at only some places on Earth?

 A folded mountain can form where plates slide toward each other.

Read a Diagram

Compare how fold mountains and fault-block mountains form.

Clue: Look for the differences between the two kinds of mountains in the diagram.

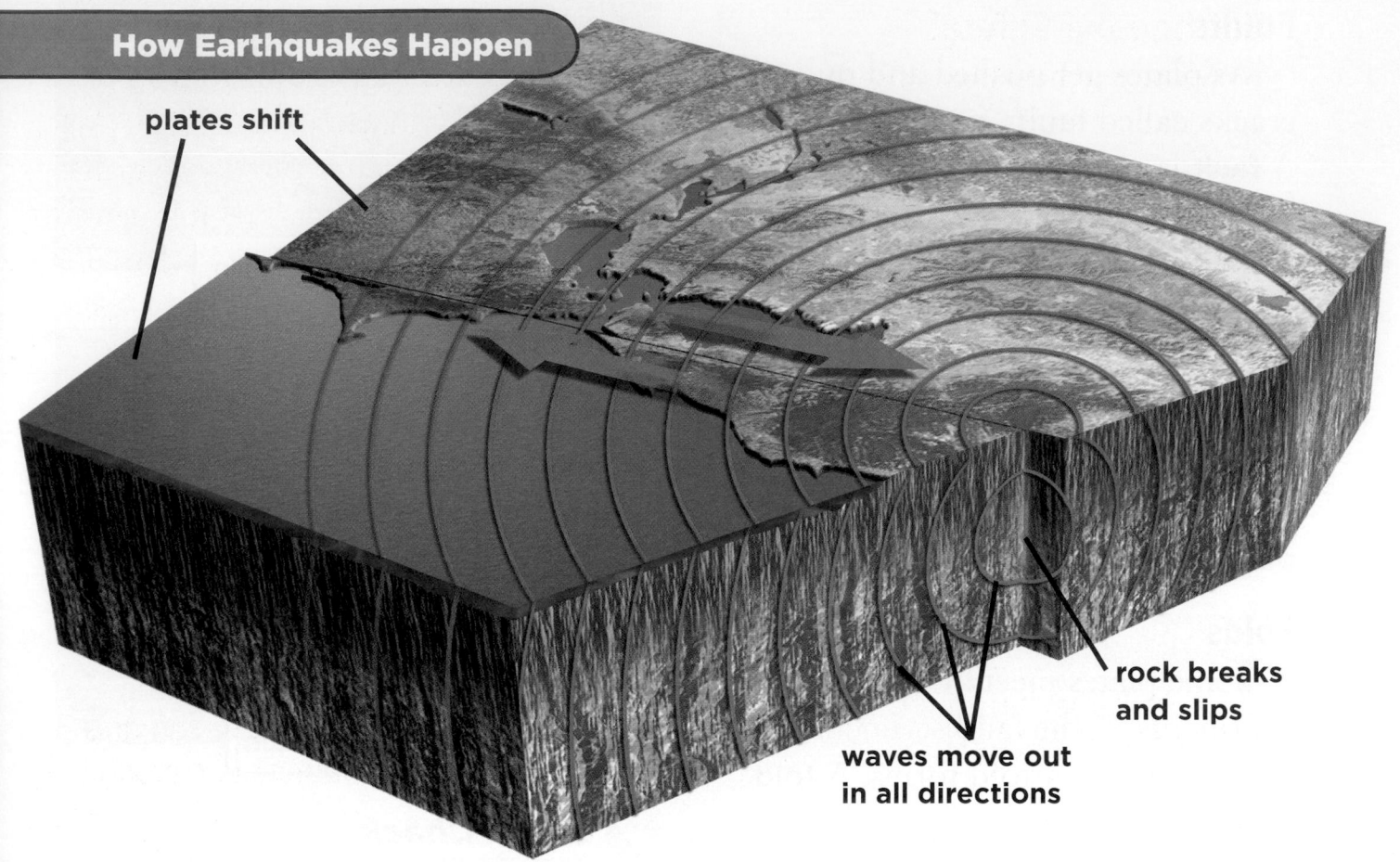

plates shift

rock breaks
and slips

waves move out
in all directions

What causes earthquakes?

An **earthquake** is a sudden shaking of Earth's crust. It is caused by sudden movement along a fault. When forces push and pull a plate, energy builds up. Earth's crust may store this energy for many years. Then suddenly, the energy is released. The rock breaks apart and moves along the fault! Earthquakes are common in places with active faults, like parts of Alaska and California.

How Earthquakes Travel

An earthquake begins below the ground. The energy released by a sudden movement shakes the crust. Vibrations, or waves, move through Earth in all directions.

Did you ever drop a pebble in water? The waves of an earthquake travel like the ripples of water. As they move away from the center of the earthquake, the waves weaken. Even so, you may feel them at the surface hundreds of miles away from the center!

FACT No scientist has ever predicted a major earthquake.

Earthquake Safety

Most earthquakes are too weak to notice. Others can cause extreme damage. During a major earthquake, buildings and roads may break apart. Bridges might collapse.

Do you know what to do if the ground below you starts to shake? You can stay safe in an earthquake by following a few simple rules. If you are indoors, duck under a table or doorway. Keep away from walls and windows. If you are outdoors, stay away from trees, power lines, and anything that might fall down.

Earthquakes in the Ocean

Some earthquakes strike below the ocean. If an earthquake is strong enough, it can cause the ocean crust to lift suddenly. When this happens, look out! A giant wave, called a *tsunami* (soo•NAH•mee), might hit the shore. Tsunamis cause great damage along coastlines. They can destroy everything in their path.

 Quick Check

Cause and Effect How can an earthquake cause a tsunami?

Critical Thinking How can you stay safe if a tsunami is coming?

How do scientists study earthquakes?

Any movement can cause a vibration. **Seismic waves** are the vibrations caused by earthquakes.

When an earthquake strikes, seismic waves travel out from the source in all directions. The waves move at different speeds. Some of the waves travel along or near Earth's surface. Others travel through Earth's interior.

Measuring Seismic Waves

Scientists measure seismic waves with an instrument called a seismograph (SIZE•muh•graf). A **seismograph** detects and records earthquakes. It shows seismic waves as jagged lines along a graph. The lines show how much the ground shakes. The stronger the quake, the longer the lines.

The Gray-Milne seismograph was made in Scotland.

Time Line of Seismic Study

Chang Hêng's seismoscope was invented in China.

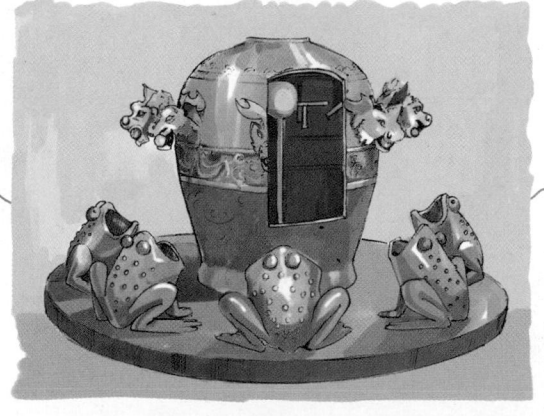

This seismoscope was invented in Italy.

A.D. 132   1856   1885

Seismic Networks

When an earthquake hits, one of the first questions is "Where was it?" Earthquake scientists have a network of seismographs around the world. They collect data from each seismic station. Then they calculate the location and depth of the quake.

 Quick Check

Cause and Effect
What causes seismic waves?

Critical Thinking Why do the seismic wave readings shown below have different heights?

≡**Quick Lab**

Hearing Clues

1 Your teacher will give you containers. Make a plan to find out what they hold without opening the containers.

2 **Observe** Carry out your plan. Be sure not to damage the containers. Record your observations.

3 **Interpret Data** Study your data. What clues do they provide?

4 What do you think is inside each container? Make a diagram to help explain your ideas.

Ocean Bottom Seismographs were invented in the United States.

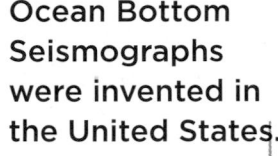

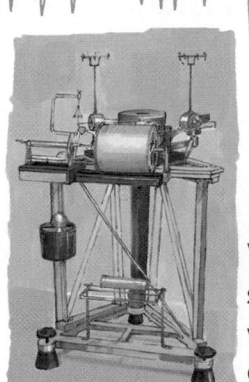

Wiechert's seismograph was invented in Germany.

Read a Diagram

How have seismographs changed over time?

Clue: Follow the time line from left to right.

1899   1937

What is a volcano?

A **volcano** is a mountain that builds up around an opening in Earth's crust. Sometimes a volcano will force materials from Earth's interior out of its opening. Scientists call this event an *eruption*.

A volcanic eruption can send out melted rock, gases, ash, or rocks into the air. Melted rock is called *magma*. Once magma reaches Earth's surface, it is called *lava*. By erupting often, a volcano can build a large mountain. Each eruption can add layers of lava and ash. The lava cools and hardens into rock.

Some volcanoes rest quietly for many years until they erupt suddenly. Others erupted often in the past but will never erupt again.

magma

Where Volcanoes Form

Most volcanoes occur at the edges of plates. When two plates meet, one can sink below the other. As it sinks, the plate gets hotter. The rock melts into magma. The magma rises and forms a volcano.

Volcanoes also form where Earth's plates move apart. The space between the moving plates allows magma to rise to the surface.

Some volcanoes form far away from plate edges. These *hot spots* are places where Earth's crust is very thin. Magma can easily break through to the surface. The islands of Hawaii formed over a hot spot in the Pacific Ocean. The islands are the tops of huge volcanoes that rose from the ocean floor.

✓ Quick Check

Cause and Effect How can volcanic eruptions build mountains?

Critical Thinking When should you stay a safe distance away from a volcano?

Volcanoes still build the Island of Hawaii.

Lesson Review

Visual Summary

Earth's crust and upper mantle are divided into slow-moving plates. **Mountains** may form where plates meet.

An **earthquake** is a sudden shaking of Earth's crust. It begins along a fault, releasing seismic waves.

A **volcano** is an opening in Earth's crust to the magma below. Volcanoes can build mountains.

Make a FOLDABLES Study Guide

Make a trifold book. Use it to summarize what you read about Earth's moving crust.

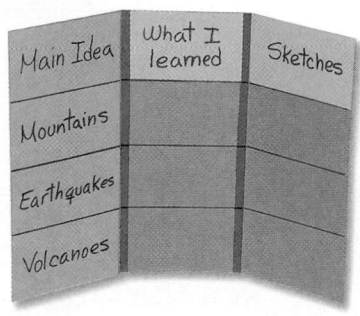

Main Idea | What I learned | Sketches
Mountains
Earthquakes
Volcanoes

Think, Talk, and Write

1 Vocabulary What do we call the vibrations from an earthquake?

2 Cause and Effect What are three ways that mountains are built?

| Cause → Effect |
| → |
| → |
| → |

3 Critical Thinking Earth's plates move slowly. Why do earthquakes happen suddenly?

4 Test Prep Which instrument measures earthquakes?

  A scientist
  B seismic wave
  C seismology
  D seismograph

5 Essential Question How can Earth's crust change?

Math Link

Measuring Distance
One kind of seismic wave moves through rock at 7 km per second. How long would it take such a wave to travel from San Francisco to New York—a distance of 4,100 km?

Social Studies Link

Research and Report
In the year A.D. 79, Mount Vesuvius erupted in Italy. Ash quickly covered the city of Pompeii. Research and report on this event. Describe the effect of the eruption.

Meet Ro Kinzler

Ro's favorite place to collect lava samples is Kilauea volcano in Hawaii.

Ro Kinzler is fascinated by volcanoes and volcanic rocks. She would go just about anywhere to find out more about them. Ro is a scientist at the American Museum of Natural History.

Ro travels to the Cascades in Northern California to collect lava samples from active volcanoes like Mount Shasta and Medicine Lake. She wants to study how magma moves through Earth. Back in the lab, Ro does experiments. She heats and squeezes the lava samples to find out how they formed deep in Earth.

Lava is melted rock that cools at Earth's surface.

▲ *Alvin* is a submersible that can take scientists to the ocean floor.

You don't find volcanoes only on land. There are lots of them on the ocean floor. Ro and other scientists have gone to the bottom of the ocean to study volcanoes. They use small underwater vehicles called submersibles.

The scientists visited the Mid-Atlantic Ridge, part of the longest volcano chain in the world. Ro is one of the few people to have ever seen it. She peered out the portholes of the submersible *Alvin* with other scientists. They made careful observations. They used these observations to make maps of the ocean floor.

Cause and Effect

▶ The cause answers the question "Why did something happen?"

▶ The effect answers the question "What happened as a result?"

 **Write About It**

Cause and Effect Read the article with a partner. Fill out a cause-and-effect chart to record why Ro visits volcanoes and collects lava samples. Tell what happens as a result of her work.

 **e-Journal** Research and write about it online at www.macmillanmh.com

Connect to
AMERICAN MUSEUM of NATURAL HISTORY
at www.macmillanmh.com

Weathering and Erosion

Look and Wonder

Once this sea arch was a continuous piece of rock. Now you can see through it. How did this arch form in the limestone cliffs of Normandy, France?

How can rain shape the land?

Make a Prediction

Water always moves down a slope. What happens when it rains? Make a prediction that tells how rainfall shapes the land.

Test Your Prediction

Materials

- potting soil
- sand
- pebbles
- shallow pan
- spray bottle with water

1 **Make a Model** Pile a mixture of potting soil, sand, and pebbles at one end of a pan. Shape the mixture into a sloping hillside.

2 Use a spray bottle to simulate rain. Spray at an even rate until the hillside is soaked.

3 **Observe** Continue the rain at the same rate. Observe what happens to the hillside. Record your observations.

Draw Conclusions

4 **Communicate** Did your results match your prediction? Explain what happened to the model land.

5 **Infer** How is your model like the real world? Use evidence from your observations.

Step 2

Explore More

Does the rate of rainfall affect the amount of land that moves downhill? What variables must you control to test a hypothesis? What variable would you change? Try it. Report your results.

Read and Learn

▶ **Essential Question**

What forces shape and change Earth's landforms?

▶ **Vocabulary**

weathering, p. 226

erosion, p. 228

deposition, p. 229

terminus, p. 231

moraine, p. 231

▶ **Reading Skill** ✔

Classify

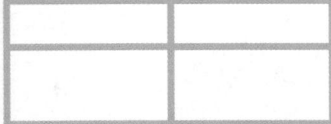
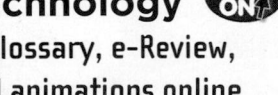

▶ **Technology** 🔵 LOG ON

e-Glossary, e-Review, and animations online at www.macmillanmh.com

What is weathering?

Every day, rocks move and change shape as the wind blows and rain falls. Look at the rock below. What happened to it?

Even the largest boulder can break apart over time. **Weathering** is the slow process that breaks rocks down into smaller pieces. Flowing water, rain, and wind are some causes of weathering.

In the distant future, the rock below will probably look very different. It may break into pieces the size of pebbles or even smaller!

What caused this boulder to break apart?

▲ Chemical weathering can form limestone caves like this one in Brazil.

Physical Weathering

Rocks can change size and shape without changing their chemical makeup. This process is called *physical weathering.*

Flowing water from streams and rivers can make sharp rocks smooth. Waves crashing onto a cliff can break off small pieces of land. Rainfall may seep through small cracks in a rock. Cycles of freezing and melting can widen the cracks.

Living things also cause physical weathering. Plant roots can force their way through cracks in a rock. As the plant grows, its roots get stronger. The strong roots can break the rock apart.

Chemical Weathering

Chemical weathering changes the minerals that make up rocks. Oxygen, acids, and carbon dioxide all cause chemical weathering. They change minerals into new substances.

Have you ever seen an iron chain get rusty? Water and air change iron into rust. Rocks that have iron in them can rust too. Water and carbon dioxide can form limestone caves. Even living things, such as lichens, can soften the rocks they grow on.

 Quick Check

Classify What are the two kinds of weathering?

Critical Thinking Where might you find examples of weathering?

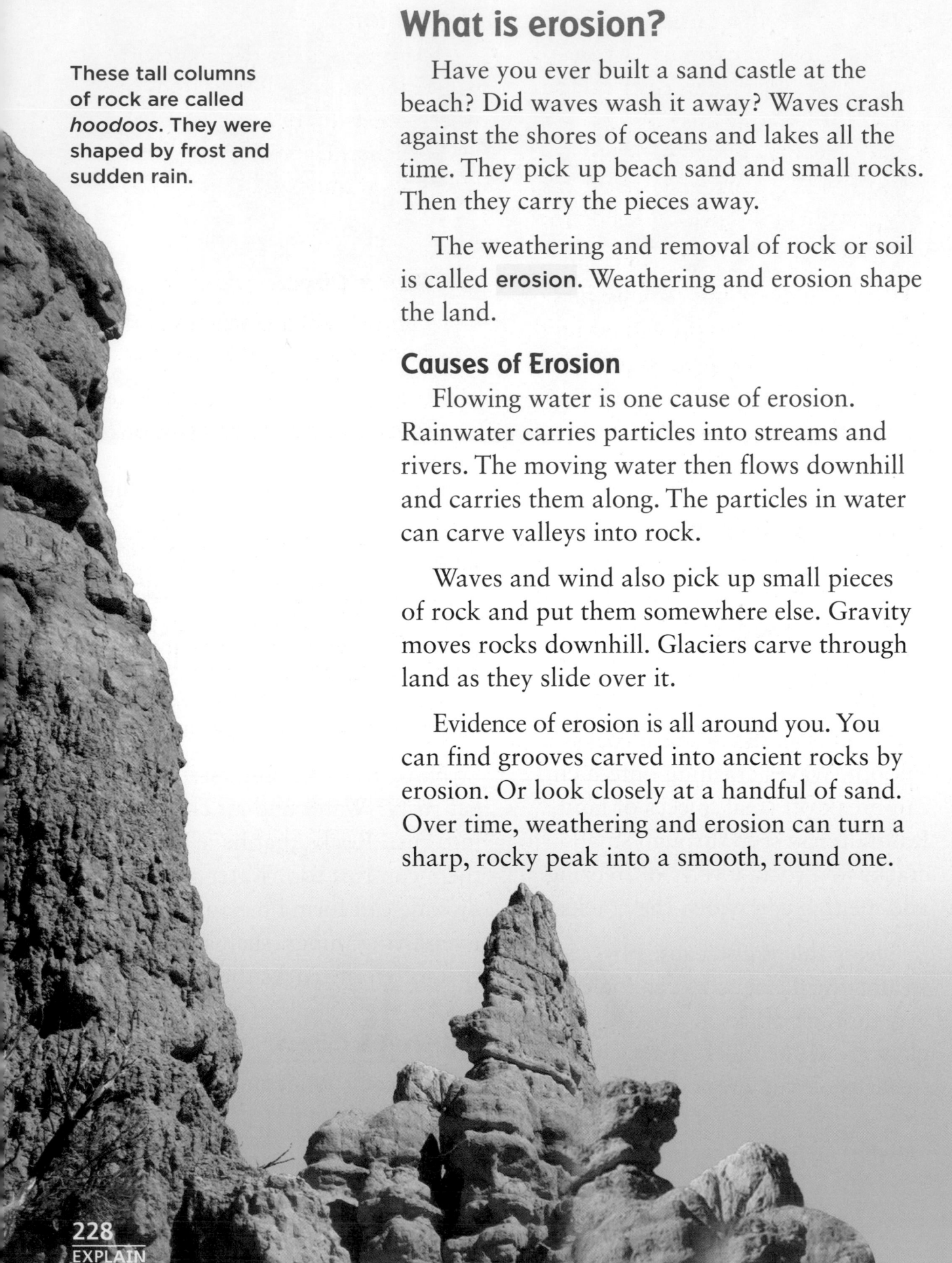

These tall columns of rock are called *hoodoos*. They were shaped by frost and sudden rain.

What is erosion?

Have you ever built a sand castle at the beach? Did waves wash it away? Waves crash against the shores of oceans and lakes all the time. They pick up beach sand and small rocks. Then they carry the pieces away.

The weathering and removal of rock or soil is called **erosion**. Weathering and erosion shape the land.

Causes of Erosion

Flowing water is one cause of erosion. Rainwater carries particles into streams and rivers. The moving water then flows downhill and carries them along. The particles in water can carve valleys into rock.

Waves and wind also pick up small pieces of rock and put them somewhere else. Gravity moves rocks downhill. Glaciers carve through land as they slide over it.

Evidence of erosion is all around you. You can find grooves carved into ancient rocks by erosion. Or look closely at a handful of sand. Over time, weathering and erosion can turn a sharp, rocky peak into a smooth, round one.

Rivers Erode the Land

The Grand Canyon in Arizona shows how powerful a river can be. The canyon is 446 km (277 mi) long. It has an average depth of 1.6 km (1 mi). This huge space was carved over millions of years by the Colorado River.

Rivers and streams pick up bits of rock and soil as they flow over land. Some of the pieces get deposited, or dropped off, on riverbanks. Others get carried to the mouth of the river.

Deposition

Deposition is the dropping off of weathered rock. Deposition by water builds up deltas, riverbanks, and beaches. Deposition by wind forms sand dunes.

✔ Quick Check

Classify What processes erode land? What processes deposit it?

Critical Thinking How do rivers cause weathering and erosion?

Rivers Shape the Land

Read a Photo

In your own words, describe how this canyon in Utah probably formed.

Clue: Observe the sides of the canyon. Look where the water is located.

 Science in Motion Watch how a river shapes the land at www.macmillanmh.com

How do glaciers shape the land?

In very cold places, thick sheets of ice called *glaciers* (GLAY•shurz) creep over the land. Over one million years ago, glaciers began to cover much of Earth. Few places are cold enough for glaciers today.

Glaciers form where snow collects quickly and melts slowly. Year after year, the snow builds higher. The weight on top of the mound puts pressure on the snow below. The bottom of the glacier slowly turns to ice. Near the ground, some ice melts.

Carving the Land

As the weight of the ice increases, the glacier begins to flow. The bottom and sides freeze onto rocks. As the glacier continues to move, it tears rocks from the ground. It scratches, flattens, breaks, or carries away the things in its path. A glacier can make a valley wider and steeper.

A glacier carved this valley in Alaska. ▼

A Glacier Deposits Land

moraine

moraine

What Glaciers Leave Behind

You have read how glaciers erode the land. Glaciers also deposit eroded rock. As glaciers melt, they leave behind the rocks they carried. The leftover rocks are called *glacial debris* (GLAY•shul duh•BREE).

Glacial debris can be made of large boulders or small particles. It can have bits of gravel, sand, and clay. The glacier drops most of this debris at its downhill end, or **terminus**.

Have you ever seen a giant boulder all by itself in a field? It was probably glacial debris. More often, glacial debris is a mix of small rocks, gravel, sand, and clay. The mixture is called *glacial till*.

Materials that a glacier picks up or pushes can forms mounds. These mounds are called **moraines**. Today, you can find glacial till and moraines across Canada and northern parts of the United States.

Read a Diagram

How does a glacier change the land as it melts?

Clue: Trace the glacier's path from the top of the hill to the terminus.

terminus

glacial till

✓ Quick Check

Classify Which landforms do glaciers erode? Which do they deposit?

Critical Thinking How do glaciers compare to other causes of weathering and erosion?

How do people shape the land?

Most processes in nature change Earth's land very slowly. People can make faster changes.

Mining

One way people change the land is by mining it. *Mining* is digging into the land for useful resources like minerals, metals, or fuel.

Landfills

Landfills are places where people pile trash. Some form large mounds or hills. Some are covered with soil and plants to blend in with the land.

Forests

People need land to build farms, and homes. Often people cut down or burn forests to clear the land. The trees are used to make products. Without the trees, erosion easily washes away the soil.

▲ You can help the land by planting and caring for trees.

✔ Quick Check

Classify List some of the ways people shape the land. Are these helpful or harmful to the land?

Critical Thinking Think of other helpful ways that people shape the land.

Part of this evergreen forest has been cut down. ▼

Lesson Review

Visual Summary

 **Weathering** is the breaking down of rock into smaller pieces. Two kinds of weathering are physical and chemical.

 **Erosion** is the removal of weathered rock. Deposition is the dropping off of eroded rock.

 **People change the land** in many ways. These changes can be helpful and harmful to the land.

Make a **FOLDABLES** Study Guide

Make a three-tab book. Use it to summarize what you read about weathering and erosion.

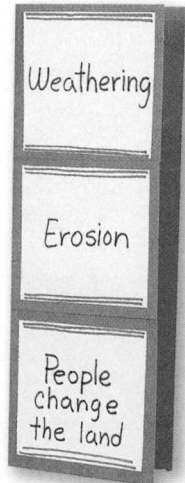

Think, Talk, and Write

1. **Vocabulary** Glacial debris collects at the _____ of a glacier.

2. **Classify** List examples of physical weathering and examples of chemical weathering.

3. **Critical Thinking** Compare natural weathering and erosion to changes that people make to the land.

4. **Test Prep** Which of the following is <u>most likely</u> to carve a canyon into Earth's crust?

   A chemical weathering

   B wind

   C moving water

   D waves

5. **Test Prep** Which is made of deposited materials?

   A a moraine

   B a mountain

   C a valley

   D a river

6. **Essential Question** What forces shape and change Earth's landforms?

 **Math Link**

Measuring Distances
A glacier advances 6 meters every year. How long will it take for the glacier to move a distance of 1,000 m?

 **Social Studies Link**

Study a Map
Study a road map of Colorado, Utah, Montana, or another state that has mountains. Compare the roads and cities to the paths of rivers.

Land over Time

Mountains may seem like mighty giants. But are they? Weathering can break down even the strongest mountains. How does this happen?

Wind carries the seeds of plants from place to place. A seed can land on a patch of soil on rock. There the seed sprouts. The roots find small cracks in the rock.

As the roots grow, rainwater fills the cracks in the rock. If it gets cold enough, the water freezes into ice. The ice expands. The cracks widen. All this time, the roots grow bigger.

At some point, the cracks widen so much that pieces of rock break off. In time, these pieces get smaller. Over millions of years, weathering can break down an entire mountain!

Expository Writing

Good expository writing

▶ presents the main idea in a topic sentence;

▶ supports the main idea with facts and details.

Write About It

Expository Writing Write a paragraph that summarizes "Land over Time." Include the main idea and the most important details.

 **-Journal** Research and write about it online at **www.macmillanmh.com**

DISAPPEARING MOUNTAINS

Mount McKinley

Mount Whitney

Mount Shasta

Wheeler Peak

This table shows the heights of some mountain peaks in the United States.

Heights of Mountain Peaks			
Mountain	State	Height in Meters	Height in Feet
Mount McKinley	Alaska	6,194	20,320
Mount Whitney	California	4,417	14,491
Mount Shasta	California	4,317	14,162
Wheeler Peak	Nevada	3,982	13,065

Mountains erode by small amounts. Suppose Mount McKinley erodes 2 m each year. How many years would it take for the mountain to be 6,174 m tall?

Solve It

If the erosion rate is 1 m each year, what will be the height of:

1. Mount Shasta in 20 years?

2. Mount Whitney in 15 years?

3. Wheeler Peak in 80 years?

Problem Solving

▶ To find the number of years, you can count backward by 2 from 6,194 m to 6,174 m.

6,192	6,190
6,188	6,186
6,184	6,182
6,180	6,178
6,176	6,174

It would take 10 years.

▶ Another way is to find the number of meters lost. Then you can divide the difference of meters by 2.
6,194 m – 6,174 m = 20 m
20 ÷ 2 = 10
It would take 10 years.

Lesson 4

Changes Caused by the Weather

Look and Wonder

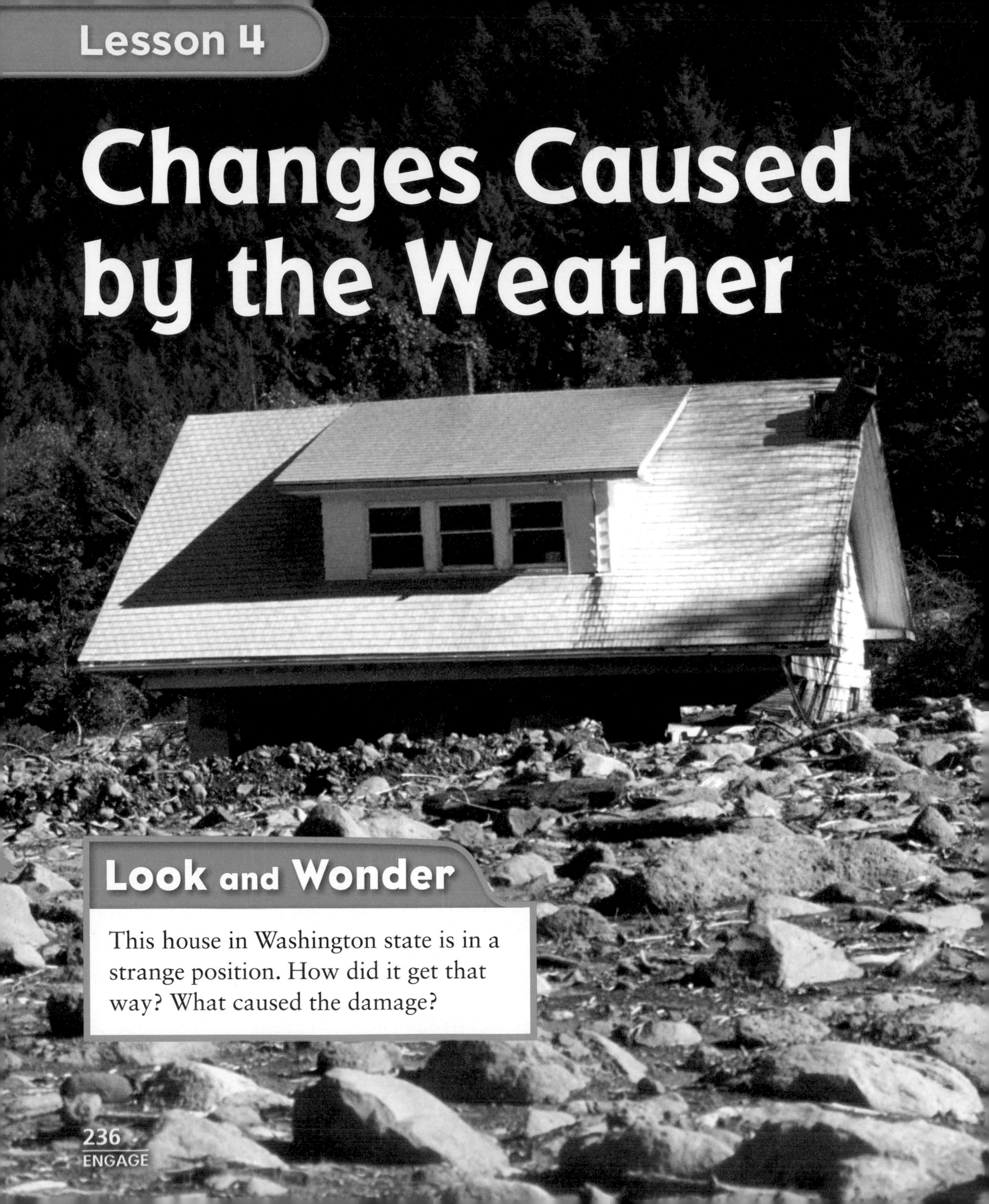

This house in Washington state is in a strange position. How did it get that way? What caused the damage?

How does steepness of slope affect the movement of Earth's materials?

Form a Hypothesis

We sometimes see evidence of sliding rocks and soil at the bottom of a hill. How does the steepness of a slope affect the downhill movement of rocks and soil? Write a hypothesis.

Test Your Hypothesis

1. Stir equal amounts of soil, gravel, and sand in the pan. Pat the mixture into a flat layer.

2. **Predict** What will happen when you raise one end of the pan? Record your prediction.

3. **Observe** Raise one end of the pan 4 centimeters. Record what happens. Continue raising that end by 4 cm at a time until the pan is nearly upright. Record your observations each time.

Draw Conclusions

4. **Interpret Data** How did raising the end of the pan affect your results?

5. What is the relationship between steepness of slope and the movement of soil and rocks?

Explore More

How do sudden downpours of rain affect very steep slopes? How could you test this? What variable would you control? What variable would you change? Try it. Report your results.

Materials

- **deep aluminum pan**
- **measuring cup**
- **potting soil**
- **gravel**
- **sand**
- **metric ruler**

Step 1

Step 3

▶ Essential Question

How does weather shape and change the land?

▶ Vocabulary

flood, p. 238

tornado, p. 240

hurricane, p. 240

landslide, p. 242

avalanche, p. 242

▶ Reading Skill ✓

Infer

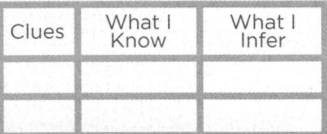

Clues	What I Know	What I Infer

▶ Technology

e-Glossary and e-Review online at www.macmillanmh.com

How do floods and fires change the land?

You have learned how erosion and deposition change the land slowly. What causes Earth's surface to change quickly?

Floods

Heavy rain or quickly melting snow can cause water to flow over the land. The water may not soak into the ground completely. It runs on top of the land. It flows into streams and rivers. The extra water may spill over the sides, or banks, of streams and rivers. Then it floods (FLUDZ) the nearby land. A **flood** is an overflow of water onto land that is normally dry.

Cities flood when water drains cannot carry water away fast enough. The drains overflow. The streets become flooded.

Floods can carry mud into homes and streets. The mud and water cause damage. Floodwaters erode the soil quickly. They can wash away trees and anything else in their path.

Floods also serve a purpose in nature. After a flood, new soil is deposited on the land. The nutrients in this soil help plants grow.

Fires

When there is too little rain, fires are likely. Many are caused by lightning. A fire can quickly change a forest into a field of charred tree trunks. Forest animals lose their habitats. Grassland fires are fueled by dry plants and spread by winds. Most places recover from natural fires.

FACT ▶ Some forests depend on fires to help plants grow.

Read a Photo

How can floods and fires change the land?

Clue: Compare the photo of a flood and the photo of a fire.

Fire Safety

Carelessness also causes fires. People can prevent wildfires by being safe around campfires and cookouts. Do not light fires in dry areas. Never play with matches.

 Quick Check

Infer What kinds of weather cause floods and fires?

Critical Thinking How can people prevent forest fires?

How do storms change the land?

Have you ever heard the saying "When it rains, it pours"? A light shower might form a few puddles here and there. A severe storm can change the land.

Tornadoes

A thunderstorm can spin off a violent storm called a tornado (tor•NAY•doh). **Tornadoes** are columns of spinning wind. They move across the ground in a narrow path. As they move, tornadoes whip up or destroy everything in their path.

Tornadoes are common in the Great Plains region of the United States. In fact, a certain path through that region is known as "Tornado Alley."

Hurricanes

If you live near an ocean or the Gulf Coast, you may have experienced a hurricane. A **hurricane** is a very large, swirling storm. At its center, or eye, is an area of very low pressure. Strong winds, walls of clouds, and pounding rains surround the eye.

A hurricane is much bigger than a tornado. It can span hundreds of kilometers. It also lasts longer.

Hurricanes form over warm oceans near the equator. They whip up large waves as they travel. When a hurricane moves toward a coast, winds and waves can force water onshore. Massive floods can occur. Heavy rains add to the flooding. The damage does not stop there. When it is over land, a hurricane can uproot trees and flatten buildings. It can change an entire ecosystem in one day.

Hurricanes are becoming more common in some places. Scientists are finding that higher temperatures are a factor.

Hurricane Damage

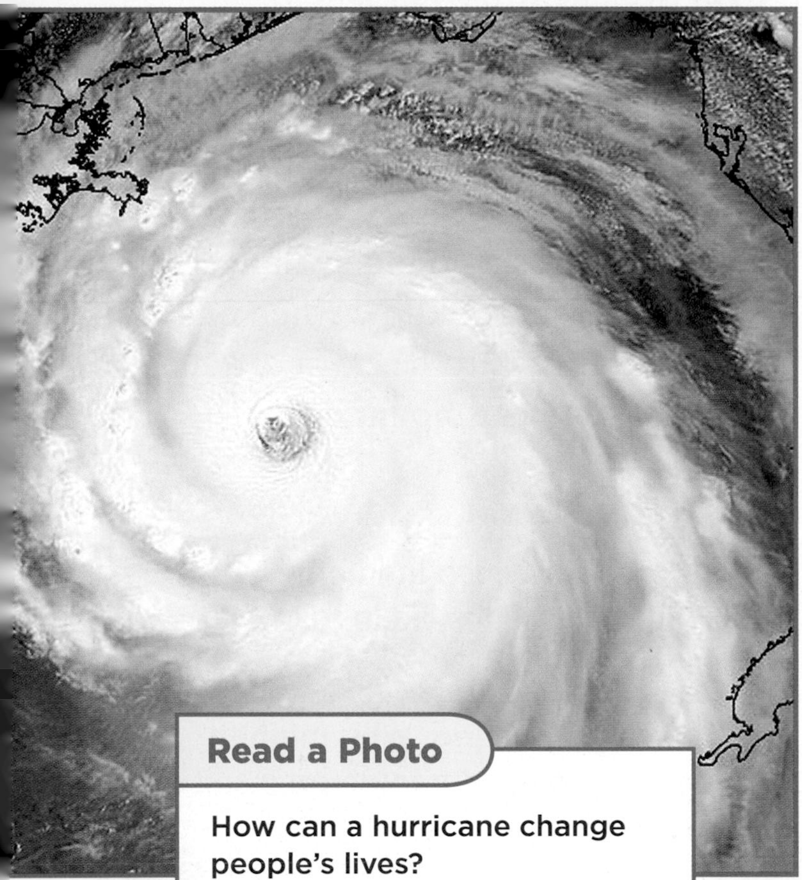

Read a Photo

How can a hurricane change people's lives?

Clue: Think about the people who live or work in the place shown below.

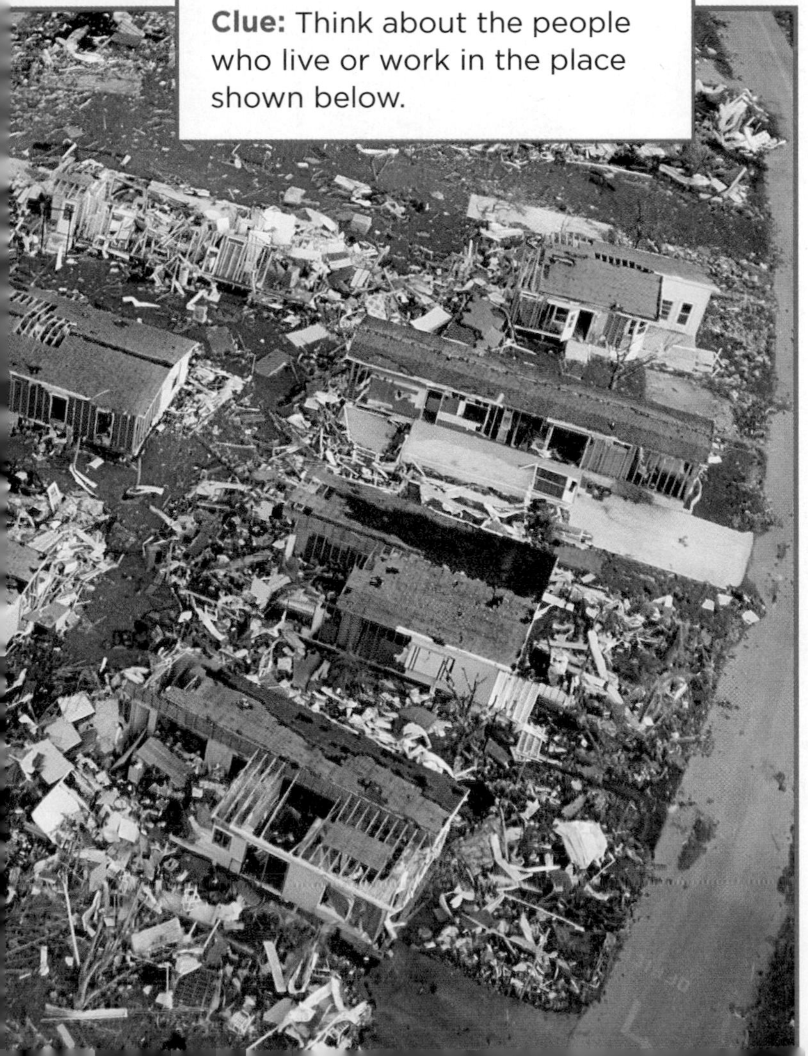

Quick Lab

Storms at the Beach

1. **Make a Model** Pour and press sand into one end of a long pan. This is your beach. Add water to the other end. The water should come up to the lower edge of the beach.
2. Make waves by moving a ruler back and forth in the water. Observe how the beach changes. Continue observing as you move the ruler more quickly, making taller waves.
3. **Infer** Storms bring taller, stronger waves. How do storms and waves affect beach erosion?

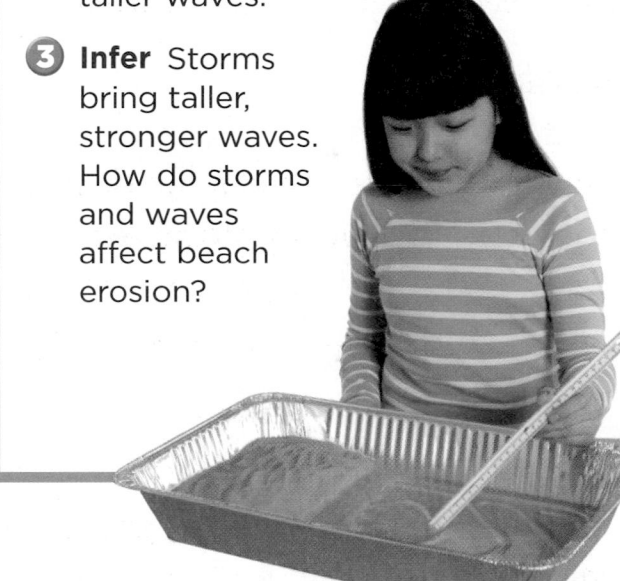

Quick Check

Infer Why is it useful to predict storms?

Critical Thinking How are tornadoes similar to hurricanes? How are they different?

How do landslides change the land?

Have you ever seen a pile of rocks at the bottom of a slope? How did they get there? Part of the answer is gravity. *Gravity* pulls rocks and other objects from high places to low places.

Heavy rains can cause loose rock and soil to move quickly down a slope. A **landslide** is the sudden downhill movement of these materials in large amounts.

An avalanche (A•vuh•lanch) is similar to a landslide. In an **avalanche**, tons of ice and snow rush down a mountain.

Scientists work to predict when and where landslides and avalanches happen. They never know when one will strike. It pays to be extra careful when you are near mountains.

✔ Quick Check

Infer Near which landforms would landslides likely take place?

Critical Thinking Why might it be risky to build a town on a soft hillside?

An avalanche like this one can shift the snowy landscape.

Lesson Review

Visual Summary

Too much rain can cause **floods.** Too little rain can lead to fires. Both can change the land quickly.

Tornadoes and hurricanes are powerful storms that shape the land quickly.

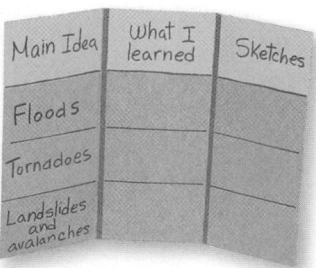
In **landslides and avalanches,** large amounts of land, ice, or snow slide downhill suddenly.

Make a FOLDABLES® Study Guide

Make a trifold book. Use it to summarize what you learned about changes caused by the weather.

Think, Talk, and Write

1 **Vocabulary** What word describes a large, sudden movement of ice or snow downhill?

2 **Infer** A photograph shows fallen palm trees along a beach. The beach is in the southeastern United States. What kind of event most likely caused this result? Fill in the graphic organizer to show your thinking.

Clues	What I Know	What I Infer

3 **Critical Thinking** Some radios are powered by batteries. How could such a radio help you prepare for a severe storm?

4 **Test Prep** Which of these events can help plants grow?

   A hurricanes

   B tornadoes

   C forest fires

   D landslides

5 **Essential Question** How does weather shape and change the land?

 **Writing Link**

Write a Newspaper Report
Research a recent hurricane or tornado. Write a newspaper report describing the storm. Include facts and details. Support your report with comments from witnesses.

 **Social Studies Link**

Research an Event
New Hampshire was once home to a rock formation called "Old Man of the Mountain." Find out what happened to the "Old Man." Describe the event that changed it.

Be a Scientist

Materials

aluminum pan

modeling clay

water

colored pencils

paper

soil and gravel

plastic cup

What happens to the environment when a river floods?

Form a Hypothesis

Rivers can move large amounts of materials from one place to another. These materials include minerals and bits of rock. When a flood erodes the sides of a river, where do the materials go? Write your answer in the form "If a river floods, the materials in the water will be deposited..."

Test Your Hypothesis

1. **Make a Model** Construct a model river and surrounding land. Mold clay inside an aluminum pan to form a flat river bottom. Flatten the land as well.

2. Pour just enough water into the river to fill it. Draw your model.

3. Make floodwater by mixing soil and gravel in a cup of water.

Step 2

4. Pour the floodwater into the river. Where does the water go? Draw the flooded river area on a sheet of paper.

5. **Observe** Let your flooded land dry overnight. Observe and record any changes the next day. Draw your river area again.

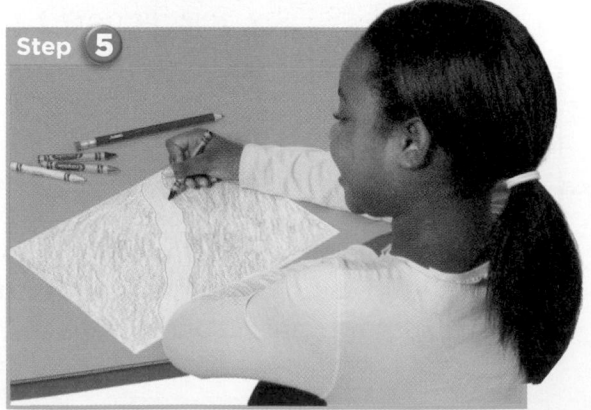
Step 5

6. **Communicate** Describe the materials that were deposited on the land.

Draw Conclusions

7 What happened to your landscape when the river overflowed?

8 **Infer** How might floods help plants, animals, or farmers on the land surrounding a river?

How does the amount of water affect damage?

Form a Hypothesis

How can the amount of water in a flood affect the amount of damage to the land? Write your hypothesis in the form "As the amount of floodwater increases, the amount of material carried away by the water..."

Test Your Hypothesis

Design an investigation to test your hypothesis. Write out the materials you need and the steps you will follow. Record your observations and results.

Draw Conclusions

Did your results support your hypothesis? Why or why not? Explain how you set up the experiment to test for only one variable.

What can you learn about floods? For example, what types of soil are most likely to wash away during a flood? Design an investigation to answer your question. Test one variable at a time. Write your procedure so that another group can repeat the investigation by following your instructions.

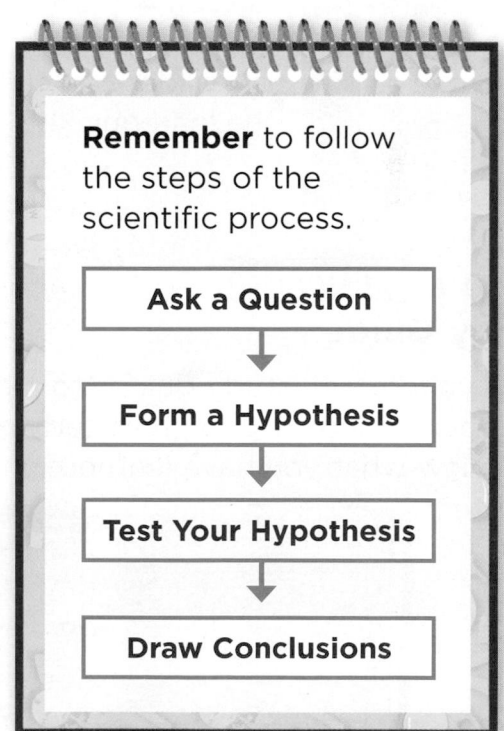

Remember to follow the steps of the scientific process.

Ask a Question

↓

Form a Hypothesis

↓

Test Your Hypothesis

↓

Draw Conclusions

Visual Summary

Lesson 1 Landforms such as mountains and valleys cover Earth's crust. Earth's interior is made of four layers.

Lesson 2 Earth's surface is broken into large plates. The plates move slowly. Many landforms form at the plate edges.

Lesson 3 Weathering, erosion, and deposition are slow processes that shape the land.

Lesson 4 Floods, fires, storms, avalanches, and landslides can change the land quickly.

Make a FOLDABLES Study Guide

Tape your lesson study guides to a piece of paper as shown. Use your study guide to review what you have learned in this chapter.

Fill in each blank with the best term from the list.

crust, p. 208	**hurricane**, p. 240
deposition, p. 229	**inner core**, p. 208
erosion, p. 228	**moraine**, p. 231
fault, p. 215	**volcano**, p. 220
flood, p. 238	**weathering**, p. 226

1. A crack in Earth's crust along which movement occurred is called a(n) _____.

2. Physical _____ is a process that changes only the size of rock.

3. The sphere of solid material at the center of Earth is the _____.

4. A mound called a(n) _____ is formed by glaciers.

5. Heavy rainfall can cause a(n) _____.

6. Earth's landforms are found on the _____.

7. A large, swirling storm with strong winds and heavy rain is called a(n) _____.

8. The _____ of weathered rock particles by a river forms a delta.

9. When gravity, waves, wind, and glaciers weather and transport rock, _____ happens.

10. When a(n) _____ erupts, magma, ash, and gases are sent into the air.

LOG ON **e-Glossary** Words and definitions online at www.macmillanmh.com

Answer each of the following.

11. Infer Why do volcanoes often appear near the edges of plates rather than at the center of plates?

12. Make a Model Design a way to show how wind causes erosion. Use a flat container, straw, and sand as your materials. How does wind move the different materials? Explain your results.

13. Critical Thinking How do scientists get information about Earth's layers?

14. Expository Writing Explain how snow can cause weathering and erosion.

15. Cause and Effect How do mountains form at faults?

16. Predict A field is plowed just before a heavy rainstorm. What effect will the rainstorm have on the recently plowed soil?

17. True or False *Canyons form by wind erosion.* Is this statement true or false? Explain.

18. True or False *The Grand Canyon reaches into Earth's mantle.* Is this statement true or false? Explain.

19. Which phrase <u>best</u> completes the following sentence? The Hawaiian Islands were formed by

A volcanoes.   **C** folding.

B glaciers.   **D** faulting.

The Big Idea

20. What causes Earth's surface to change?

Make a Tornado

1. Place rock, soil, and other heavy materials in the bottom of a bucket. Fill the bucket with water.

2. Place a large spoon or stick in the bucket. Swirl it around rapidly to model a tornado in the water.

3. Describe what happens to the materials you placed in the bottom of the bucket.

Analyze Your Results

Explain how the objects in the water reacted to the model tornado. Were there any differences between the materials? How does the tornado you made compare to an actual tornado?

1 The picture below shows rocks and sand along the side of a stream.

The settling of rocks and sand is an example of which process?

A deposition

B weathering

C eruption

D transport

DOK 1

2 What <u>most likely</u> caused these desert landforms?

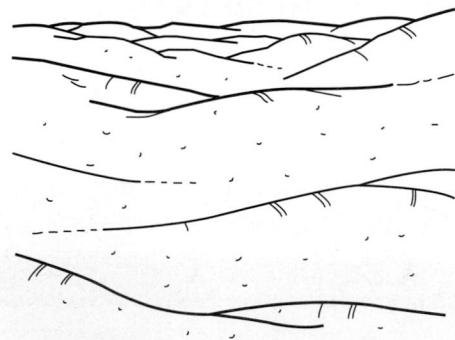

A rain   **C** flowing water

B wind   **D** freezing water

DOK 2

3 Which <u>most likely</u> shows the oldest river?

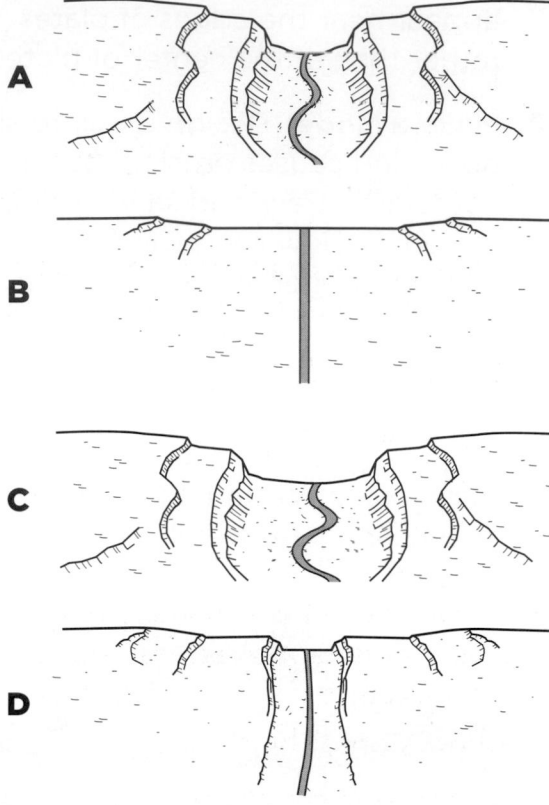

A

B

C

D

DOK 2

4 Over time, which of these can form over a hot spot?

A a delta

B a fault

C a fold

D an island

DOK 1

5 All of these are causes of erosion **except**

A gravity.

B moving water.

C wind.

D sunlight.
DOK 1

6 Each of the following causes rapid changes to Earth's surface **except**

A landslides.

B weathering.

C an earthquake.

D a volcanic eruption.
DOK 1

7 All of the following are a result of a glacier's movement across land **except**

A erosion of land.

B volcanic eruption.

C making a valley wider.

D deposition of large boulders and smaller rocks.
DOK 1

8 A tornado can cause all of the following **except**

A uprooted trees.

B flattened buildings.

C destroyed habitats.

D formation of moraines.
DOK 1

9 Which event would <u>most likely</u> cause a flood?

A volcano

B tornado

C hurricane

D forest fire
DOK 1

10 Look at the picture below.

Describe two different ways that mountains can form.
DOK 1

Describe two forces that can change the shape of a mountain.
DOK 2

Check Your Understanding

Question	Review	Question	Review
1	p. 229	6	pp. 216–220, 226–227, 238, 242
2	pp. 204, 226–229	7	pp. 230–231
3	pp. 228–229	8	p. 240
4	p. 220	9	pp. 220, 238, 240
5	pp. 228–229, 242	10	pp. 215, 220, 226–232

CHAPTER 6

Saving Earth's Resources

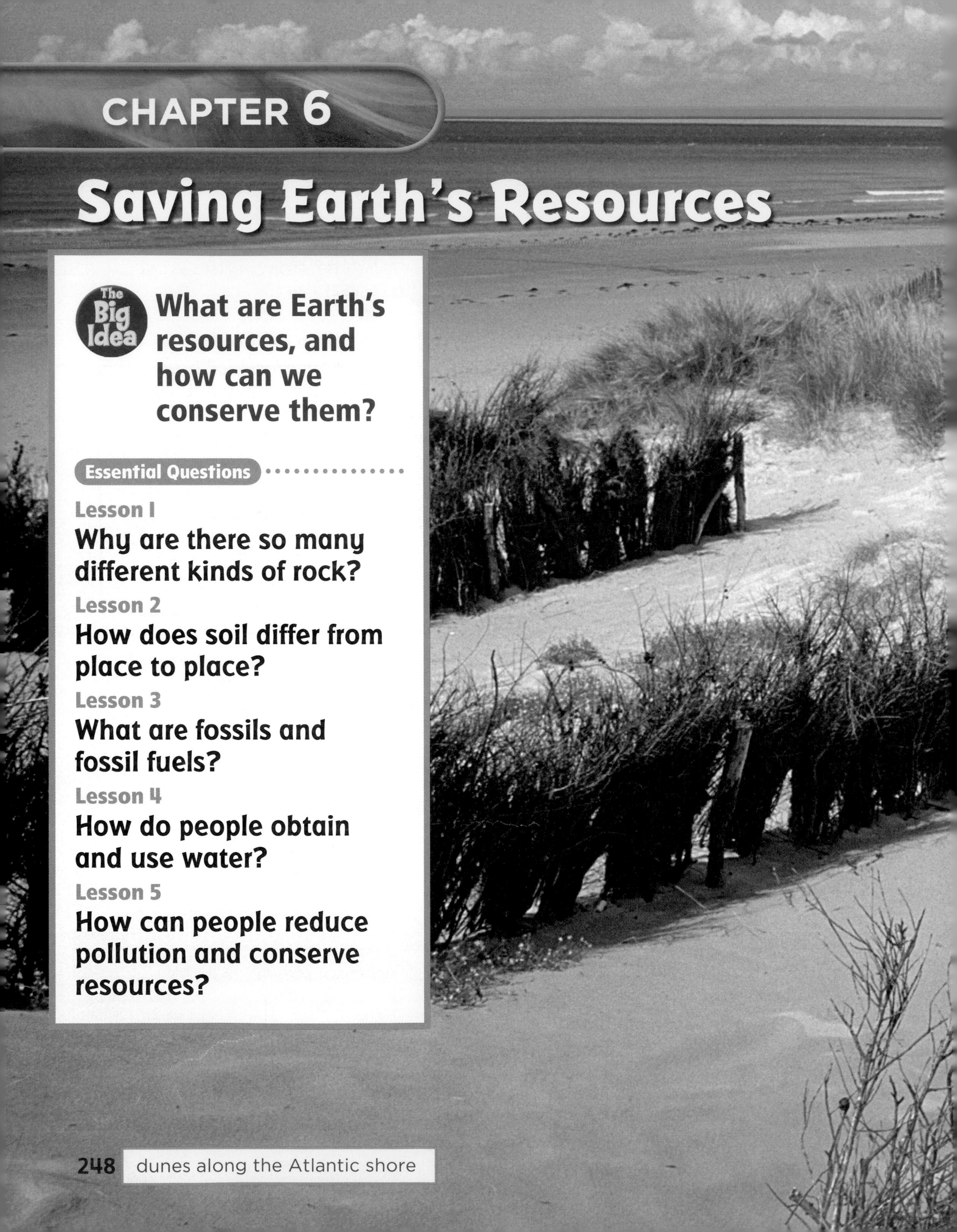

 **The Big Idea** **What are Earth's resources, and how can we conserve them?**

Essential Questions ·················

Lesson 1
Why are there so many different kinds of rock?

Lesson 2
How does soil differ from place to place?

Lesson 3
What are fossils and fossil fuels?

Lesson 4
How do people obtain and use water?

Lesson 5
How can people reduce pollution and conserve resources?

dunes along the Atlantic shore

 **Big Idea Vocabulary**

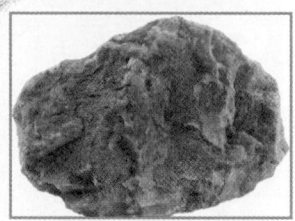

mineral a natural, nonliving, usually solid material from Earth's crust (p. 252)

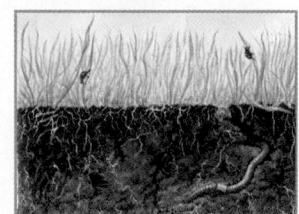

topsoil surface soil layer rich in nonliving plant and animal matter and minerals (p. 265)

fossil fuel an energy source from the remains of an organism that lived millions of years ago (p. 278)

reservoir a storage area for holding and managing freshwater (p. 288)

pollution harmful or unwanted material that has been added to the environment (p. 296)

conservation the wise use of resources (p. 298)

 Visit www.macmillanmh.com for online resources.

Lesson 1

Minerals and Rocks

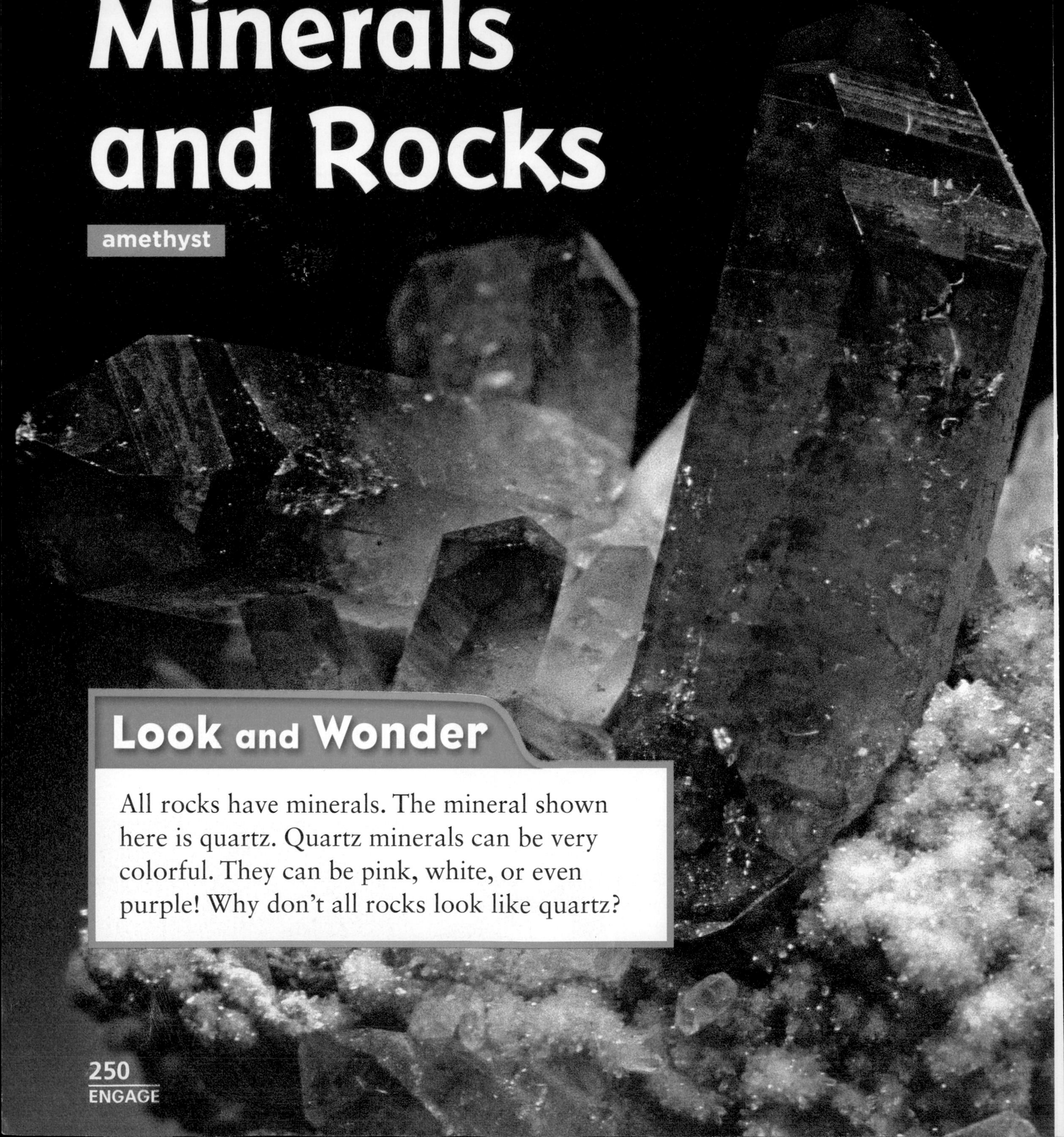

amethyst

Look and Wonder

All rocks have minerals. The mineral shown here is quartz. Quartz minerals can be very colorful. They can be pink, white, or even purple! Why don't all rocks look like quartz?

What makes rocks different from one another?

Purpose

Explore the properties of different rocks.

Materials

- several different rocks
- hand lens

Procedure

1. Look at each rock. What color is the rock? What is its shape? How does it feel?

2. **Communicate** Make a chart to record all your observations.

3. **Observe** Choose a rock that has more than one color. Using a hand lens, compare the parts that are the same color. Are those parts shiny or dull? Rough or smooth? Record your observations in your chart.

4. Choose another color in the same rock. How do the parts with this color compare?

Draw Conclusions

5. **Infer** Are the differently colored parts of the rock made of the same or different materials? Explain your answer.

6. What do you think makes these rocks different from one another?

Explore More

Choose one of the rocks. How could you identify the rock and tell what it is made of? Do some research. Report your findings.

Step 1

Step 3

▶ **Essential Question**

Why are there so many different kinds of rock?

▶ **Vocabulary**

mineral, p. 252

igneous rock, p. 254

sedimentary rock, p. 255

relative age, p. 255

metamorphic rock, p. 256

rock cycle, p. 256

resource, p. 258

▶ **Reading Skill** ✔

Sequence

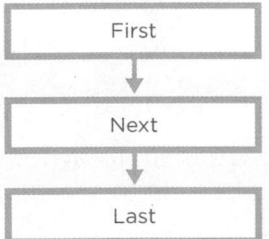

First
Next
Last

▶ **Technology** 🔵 **LOG ON**

e-Glossary and e-Review online at www.macmillanmh.com

What is a mineral?

Why are rocks so different from each other? To answer that question, it helps to know about minerals (MIH•nuh•rulz). A **mineral** is a natural, nonliving substance that makes up rock. In fact, minerals are the building blocks of rocks.

Scientists have identified more than 3,000 kinds of minerals. How? They observe minerals' properties. A *property* is a characteristic that describes something. Look at the minerals shown on these pages. Each has different properties.

Color

One property is color. Talc, for example, is white. Topaz can be blue. Color alone, however, cannot identify minerals. Many minerals have the same color. Quartz can be many different colors!

Hardness

Hardness is a property that refers to a mineral's ability to scratch another mineral or be scratched by another mineral. Mohs' scale shows the hardness of a few common minerals. Each mineral has a number from one to ten. Ten is the hardest kind. The harder a mineral is, the more it resists scratching. As the scale shows, diamond is the hardest mineral of all.

Mohs' Hardness Scale

talc	gypsum	calcite	fluorite	apatite
softest 1	2	3	4	5

Properties of Minerals

Mineral	mica	pyrite	feldspar	hematite
Color	white, green, silver, or brown	gold or brassy yellow	white, pink, gray, or smoky black	gray or brown
Luster	pearly	metallic	glassy	metallic or dull
Streak	white	green-black 	white	red
Hardness	2-2.5	6-6.5	6-6.5	5-6

Luster

Luster refers to the way light bounces off the surface of a mineral. Some minerals have a *metallic,* or shiny, luster. Minerals that have no shine at all have a dull luster. As the table shows, other minerals have glassy or pearly lusters.

Read a Table

Which is harder—feldspar or calcite?

Clue: Compare the values for these minerals on Mohs' scale.

Streak

When you scratch a mineral along a white tile, it leaves behind a powder. *Streak* is the color of that powder. The streak may be different from the color of a mineral's surface.

✓ Quick Check

Sequence What steps would you follow to identify a mineral?

Critical Thinking Why do scientists use several properties to identify a mineral?

feldspar	quartz	topaz	corundum	diamond
6	7	8	9	10 hardest

What are igneous and sedimentary rocks?

Have you ever wondered how a rock forms? Its minerals offer clues.

Igneous Rocks

Below Earth's surface are areas of melted rock called magma. When magma cools and hardens, igneous (IG•nee•us) rocks form. *Igneous* is a Latin word meaning "fire." Igneous rocks form from melted rock—either from magma below Earth's crust or from lava above it.

When a rock cools slowly, large mineral grains can form. The cooled rock looks rough or coarse. *Texture* is a property that describes the appearance of a rock. It is related to the size of the mineral grains. If a rock cools quickly, there is no time for large grains to form.

obsidian

Ancient North American hunters used obsidian to make spears.

Examples of Igneous Rocks

Both obsidian (ub•SIH•dee•un) and basalt (buh•SAWLT) cooled quickly. Obsidian is smooth and glassy. It cooled so quickly that mineral grains did not have time to form. Basalt has small mineral grains because it cooled more slowly than obsidian. Granite (GRA•nut) is different. It cooled slowly underground. It had enough time to form large mineral grains.

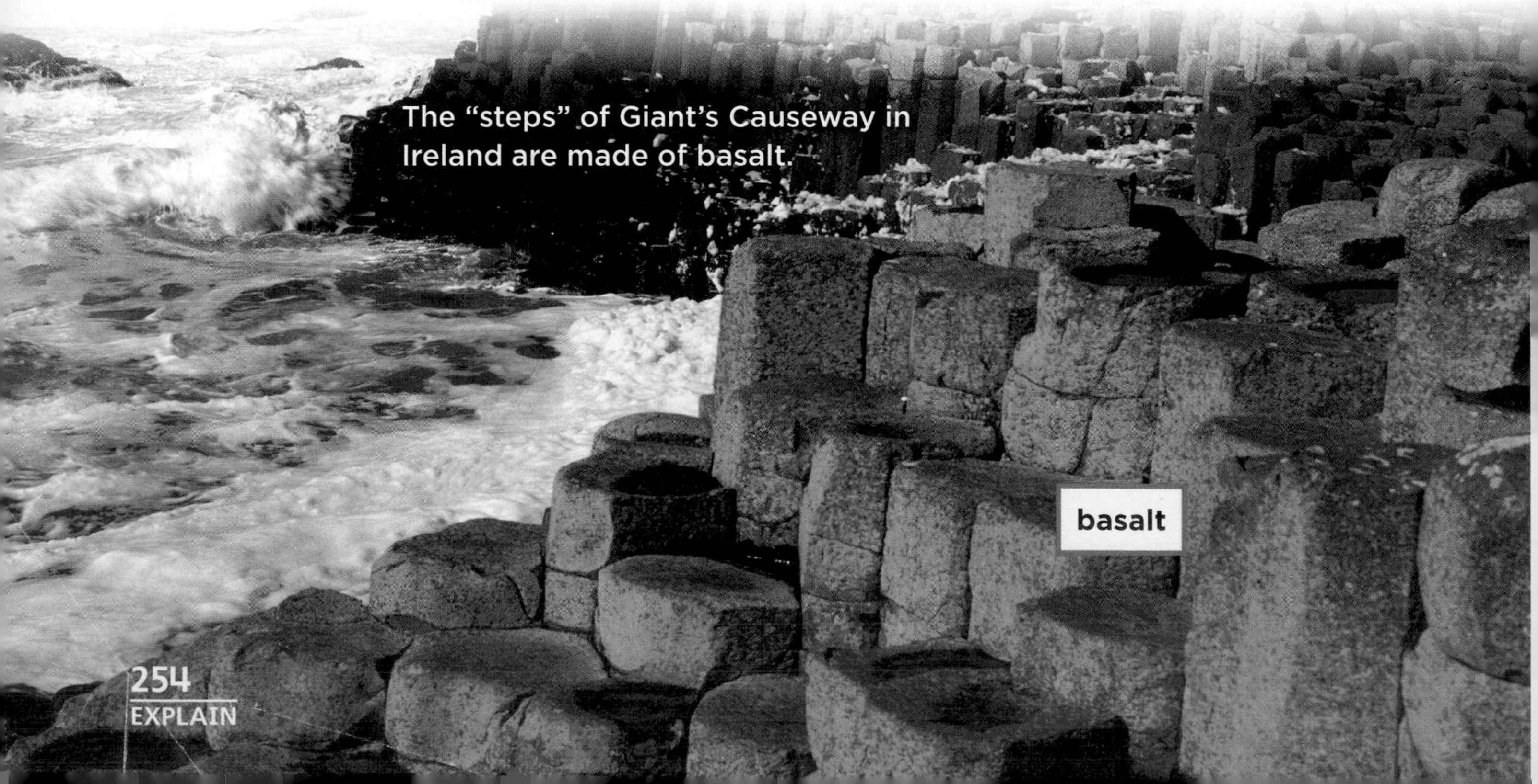

The "steps" of Giant's Causeway in Ireland are made of basalt.

basalt

Sedimentary Rocks

Look at the picture of sandstone. Do you see its layers? These layers are made of tiny pieces called *sediment* (SE•duh•munt). Some sediments are from rocks or minerals. Others are bits of plants, shells, or other hard materials.

Sedimentary rocks form from sediments that are cemented or pressed together. Wind and water deposit most of the sediments. Over time, new sediments are deposited on top of older layers. Sometimes dissolved minerals will cement the sediments together. Other times the weight of the top layers presses the sediment together. It can take millions of years for sediment to become rock.

Relative Age

The layers in sedimentary rock are stacked in order of their relative ages. **Relative age** is the age of one thing compared to another. The older the relative age of a rock layer, the lower it is found. Relative age also applies to any fossils in the rock layers. A *fossil* is a trace of something that was once alive.

✔ Quick Check

Sequence How do sedimentary rocks form?

Critical Thinking Can you observe sedimentary rocks forming? Explain.

FACT▶ A rock can be made from more than one mineral.

≡Quick Lab

Observing Igneous Rocks

1. Obtain a piece of pumice and a piece of granite. How do these two igneous rocks compare in size and weight?

2. **Predict** Will the rocks sink or float? Explain your prediction.

3. Place both rocks in water. What happens?

4. **Infer** What property might have contributed to whether the rocks sink or float?

sandstone

Sandstone has the minerals quartz and feldspar. It is often used as a building stone.

What are metamorphic rocks?

Temperatures below Earth's surface can be very high. Pressure there is high too. When rocks are under so much heat and pressure, their chemical properties can change.

Rocks formed from other rocks by extreme heat and pressure are called **metamorphic** (me•tuh•MOR•fik) **rocks**. These rocks can form from igneous, sedimentary, or even other metamorphic rocks. The chart shows some metamorphic rocks and the rocks they formed from.

Before Original Rock	**After** Metamorphic Rock
granite (igneous rock)	gneiss
limestone (sedimentary rock)	marble
sandstone (sedimentary rock)	quartzite

How Metamorphic Rocks Form

The properties of metamorphic rocks depend on the amounts of heat and pressure. In some rocks, the minerals get rearranged and pressed into thin layers. These layers are called *bands*. Bands may be straight or wavy.

You can see an example of banding in a metamorphic rock called gneiss (NISE). Gneiss starts out as granite—an igneous rock. It takes lots of heat and pressure for those colorful bands to form.

Some metamorphic rocks form from sedimentary rocks. Marble, for example, forms from limestone. Unlike gneiss, marble does not have bands. Quartzite is another metamorphic rock that does not have bands. It forms from sandstone.

The Rock Cycle

You have learned how igneous, sedimentary, and metamorphic rocks form. Does it surprise you that they can all change from one type to another?

The **rock cycle** shows how rocks change from one form to another. It shows how all rocks are related to one another. As you can see on the next page, the rock cycle has many paths. It also takes millions of years!

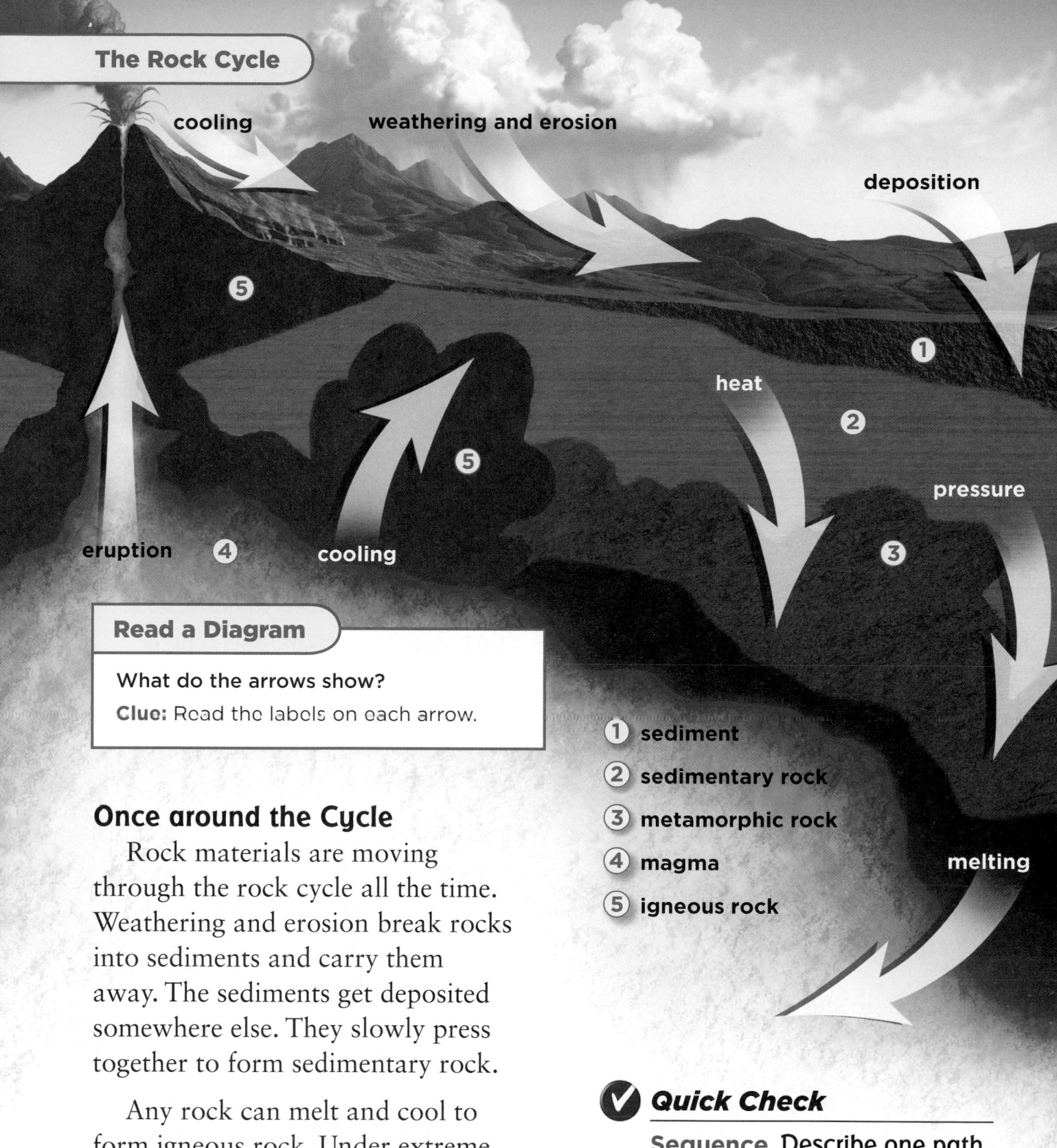

The Rock Cycle

cooling

weathering and erosion

deposition

5

heat

2

pressure

eruption 4 cooling 5 3

Read a Diagram

What do the arrows show?

Clue: Read the labels on each arrow.

① sediment
② sedimentary rock
③ metamorphic rock
④ magma
⑤ igneous rock

melting

Once around the Cycle

Rock materials are moving through the rock cycle all the time. Weathering and erosion break rocks into sediments and carry them away. The sediments get deposited somewhere else. They slowly press together to form sedimentary rock.

Any rock can melt and cool to form igneous rock. Under extreme heat and pressure, some rocks become metamorphic. In time, weathering and erosion break those rocks apart, and the cycle goes on.

✔ Quick Check

Sequence Describe one path through the rock cycle.

Critical Thinking Compare slow and fast events in the rock cycle.

Quartzite is used to make glass.

▼ This guardian lion in Thailand is made of marble.

How do we use rocks?

Rocks and minerals are **resources**—materials from Earth that have useful properties. You can see examples all around you.

Uses of Igneous Rocks

Granite is strong and long lasting. These qualities make it ideal for building schools and other structures. Pumice is found in some soaps and cleansers. Its rough texture helps to scrub off dirt.

Uses of Sedimentary Rocks

Limestone is often used to make glass. Shale is used to make bricks, china, and pottery. When shale is combined with limestone, it can be used to make cement. Scientists use the layers in sedimentary rocks to piece together Earth's history.

Uses of Metamorphic Rocks

Slate is waterproof. It is a good choice as tiles for roofs, billiard tables, and walkways. Marble is valued for its beauty and strength. We use it in flooring, hearths, monuments, and statues. It is easy to carve and it resists fire.

 Quick Check

Sequence How can a rock end up in a building?

Critical Thinking How have you used rocks today?

Lesson Review

Visual Summary

Minerals are the building blocks of rocks. Scientists use several properties to identify minerals.

Rock classification includes igneous, sedimentary, and metamorphic rocks.

Rocks change form slowly during **the rock cycle.** Many rocks make useful resources.

Make a FOLDABLES® Study Guide

Make a three-tab book. Use it to summarize what you read about rocks and minerals.

Minerals

Rock Classification

The rock cycle

Think, Talk, and Write

1 Vocabulary Rocks that have been changed into new rock by heat and pressure are called _____.

2 Sequence What happens in the rock cycle?

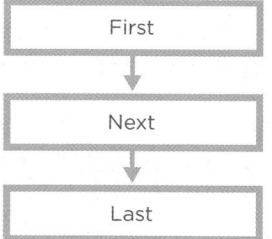

First

Next

Last

3 Critical Thinking Would you be more likely to find a fossil in a metamorphic rock or a sedimentary rock? Explain.

4 Test Prep Which property of a mineral does Mohs' scale measure?

A luster

B texture

C color

D hardness

5 Essential Question Why are there so many different kinds of rock?

 **Math Link**

Solve a Problem

José has 33 rock samples. Exactly $\frac{1}{3}$ are igneous, $\frac{1}{3}$ are sedimentary, and $\frac{1}{3}$ are metamorphic. How many samples of each type does he have?

 **Social Studies Link**

Learn about Rocks in Your Area

What types of rock make up the land where you live? Research the answer at the library. Write a report about your findings.

Focus on Skills

Inquiry Skill: Communicate

Metamorphic rocks have many grains of minerals. By observing these minerals, scientists can tell what transformed one rock type into another. They make models to show how the size and shape of the minerals change. You **communicate** to tell others your findings.

▶ Learn It

When you **communicate**, you share information with others. In science, it is important to be as clear as you can be about your results. Then people can understand what you did and what you found. It is a good idea to communicate in more than one way. You can show your results as a diagram, chart, or table. You can write a report too.

▶ Try It

Model the effect of pressure on metamorphic rock. Then **communicate** your results.

> **Materials** clay, mat or tray, ruler, wood block

1 Roll modeling clay into three balls on a mat or tray. Make them equal in size. Flatten the clay balls slightly so they have two sides. Smooth the sides so you can stack the balls on top of one another. These model the grains of minerals in rock.

2 Make a data chart like the one shown on the opposite page.

3 Observe the shape of your model grains. Draw their shapes in your data chart.

4 Measure the height of the grains in centimeters. Record the measurement in your data chart. Do the same for the width of the grains.

5 Place the flat part of the wood block at the top of the stack. Slowly but firmly push down on the block. This models how pressure squeezes the grains of minerals from above.

6 Repeat steps 3 and 4. Enter your results in the chart where it says *After Squeezing*.

	Drawing of Grains	Height of Grains (cm)	Width of Grains (cm)
Before Squeezing			
After Squeezing			

▶ **Apply It**

Using your data chart, **communicate** your results in a report.

1 Write a summary sentence describing how the grains changed.

2 How did the height and width change? Did your measurements increase or decrease? Write two sentences explaining how the measurements of your model changed.

3 Write a short paragraph explaining how your model is like a real metamorphic rock below the ground. Communicate your conclusions.

4 What would happen if you squeezed the model grains from side to side? How would they change? Finish your report with your prediction.

Soil

Look and Wonder

Farmers depend on soil to grow healthy crops. That means you depend on soil too! Just what is soil? How does it form?

What is soil made of?

Purpose

Compare the parts and properties of different soil samples.

Procedure

1. Spread newspaper over a desk or table. Then spread three sheets of paper towel on the newspaper. Place one soil sample on each towel.

2. **Observe** Use the pencil to separate pieces from each sample. Observe the pieces closely with the hand lens. Record your observations.

3. Place ten drops of water on each sample. After a few minutes, lift the paper towels. Observe any water stains on the newspaper.

Draw Conclusions

4. **Interpret Data** Which sample takes up the most water? Explain your evidence.

5. How are the soil samples alike? How are they different?

6. What kind of materials do you think make up each soil sample?

7. **Infer** Why is the ability to hold water an important property of soil?

Explore More

Exactly how much water is in each soil sample? Form a hypothesis. Design an experiment to test your idea. Try it and report your results.

Materials

- newspaper
- paper towels
- 3 soil samples
- pencil
- hand lens
- eyedropper
- water

Step 3

Essential Question

How does soil differ from place to place?

Vocabulary

humus, p. 264

horizon, p. 265

soil profile, p. 265

topsoil, p. 265

subsoil, p. 265

pore spaces, p. 266

porous, p. 266

permeability, p. 266

Reading Skill
Draw Conclusions

Text Clues	Conclusions

Technology
e-Glossary, e-Review, and animations online at www.macmillanmh.com

What is soil made of?

If you look at soil with a hand lens, you find many different things. You find small pieces of rocks and minerals. You also find humus (HYEW•mus). **Humus** is nonliving plant or animal matter. What else is in soil? Some things you may not see are water, air, and living things.

How Soil Forms

Soil can take hundreds or thousands of years to form. Through weathering, rock becomes sediment. The sediment gets deeper the longer the rock is weathered. Plants take root in the sediment and weather more of the rock. Animals move and mix the sediment.

When plants and animals die, bacteria and fungi decompose them. Humus forms. Humus has nutrients for new plants to grow. In this way, living things renew the soil year after year.

Weathering Caused by Living Things

Read a Photo

How can animals contribute to the soil?

Clue: Look closely at where the rabbit is.

LOG ON *Science in Motion* Watch animals in the soil at www.macmillanmh.com

Soil Horizons

Soil forms in layers called **horizons** (huh•RI•zunz). Each horizon has a different amount of sediment, rock, and humus. A **soil profile** shows these horizons. In some places, the soil profile might look like the one on this page.

The layer of soil at the surface is called the A horizon. It is rich in humus and minerals. Another name for the A horizon is **topsoil**. Topsoil is home to many living things.

The next layer down is the B horizon, or **subsoil**. It is often lighter and harder than topsoil. It has bits of clay and minerals that trickle down from the topsoil. The roots of strong plants grow down into the subsoil.

At the bottom of most profiles lies bedrock. The C horizon is above the bedrock and below the subsoil. It is made up of weathered bedrock.

The rock and humus that make up soil are not the same everywhere. That is why soil profiles are different from place to place.

✔ Quick Check

Draw Conclusions How does bedrock change as soil forms?

Critical Thinking How might cold winters help form soil?

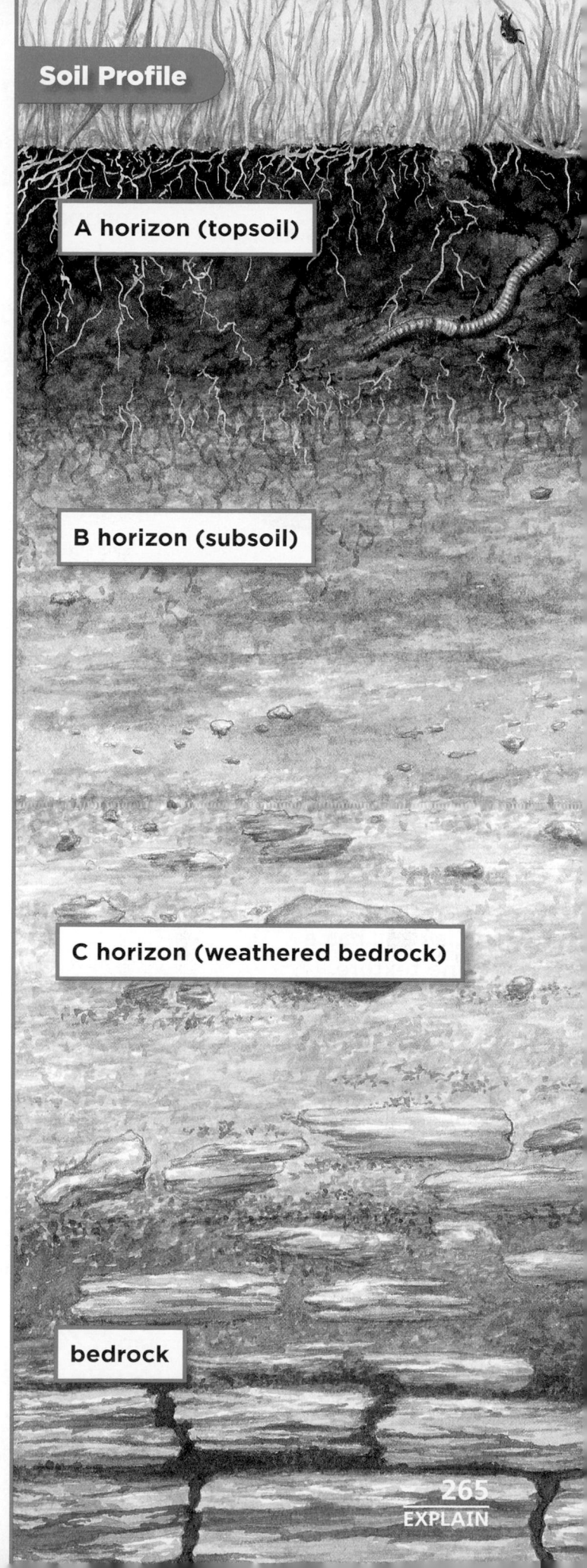

Soil Profile

A horizon (topsoil)

B horizon (subsoil)

C horizon (weathered bedrock)

bedrock

▲ Sandy soil has a coarse texture.

▲ Silty soil has a medium texture.

▲ Clay soil has a fine texture.

What are some properties of soil?

There are dozens of different kinds of soils. Each has its own set of properties. One soil property is color. Another is texture. Texture refers to the size of the particles of soil.

Pore Spaces

The spaces between particles of soil are called **pore spaces**. The pore spaces in soil act like filters. They remove certain substances from the water as it moves through. This keeps the water clean. Materials with pore spaces are said to be **porous** (POR•uhs). They hold air and water.

Permeability

The sizes and numbers of pore spaces affect a soil's permeability (pur•mee•uh•BIH•luh•tee). **Permeability** describes how fast water passes through a porous material. Sandy soil has high permeability. The size and shape of the sand lets water move freely between the pore spaces.

Permeability of Soils

Read a Photo

How does the permeability of fine soil differ from coarse soil?

Clue: Look at the pore spaces shown in the circles.

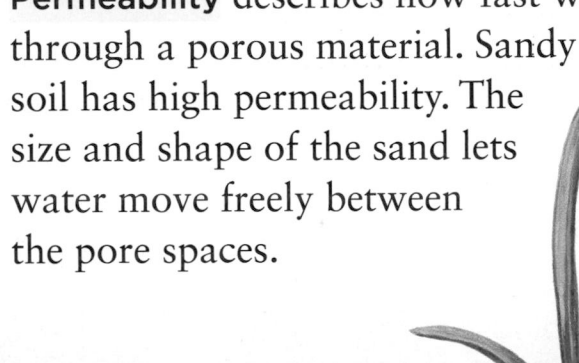

fine soil

cattail

Factors That Affect Soils

The properties and thicknesses of soils depend on the climate, the ecosystem, and the bedrock. The steepness of the land is a factor too. Time also affects soils.

Steep slopes often have thin soils because they erode quickly. Thin soils are poor for growing crops. Thicker soils can build up on flat lands. Any soil can thicken over time if it is left alone.

Water, wind, and ice can erode soil and move it from place to place. This kind of *transported soil* covers large parts of the central United States. Minerals in transported soil may be very different from those in the bedrock.

mallee tree

≡Quick Lab

Rate of Flow

1 Make two containers like the one shown. Fill each with a different type of soil.

2 Hold one container over a measuring cup. Begin timing as you slowly pour one cup of water into the soil.

3 Record the amount of time it takes for the water to stop flowing into the cup.

4 **Use Numbers** Calculate how much water remains in the soil. Repeat steps 2 and 3 with the second soil sample.

5 **Interpret Data** Which soil type has a higher permeability? Why?

✔ Quick Check

Draw Conclusions How does the size of the pore spaces affect the permeability of soil?

Critical Thinking A farmer wants to grow a crop on flat land. The soil has lots of humus. Is this a good idea? Why or why not?

coarse soil

Why is soil type important?

Soil permeability is important to plants that live on land. That means the type of soil in which plants grow is also important.

Topsoil is home to a large variety of living things. All living things need at least a little water. They need air too. Plants and animals can survive in soil only if the soil is porous enough.

Soil permeability is especially important for farmers. Sandy soil is very porous. Water moves through quickly, carrying minerals as it goes. The minerals often travel below the reach of plant roots. If the soil holds too little water, crops dry up.

Fine soil is porous, but it has low permeability. Water soaks into it slowly. The water may stay in the pore spaces for a long time. This is not good for plants either. A crop can drown from too much water.

✔ Quick Check

Draw Conclusions How does soil permeability affect plants?

Critical Thinking Why might farmers grow different kinds of plants in different kinds of soil?

Desert plants are adapted to grow in sandy soils.

Medium-textured soils are good for many crops.

Some kinds of grapes grow well in clay soils.

 **FACT** Sandy deserts are not lifeless.

Lesson Review

Visual Summary

Soil is made of humus, weathered rock, and minerals. **Soil horizons** show the different layers of soil.

Soil properties include color, texture, and the number of pore spaces. Permeability is another property of soil.

Soil **permeability** affects the organisms that live in soil. Plants are adapted to certain kinds of soils.

Make a FOLDABLES Study Guide

Make a trifold book. Use it to summarize what you read about soil.

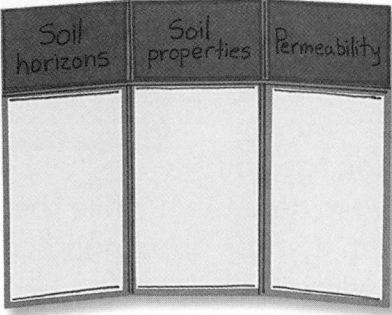

Think, Talk, and Write

1. **Vocabulary** Soil forms in layers called _____.

2. **Draw Conclusions** What can you conclude about a soil's permeability by observing its texture?

Text Clues	Conclusions

3. **Critical Thinking** Why might a desert plant grow poorly in clay soils?

4. **Test Prep** What can pore spaces hold in soil?

   A air only

   B water only

   C air and water

   D humus

5. **Test Prep** Which soil is likely to have the thinnest layers?

   A soil along a steep slope

   B soil on flat land

   C any transported soil

   D any clay soil

6. **Essential Question** How does soil differ from place to place?

 **Writing Link**

Write a Report
What kind of soil is common in your area? What crops does it support? Research the answers. Report your findings.

 **Math Link**

Solve an Equation
A student adds 35 milliliters of water to a soil sample. Then 28 mL drip out. How much water is the soil still holding? Write a math sentence.

Be a Scientist

Materials

filter paper, funnel

measuring cup, teaspoon

500 mL beaker, 1,000 mL beaker

drink mix

water

potting soil, sand, clay soil

Structured Inquiry

How do different soil types hold minerals when it rains?

Form a Hypothesis

Does soil type affect the amount of nutrients that the soil can hold? Consider three types—potting soil, sand, and clay. Write your answer in the form "In a heavy rain, the soil type that loses the most nutrients is..."

Test Your Hypothesis

1 Place a piece of filter paper in a funnel. Put the funnel in a 1,000 mL beaker.

2 **Make a Model** Mix one cup of potting soil with one teaspoon of drink mix. The drink mix models the nutrients that plants need for growth. Pour the mixture into the funnel.

Step 3

3 **Measure** Put 250 mL of water in a beaker. Slowly pour the water on the soil mixture. Make your data table while the water drips into the jar. When it stops dripping, record the color and amount of water in your table.

4 **Use Variables** Repeat steps 1 through 3 with sand. Repeat again with clay soil. Record your observations.

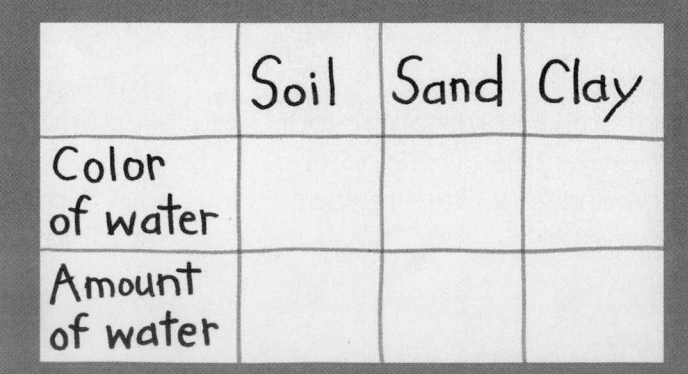

	Soil	Sand	Clay
Color of water			
Amount of water			

Draw Conclusions

5 What differences did you observe among the types of soils you tested?

6 Which type of soil lost the most nutrients? Which type held the least amount of nutrients?

7 **Infer** Why might some plants have trouble growing in sandy soil?

Guided Inquiry

How does soil permeability affect plant growth?

Form a Hypothesis

How does the permeability of soil affect the kinds of plants that grow in it? Write a hypothesis.

Test Your Hypothesis

Design an experiment to find the best soil permeability for growing lima beans. Which variable will you test? Which variables will you control? Write out the materials you need and the steps you will follow. Then try it. Record your observations in a data table. Explain your results.

Draw Conclusions

Did your results support your hypothesis? Which soil type is best for growing lima beans? Which type is worst?

Open Inquiry

What else would you like to learn about soil types? Design an investigation to answer your question. Write the steps of your investigation so that another group can complete it by following your instructions.

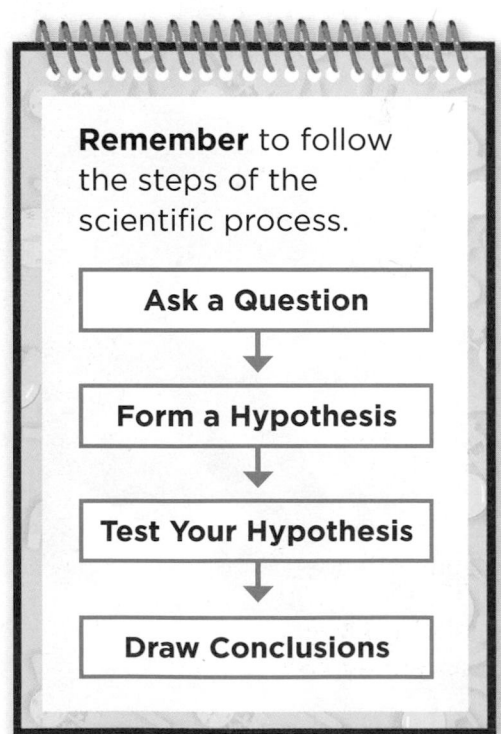

Remember to follow the steps of the scientific process.

Ask a Question

↓

Form a Hypothesis

↓

Test Your Hypothesis

↓

Draw Conclusions

Resources from the Past

Look and Wonder

Can you identify this creature? Where and when did it live? Scientists use evidence like this to find out about ancient life. What other evidence provides clues about Earth's past?

What can you learn from footprints?

Purpose

Make inferences from observations of footprints.

Procedure

1. **Make a Model** Flatten some modeling clay. Make model footprints by pressing small objects into the clay. Model different animals—heavy, light, walking, and running.

2. **Communicate** Write a brief story about two or more different animals. Tell how they interact.

3. Flatten the clay to form a smooth surface. Use the objects to make model footprints that show your story.

4. **Infer** Exchange clay models with another group. Try to infer the story that the other group modeled.

Draw Conclusions

5. What do prints in clay reveal about the objects that made them?

6. **Interpret Data** How can footprints show whether an animal was walking or running?

7. What do you think scientists can infer from footprints made long ago?

Explore More

How can you tell whether footprints were made by an animal walking on two legs? On four legs? Do some research and make observations. Share your findings with the class.

Materials

- modeling clay
- small common objects such as pencil, shell, eraser, coin, paper clip

Step 1

► **Reading Skill** ✔
Fact and Opinion

Fact	Opinion

► **Technology**
e-Glossary and e-Review online at **www.macmillanmh.com**

What are fossils?

Scientists use clues from fossils to learn about the past. A **fossil** is evidence of an organism that lived long ago. For instance, scientists have found dinosaur tracks. By observing the patterns that the footprints made, scientists learned how the dinosaurs walked.

How Fossils Form

Most fossils are found in sedimentary rocks. Sediments bury whatever remains from a plant or animal that once lived. As the sediments turn to rock, the buried remains can become fossils.

When a plant or animal dies, the soft parts quickly decay or are eaten. Bones, teeth, and shells last longer. These hard parts are more likely to become fossils.

Sometimes entire organisms are preserved. Large animals called mammoths have been found frozen in the tundra ice. Insects and spiders can become trapped in sticky tree sap. The sap can harden into **amber**—a hard, smooth material.

◄ The body of this spider is preserved in green amber.

The ants in this amber look a lot like today's ants. ►

Kinds of Fossils

foot imprint

trilobite cast

petrified wood

Molds and Casts

Shells often leave behind fossils known as molds. A **mold** is a hollow form with a certain shape. How does a mold form? Water can seep into the spaces in the rock where an organism is buried. Slowly, the water washes away the shell. It leaves a hollow space, or mold, where the shell was.

If minerals build up inside a mold, another kind of fossil may form. A **cast** is a fossil that is formed or shaped in a mold. Have you ever made gelatin in a shaped cup? The cup was a mold. The hardened gelatin was a cast.

Imprints

Sometimes a shallow print is the only fossil evidence we have. Tracks, body outlines, and leaf prints are called imprint fossils. An **imprint** is a mark made by pressing.

Stony Fossils

Wood and bones can become *petrified,* or turned to stone. As minerals slowly seep inside a dead tree or animal, the minerals replace its insides. The organism becomes a fossil of rock!

✔ Quick Check

Fact and Opinion How do we know about dinosaurs?

Critical Thinking How could you model a cast fossil and its mold?

How do we study fossils?

Scientists use powerful computers and microscopes to learn about ancient life. When a new fossil is discovered, scientists compare it to similar living organisms. In doing so, scientists must consider that organisms change over time.

Sometimes a fossil's location is more puzzling than the fossil itself. For example, fossils of ferns have been found in icy Antarctica. Antarctica is a polar biome. It is much too cold for ferns to grow!

Geologic Time

Organisms are not the only things that change over time. Earth's land changes too. You are familiar with fast changes such as hurricanes and landslides. However, most of Earth's changes are very, very slow.

Scientists measure Earth's history in millions—even billions—of years. We call such long spans *geologic time*. When scientists study fossils, they are also studying geologic time.

Examining Rock Layers

How have Earth's land and living things changed over geologic time? To find out, scientists examine rock layers. They look for fossils. Then they compare those fossils with other fossils in the same rock.

The oldest fossils are in the oldest rock layers at the bottom. Younger fossils are found in upper rock layers. Those layers formed later.

Scientists must work carefully to clean and prepare fossils for study.

Finding Evidence

Rock layers and fossils are evidence of Earth's changes over time. In the past, Earth's climate was warmer than it is now. At other times in the past, it was cooler. If you find a fish fossil on land, you know that water once covered that land.

▲ Ammonites lived in water. These fossil ammonites were found on land.

▼ This fossil fern was found where it is too cold for ferns to grow.

≡ Quick Lab

Older and Younger

1. Cut a piece of paper into four pieces. Draw a fossil on each.

2. **Make a Model** Have a partner place each fossil inside the front cover of four different books. Stack the books. The stack models Earth's rock layers.

3. Find the fossils. Arrange them from oldest to youngest.

4. **Communicate** How did you decide which fossil was oldest and which was youngest? Explain this to your partner.

✓ Quick Check

Fact and Opinion Scientists are certain that dinosaurs and mammals once lived together. How can they know this?

Critical Thinking How could ferns have lived on Antarctica?

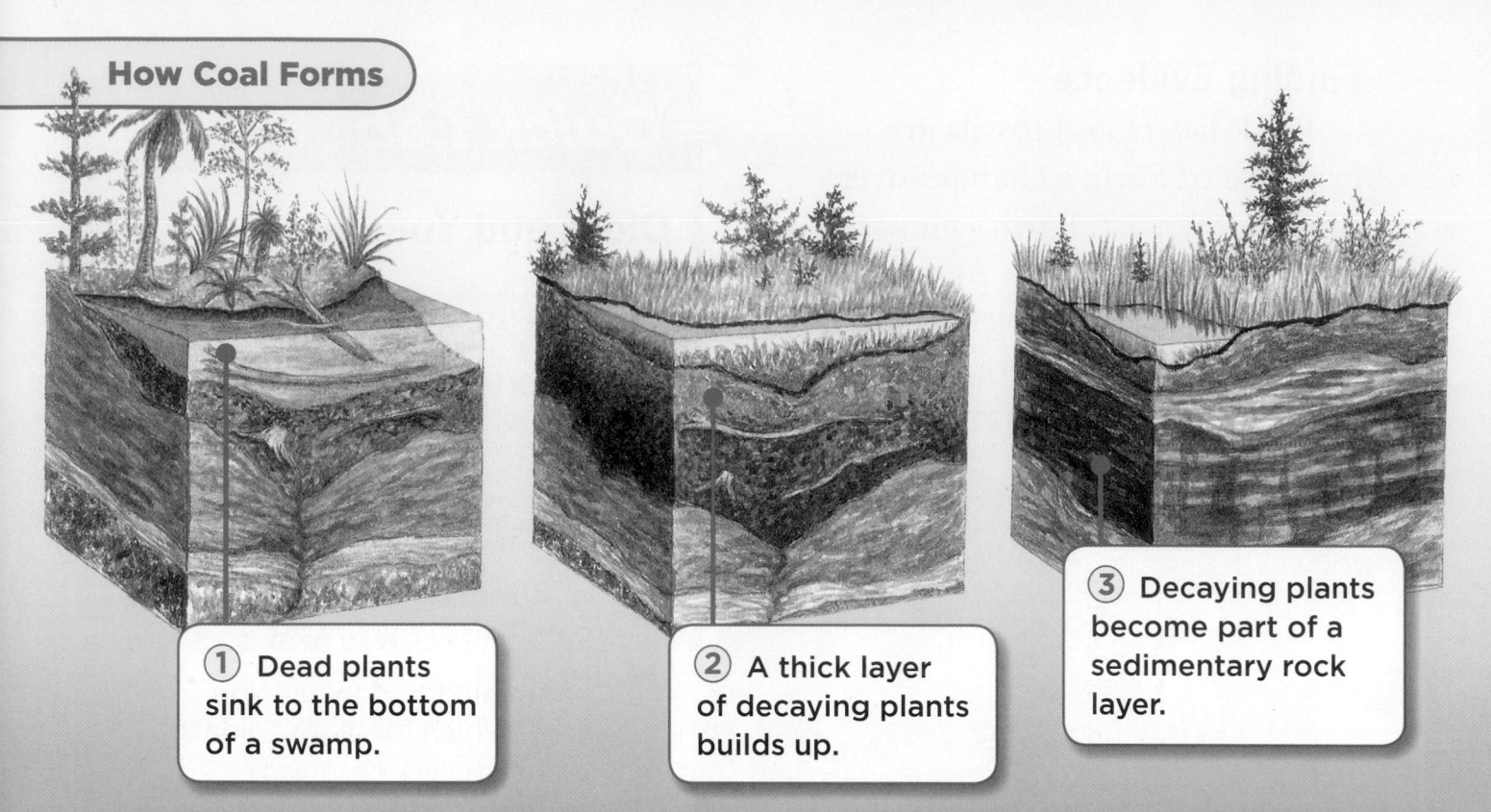

① Dead plants sink to the bottom of a swamp.

② A thick layer of decaying plants builds up.

③ Decaying plants become part of a sedimentary rock layer.

What are fossil fuels?

Where do people get energy to make electricity and drive vehicles? Most of it comes from fossil fuels such as coal, oil, and natural gas. A **fossil fuel** is an energy source that formed millions of years ago. Fossil fuels form from the remains of plants and animals.

Fossil fuels are nonrenewable resources (NON•ree•NEW•uh•bul REE•sors•ez). A **nonrenewable resource** is a useful material that cannot be replaced easily. Once it is used up, it is gone forever. To release the energy stored in fossil fuels, we have to burn them. When we burn the fuels, we destroy the resource.

Finding Fossil Fuels

Searching for fossil fuels can be difficult. It involves mining and drilling deep below Earth's surface. We need large power plants to get the energy from fossil fuels. We also need ways to deliver that energy. These are all expensive processes.

Using Fossil Fuels

One liter of natural gas takes millions of years to form. It burns within a few seconds! Burning fossil fuels causes air pollution. People who breathe the polluted air can become sick. Polluted air may cause acid rain. Acid rain can harm other organisms as well. Do we have other choices? Yes, we do.

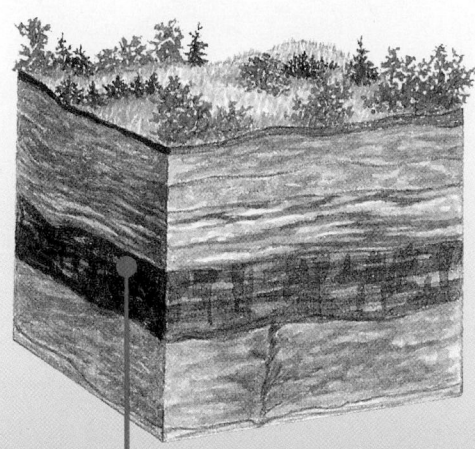

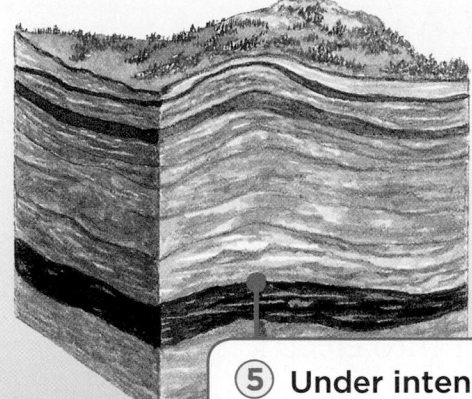

④ The rock layer is pressed into soft coal.

⑤ Under intense heat and pressure, the soft coal turns to hard coal. It is now a fossil fuel.

Read a Diagram

How can dead plants or animals become a fossil fuel?

Clue: Follow the steps in the diagram.

Alternative Energy

No one knows how long our fossil fuel supply will last. That is why scientists are always looking for other ways to produce energy. These are called *alternative energy sources*. Can you think of some?

This machine is an oil pump. It gets oil from below the ground.

✔ Quick Check

Fact and Opinion *Fossil fuels store energy.* Is this statement a fact or an opinion? Explain.

Critical Thinking How do you and your family rely on fossil fuels?

What can we use instead of fossil fuels?

Earth and the Sun supply us with renewable resources. A **renewable resource** is a useful material that is replaced quickly in nature.

The Sun provides us with a source of energy every day. A tool called a solar cell can change the energy from sunlight into electricity.

Windmills can harness the wind. We can get energy from flowing water too. Ocean tides carry usable energy. In some places, people get energy from the heat inside Earth.

The pie chart shows where the United States got energy for electricity in 2007. You can see that 8 percent of the energy came from renewable resources. The rest were nonrenewable resources.

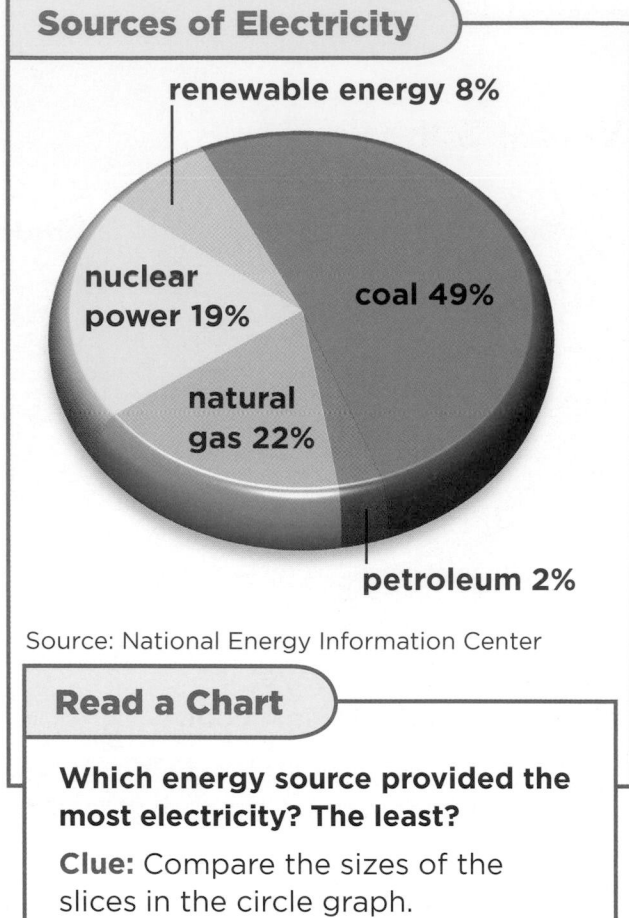

Sources of Electricity

- renewable energy 8%
- nuclear power 19%
- coal 49%
- natural gas 22%
- petroleum 2%

Source: National Energy Information Center

Read a Chart

Which energy source provided the most electricity? The least?

Clue: Compare the sizes of the slices in the circle graph.

Quick Check

Fact and Opinion *We should not use coal.* Is this statement a fact or an opinion? Explain.

Critical Thinking Should scientists try to develop new ways to use renewable energy resources? Why or why not?

Wind farms gather the energy from wind and turn it into electricity.

Lesson Review

Visual Summary

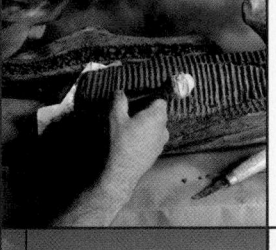 **Fossils** provide clues about what Earth was like in the past. There are different kinds of fossils.

 **Fossil fuels** such as coal, oil, and natural gas are nonrenewable resources. They cannot be replaced quickly.

Renewable resources include wind, water, and sunlight. They are replaced quickly.

Make a FOLDABLES® Study Guide

Make a three-tab book. Use it to summarize what you read about fossils and fossil fuels.

Fossils

Fossil fuels

Renewable resources

Think, Talk, and Write

1 **Vocabulary** The energy from wind can be used again and again. This makes wind a(n) _____.

2 **Fact and Opinion** List facts about the use of fossil fuels. Write your opinion about each use.

Fact	Opinion

3 **Critical Thinking** How do the fossils displayed in museums compare with those in the ground?

4 **Test Prep** Which resource is renewable?

A coal

B diamonds

C silver

D wind

5 **Essential Question** What are fossils and fossil fuels?

 **Writing Link**

Write a Short Story
Write a short story in which a fossil plays an important role. Describe the organism that left the fossil. Tell about the people who find or study the fossil.

 **Social Studies Link**

Write a Report
What is the price of gasoline in your community? Ask adults what they think about energy costs. How do these costs affect them? Write a report about what you learned.

Be a Scientist

model of fossil
T. rex tooth

model of fossil
Edmontosaurus tooth

model of fossil
shark tooth

horse tooth

colored pencils

Structured Inquiry

How do scientists learn about dinosaurs?

Form a Hypothesis

Scientists use fossils to infer things about dinosaurs. For example, scientists look at fossil teeth to infer what a dinosaur would have eaten. What can you learn from teeth? Write your answer in the form "If a dinosaur's tooth is flat, the dinosaur would have eaten..."

Test Your Hypothesis

1. Create a data table that includes rows for length and width. Make a row in your data table for drawings.

2. **Observe** Look closely at each model tooth. Draw its picture in your table.

3. **Measure** Find the length and width of each tooth. Record each measurement in your table.

	T. rex	Edmontosaurus	Shark	Horse
Drawing				
Length (cm)				
Width (cm)				

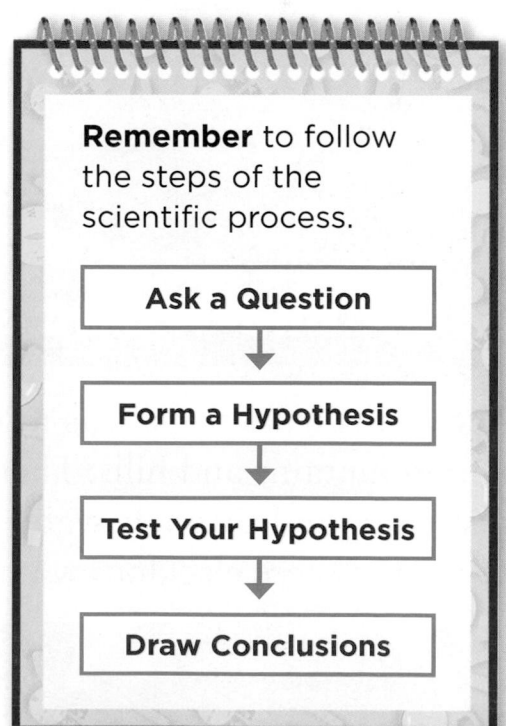

Draw Conclusions

4 **Interpret Data** Compare and contrast the teeth you examined. Which dinosaur tooth is more like the horse tooth? Which dinosaur tooth is more like the shark tooth? Explain your answers.

5 The diet of modern horses includes grasses, hay, and wheat. Modern sharks eat fish and other animals. Based on your results, what do you think *Edmontosaurus* ate? Why?

6 **Infer** What did *T. rex* eat? How do you know?

7 Was your hypothesis correct? Explain.

Guided Inquiry

What else can you learn from fossil teeth?

Form a Hypothesis

What other information can scientists infer from animal teeth? Write your answer as a hypothesis.

Test Your Hypothesis

Design an investigation to find out whether tooth size can tell you the size of an animal. Write out the materials you need and the steps you will follow. Record your results and observations.

Draw Conclusions

Did your results support your hypothesis? Why or why not? Form a new hypothesis if yours was not supported.

Open Inquiry

What else would you like to learn about dinosaur fossils? Design an investigation to answer your question.

Remember to follow the steps of the scientific process.

> **Ask a Question**
> ↓
> **Form a Hypothesis**
> ↓
> **Test Your Hypothesis**
> ↓
> **Draw Conclusions**

Water

Look and Wonder

Water falls from the sky as rain. It flows over mountains and hills. Then it collects in streams and rivers. Is water always moving? Where else does water collect?

Does water flow faster through soil or gravel?

Form a Hypothesis

Will water flow faster through a cup of soil or a cup of gravel? Recall what you know about soil and rock. Then write a hypothesis.

Test Your Hypothesis

Materials

- pencil
- two 12-oz. paper cups
- perlite or soil
- plastic container
- 200 mL of water
- measuring cup
- stopwatch
- gravel

1. With a pencil tip, make a small hole in the bottom of one paper cup. Make a mark on the inside of the cup, close to the top.

2. **Measure** Place your finger over the hole. Fill the cup to the mark with perlite or soil. Hold the cup over a plastic container. Have your partner pour in 100 mL of water.

3. Remove your finger. Time how long it takes the water to drain. Record the time in a data table.

4. Repeat steps 1, 2, and 3 using gravel.

Draw Conclusions

5. **Interpret Data** Which material lets water soak through faster?

6. What might happen to rainwater when it falls on soil? On gravel?

7. **Infer** Which material might support more plant growth—soil or gravel? Explain.

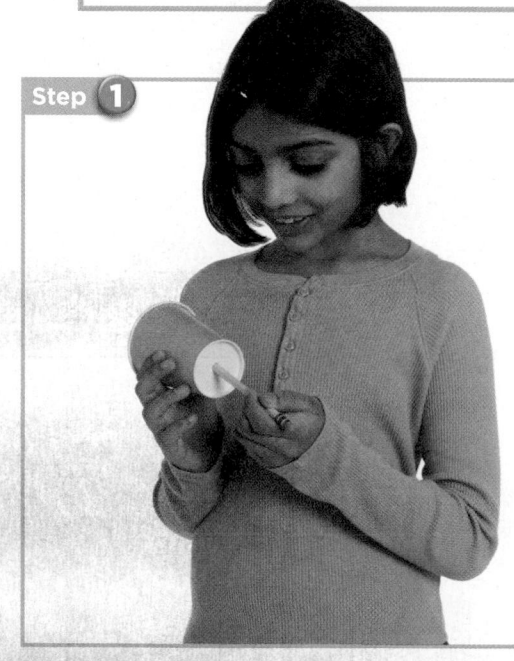

Step 1

Explore More

Which holds more water after the water first flows through—soil or gravel? Design a test of your hypothesis. Use evidence to support your conclusion.

▶ Essential Question

How do people obtain and use water?

▶ Vocabulary

soil water, p. 287

groundwater, p. 287

watershed, p. 287

reservoir, p. 288

well, p. 288

runoff, p. 288

irrigation, p. 290

▶ Reading Skill ✔

Problem and Solution

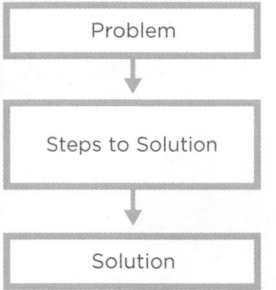

Problem
↓
Steps to Solution
↓
Solution

▶ Technology

e-Glossary and e-Review online at www.macmillanmh.com

Where is Earth's water found?

Have you ever wondered whether some places have more water than others? Look at a globe. You will see that we live in a watery world.

Salt Water

Oceans and seas cover almost three fourths of Earth's surface. That is a lot of water! Can people drink it? Can we use it to grow plants? Ocean water, or *seawater*, contains a great deal of salt. We cannot drink it or use it on soil.

Freshwater

Freshwater is water without much salt. Most streams, rivers, lakes, wells, and ponds contain freshwater. However, most of Earth's freshwater is not in a liquid state. It is solid!

Glaciers and ice caps hold most of Earth's freshwater. *Ice caps* are thick layers of ice on land. Giant ice caps cover Greenland and Antarctica— the continent at the South Pole.

Most of Earth's fresh water is solid ice.

Read a Map

How could a ship travel from Chicago, Illinois, to the Atlantic Ocean?

Clue: Trace a path on the map through water.

Below the Ground

When water soaks into soil, it becomes **soil water**. Plants use some of the soil water. The rest travels farther down below the surface. It seeps downward until it reaches a layer without cracks or pore space. Then the water collects in the spaces above. **Groundwater** is the term for water that fills the cracks and spaces of rocks under the ground.

Watersheds

On land, water may flow downhill into a common stream, lake, or river. Such areas are known as **watersheds**. People who live in a watershed tend to use the water that drains through it. Laws and government agencies help us protect watersheds.

 **Quick Check**

Problem and Solution Where would you go to find water?

Critical Thinking How could we use salt water?

A wetland can contain freshwater, salt water, or a mixture of both.

How is freshwater supplied?

Most large towns and cities get their water from reservoirs (REH•zuh•vworz). A **reservoir** is a storage area for holding and managing freshwater. Some are natural lakes or ponds. Others are built by people. Pipelines supply people with the water in reservoirs.

Groundwater is another source of freshwater. A well is the most common way of getting groundwater. **Wells** are deep holes drilled or dug below the ground. Pumps get the water to the surface.

Freshwater is rarely pure. It may have bacteria or harmful chemicals in it. Such substances are often carried to a water source by runoff. **Runoff** is water that flows over the land without evaporating or soaking into the ground.

Lake Mead is a reservoir shared by Arizona and Nevada.

Water Treatment Plants

Water cannot be supplied to people before it is safe to use. A *water treatment plant* is a place where water is made clean and pure.

First, the water passes through a filter. The filter removes trash and other large objects. Next, chemicals are added to kill harmful organisms.

Water Treatment

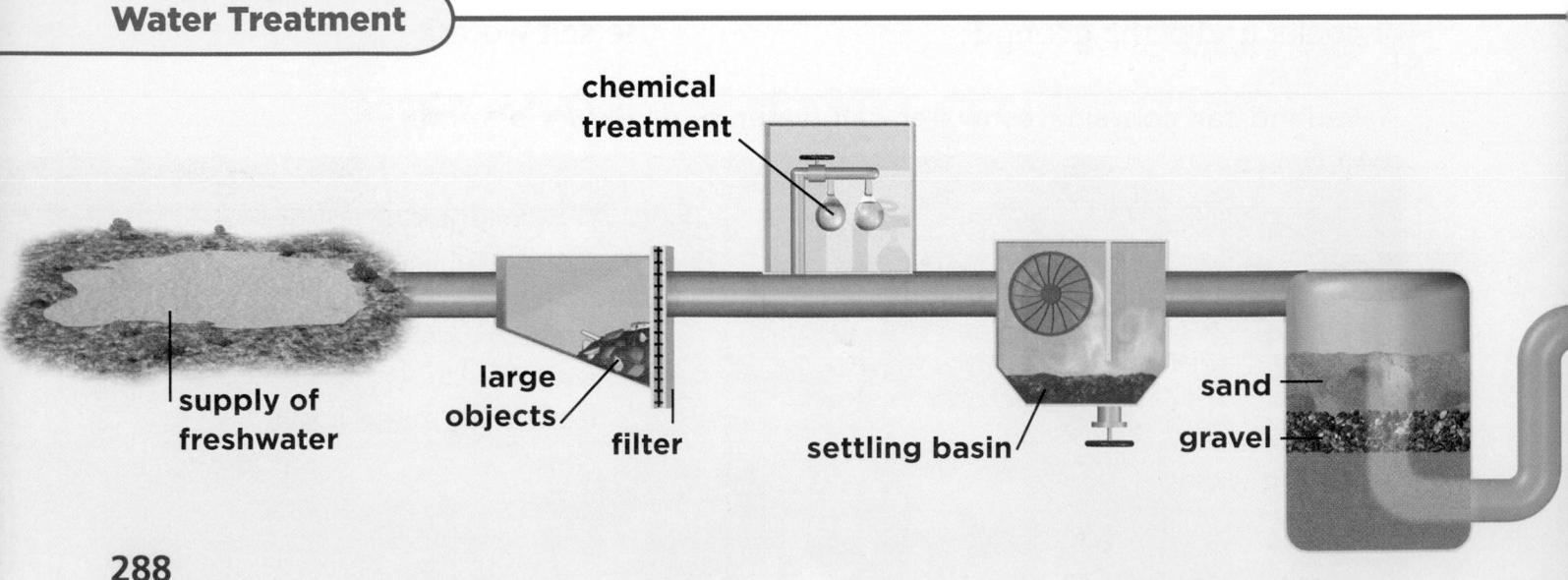

chemical treatment

supply of freshwater

large objects

filter

settling basin

sand

gravel

Look at the diagram below. It shows the sequence of events at a water treatment plant. After the water is cleaned, it is stored in reservoirs until it is needed.

Read a Diagram

How does water reach homes and other buildings?

Clue: Trace a path from the water supply to the end of the diagram.

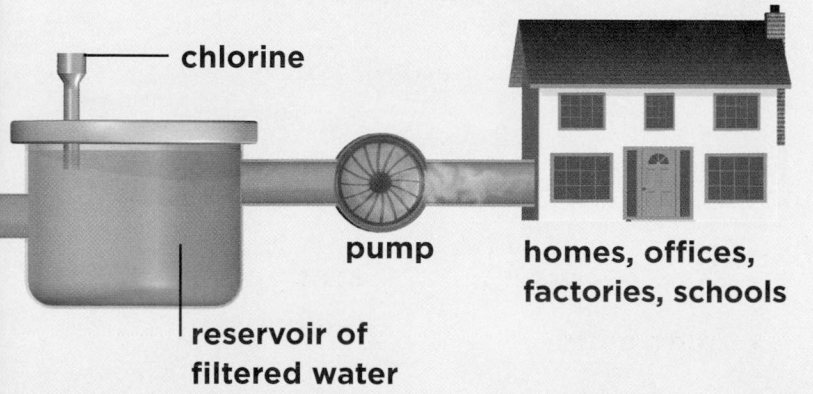

chlorine

pump

reservoir of filtered water

homes, offices, factories, schools

≡*Quick Lab*

Freshwater in Plants

1 **Measure** Using a balance, measure the mass of some apple slices.

2 Leave the apple slices on an open tray. When they are completely dry, measure their mass again.

3 **Use Numbers** Calculate the difference between the masses of the apple slices before and after you dried them. What does this difference show?

4 Repeat with another kind of fruit. Compare the results from this fruit and the apple.

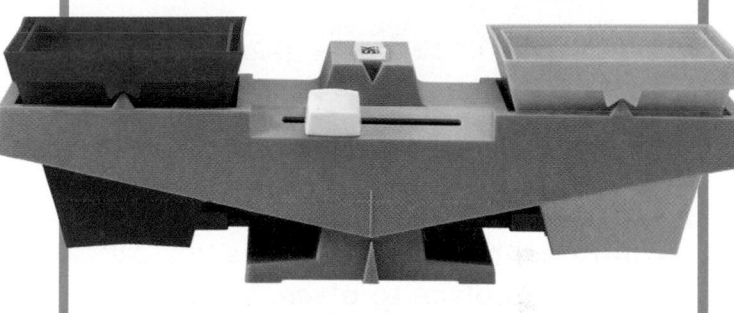

✓ *Quick Check*

Problem and Solution How do people make water safe for drinking?

Critical Thinking Why should you not drink water directly from streams or lakes?

▲ Some farms grow plants in water instead of soil.

▲ Waterways help people move things from place to place.

How else do we use water?

People use Earth's water in all sorts of ways. Freshwater is used in farming. In some places, irrigation (ihr•uh•GAY•shun) supplies the water for growing crops. **Irrigation** is a way to bring water into the soil through pipes or ditches.

Water is important to industry too. It is used to generate electricity. Ships need water to transport goods.

What are other ways people use freshwater? They use it to have fun! Swimming, boating, and fishing are some of the many examples.

 Quick Check

Problem and Solution What problem does irrigation solve?

Critical Thinking Describe three ways that people could use a river.

Many people use water for fun and recreation.

Lesson Review

Visual Summary

Forms of Earth's water include oceans, lakes, rivers, streams, groundwater, and watersheds.

People get drinking water from wells and reservoirs. The water must be cleaned before it is used.

Uses of freshwater include farming, irrigation, transportation, and recreation.

Make a FOLDABLES Study Guide

Make a trifold book. Use it to summarize what you read about Earth's water.

Think, Talk, and Write

1 Vocabulary All the water in a(n) _____ drains into one river or stream.

2 Problem and Solution How can people make sure their water is safe to drink?

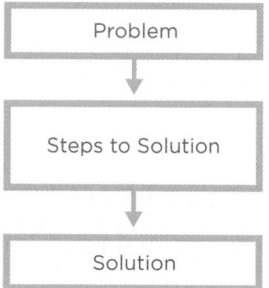

3 Critical Thinking How does the Sun help provide you with freshwater?

4 Test Prep Where is <u>most</u> of Earth's freshwater found?

A in lakes, rivers, and streams

B in glaciers and ice caps

C in the atmosphere

D below the ground

5 Essential Question How do people obtain and use water?

+6 Math Link

Solve a Problem
A leaky faucet wastes about 300 milliliters of water each day. About how much water does this leaky faucet waste in one year?

Social Studies Link

Learn about Water Use
The Southwest United States is hot and dry. It is home to many growing cities, like Phoenix, Tucson, and Las Vegas. Research how these cities are meeting their water needs.

Writing in Science

Saving Water

Dear Editor,

All life depends on water. We need clean water to drink. Plants and animals need water to survive. We also use water to make food.

Every year the population grows, so every year the amount of water we need increases. But the water supply doesn't always increase.

I believe each of us can make a difference. Here are some simple actions everyone can take right now.

- Fix any leaky faucets.
- Grow only plants that are adapted to local climate and rainfall, rather than plants that must be watered often.
- Run the dishwasher only when it is full.

All of us can help. We cannot let this important resource drip away!

Persuasive Writing

Good persuasive writing

▶ states an opinion about a topic;

▶ uses convincing reasons for that opinion;

▶ includes a call to action.

You can save water by turning off the faucet while brushing your teeth.

Write About It

Persuasive Writing Write a letter to the editor of your local newspaper. Your letter should inform people about the need to keep the groundwater clean. Include facts and details to make your letter persuasive.

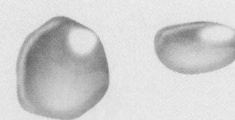 **e-Journal** Research and write about it online at www.macmillanmh.com

How much water do you use?

How much water do you use each day? You probably use more than you think you do.

The table below shows how much water people use doing different tasks each day. Keep a journal of your water use for one day. Record every time you do one of the tasks below. Then use the table to calculate how much water you used.

Daily Water Use

Task	Average Number of Gallons Used
take a bath	50
use a dishwasher	20 per load
use a washing machine	10 per load
wash dishes by hand	5 per load
flush a toilet	1.6 per flush
take a shower	2 per minute
brush your teeth	1
wash your hands	1
drink a glass of water	0.06

Add Decimals

▶ Adding decimals is like adding money amounts. Write the numbers in a column. Line up the decimal points.

$$\begin{array}{r} 50 \\ 0.06 \\ +1.6 \\ \hline \end{array}$$

▶ If a number does not have a decimal point, you can give it one. Insert zeros after the decimal point so the amounts line up correctly.

$$\begin{array}{r} 50.00 \\ + 0.06 \\ + 1.60 \\ \hline \end{array}$$

▶ Add each column. Remember to write the decimal point in the sum.

$$\begin{array}{r} 50.00 \\ + 0.06 \\ + 1.60 \\ \hline 51.66 \end{array}$$

 Solve It

1. How much water did you use in one day?

2. Try to conserve water for one day. What are some ways you can cut down on the amount of water you use? Keep a journal on your water conservation day. How many gallons did you save?

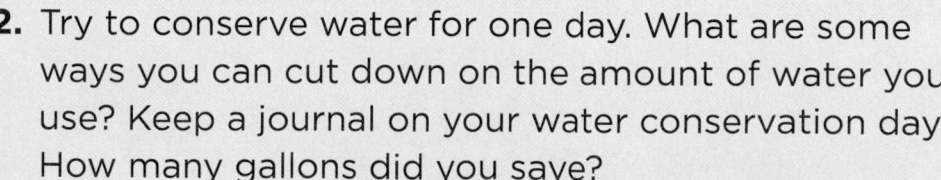

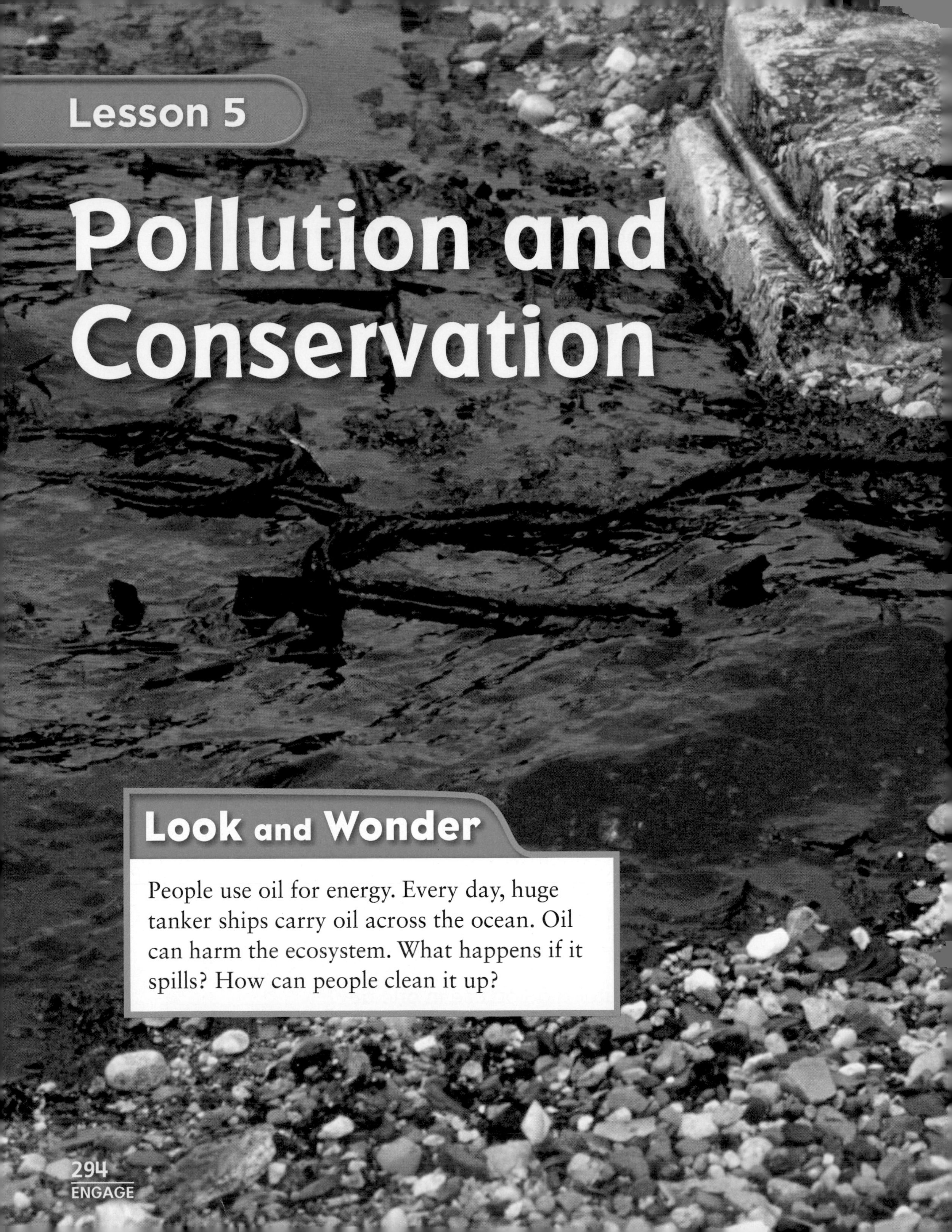

Pollution and Conservation

Look and Wonder

People use oil for energy. Every day, huge tanker ships carry oil across the ocean. Oil can harm the ecosystem. What happens if it spills? How can people clean it up?

How can you clean an oil spill?

Make a Prediction

Oil and water do not mix. How could you separate oil from the surface of water? From the surface of a solid? Make a prediction.

Test Your Prediction

1. Fill a plastic container halfway with water. Float a cork in the water.

2. **Make a Model** Using an eyedropper, carefully drip 6–7 drops of oil onto the water's surface.

3. **Observe** Watch the oil, water, and cork for about 30 seconds. Record your observations.

4. Based on your observations, make a plan to test your prediction. Use only the materials your teacher gives you.

5. Carry out your plan to clean the oil from the water and cork. Record your results.

Draw Conclusions

6. **Communicate** How well were you able to clean the oil from the water? From the cork? Describe your findings.

7. Was your prediction correct? Explain. What other materials do you think might have worked?

8. **Infer** What can be done to clean up an oil spill in the ocean?

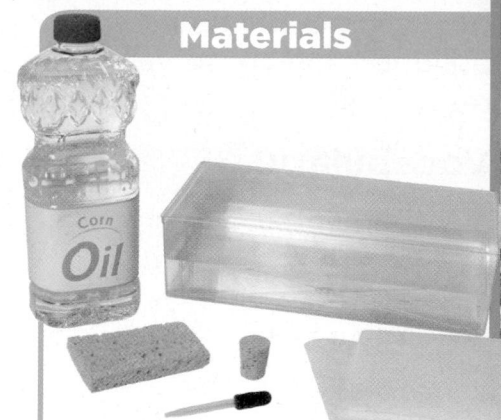

Materials

- plastic container
- water
- cork
- eyedropper
- vegetable oil
- paper towels
- sponge

Step 2

Explore More

Research the 1989 oil spill from the *Exxon Valdez.* Where did the oil spill? How far did it travel? How was it cleaned? Report your findings.

Read and Learn

▶ **Essential Question**

How can people reduce pollution and conserve resources?

▶ **Vocabulary**

environment, p. 296

pollution, p. 296

acid rain, p. 296

conservation, p. 298

compost, p. 298

reduce, p. 300

reuse, p. 300

recycle, p. 300

▶ **Reading Skill** ✔

Main Idea and Details

Main Idea	Details

▶ **Technology**

e-Glossary and e-Review online at www.macmillanmh.com

What is pollution?

The living and nonliving things that make up an area form an **environment**. All living things need a healthy environment. A healthy environment has clean air, water, and land.

When a harmful substance is added to the environment, it causes **pollution** (puh•LEW•shun). Some pollution comes from natural sources, like forest fires and volcanoes. Most pollution comes from human activities.

Air Pollution

When we burn fossil fuels, gases and bits of dust go into the air. Some of the gases combine with water droplets in the air. When this happens, **acid rain** forms. Acid rain can harm living things and damage buildings.

Water vapor, other gases, and dust can form smog, which hangs in the air like fog. Smog makes the air dangerous to breathe.

Water and Land Pollution

What happens when people dump wastes from homes or factories into oceans, lakes, or rivers? They cause water pollution. Polluted water can kill plants and animals. It can make people sick.

Even chemicals that help people can pollute the water. Fertilizers and pesticides can be washed into nearby waters during a storm. Storm drains lead to an ocean or stream. Oil spills from ships can pollute water and beaches. Many fish, birds, and mammals are harmed in this way.

You know better than to *litter,* or throw trash on the ground. Littering is ugly. It is also against the law. Littering, chemicals, solid wastes, and acid rain pollute the land. They can harm or kill plants and animals.

People cause land pollution by leaving tires, paper, and other trash on the ground.

 Quick Check

Main Idea and Details How can pollution harm the environment?

Critical Thinking Are people always aware of the pollution they cause? Give examples.

Cars, trucks, and buses are major sources of air pollution in cities.

Read a Photo

Why would a farmer use this method of planting?

Clue: Look at the relationship of the crop rows to the slope of the land.

How can we protect the soil and water?

Everyone can conserve resources. **Conservation** (kahn•sur•VAY•shun) means using resources wisely.

Soil Conservation

You have learned how easily soil can be carried away, or eroded. Farmers use methods to slow erosion. These methods are called soil conservation. They keep the soil in place to support plants. One method is to plant a row of trees that slows down the wind.

On sloping lands, farmers plow fields in curved rows that follow the shape of the land. This method is called *contour plowing*. When farmers change crops every year, it is called *crop rotation*. Crop rotation conserves the nutrients in soil.

People can conserve soil by spreading compost in their gardens. **Compost** is a mix of dead or decaying matter, such as food scraps, fallen leaves, and cut grass.

Water Conservation

No one has an unlimited water supply. How can we conserve water?

Many towns and cities collect used water, or *wastewater*, from homes and businesses. They pass it through a system of pipes called *sewers*. The pipes lead to a sewage treatment plant. There the water is cleaned and released.

How can you conserve water? Turn off the faucet when you are not using it. Run only full loads in laundry machines and dishwashers. Ask your family to fix leaky toilets and faucets. By saving water, you conserve it!

People in desert regions can conserve water by growing native plants instead of grass. ▼

Quick Lab

Conservation Plan

1. **Observe** How does your school use resources? Find out. Consider water use, energy use, and garbage disposal.

2. Think of ways your school could conserve resources or produce less waste. Write down your ideas.

3. **Communicate** Share your ideas with your classmates. As a class, write a plan to present to your principal.

✔ Quick Check

Main Idea and Details Describe some methods of conservation.

Critical Thinking What kinds of things can you do to conserve soil and water?

What are the 3 Rs?

Three ways to conserve resources start with the letter *R*.

Reduce

To **reduce** means to use less of something. This is the simplest way to conserve. How do you reduce your use of paper? Write on both sides of the sheet. How do you reduce your energy use? Turn off the lights when you do not need them.

Reuse

To **reuse** means to use something over again. You can reuse glass or plastic cups instead of throwing away paper cups. You can reuse old clothes as cleaning rags.

Recycle

To **recycle** means to make a new product from old materials. Recycling keeps materials in use and out of landfills. Many communities recycle paper, plastic, glass, and metal. Can you think of other ways to recycle?

 Quick Check

Main Idea and Details What are the 3 *R*s of conservation?

Critical Thinking How can the 3 *R*s help us conserve fossil fuels?

Read a Photo

Which of the 3 *R*s does each picture show?

Clue: Look closely at what is being

The 3 Rs in Action

WE RECYCLE

Lesson Review

Visual Summary

Human activities can cause **pollution** of Earth's land, air, and water.

Conservation means using resources wisely. There are many ways to conserve the soil and water.

Anyone can practice **the 3 Rs** of conservation—reduce, reuse, and recycle.

Make a FOLDABLES Study Guide

Make a trifold table. Use it to summarize what you read about pollution and conservation.

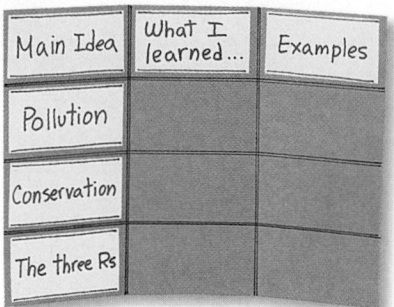

Main Idea	What I learned...	Examples
Pollution		
Conservation		
The three Rs		

Think, Talk, and Write

1 **Vocabulary** You can form _____ from table scraps and grass cuttings.

2 **Main Idea and Details** Give one example for each of the 3 Rs of conservation.

Main Idea	Details

3 **Critical Thinking** Write a plan that your family can follow to conserve water. Which of the steps do you think will help conserve the most water? Explain.

4 **Test Prep** The burning of which product can cause acid rain?

   A composts

   B fertilizers

   C pesticides

   D fossil fuels

5 **Essential Question** How can people reduce pollution and conserve resources?

 **Writing Link**

Write a Report

Research how your community treats wastewater. Do people pay for this service? How much? Write a report to show what you found out.

 **Art Link**

Make a Poster

Make a poster that teaches people about one or more of the 3 Rs. Use bright colors or good humor to express your message.

SAVING THE SOIL

Much of the food we eat comes from plants grown in soil. Sometimes the way we grow our food can harm the soil. It's important that we have good soil to grow crops. Farmers can use different methods to protect the soil now and in the future.

Plowing is pulling a blade that turns over the soil. Some farmers use a method called contour plowing. They follow the curves of a hillside instead of plowing straight up and down. This method makes "steps" that stop or channel the flow of rainwater. Then the water soaks into the soil. Without contour plowing, the water runs downhill, carrying loose soil and nutrients with it. The problem is that contour plowing takes longer than plowing in straight lines. It also uses more fuel.

Many farmers use tractors during planting season.

Step farming is a type of contour plowing.

Connect to

AMERICAN
MUSEUM of
NATURAL
HISTORY

302
EXTEND

at www.macmillanmh.com

Other farmers do not plow their fields after a harvest. A protective cover forms over the soil. This cover lasts through the winter. When it's time to plant, the farmers don't need to plow. They just dig holes in the field and place seeds in the holes. This method is called no-till planting. No-till planting also has its problems. Farmers may need to use chemicals to kill weeds that plowing would have removed.

Soybeans can be used as a cover crop. ▲

Some farmers plant seeds at the end of a harvest to protect the soil. The plants that grow are called cover crops. They add nutrients to the soil. Common cover crops include rye, clover, and fava beans. When farmers are ready to plant again, they plow the cover crop into the soil. Then they plant seeds for the new crop.

Wheat also makes a good cover crop. ▶

Main Idea and Details

▶ The main idea is the focus of the entire article.

▶ Details support and explain the main idea.

Write About It

Main Idea and Details

1. Why do farmers need to protect the soil?
2. What are some ways that farmers protect the soil? List the advantages and disadvantages of these methods.

 **e–Journal** Research and write about it online at **www.macmillanmh.com**

Visual Summary

Lesson 1 Rocks are made up of minerals. The rock cycle describes how rocks form and change.

Lesson 2 Soil is made of weathered rock and other materials. Soil forms slowly in layers.

Lesson 3 Fossils give us clues about Earth's past. Fossil fuels are nonrenewable resources.

Lesson 4 Water collects on Earth's surface and below the ground. It is stored and used in many ways.

Lesson 5 People may pollute the land, air, and water. We can protect our resources through conservation.

Make a **FOLDABLES** Study Guide

Tape your lesson study guides to a piece of paper as shown. Use your study guide to review what you have learned.

Fill each blank with the best term from the list.

environment, p. 296　　**mineral,** p. 252

fossil, p. 274　　**permeability,** p. 266

groundwater, p. 287　　**renewable resource,** p. 280

humus, p. 264　　**resource,** p. 258

irrigation, p. 290　　**rock cycle,** p. 256

1. All the living and nonliving things in an area make up a(n) _____.

2. The nonliving substance that makes up rock is called a(n) _____.

3. Nature can quickly replace a(n) _____, such as air or water.

4. Many farmers depend on _____ to bring water to their crops.

5. Sandy soils have a higher _____ than clay soils.

6. To tap into _____, people drill or dig a deep hole called a well.

7. A material or object that people can use is called a(n) _____.

8. The evidence of a living thing from long ago is a(n) _____.

9. Rocks change form through the _____.

10. Nonliving plant and animal matter, called _____, has nutrients for plants to grow.

Answer each of the following.

11. Main Idea and Details How are the three types of rock formed? Provide an example for each.

12. Communicate Write a public notice. Explain how a water treatment plant works. Describe how it makes water safe for drinking.

13. Critical Thinking Which renewable energy source do you think will be most important in the future? Explain your answer.

14. Persuasive Writing Write a letter to your school newspaper convincing other students to conserve resources. Suggest ways that they can reduce, reuse, and recycle.

15. The properties of soil

    A vary from place to place.

    B are the same in all climates.

    C are similar in all horizons.

    D cannot be observed easily.

16. Problem and Solution How can drinking water become polluted? What can be done to keep it safe?

17. True or False *Fossil fuels are a renewable resource.* Is this statement true or false? Explain.

18. True or False *Heat and pressure can change the properties of a rock.* Is this statement true or false? Explain.

19. Critical Thinking What do fossils tell scientists about how life on Earth has changed?

20. What are Earth's resources, and how can we conserve them?

Marvelous Minerals

Learn more about the properties and uses of different minerals.

What to Do

1. Use resources to find information about diamond, quartz, chromite, and copper. What properties does each mineral have?

2. Find out how each mineral is used. In what common objects can each mineral be found?

3. Copy the chart below. Use your chart to record your findings.

Mineral	Properties	Uses
diamond		
quartz		
chromite		
copper		

Analyze Your Results

Explain how the uses of each mineral are related to its properties.

1 Which mineral is softest?

Moh's Hardness Scale	
Mineral	**Hardness**
gypsum	2
calcite	3
quartz	7
diamond	10

A diamond

B gypsum

C quartz

D calcite
DOK I

2 The fossil shown below was found in a tundra biome.

What can you <u>most likely</u> infer from this finding?

A The climate has gotten colder.

B The climate has gotten warmer.

C Fossil fuels can be found nearby.

D Fossil fuels cannot be found here.
DOK 2

3 Which properties are <u>most</u> helpful in identifying minerals?

A luster, streak

B size, ability to float

C weight, shape

D shape, width
DOK I

4 The properties of soils

A vary from place to place.

B are the same in all biomes.

C are similar in all horizons.

D can only be observed with scientific equipment.
DOK I

5 Which human activity <u>most likely</u> has a negative impact on the environment?

A composting

B conserving resources

C recycling notebook paper

D burning fossil fuels
DOK I

6 Most fossils are found in

A minerals.

B igneous rocks.

C sedimentary rocks.

D metamorphic rocks.
DOK I

7 Look at the diamond below.

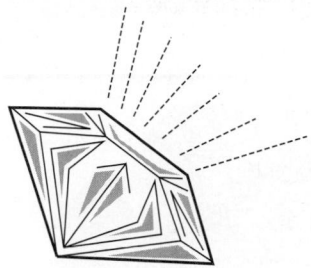

In which group does this diamond belong?

A renewable resource

B fossil fuel

C building material

D mineral resource
DOK I

8 A soil sample has a coarse texture and high permeability. It is __most likely__

A clay.

B sand.

C humus.

D bedrock.
DOK I

9 A rock is soft and has layers. The rock is __most likely__

A sedimentary.

B igneous.

C metamorphic.

D granite.
DOK I

10 Windmills are a renewable source of electricity.

Name another renewable energy source.
DOK I

Explain how using this renewable energy source can reduce pollution.
DOK 2

Check Your Understanding

Question	Review	Question	Review
1	pp. 252–253	6	pp. 274-277
2	pp. 276–277	7	pp. 252-253, 278–280
3	pp. 252–253	8	pp. 267–268
4	pp. 266–268	9	pp. 254-255
5	pp. 278–280	10	pp. 278–280

Survey Technician

How can you find out exactly where your front yard ends and your neighbor's begins? You need the help of a survey technician! Survey technicians use certain instruments to find the borders of lands. They help plan where new roads, bridges, and homes are built.

To join a surveying crew, you will probably need a technical degree in surveying. You will also need a steady hand and strong math skills. Computer skills are especially useful.

▲ Surveyors use high-tech tools to measure land area.

Geologist

If you are curious about planet Earth, you may want to become a geologist. Geologists are also called Earth scientists.

Not all geologists study rocks. Some work with businesses to locate oil or other resources in the ground. Others study earthquakes or volcanoes. Geologists also look at how Earth has changed over time. They make predictions about the future.

To be a geologist, you need a college degree. Most geologists go to school for several more years after college. If you want to be a real "rock hound," you'd better bone up on science!

▲ Geologists who study volcanoes wear thermal suits for protection.

Weather and Space

Satellites can give us pictures of the weather anywhere on Earth!

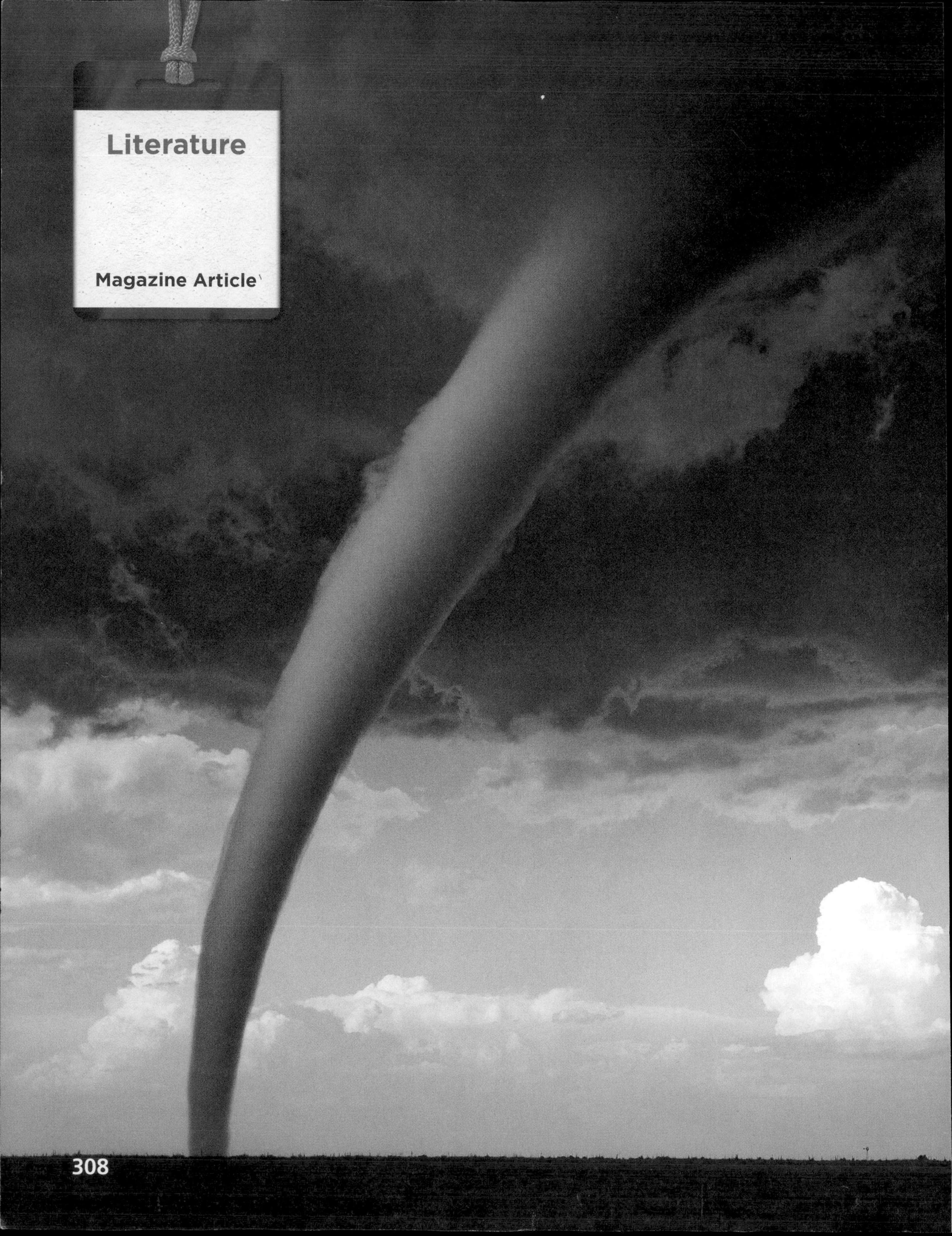

Literature

Magazine Article

Tornado
Tears through Midwest

FUJITA SCALE

F0	branches broken off some trees
F1	surfaces peeled off roofs
F2	whole roofs torn from some houses
F3	most trees in the forest uprooted
F4	well-constructed houses destroyed
F5	houses and trucks hurled through the air

Weather changes all of the time, but sometimes it is more powerful than others! A tornado can form and the effects can last a long time.

In 2005, a very powerful tornado hit Indiana and Kentucky. It was very destructive. Officials believe that the path of the tornado was three-fourths of a mile wide and 20 miles long.

Do you know how a tornado is formed? They are made up of strong, spinning winds that are created when the warm air of a giant storm system rises and hits a current of downward-moving cool air. This can cause the wind to start spinning, resulting in a tornado. The vortex, or center of a tornado, sucks in air and carries it upward. Tornadoes can often be black because they pick up gust.

The tornado that hit Indiana and Kentucky was rated a F# on the Fujita Scale. This is very severe because the winds ranged from 158 miles per hour (mph) to 206 mph. The Fujita scale ranges from the weakest, F0, to the strongest, F5. The Fujita scale got its name from its creator, Dr. Theodore Fujita.

Write About It

Response to Literature What would happen if a tornado struck your community? Write a fictional story. Describe how your community would stay safe. How would it rebuild after the disaster?

e-Journal Write about it online at **www.macmillanmh.com**

CHAPTER 7

Weather and Climate

What are weather and climate?

Essential Questions ·

Lesson 1
How can you tell that air is around you?

Lesson 2
How is water recycled?

Lesson 3
How do fronts and air masses change the weather?

Lesson 4
Why do weather patterns change?

 # Big Idea Vocabulary

atmosphere the blanket of gases that surrounds Earth (p. 314)

condensation the process of a gas changing into a liquid (p. 325)

cloud a collection of tiny water droplets or ice crystals that hangs in the air (p. 325)

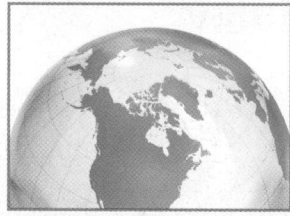

air mass a large region of the atmosphere where the air has similar properties throughout (p. 336)

front a boundary between air masses with different temperatures (p. 337)

climate the average weather pattern of a region over time (p. 346)

 Visit www.macmillanmh.com for online resources.

Air and Weather

Look and Wonder

Pinwheels spin wildly in a strong wind. What makes the wind blow strongly? Why does it blow from different directions?

How does the wind move?

Make a Prediction

Air can move from place to place. When you open a sealed bottle of liquid that is under pressure, air moves. Does the air move into or out of the bottle? Why? Make a prediction.

Test Your Prediction

1 **Make a Model** Fill an empty plastic bottle halfway with very warm water from a faucet.

2 ⚠ **Be Careful.** Pour warm liquids carefully. Place the cap on the bottle. Shake the bottle several times. Pour the water out. Replace the cap and set the bottle on a table. Observe it for several minutes.

3 **Observe** Hold the bottle near your ear. Remove the cap slowly. Listen carefully.

Draw Conclusions

4 Did air move into or out of the bottle? What happened to the pressure inside the bottle before the cap came off? After it came off?

5 **Infer** How might air pressure affect the direction from which winds blow? Use evidence from your model in your answer.

Explore More

Suppose you warm the air inside a capped bottle. What will happen to the air pressure inside the bottle? Write a prediction. Try it!

Materials

- bottle with cap
- funnel
- very warm water

Step **1**

Read and Learn

▶ **Essential Question**

How can you tell that air is around you?

▶ **Vocabulary**

atmosphere, p. 314

temperature, p. 316

humidity, p. 316

air pressure, p. 317

thermometer, p. 318

wind vane, p. 318

barometer, p. 318

rain gauge, p. 318

▶ **Reading Skill** ✔
Summarize

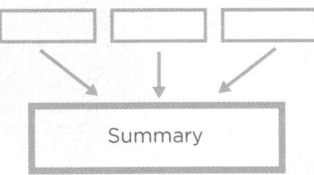

Summary

▶ **Technology** LOG ON
e-Glossary and e-Review online at www.macmillanmh.com

What is in the air?

Air surrounds Earth like a thin blanket. This blanket of air is the **atmosphere** (AT•muh•sfeer). How is the atmosphere important to people and other living things?

Gases

The atmosphere is a mix of different gases. You can tell from the circle graph that most of the atmosphere is made of nitrogen (NI•truh•jun) and oxygen. The atmosphere also has carbon dioxide and other important gases.

Animals and most other organisms need oxygen to live. Plants also need carbon dioxide. The atmosphere allows living things to survive on Earth.

Layers of Earth's Atmosphere

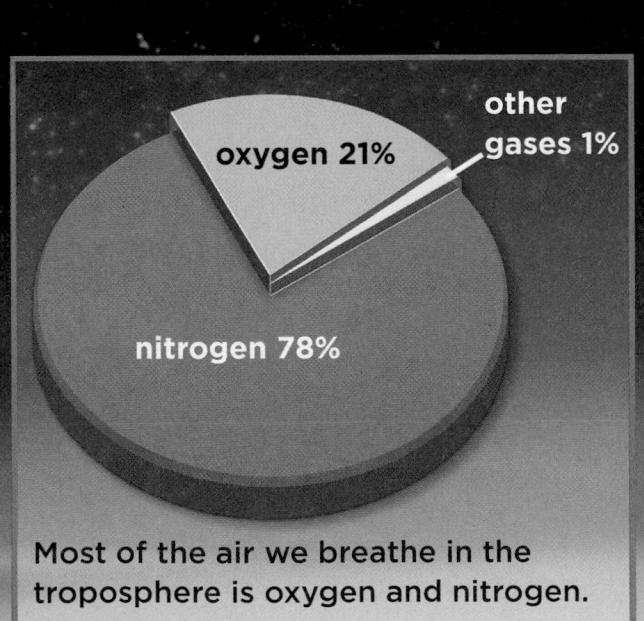

other gases 1%

oxygen 21%

nitrogen 78%

Most of the air we breathe in the troposphere is oxygen and nitrogen.

The Troposphere

Earth's atmosphere is made up of layers. The layer closest to Earth's surface is the *troposphere* (TROH•puh•sfeer). Compared to the rest of the atmosphere, the troposphere is very thin. Yet all of Earth's life exists here.

The troposphere is also where all of Earth's weather takes place. Here the air is always on the move. Air that moves from place to place is called *wind*. Wind can be as gentle as a light breeze. It can be as fierce as a tornado. Any change in the wind brings a change in the weather.

Other Layers of the Atmosphere

The diagram shows three other layers of Earth's atmosphere. The stratosphere (STRA•tuh•sfeer) is the layer above the troposphere. The troposphere has few air particles. The air gets even lighter in the mesosphere (ME•zuh•sfeer) and thermosphere (THUR•muh•sfeer).

 Quick Check

Summarize How are the troposphere and the atmosphere related?

Critical Thinking In what way is Earth's atmosphere like an orange peel? How is it different?

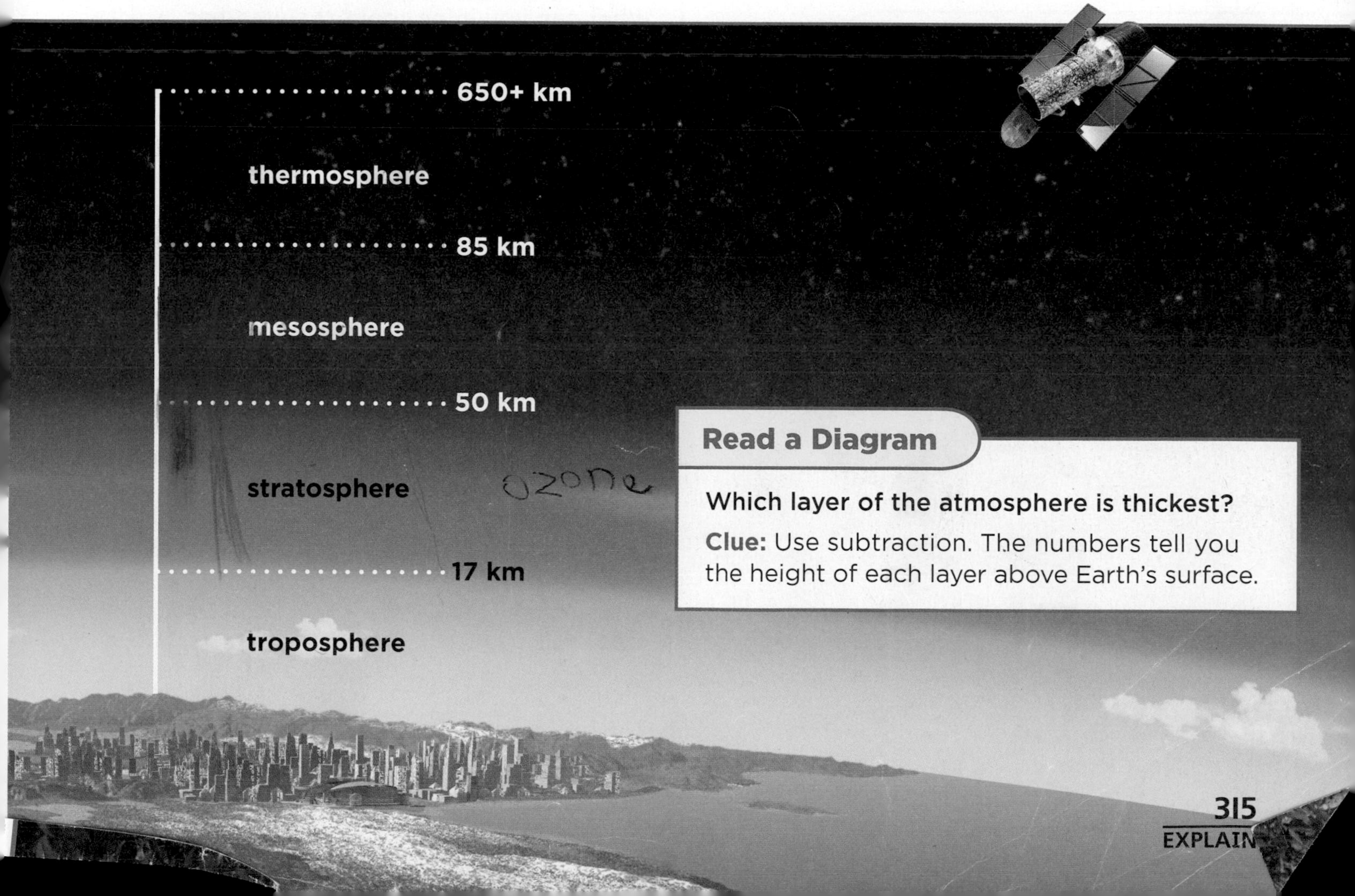

650+ km

thermosphere

85 km

mesosphere

50 km

stratosphere   ozone

17 km

troposphere

Read a Diagram

Which layer of the atmosphere is thickest?

Clue: Use subtraction. The numbers tell you the height of each layer above Earth's surface.

What is weather?

Weather is the condition of the atmosphere at a given time and place. Weather can vary depending on the time of day, season, or place.

Air Temperature

Temperature (TEM•puh•ruh•chur) describes how hot or cold something is. When the Sun's energy heats Earth's surface, the surface warms the air above it. The air moves.

Some parts of Earth's surface heat up more than other parts. The uneven heating of Earth's surface causes air to move at different speeds. Moving air is called wind.

Humidity

If the air around us feels damp and sticky, we call the weather humid (HYEW•mud). **Humidity** (hyew•MIH•duh•tee) is a measure of how much moisture is in the air. Deserts usually have very low humidity. Rain forests have very high humidity.

Air always has some amount of moisture. Most of the moisture comes from ocean water that changes into a gas. The rest comes from bodies of water, soil, and plants.

Read a Photo

What can you infer about the weather in a tropical rain forest?

Clue: Look for clues that show humidity and air temperature.

Humidity in a Rain Forest

Mountain climbers use special equipment to deal with low temperature and low air pressure.

Air Pressure

We live at the bottom of the troposphere. Here, the weight of the entire atmosphere pushes down on us. The force of air pushing on an area is called **air pressure**.

Particles of cool air are closer together than particles of warm air. In the same amount of space, cool air weighs more than warm air. Warm air is less *dense*, or packed together, than cold air. As air warms, its pressure decreases. Air moves from an area of high pressure to an area of low pressure.

Precipitation

Any form of water that falls from clouds is *precipitation* (prih•sih•puh•TAY•shun). The term includes rain, snow, sleet, and hail.

≡Quick Lab

Humidity in a Cup

1. Pour 5 milliliters of water in each of two cups. Cover each cup with plastic wrap.

2. Place one cup in the refrigerator for ten minutes. Keep the other cup on a flat surface.

3. **Observe** Remove the cup from the refrigerator. Set it beside the other cup. Observe and compare the water in both cups. What differences do you notice?

4. **Infer** Which cup do you think has greater humidity—the warm cup or the cold cup? How do you know?

✔ Quick Check

Summarize What properties can you use to describe the weather?

Critical Thinking What role does the Sun play in Earth's weather?

FACT ▶ Humidity on Earth's surface never reaches zero.

How can you measure weather?

Weather scientists often collect data from a place called a <u>weather station</u>. You can set up your own weather station. All you need are a few of the tools shown on this page.

A *hygrometer* (hi•GRAH•muh•tur) measures humidity. ▲

A wind vane points in the direction from which the wind is blowing. ▶

◀ A thermometer measures air temperature in degrees Celsius (°C) or degrees Fahrenheit (°F).

A barometer measures air pressure. ▶

An *anemometer* (a•nuh•MAH•muh•tur) measures wind speed. The faster the wind blows, the faster the cups spin. ▼

◀ A rain gauge (GAYJ) is a tube that collects water. It shows the amount of rainfall.

✔ Quick Check

Summarize What tools could you use to measure the weather?

Critical Thinking Why do scientists use different tools to measure weather?

Lesson Review

Visual Summary

Earth's atmosphere is made of gases. It has several layers. The troposphere is the layer where weather forms.

We can describe the **properties of weather** using air temperature, humidity, air pressure, precipitation, and wind.

Scientists use many different **tools to measure weather,** such as hygrometers and thermometers.

Make a FOLDABLES Study Guide

Make a three-tab book. Use it to summarize what you read about air and weather.

Earth's atmosphere

Properties of weather

Tools to measure weather

Think, Talk, and Write

1 **Vocabulary** A(n) _____ measures the speed of the wind.

2 **Summarize** What are the different parts of Earth's atmosphere?

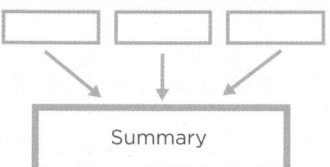

Summary

3 **Critical Thinking** Compare and contrast two examples of weather that you have experienced. Your comparison should include the vocabulary terms from this lesson.

4 **Test Prep** In which layer of the atmosphere do we experience weather?

   A thermosphere

   B stratosphere

   C mesosphere

   D troposphere

5 **Essential Question** How can you tell that air is around you?

Math Link

Find the Average Rainfall
It rained 4 centimeters on Monday, 8 cm on Tuesday, and 6 cm on Wednesday. What is the average rainfall for the three days?

Health Link

Report on Staying Healthy
How do people stay healthy when the air temperature is very cold or very hot? Research the answers. Report on your findings.

Writing in Science

WATCHING SPRING WEATHER

Spring weather can be very different from day to day. Last week, we had a stretch of sunny and mild spring weather. Temperatures were in the seventies. At night, they dropped to the mid-60s. The air was mostly still, with a gentle breeze moving in every now and then.

Then the barometer started to fall rapidly. This signaled an approaching storm.

Yesterday strong winds swept in from the northwest. A heavy rain began to fall. The temperature was 41°F at noon. At night, it fell to the low 30s.

Today it is cloudy and overcast. The temperatures are in the high 40s. The wind speed is 23 miles per hour.

Don't put away your winter coat yet. The weather forecasters predict more cold weather to come. We might have a light snowfall tonight.

 **Write About It**

Expository Writing Observe the weather in your area every day for two weeks. Record the temperature, air pressure, precipitation, clouds, and wind speed. Write a newspaper article about the changes you observe.

**-Journal** Research and write about it online at www.macmillanmh.com

Graphing Weather Changes

You can use line graphs to show how things change over time. Record the high and low temperatures in your area every day for seven days. Use newspapers, television, or radio broadcasts to collect your data. Then plot the data on a line graph.

First, title your graph *Temperature Changes*. Label the bottom and left side as shown below. Start the temperature scale with a lower number than the lowest temperature you recorded. Then, mark off equal spaces in intervals of 5. Write the days of the week across the bottom.

Plotting Points on a Line Graph

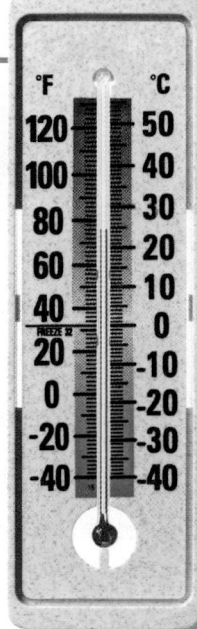

► Use different colors for high and low temperatures.

► Find the high temperature for your first day. If it is between two markings, make an estimate. Slide your finger over to the day. Mark that spot with a point.

► Continue plotting all the high and low temperatures. Use straight lines to connect all the highs. Use another line to connect all the lows.

Temperature Changes

Temperature (F°): 50, 45, 40, 35, 30, 25, 20

Sun | Mon | Tues | Wed | Thurs | Fri | Sat

Solve It

Plot your data on the graph you made. Describe the temperature pattern shown on your graph.

The Water Cycle

Look and Wonder

Earth has had the same amount of water for billions of years. But not all of that water is in the liquid state. Some is solid ice. Some is even in the gas state. How can this be so?

How does water change from a liquid to a gas?

Form a Hypothesis

What variables affect how water changes from a liquid to a gas? Write a hypothesis.

Test Your Hypothesis

1 Communicate Work in a small group. Discuss examples of water changing from a liquid to a gas. What might affect how fast this change occurs? Consider temperature, wind, area, and volume of water.

2 Use Variables Using the materials, design an experiment to test one of the variables you discussed. Use two water samples. One will test the independent variable. The other water sample is your control.

3 Experiment Conduct your experiment. Record your observations at each step.

Draw Conclusions

4 Was your hypothesis correct? Does the variable you tested affect how water changes from a liquid to a gas? Give evidence to support your conclusion.

5 Classify Share your results as a class. Classify the variables you tested into those that affect the change and those that do not.

Explore More

Choose a different variable that might affect how liquid water changes to a gas. Form a new hypothesis. Design an experiment to test it. Then conduct your experiment. Share your findings with the class.

Materials

- water
- 2 plastic containers with lids
- spoon
- salt

Step 3

▶ **Reading Skill** ✔
Sequence

First

↓

Next

↓

Last

▶ **Technology** 🔵 LOG ON
e-Glossary, e-Review, and animations online at www.macmillanmh.com

Why does water change state?

Water moves from Earth's surface into the atmosphere. Then it moves back to the surface. Water changes state, or form, as it moves.

Evaporation

Water seems to disappear when it evaporates (ih•VA•puh•rayts). **Evaporation** occurs when a liquid changes slowly into a gas. Liquid water does not really disappear. It just changes to a gas.

Water vapor is water in the gas state. You cannot see water vapor, but it is part of the air around you.

Water is always evaporating from oceans, streams, lakes, rivers, and ponds. The Sun's heat causes particles of water at the surface to move rapidly. The more heat they take in, the faster and farther apart they move. Some of the particles rise into the air as a gas—water vapor.

❶ The Sun's energy heats the surface of the water.

❷ Particles of water evaporate from the surface. They rise into the air as water vapor.

Condensation

As particles of water vapor rise into the air, they cool. The particles lose energy. They move more slowly. High in the atmosphere, the water vapor *condenses* (kun•DENS•ez) to liquid water. **Condensation** occurs when a gas changes to a liquid.

Dew is a familiar kind of condensation. Dew forms when water vapor cools and condenses onto a surface. Have you ever seen drops of water cover the grass on a cool morning? Those drops are dew.

Water vapor can also condense onto dust particles in the air. The tiny drops, or *droplets,* form clouds. A **cloud** is a group of water droplets in the atmosphere. The droplets are pure water in liquid form.

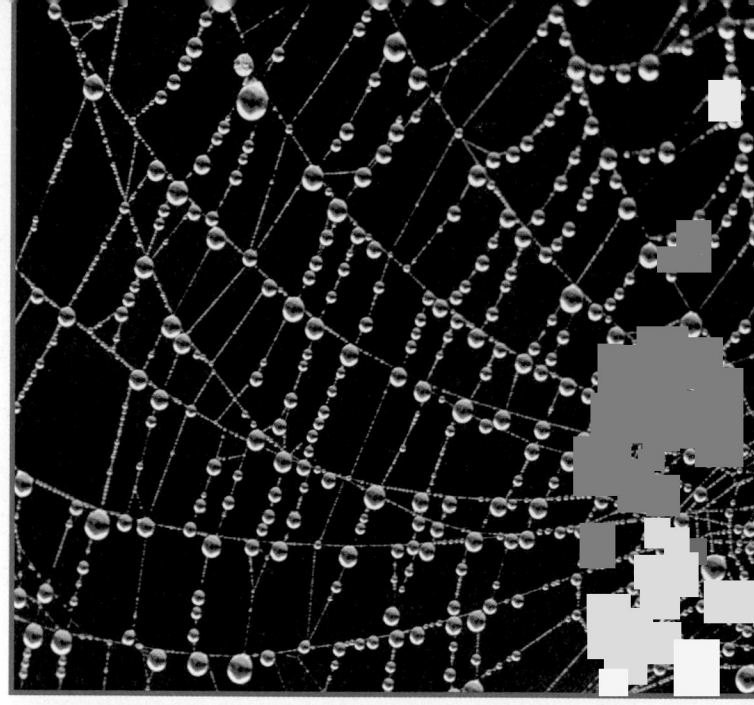

Dew can form on spiderwebs in the early morning.

Precipitation

Inside a cloud, small water droplets may join together and form larger ones. If it is very cold, some droplets freeze into ice. To **freeze** is to change from a liquid to a solid.

The droplets and bits of ice grow larger and heavier. When they are too heavy, they fall to Earth's surface. **Precipitation** (prih•sih•puh•TAY•shun) is the water that falls from clouds down to Earth.

3 As they rise higher, the particles of water vapor cool and condense.

4 Clouds form from droplets of liquid water.

5 When droplets in the clouds grow large and heavy, they fall to Earth.

✓ Quick Check

Sequence Explain the steps in evaporation and condensation.

Critical Thinking What happens to a puddle of water on a sunny day? Why?

Where does water go?

By now you know a lot about water. You know that water can be found in many places. You know it has three different states.

Water is always moving from place to place, in one form or another. The **water cycle** is the movement of water between Earth's surface and the air. Evaporation, condensation, and precipitation help water move through the cycle. The diagram shows you how.

In the Air

In the water cycle, water changes state among liquid, gas, and solid. The Sun is the energy source for this cycle. The Sun's energy causes water to evaporate from lakes, oceans, and other bodies of water. Water also evaporates from the leaves of plants. This is called *transpiration* (trans•puh•RAY•shun). As it rises in the air, the water vapor condenses. Clouds form. During precipitation, water falls from the clouds over land and water.

The Water Cycle

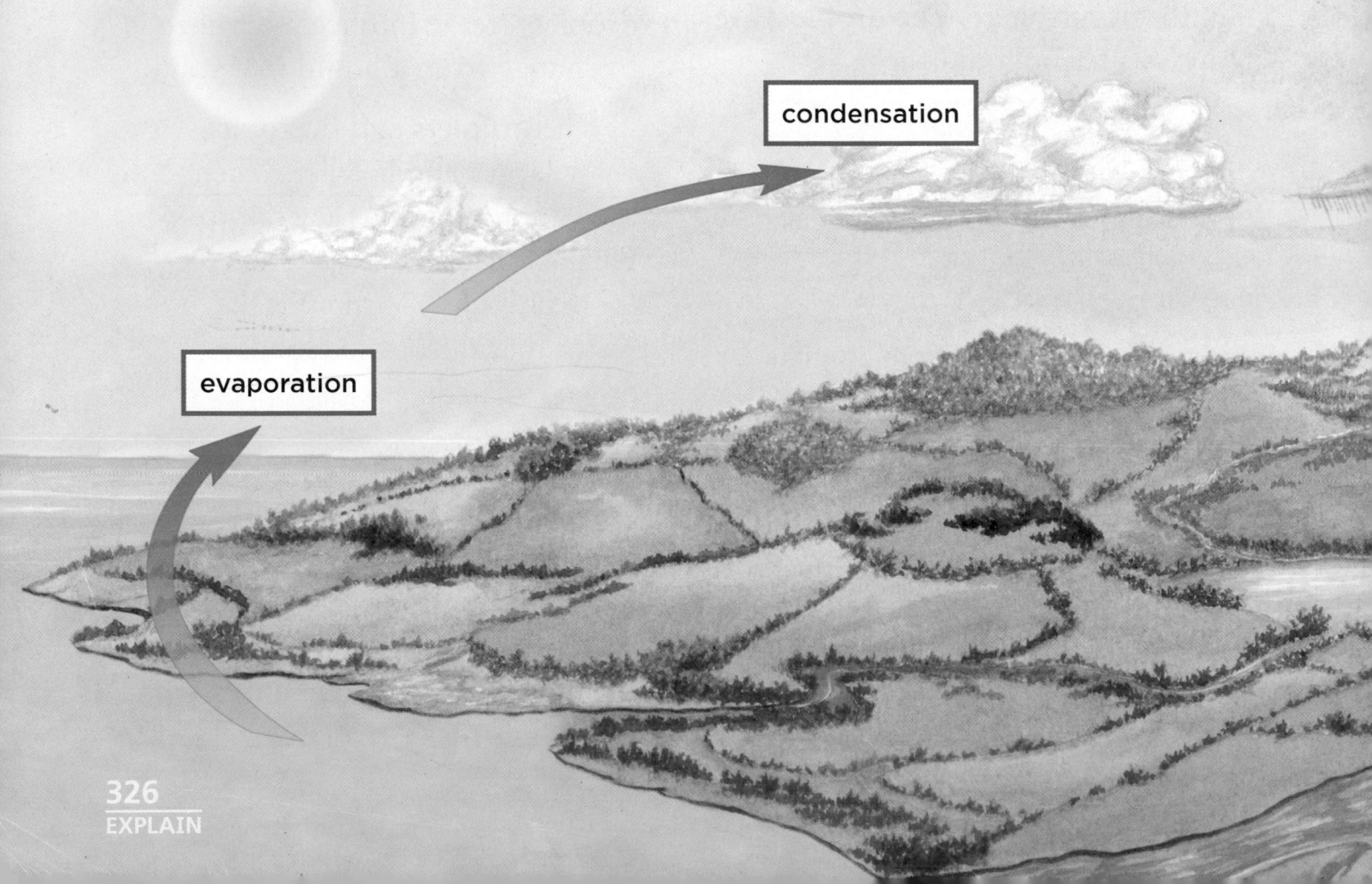

condensation

evaporation

On and Below the Ground

Precipitation can fall as rain, snow, sleet, or hail. When it rains, water flows over Earth's surface as *runoff*. Runoff gathers in lakes, oceans, and streams. Water also collects in glaciers and ice caps.

Water that soaks into the ground moves downward through small cracks and spaces. Some of this water becomes groundwater that flows to wells, rivers, and lakes. Plants take up some water and some evaporates. Water is always moving and recycling.

Quick Check

Sequence How does water enter and leave the atmosphere?

Critical Thinking How does the Sun's energy affect Earth's weather?

Read a Diagram

Describe one path through the water cycle.

Clue: Follow the arrows.

 Science in Motion Watch how the water cycle works at www.macmillanmh.com

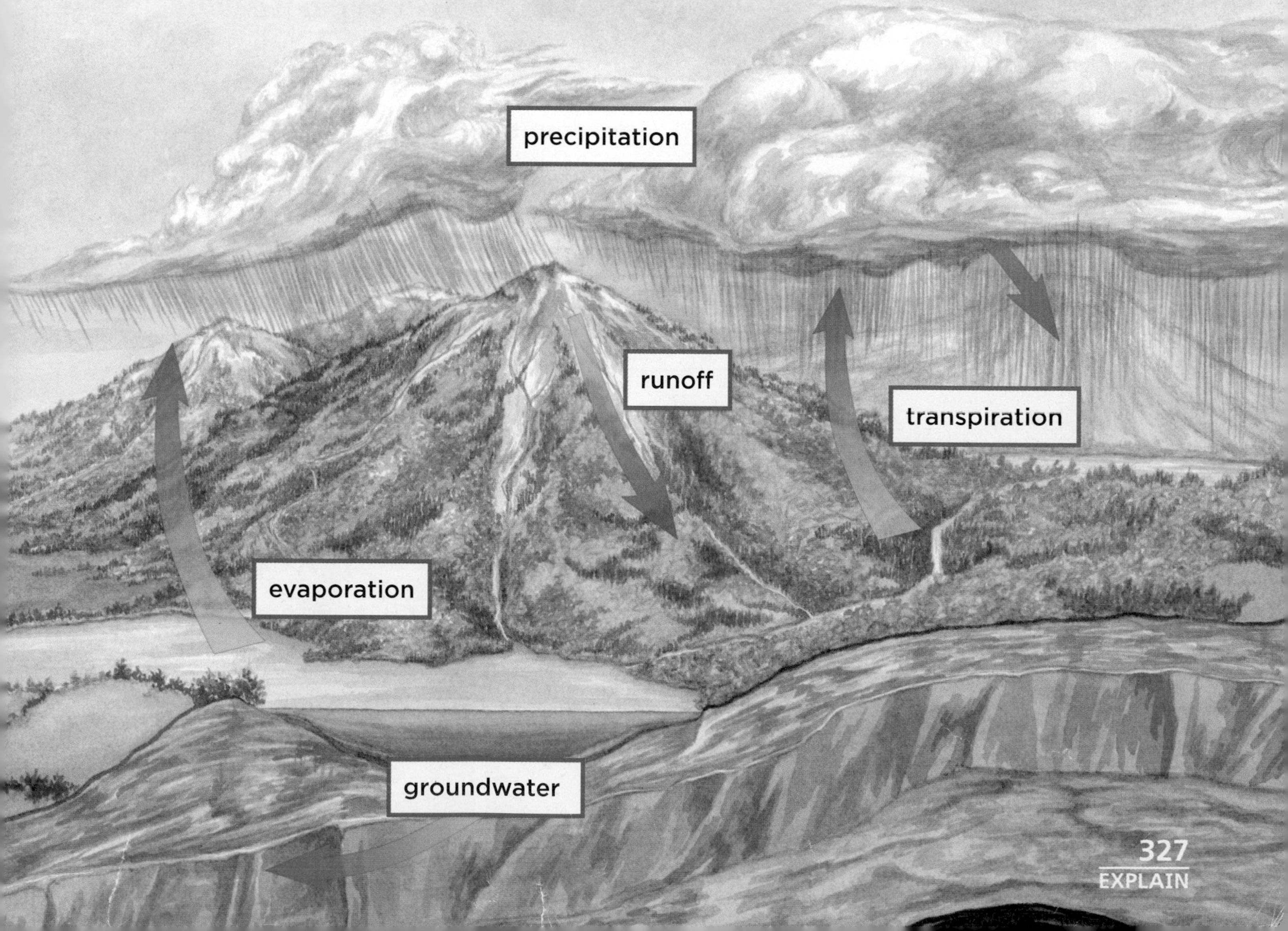

precipitation

runoff

transpiration

evaporation

groundwater

Cloud in a Jar

1. Pour very warm water into a jar until it is about 1 cm deep. Seal the jar tightly. Then shake it several times.

2. Open the jar and quickly place a plastic sandwich bag inside it. Using a rubber band, seal the bag tightly around the mouth of the jar.

3. **Observe** Reach into the bag. Gently pull it up. Then release the bag. Observe and describe what happens in the jar. Repeat this step several times.

4. **Interpret Data** When does the cloud form? When does it disappear? Why do you think this happens?

What are some types of clouds?

Clouds form at different heights above Earth's surface. Scientists classify clouds into three main types based on how and where they form.

Cumulus

Cumulus (KYEW•myuh•lus) clouds are puffy, white clouds that look like cotton balls. They often have a flat bottom.

You have probably seen clouds grow dark before a rainstorm. If a cumulus cloud becomes dark and thick, it is called a *cumulonimbus* (kyew•myuh•loh•NIM•bus) cloud. This kind of cloud produces precipitation.

Cloud Types

cumulus

stratus

cirrus

Stratus

Stratus (STRA•tus) clouds form in layers. The layers look like sheets or blankets. Stratus clouds are often the lowest clouds in the sky. What we call fog is really a stratus cloud near Earth's surface. Like cumulonimbus clouds, stratus clouds can form precipitation.

Cirrus

Cirrus (SEER•us) clouds look thin, wispy, or feathery. They are made of tiny bits of ice. Cirrus clouds are usually found very high in the sky.

Observing Clouds

In the diagram at right, you can see other cloud types. Often, you can find more than one cloud type in the sky at one time.

 Quick Check

Sequence How might clouds change as a morning rain shower turns into a sunny day?

Critical Thinking Classify the types of clouds you see in the sky today.

Many Kinds of Clouds

cirrus

cirrocumulus

altocumulus

cumulonimbus

Read a Diagram

Which cloud types are related to one another?

Clue: Compare the word parts and pictures for the different cloud types.

What are other forms of precipitation?

Rain is just one form of precipitation. Water can change state as it moves through the air. When this happens, other types of precipitation may fall.

Snow

When water reaches a temperature below 0°C (32°F), it freezes into ice. Remember, to freeze is to change from a liquid to a solid. Bits of ice can collect in a cloud. If they get too heavy, they fall as snow.

Snow may melt as it falls to the ground. To **melt** is to change from a solid to a liquid. Melting happens when sunshine or warm air heats the snowflakes. The heat makes the snow change to rain.

Sleet and Hail

Sometimes rain falls from clouds as a liquid but freezes along the way. The rain turns into small chunks of ice. The ice that falls to the ground is called *sleet*.

Hail is made of ice too. The ice chunks are much larger than sleet. Hail forms inside the tall clouds of a thunderstorm. Most hailstones are the size of peas. However, some are bigger than baseballs!

 Quick Check

Sequence How does snow form?

Critical Thinking Do all pieces of ice that fall to the ground come from icy clouds? Explain.

Most hailstones are small. Large ones can be dangerous! How big is the hailstone to the left?

FACT ▶ Hail can fall in spring and summer.

Lesson Review

Visual Summary

 Water changes from a liquid to a gas through **evaporation.** It changes from a gas to a liquid through **condensation.**

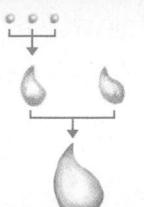 In the **water cycle,** water travels by runoff, evaporation, condensation, and **precipitation.**

 **Clouds** form at different heights above Earth's surface. They are classified by how and where they form.

Make a FOLDABLES® Study Guide

Make a layered-look book. Use it to summarize what you read about the water cycle.

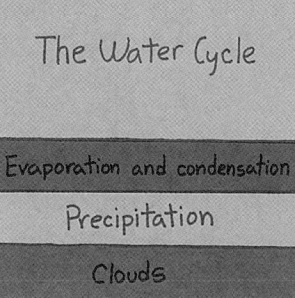

The Water Cycle

Evaporation and condensation

Precipitation

Clouds

Think, Talk, and Write

1. **Vocabulary** Water vapor becomes liquid water through _____.

2. **Sequence** Describe the path of water from the ocean to a raindrop.

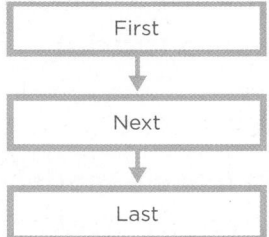

First

↓

Next

↓

Last

3. **Critical Thinking** How are hail and sleet alike? How are they different?

4. **Test Prep** Clouds form when water vapor
   A evaporates.
   B condenses.
   C precipitates.
   D transpires.

5. **Test Prep** Puffy, white clouds with flat bottoms are
   A cumulus clouds.
   B cirrus clouds.
   C stratus clouds.
   D cirrostratus clouds.

6. **Essential Question** How is water recycled?

Writing Link

Write a Cloud Poem
Write a poem about clouds. Choose ones you have seen or ones you would like to see. Include several different cloud types in your poem.

Art Link

Water Cycle Diorama
Make a diorama that shows how the water cycle works. Label the places where water goes. Write captions to describe how water changes state.

Focus on Skills

Inquiry Skill: Make a Model

You have seen water collect in puddles during a heavy rainstorm. You have learned that evaporation causes puddles to dry up. Does the size of a puddle affect how fast it evaporates? To answer this question and still stay dry, you can **make a model.**

▶ Learn It

When you **make a model,** you build something to represent an object or event. A model helps you learn more about the real object or event you are investigating. It is important to record your observations about your model. Then you can make inferences about the real thing.

▶ Try It

Make a model to study how the size of a puddle affects evaporation.

> **Materials** whole kitchen sponge, half kitchen sponge, two-pan balance, paper clips, water, measuring cup, lamp

1. Place the whole sponge in one balance pan and the half sponge in the other. The sponges represent puddles.

2. Add paper clips to the pan with the half sponge until both sides of the balance are equal in mass.

3. Add equal amounts of water to each sponge.

4. Place the lamp so it will shine on both puddles. Turn on the lamp. This models the Sun.

5. Observe the sponges after 5 minutes. Read the measurement on the balance. Record your observations in a data table like the one shown.

6 Continue to read the balance every 5 minutes for 15 more minutes. Record your observations.

7 Look at your results. Which sponge became lighter first? Why do you think it did?

8 How are your model puddles similar to real rain puddles? How are they different?

My Observations

	Whole Sponge	Half Sponge
After 5 minutes	_____	_____
After 10 minutes	_____	_____
After 15 minutes	_____	_____
After 20 minutes	_____	_____

▶ **Apply It**

Now **make a model** to test the effect of wind on evaporation. Use two rectangular plastic containers.

1 Pour the same amount of water into each container. Place a fan so it will blow across the surface of only one container. Turn the fan on. Use a low setting.

2 Wait 10–15 minutes. Then measure the amount of water in each container.

3 How much water evaporated from each container? What does this tell you about wind and evaporation?

Tracking the Weather

Look and Wonder

Suppose you have tickets for an outdoor event. The event will be held tomorrow. Should you bring an umbrella? How can you predict the rain?

How do raindrops form?

Form a Hypothesis

How do changes in air temperature affect water in the liquid and gas states? Write a hypothesis.

Test Your Hypothesis

1 Pour just enough water into each jar to cover the bottom of the jars.

2 **Use Variables** Place one lid upside down on one jar. Put three or four ice cubes in that lid. Place the other lid upside down on the second jar. Do not add ice cubes to that lid.

3 **Observe** Wait two minutes. Then look closely at the parts of the lids inside the jars. Record your observations every two minutes over the next ten minutes.

4 Draw a diagram that shows what happened to the water inside the jars. Add labels and arrows to explain how the water changed.

Draw Conclusions

5 Why did water droplets form mostly underneath the lid? Why didn't they form inside the jar or on the upside-down lid?

6 **Predict** What if you shined a heat lamp on the water in the jars before step 3? Predict how your results would change.

Materials

- 2 jars with lids
- water
- ice cubes

Step 2

Explore More

What would happen if you used ice instead of water in step 1? Make a prediction. Then repeat the activity with the ice. Explain your results.

▶ **Essential Question**

How do fronts and air masses change the weather?

▶ **Vocabulary**

air mass, p. 336

front, p. 337

warm front, p. 337

cold front, p. 337

stationary front, p. 337

forecast, p. 339

▶ **Reading Skill** ✔
Predict

My Prediction	What Happens

▶ **Technology** LOG ON ℮

e-Glossary and e-Review online at www.macmillanmh.com

What are air masses and fronts?

The wettest place on Earth is in the state of Hawaii. Rain falls over one of the islands about 350 days of the year. One of the driest places in the world is a desert in South America. Why is it rainy in some places and dry in others?

Air Masses

The properties of the air in different places on Earth vary. Some large regions of air have nearly the same properties throughout. These regions are called **air masses**. Weather in one part of an air mass is like the weather throughout the rest of the air mass.

Air masses form all the time, usually near the poles or the equator. They move across Earth, covering it like an ever-changing blanket. The map shows some of the common paths they take.

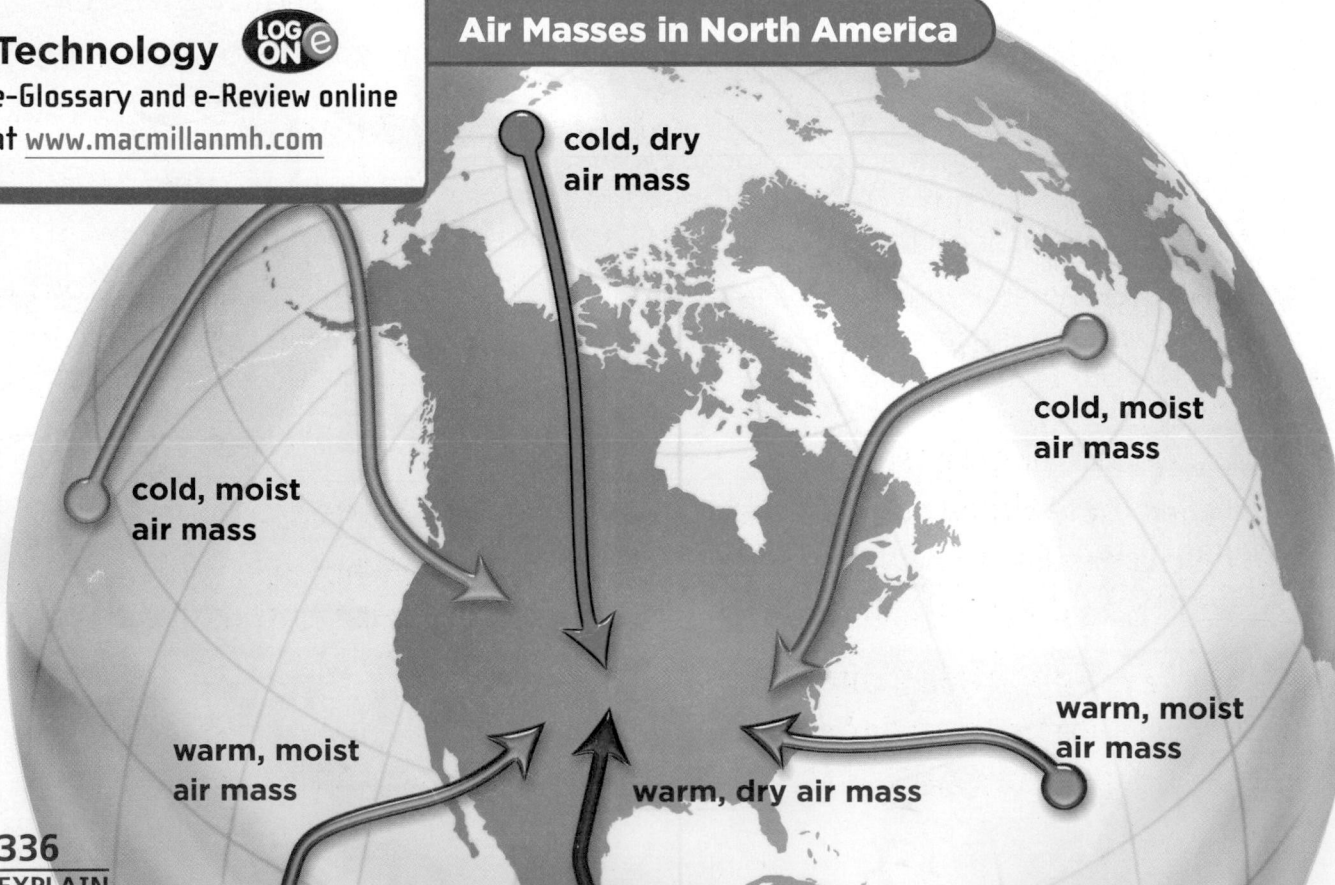

Air Masses in North America

cold, dry air mass

cold, moist air mass

cold, moist air mass

warm, moist air mass

warm, dry air mass

warm, moist air mass

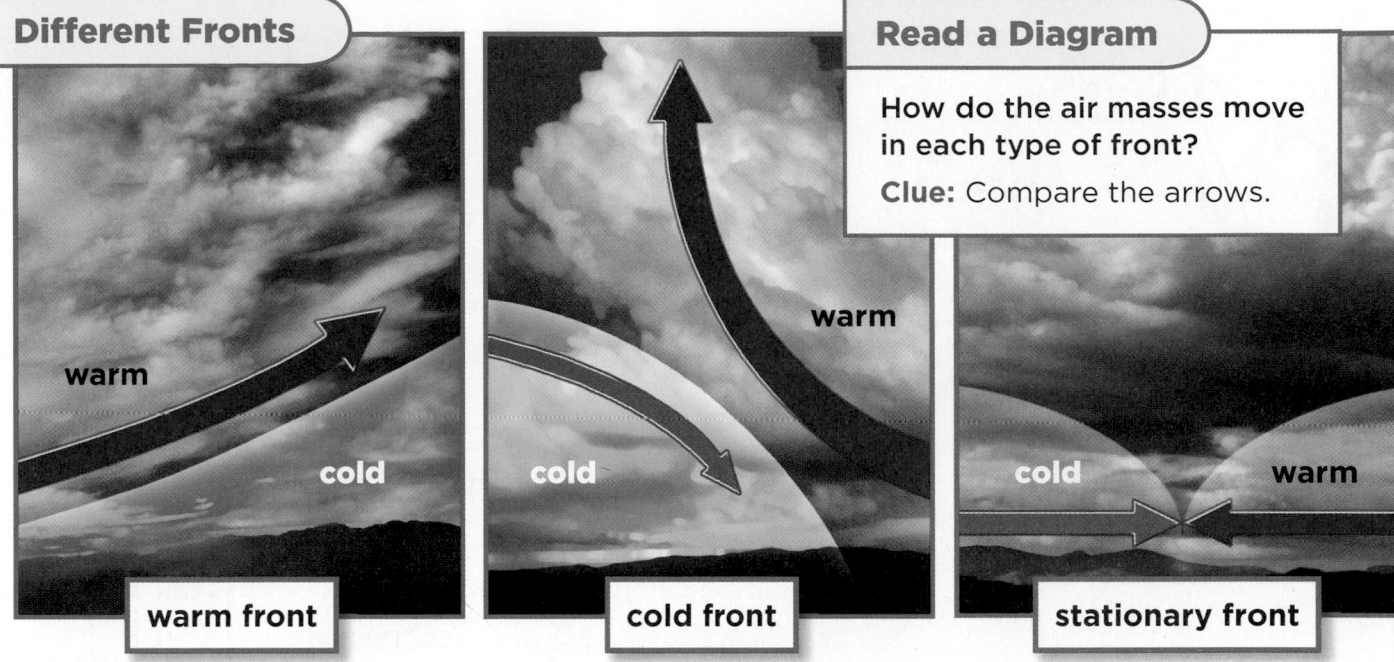

warm

cold

warm front

cold

warm

cold front

cold   warm

stationary front

Fronts

As an air mass moves, it brings weather with it. What happens when different air masses meet? Like two cars in a crash, the air masses slam into each other. The area where they meet is called a front.

A **front** is the boundary between two air masses that have different temperatures. Fronts usually cause a change in the weather.

Warm Fronts

When a warm air mass pushes into a cold air mass, a **warm front** forms. As the diagram shows, the warm air mass slides up and over the cold air mass. Layers of clouds form. The cold air retreats.

A warm front often brings light, steady rain. After the front passes, the air temperature rises.

Cold Fronts

A **cold front** forms when a cold air mass pushes under a warm air mass. The cold air mass forces the warm air mass upward quickly. Thick clouds form as the warm air rises and cools. Cold fronts often bring stormy weather.

Stationary Fronts

Sometimes rainy weather lasts for days. This can be caused by a stationary front. A **stationary front** is a boundary between air masses that are not moving.

 **Quick Check**

Predict What will happen if a cold air mass pushes into a warm air mass?

Critical Thinking How do warm fronts differ from cold fronts?

Seattle
L
Billings
Minneapolis
Detroit
Boston
Salt Lake City
Omaha
Chicago
Columbus
New York
San Francisco
L
Denver
H
Kansas City
Washington D.C.
Las Vegas
St. Louis
Raleigh
Los Angeles
Oklahoma City
Little Rock
Nashville
Atlanta
Santa Fe
New Orleans
H
Dallas
El Paso
Houston
Miami

TODAY'S HIGH TEMPERATURES (F°)

	70s
	60s
	50s
	40s
	30s
	20s
	10s

FRONTS			PRESSURE		PRECIPITATION			
Cold	Warm	Stationary	High	Low	T-storms	Rain	Snow	Sunny
			H	L				

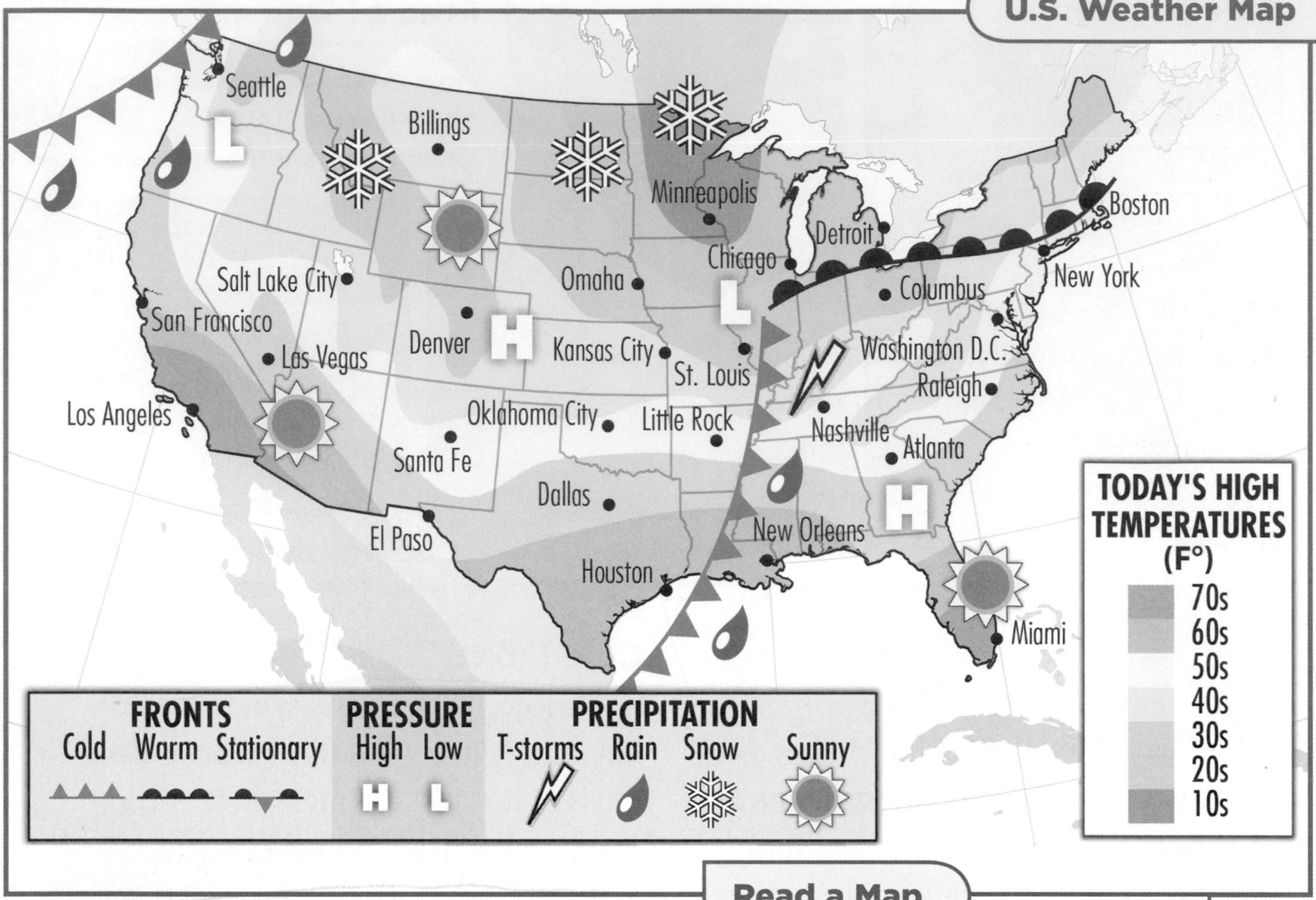

Read a Map

What does a weather map show?

Every day, scientists make and share weather maps like the one above. Weather maps show weather conditions at a certain time and place. They tell about air temperature, pressure, precipitation, and winds.

Weather maps might also show the locations of fronts. The fronts appear as a line of triangles or half circles. In the map above, rain and thunderstorms have formed along the two cold fronts.

What does this map show about the weather in Nashville?

Clue: Use the legends to find the meanings of the colors and symbols.

338
EXPLAIN

Forecasting

Maps can help us answer questions. Scientists use weather maps to make forecasts. To **forecast** is to predict weather conditions.

Temperature, air pressure, and the direction of moving fronts give important clues for forecasts. Look at the map again. Do you see the cold front from St. Louis to Houston? The triangles point toward the east. Like most fronts in the United States, this one is moving from west to east. A forecast based on this map might predict a chance of rainy weather for New Orleans.

Scientists use many technologies in forecasting. Satellites in orbit around Earth take pictures of the atmosphere. Computers help scientists analyze weather data and produce better weather maps.

Quick Lab

Weather Forecast

1. Study a weather map from today's newspaper. Compare it to maps from yesterday and the day before, if they are available.

2. **Communicate** Describe today's weather in your region and in surrounding regions.

3. **Predict** Use the weather map to predict tomorrow's weather. Explain your prediction.

4. Study the weather map tomorrow. Compare it to your prediction. How close was your forecast to the actual weather?

BASYANG" HAS FURTHER INTENSIFIED AS IT MOVES TOWARDS ISABELA-AURORA AREA.
SASA MTSAT IR1 23:32:00 12/07/2010
8 AM UPDATE   13 JULY 2010   TUESDAY

✔ Quick Check

Predict How can weather maps be used to predict weather?

Critical Thinking How likely are you to see the same cold front for several days in one place? Why?

Strong winds and lightning can make a storm dangerous.

What are the signs of severe weather?

Have you ever heard a loud clap of thunder just before a storm? Thunder is the booming sound made when lightning quickly heats the air around it. Thunder tells you that a storm is nearby.

If you see a tall, swirling mass of air shaped like a funnel, take cover! It could be a tornado. A *tornado* is a rotating column of air that touches the ground during a thunderstorm. Tornadoes can reach speeds of 400 km per hour (250 mph) or more.

A *hurricane* is a very wide storm. A typical one spans about 480 km (300 mi) across. Hurricanes form over warm water in the ocean. They bring very heavy rains and strong winds. If a hurricane moves across land, it can cause severe damage.

Storm Safety

Scientists pay close attention to signs that severe storms are forming. If one appears in their forecast, they alert the government and the public.

Do you know how to stay safe in severe weather? If thunderstorms are predicted, stay away from water and trees. When tornadoes are predicted, head for a sturdy shelter, such as a basement. To avoid a hurricane, you might need to move inland.

In any storm, always listen for directions. Seek out a trusted adult if a severe storm strikes. Be sure to follow warnings on the radio and television.

 Quick Check

Predict What might happen if a hurricane strikes the land?

Critical Thinking Why should you stay inside during a storm?

Lesson Review

Visual Summary

When two **air masses** meet, a **front** forms between them. Fronts usually bring a change in the weather.

Scientists use weather maps to make **forecasts** about the weather to come.

It is important to know about **severe storms** so you can stay safe.

Make a FOLDABLES Study Guide

Make a three-tab book. Use it to summarize what you read about tracking the weather.

Think, Talk, and Write

① Vocabulary To _____ is to predict the weather.

② Predict Study today's weather map. Forecast the weather for tomorrow.

My Prediction	What Happens

③ Critical Thinking How can a battery-powered radio help you stay safe during a storm?

④ Test Prep A storm usually forms

  A inside an air mass.

  B along a front.

  C over tall buildings.

  D over a river.

⑤ Test Prep Which term describes a very tall, gray funnel shape?

  A hurricane

  B tornado

  C cold front

  D thunderhead

⑥ Essential Question How do fronts and air masses change the weather?

Writing Link

Write a Short Essay

Write an essay about a storm that you experienced. Or ask an adult to tell you about a storm he or she remembers. Include facts about the storm and how it affected people.

Social Studies Link

History Report

Research and write a report on a severe weather event in history. If possible, write about an event in your area. Include information about how people solved problems caused by the event.

Hurricane Season

June is the beginning of a busy time for the National Hurricane Center in Miami, Florida. That's when hurricane season begins, and the scientists at the center are ready for action.

Hurricanes develop at sea under particular conditions. These include warm ocean water, low pressure, moist air, and light winds. They usually happen in the Atlantic and northeast Pacific Oceans from June through November. When a hurricane forms, it can bring violent winds, large waves, floods, and lots of damage.

To study a hurricane, scientists gather large amounts of data. Satellites that orbit Earth collect information about cloud patterns. They record temperatures on top of clouds and at the sea surface. Satellites also measure the direction and speed of winds above the ocean. This information helps scientists track the size, path, and intensity of a storm.

Connect to

AMERICAN
MUSEUM ŏ
NATURAL
HISTORY

at **www.macmillanmh.com**

Doppler radar is another tool that hurricane scientists use. It sends out radio waves from an antenna. Objects in the air, like raindrops, reflect the waves back to the antenna. Doppler radar can measure the direction and speed of a moving object, like a hurricane moving toward land.

Buoys spread across the ocean measure conditions like surface wind, waves, temperature, and fog. Planes fly to the center of a hurricane to gather data about wind, pressure, temperature, and humidity.

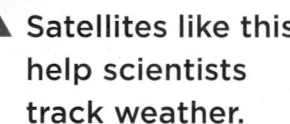

▲ Satellites like this help scientists track weather.

Scientists enter all of this data into supercomputers to create a model of the hurricane. This model helps them predict the wind speed, size, and direction of the hurricane, and where and when it might hit land. Accurate predictions of a hurricane's path can reduce the loss of life and property.

Fact and Opinion

▶ Facts tell you about something that has really happened.

▶ Opinions are what someone thinks about facts or events.

 **Write About It**

Fact and Opinion

1. What technologies help scientists study hurricanes?

2. What do you think would happen during a hurricane in your neighborhood?

 **-Journal** Research and write about it online at **www.macmillanmh.com**

Climate

Look and Wonder

It is a cool, clear day in October. The leaves have changed color to gold, orange, and red. Somewhere else on Earth, the leaves are green. There, flowers bloom under the Sun's warmth. How can the same time of year be different from place to place?

What affects weather patterns?

Purpose

Explore the factors that determine the weather patterns in different places.

Procedure

① Locate the cities of Chicago, Miami, Phoenix, and Seattle on a map.

② **Predict** The data table shows the yearly temperature and precipitation for these four cities. Predict where each belongs in the table.

③ **Classify** Copy the table. Research the weather patterns of the four cities. Fill in the cities where they belong.

④ Find out the yearly temperature and precipitation for the place where you live. Add this data to your table.

Draw Conclusions

⑤ Compare the table to your prediction in step 2. How does it compare?

⑥ **Interpret Data** Which cities are near oceans? How does their data compare to the other cities? Which cities are farthest south? How do they compare to the northern cities?

City	Yearly Temperatures	Yearly Precipitation
1	Hot summers mild winters	Very little rain
2	Hot summers warm winters	A lot of rain
3	Hot or warm summers cold winters	Much rain and snow
4	Warm summers mild winters	A lot of rain
5 My Community		

Explore More

Look at today's weather map. Compare the weather in each of the four cities with your data table. Is today's weather similar to or different from yearly patterns? Can you explain any differences?

Read and Learn

▶ **Essential Question**

Why do weather patterns change?

▶ **Vocabulary**

climate, p. 346

current, p. 348

▶ **Reading Skill** ✔

Fact and Opinion

Fact	Opinion

▶ **Technology** 🔵 LOG ON

e-Glossary and e-Review online at www.macmillanmh.com

What is climate?

The weather where you live might change from day to day. Yet you can predict what the weather will be like each season. The pattern of seasonal weather that happens year after year is called **climate** (KLI•mut).

Climate is not the same everywhere on Earth. The city of Phoenix is in the southwestern United States. The climate there is warm and dry all year. Snow and rain rarely fall. Seattle is in the northwestern United States. There the climate is cool and wet.

Farmers depend on climate to grow their crops. Some crops grow well in cool climates with steady rain. Other crops need dry climates. Still others need warm, humid climates.

Canada

temperate

Antarctica

polar

Arizona

dry

Climate Regions

Think of climate as the average weather in a certain place for a long period of time. It has similar patterns of temperature, humidity, precipitation, and wind. We can call such an area a *climate region*.

Polar regions have cold climates with low precipitation. Tropical regions are near the equator. There, the climate is warm, humid, and rainy. *Temperate* regions lie between polar and tropical regions. Temperate climates often have four seasons. Some have just two seasons—a dry one and a rainy one. Still other regions are dry or cool.

✔ *Quick Check*

Fact and Opinion *Cool climate regions are best.* Is this statement a fact or an opinion? Explain.

Critical Thinking Describe the climate of your region.

Ecuador

tropical

Alaska

cool

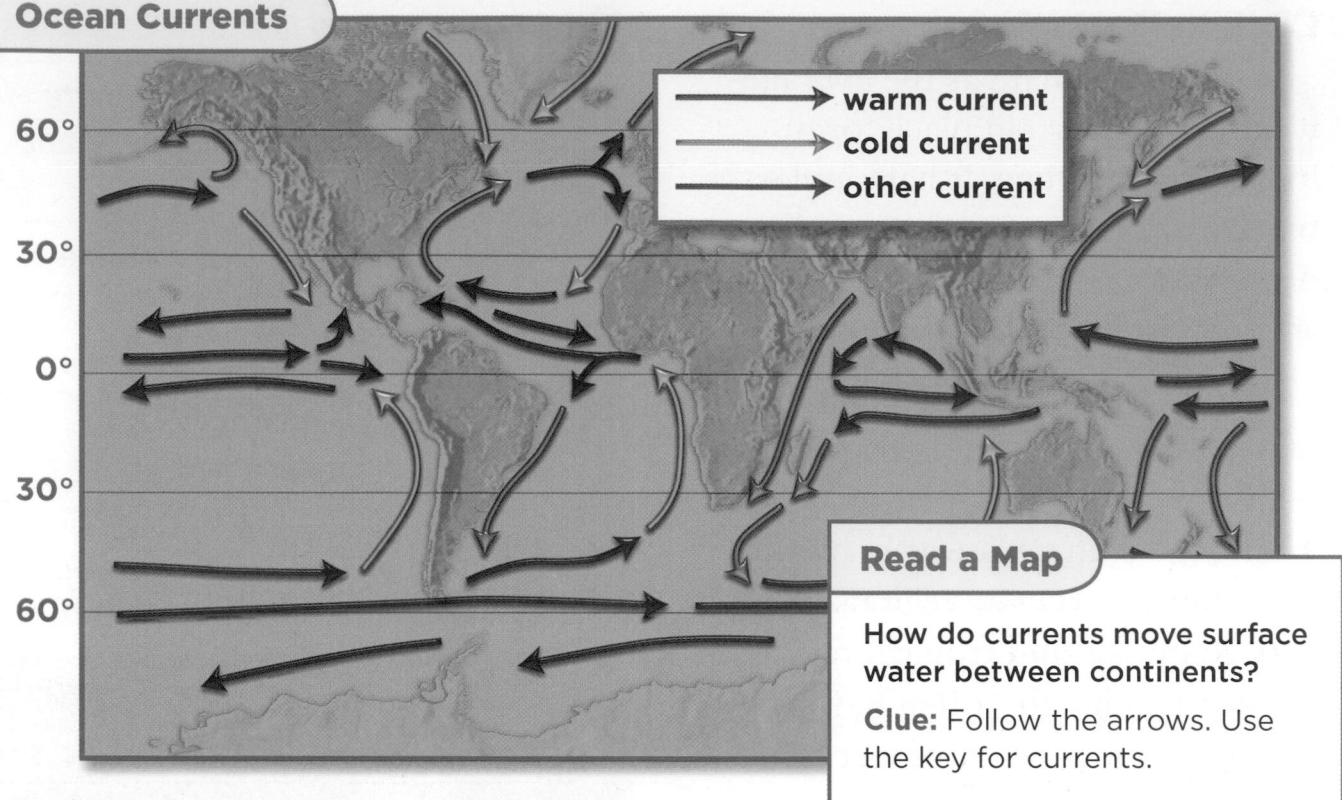

60°
30°
0°
30°
60°

→ **warm current**
→ **cold current**
→ **other current**

Read a Map

How do currents move surface water between continents?

Clue: Follow the arrows. Use the key for currents.

What determines climate?

Several things affect a region's climate over time. These include latitude, winds, and ocean currents.

Latitude

The thin lines that run east and west across some maps are lines of latitude. *Latitude* is a measure of how far a place is from the equator. The equator's latitude is set at zero degrees. Latitude increases as you move north or south from there. The highest latitude is at the North and South Poles. Both are 90 degrees.

Climates near the equator are warm and rainy. Between the equator and the poles, the climate is mild or temperate. Near the poles, the climate is cold all year.

Global Winds

Temperature differences between latitudes cause *global winds*. These are winds that move air between the equator and poles. Warm air near the equator rises and moves toward the poles. Cold air near the poles sinks and moves toward the equator.

Ocean Currents

A **current** is a directed flow of a gas or a liquid. Some ocean currents move warm water from the equator to the poles. Others move cold water from the poles toward the equator. There are also currents that move along lines of latitude. Together, these currents form circular patterns in the oceans.

Distance from Water

Do you like to swim at the beach in summer? You might have noticed that the water stays cool even on the hottest days. That is because water heats up more slowly than land does. Water cools more slowly too.

Remember that more than 70 percent of Earth's surface is covered by water. Land and water heat and cool at different rates. These differences affect the air temperature and precipitation nearby.

Climates near lakes and oceans are cloudier and rainier than regions farther inland. Summers are cooler. Winters are warmer. Nearness to water reduces temperature extremes. It also increases moisture in the air.

Indiana is an inland state. Winters there are cold and snowy.

Quick Lab

Climate in Two Cities

1. Study the data table. It shows climate information for Seattle, WA, and Fargo, ND. Locate these two cities on a map.

2. **Communicate** Describe the climates of the two cities. How do the climates compare?

3. **Infer** What factor best explains the differences between the two climates? Why do you think so?

Month	Property	Seattle	Fargo
July	high temperature	75°F	83°F
July	precipitation	19 mm	69 mm
December	high temperature	45°F	20°F
December	precipitation	150 mm	17 mm

Quick Check

Fact and Opinion *The equator has a warm climate.* Is this statement a fact or an opinion? Explain.

Critical Thinking How might sailors in the past have studied ocean currents and global winds?

An air mass loses moisture as it moves over a mountain.

How do mountains affect climate?

Latitude, water, and winds are not the only factors that affect climate. Mountains also have an effect.

Altitude

Climate at the base of a mountain is always warmer than at its peak. The higher the altitude, the lower the air temperature. *Altitude* is a measure of the height of a place above sea level.

What happens when an air mass meets a mountain? The air rises up the side of the mountain. As the altitude gets higher, the temperature gets cooler. Water vapor in the air condenses into clouds.

Clouds and Precipitation

As a cloud moves up a mountain, its water droplets get heavy. Precipitation falls. By the time the air mass passes over the mountain, the air is dry. For this reason, the climate on one side of a mountain tends to be wet. The climate on the other side is often dry.

 Quick Check

Fact and Opinion State one fact and one opinion about mountains and climate.

Critical Thinking How can a mountain "dry out" the air?

Lesson Review

Visual Summary

Climate regions have regular patterns of air temperature, humidity, precipitation, and wind.

Factors that affect climate are latitude, global winds, ocean currents, and distance from oceans and lakes.

Altitude affects **mountain climates.** The air temperature gets lower as you move up a mountain.

Make a **FOLDABLES®** Study Guide

Make a trifold chart. Use it to summarize what you read about climate.

Think, Talk, and Write

1. **Vocabulary** Ocean _____ move heat from one place to another.

2. **Fact and Opinion** Choose a climate. Why would you enjoy living in this climate? Why would you not enjoy this climate? Include facts from this lesson.

Fact	Opinion

3. **Critical Thinking** How is climate different from weather?

4. **Test Prep** Latitude is a measure of distance from

   A an air mass.

   B an ocean current.

   C a mountain.

   D the equator.

5. **Test Prep** Where is altitude highest?

   A on a mountaintop

   B at the base of a mountain

   C at sea level

   D in a valley

6. **Essential Question** Why do weather patterns change?

 **Math Link**

Find the Average Temperature
For five years, a weather station recorded high temperatures of 86°F, 89°F, 90°F, 92°F, and 88°F for the same date. What was the average for that date over the five years?

 **Social Studies Link**

Learn about Climate
Choose another country or region. Research and report on its climate. Show how climate affects the people who live there. Find out about the crops they grow.

Materials

paper

scissors

string

heat source

How does warmed air affect the weather?

Form a Hypothesis

Large masses of warm air can affect the climate of a region. You can model how warm air moves. What do you think will happen if you hold a spiral of paper over a heat source? Write your answer in the form "If the air warms, then the paper spiral will…"

Test Your Hypothesis

1. ⚠️ **Be Careful.** Cut a circle of paper to form a spiral.

2. Tie a piece of string to one end of the paper.

3. Have your teacher turn on a heat source, such as a lamp. Carefully hold or hang the spiral about 15 centimeters above the heat source.

Step 1

4. **Observe** Describe what the spiral does.

5. While holding the spiral above the heat source, turn the heat off. Describe what happens to the spiral.

Step 2

Draw Conclusions

6 Was your hypothesis correct? How did the paper spiral move when it was heated?

7 **Communicate** What happened to the paper spiral when you turned the heat off? How can you explain this?

8 **Infer** What happens to air over ground that is warmed throughout the day?

Which type of land changes temperature fastest?

Form a Hypothesis

Air is warmed by heat released from the land or water. Of soil, sand, or rock, which type of land holds heat longest? Write your answer in the form of a hypothesis.

Test Your Hypothesis

Design an investigation to find out which type of land holds heat longest. Write out the materials you will need and the steps you will follow. Record your results and observations.

Draw Conclusions

Did your results support your hypothesis? Why or why not?

What else would you like to learn about air, heat, and climate? Design an investigation to answer your question. Your investigation must be written so that another group can repeat the investigation by following your instructions.

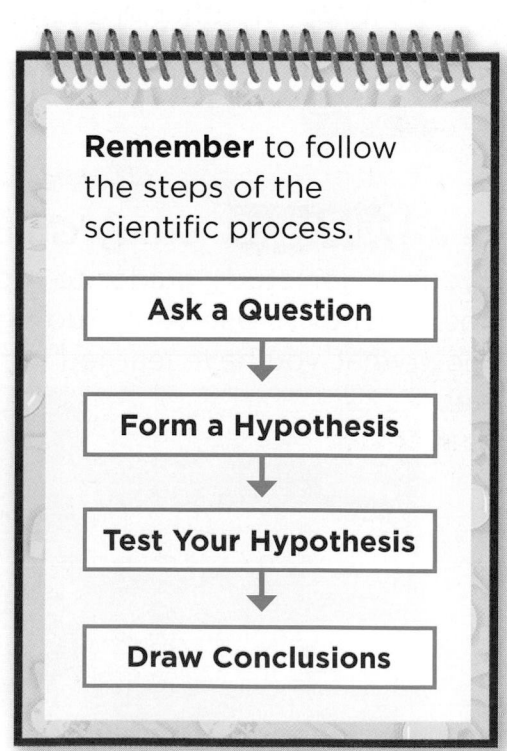

Remember to follow the steps of the scientific process.

Ask a Question

↓

Form a Hypothesis

↓

Test Your Hypothesis

↓

Draw Conclusions

Visual Summary

Lesson 1 Scientists measure the properties of Earth's atmosphere to describe weather.

Lesson 2 Water changes state as it moves between Earth's surface and atmosphere.

Lesson 3 We can predict the weather by observing air masses and fronts.

Lesson 4 Climate is the pattern of seasonal weather in a region. Latitude and other factors affect climate.

Make a FOLDABLES Study Guide

Tape your lesson study guides to a piece of paper as shown. Use your study guide to review what you have learned in this chapter.

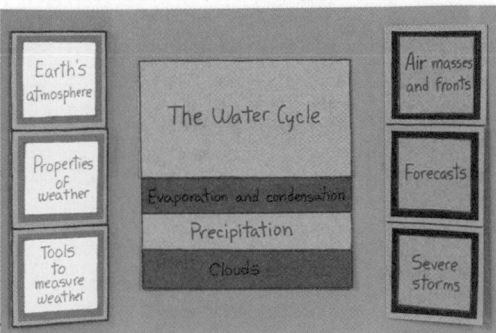

Fill each blank with the best term from the list.

air mass, p. 336   evaporation, p. 324

atmosphere, p. 314   forecast, p. 339

climate, p. 346   front, p. 337

condensation, p. 325   humidity, p. 316

current, p. 348   precipitation, p. 325

1. During the process of _____, a liquid changes slowly into a gas.

2. A large region of air with nearly the same temperature and water vapor throughout is called a(n) _____.

3. The blanket of air surrounding Earth is called the _____.

4. The pattern of seasonal weather in a region over many years is called _____.

5. A measurement of the amount of water vapor in the air is _____.

6. A directed flow of a gas or liquid is called a(n) _____.

7. A boundary between two air masses that have different temperatures is called a(n) _____.

8. A gas changes to a liquid during _____.

9. Water that falls from clouds to Earth is called _____.

10. A(n) _____ is a prediction of weather conditions.

LOG ON e-Glossary Words and definitions online at www.macmillanmh.com

Answer each of the following.

11. Summarize Describe the different kinds of fronts.

12. Make a Model Construct a simple rain gauge. On an index card, write a short explanation of how it works.

13. Critical Thinking A mountain climber goes up a tall peak. At what point in the climb would you expect the air pressure to be the strongest?

14. Expository Writing Write a paragraph describing the impact of oceans on climate.

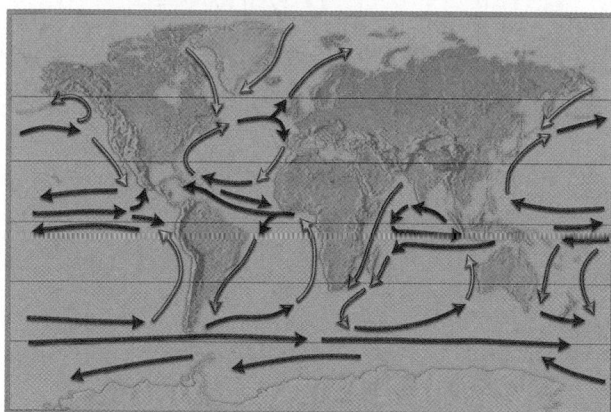

15. Sequence What happens to water in a lake during the changing seasons?

16. Infer What type of front would you infer from high cirrus clouds?

17. Critical Thinking Why can you see your breath outside on a cold winter day but not on a warm summer day?

18. True or False *A barometer measures wind speed.* Is this statement true or false? Explain.

19. Look at the picture below. What does this tool measure?

A precipitation **C** wind speed

B wind direction **D** air pressure

20. What are weather and climate?

Weather Words

1. Observe the weather at three different points during the day—morning, afternoon, and evening. Write a description of what you observe at each time of day.

2. Look at a weather report for the same time period. Create a chart comparing your observations to those of weather forecasters.

Analyze Your Results

Write a paragraph describing your results. How well did your weather observations compare to reports by weather forecasters? How would you explain any differences?

1 Which tool can be used to measure air temperature?

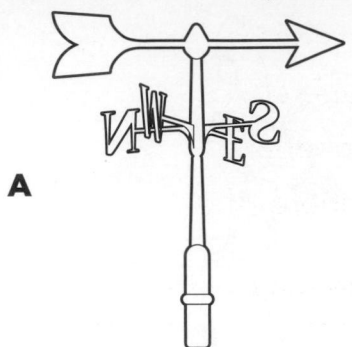

A

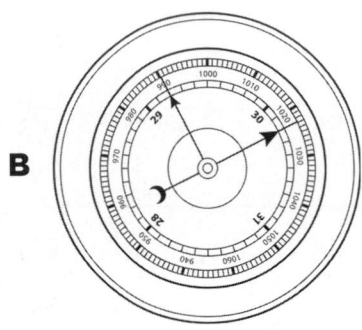

B

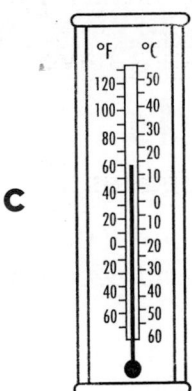

C

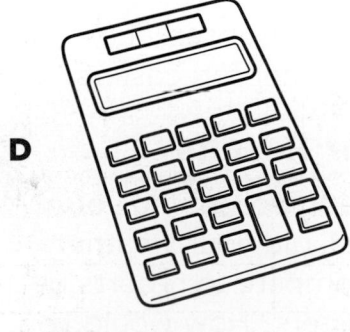

D

DOK I

2 Look at the picture below.

Which type of weather will you most likely find when these clouds are in the sky?

A stormy   **C** wet

B fair   **D** snowy

DOK 2

3 Why does precipitation occur at cold fronts?

A Warm air cools and water vapor condenses.

B Warm air cools and water evaporates.

C Cold air warms and water vapor forms.

D Cold air warms and water condenses.

DOK 2

4 If sleet falls, what can you infer?

A The air below the cloud is freezing.

B A thunderstorm is occurring.

C The ice in a cloud became too heavy and fell.

D Air temperatures are above freezing.

DOK I

5 Which of these is the source of energy for the water cycle?

A
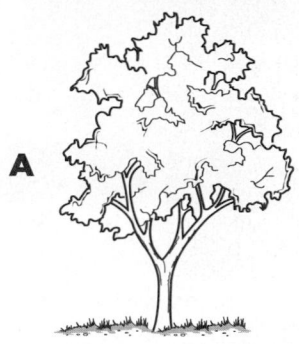

B

C

D

DOK I

6 What determines whether a storm is a hurricane?

A the amount of precipitation

B the wind direction

C the presence of lightning

D the location where it started

DOK I

7 In which layer of the atmosphere do organisms live?

A thermosphere

B mesosphere

C stratosphere

D troposphere

DOK I

8 Look at the map below.

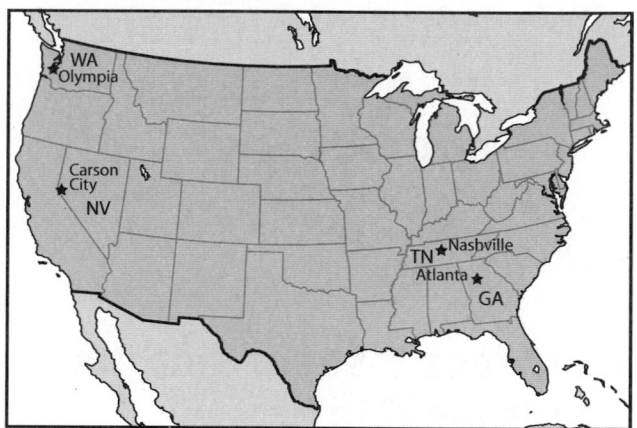

Which cities <u>most likely</u> have similar climates?

DOK I

Explain why their climates are similar.

DOK 2

Check Your Understanding

Question	Review	Question	Review
1	pp. 316–318	5	pp. 324–327
2	pp. 328–329	6	p. 340
3	pp. 336–337	7	pp. 314–315
4	pp. 324–325, 330	8	pp. 346–350

CHAPTER 8

The Solar System and Beyond

The Big Idea

What objects are in the solar system and beyond?

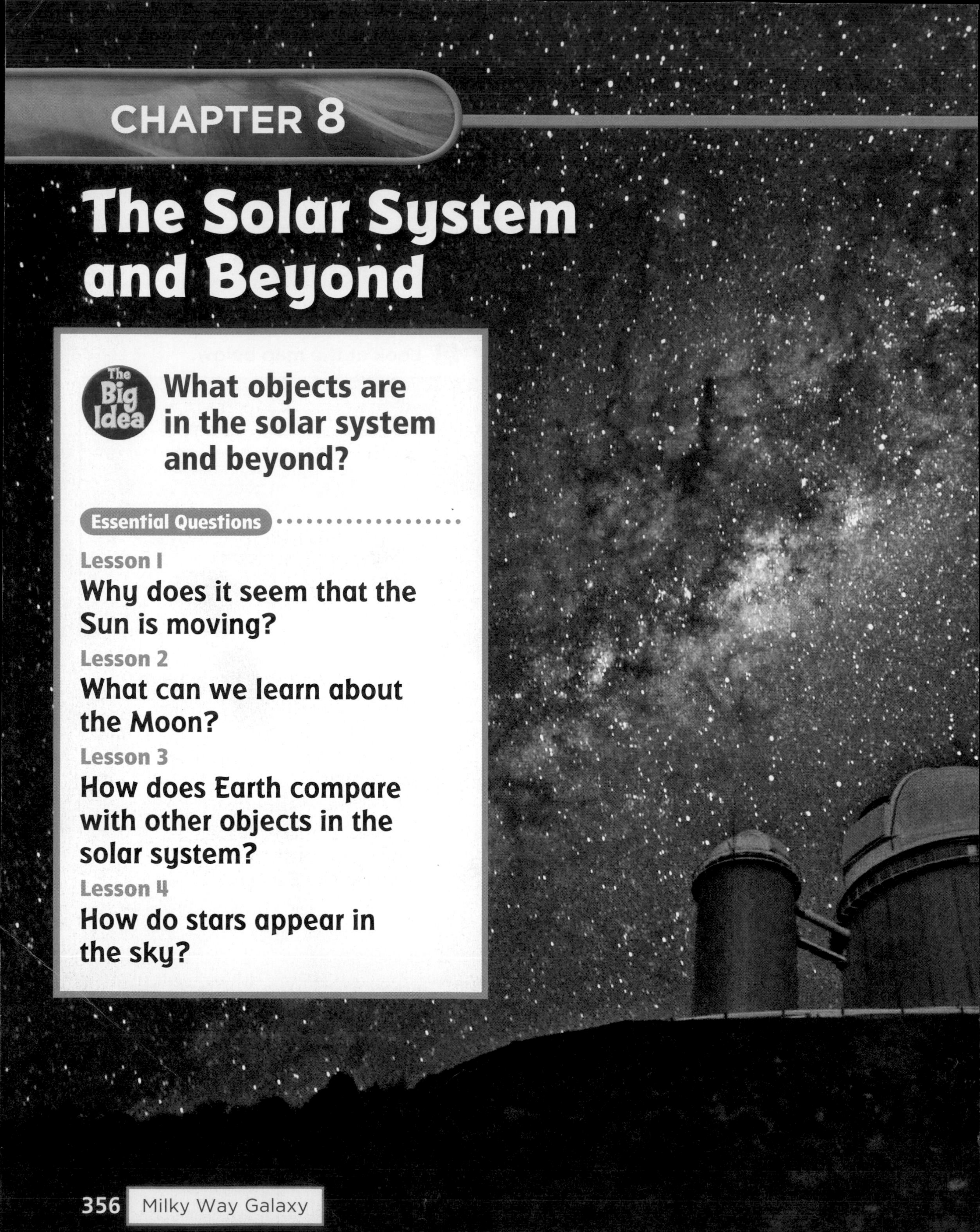

 **Big Idea Vocabulary**

 **rotation** the complete spin of an object around its axis (p. 360)

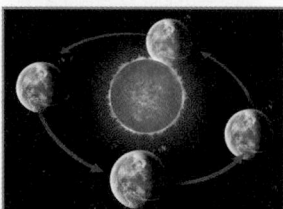 **revolution** one complete trip around an object in a circular or nearly circular path (p. 362)

 **phase** an apparent change in the Moon's shape (p. 373)

 **solar system** the Sun and all the objects that travel around it (p. 380)

 **comet** a chunk of ice, rock, and dust that moves around the Sun (p. 388)

 **constellation** a group of stars that appear to form a pattern in the night sky (p. 396)

 Visit www.macmillanmh.com for online resources.

Earth and Sun

Look and Wonder

Every day the Sun rises and sets. At dawn it appears in the east. You will find it in the west by sunset. Is the Sun really moving across the sky? Is Earth moving?

What causes day and night?

Purpose

Explore why Earth has both day and night.

Procedure

1. Write *I live here* on a self-stick note. Place the note over your home on the globe.

2. **Make a Model** Darken the room. Shine the flashlight on the self-stick note. The flashlight models the Sun.

3. **Observe** What part of the globe is lit? What part is dark? Record your observations.

4. **Form a Hypothesis** What do you think causes Earth's cycle of day and night? Write a hypothesis that you can test.

5. Make two plans to test your hypothesis. You can move the flashlight, the globe, or both. Carry out your plans to test your idea.

Draw Conclusions

6. **Communicate** Describe how you modeled day and night. How did your tests differ?

7. Do you think one of your models is correct? Which one? Why?

8. How much of Earth is lit during the day? How much is lit at night?

Explore More

The Sun rose at a certain time this morning. It will set at a certain time tonight. Does the Sun rise and set at the same time everywhere on Earth? Use your model to support your answer.

Materials

- self-stick notes
- globe
- flashlight

Step 2

Read and Learn

Essential Question
Why does it seem that the Sun is moving?

Vocabulary
rotation, p. 360
axis, p. 360
revolution, p. 362
orbit, p. 362

Reading Skill ✓
Cause and Effect

Cause → Effect
→
→
→
→

Technology LOG ON
e-Glossary, e-Review, and animations online at www.macmillanmh.com

What causes day and night?

How can it be afternoon where you live and nighttime in Asia? The answer is that you and Asia are on opposite sides of Earth. When your side of Earth faces the Sun, Asia is facing away from the Sun.

Earth Rotates

As Earth moves around the Sun, it also spins. **Rotation** (roh•TAY•shun) is the act of spinning. The diagram shows how Earth rotates.

The dotted line between the North Pole and the South Pole is Earth's axis (AK•sus). An **axis** is a real or imaginary line that an object spins around. Every day, Earth completes one rotation. One rotation takes 24 hours. We divide each hour into 60 minutes. Every minute has 60 seconds.

Earth's Rotation

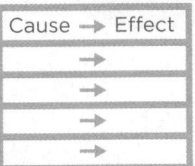

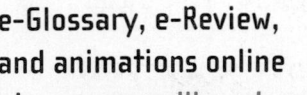

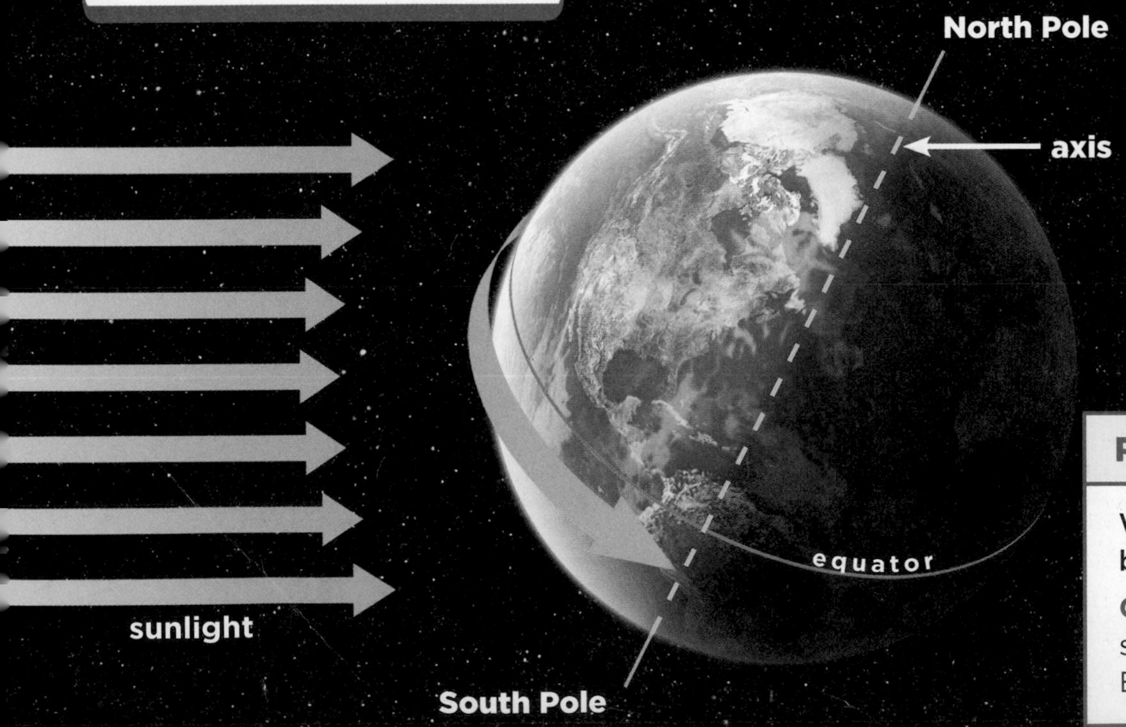

North Pole
axis
equator
sunlight
South Pole

Read a Diagram

Where on Earth will it be night next?

Clue: The green arrow shows the direction of Earth's rotation.

▲ When the Sun is high in the sky, this antelope has a shorter shadow.

▲ When the Sun is low in the sky, the antelope has a longer shadow.

Apparent Motion

As Earth rotates, you see different parts of space. During the day, the side of Earth where you live faces the Sun. As that part turns away from the Sun, it becomes night. The rotation of Earth changes day into night and night into day again.

Apparent motion is the way something appears, or seems, to move. The Sun appears to rise in the east. It seems to set in the west. Apparent motion is not real motion.

Earth's rotation causes the apparent motion of many objects in space. Stars only seem to move. The Moon and planets do not always move in the same direction as their apparent motion.

Shadows

Have you ever made a shadow puppet? A shadow forms when light is blocked. The light strikes an object but cannot pass through.

You cast a shadow when your body blocks the sunlight. Your shadow always points away from the Sun. As the position of the Sun changes, your shadow changes too. Early in the morning, your shadow is long. It shrinks until midday. Then it grows longer again until sunset.

✔ Quick Check

Cause and Effect What causes Earth's cycle of day and night?

Critical Thinking How might you use the Sun to estimate the time of day?

FACT The Sun is not always at its highest point in the sky at noon.

What causes seasons?

Not only does Earth rotate around its axis, it also revolves (rih•VAHLVZ) around the Sun. **Revolution** is when one object travels around another.

The path a revolving object takes is its **orbit**. Earth's orbit is shaped like an *ellipse* (ih•LIPS), or flattened circle. Earth's orbit around the Sun takes $365\frac{1}{4}$ days, or one year.

Earth's Tilted Axis

Earth's axis is not straight up and down. It is tilted at an angle of 23.5°. The tilt points in the same direction throughout Earth's orbit.

Earth's tilt causes sunlight to strike Earth at different angles. At any given time, each *hemisphere* (HE•muh•sfeer), or half, of Earth gets more or less sunlight. The seasons result from both Earth's tilted axis and its revolution around the Sun.

Earth's Revolution

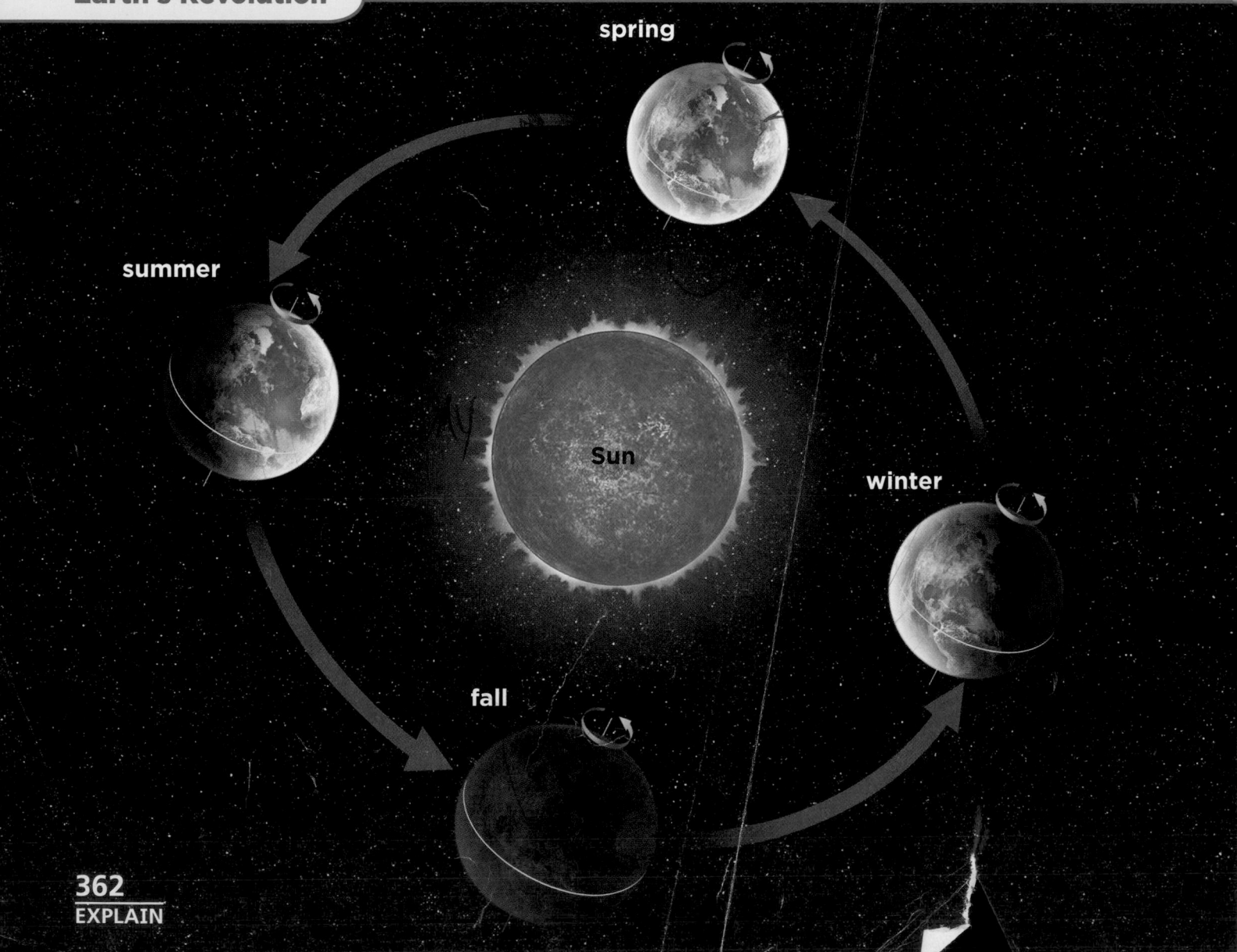

spring

summer

Sun

winter

fall

The Four Seasons

How does Earth's tilt cause summer, fall, spring, and winter? In June, the North Pole tilts toward the Sun. Sunlight hits the Northern Hemisphere at steep angles. The light is more intense. It is summer.

In December, the North Pole tilts away from the Sun. Sunlight strikes the Northern Hemisphere at low angles. It is winter in the northern part of the world. In the Southern Hemisphere, however, it is summer.

≡Quick Lab

Sun and Seasons

1. Hold the bottom of a flashlight 5 centimeters above a piece of graph paper. Trace the circle of light on the graph paper. Label the circle *A*.

2. Tilt the flashlight as shown. Keep it the same distance above the paper. Trace the circle of light. Label it *B*.

3. **Use Numbers** Count the squares on the graph paper that fall inside or mostly inside each circle.

4. Did tilting the flashlight change the number of squares? How?

5. **Infer** How do your results help explain the seasons?

summer
June 21–Sept. 22

fall
Sept. 22–Dec. 21

winter
Dec. 21–March 20

spring
March 20–June 21

Read a Diagram

Describe how sunlight changes in the Northern Hemisphere in one year.

Clue: Follow the red arrows. Look for the shadow.

 Science in Motion Watch how Earth revolves around the Sun at www.macmillanmh.com

Quick Check

Cause and Effect What causes the seasons?

Critical Thinking What would happen to the seasons if Earth's axis were not tilted?

How does the Sun's apparent path change over the seasons?

The diagram shows the apparent path of the Sun during the day. Each yellow circle represents the Sun's position at noon. How does that position change from winter to summer? The Sun rises much higher in the sky during a summer day. It also rises earlier and sets later.

At the Equator

The diagram does not apply to all parts of the world. Near the equator, the Sun's apparent path changes much less during the year. Temperatures there change little from season to season. All year long, sunlight strikes at similar angles.

At the Poles

Near the poles, the Sun's apparent path is very different between seasons. In northern Alaska, for instance, summer nights are very short. During winter, the Sun hardly appears.

Making Predictions

The Sun's apparent path changes in the same pattern every year. Scientists use these patterns to make predictions. They can predict the exact times the Sun will rise and set.

✔ Quick Check

Cause and Effect How does the Sun's apparent path change over the year?

Critical Thinking Why are the differences so large near the poles?

Apparent Path of the Sun

summer

spring and fall

winter

Lesson Review

Visual Summary

Earth's rotation causes day and night. Shadows change with the Sun's apparent motion across the sky.

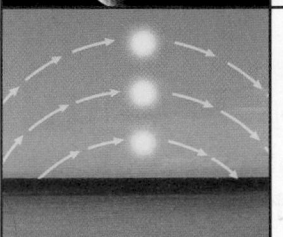

The tilt of Earth's axis and **Earth's revolution** around the Sun cause seasons to change during the year.

The Sun's apparent path depends on the tilt of **Earth's axis.** The path is different near the equator and poles.

Make a FOLDABLES Study Guide

Make a three-tab book. Use it to summarize what you learned about the Sun and Earth.

Earth's rotation

Earth's revolution

Earth's axis

Think, Talk, and Write

1 Vocabulary Earth's _orbit_ is the path it takes during its revolution.

2 Cause and Effect List the different effects caused by Earth's motion.

Cause	→	Effect
	→	
	→	
	→	
	→	

3 Critical Thinking How would Earth be different if its axis were not tilted?

4 Test Prep When does the Sun rise highest in the sky in the Northern Hemisphere?

  A March
  B June
  C September
  D December

5 Test Prep Which process takes Earth 24 hours to complete?

  A rotation
  B revolution
  C shadows
  D seasons

6 Essential Question Why does it seem that the Sun is moving?

 **Math Link**

Use Multiplication
A tree is 9 meters tall. In the morning, the tree casts a shadow 3 times its height. How long is the tree's shadow in the morning?

Social Studies Link

Learn about Seasons in Other Places
Find your state on a globe. Where on Earth is the cycle of day and night the opposite from where you live? Where is the cycle of seasons the opposite?

Without the Sun

It is the year 3528. Planet Mungo is in conflict with Earth. Mungo's scientists have built a huge device that blocks the Sun's light from reaching Earth.

It happened 14 days ago. First, the sky darkened. Then, the air cooled. It became very still. Rain began to fall. It has been raining for 13 days.

The High Global Commission met. "We must resolve our conflict with Mungo," said the commission chief. "Without the Sun, plants cannot make food. The plants are dying. With no plants, all the animals will die!"

"If this continues," argued the vice chief, "water won't evaporate. We'll face floods and freezing."

"Get up, Lisa," shouted Mom.

Lisa opened her eyes. "Mom, I just had the weirdest dream." She looked out the window and smiled in the sunlight.

Fictional Story

A good fictional story

▶ has an interesting beginning, middle, and end;

▶ describes a setting, telling when and where a story happens.

Write About It

Fictional Story Write your own story about what would happen if sunlight could not reach Earth.

 **e-Journal** Research and write about it online at **www.macmillanmh.com**

Light Speed!

Light travels about 10 trillion kilometers in one year. Light travels about 18 million km in one minute. The Sun is about 150 million km away from Earth.

Using this information, you can figure how many minutes it takes for sunlight to reach Earth. Divide 150 million by 18 million to get the answer.

Instead of using long division to get an exact answer, you can make an estimate. Use numbers that are close to the ones in the problem but are easier to divide.

Estimate Quotients

▶ You can use compatible numbers to estimate $150 \div 18$. What numbers close to these are easier to divide?

▶ 18 is close to 20. 150 is between 140 and 160. What is 140 divided by 20?

Think: $14 \div 2 = 7$
So: $140 \div 20 = 7$

What is 160 divided by 20?

Think: $16 \div 2 = 8$
So: $160 \div 20 = 8$

It takes 7 to 8 minutes for the Sun's light to reach Earth.

Solve It

Mars is about 230 million km away from the Sun. Estimate the number of minutes it takes for the Sun's light to reach Mars.

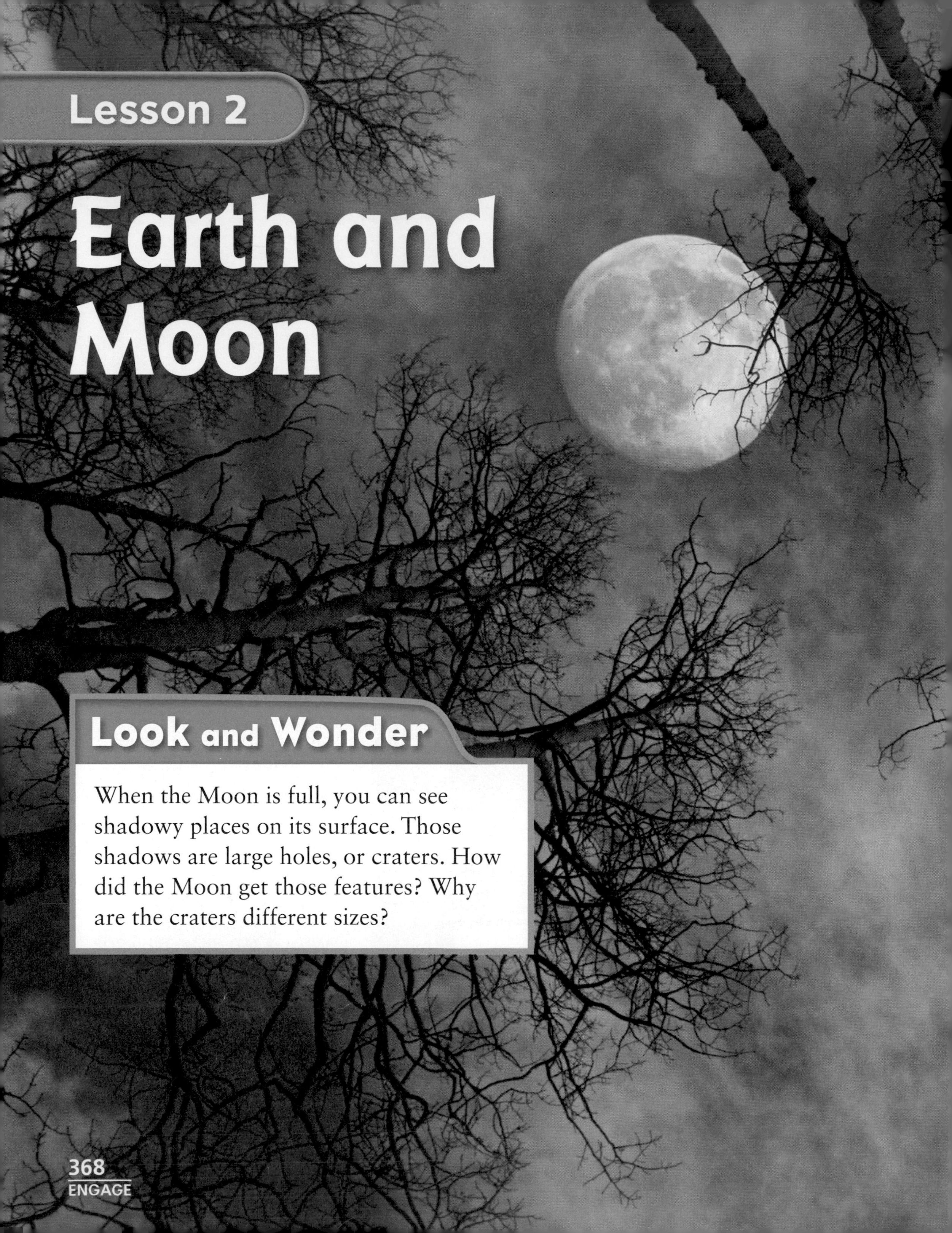

Earth and Moon

Look and Wonder

When the Moon is full, you can see shadowy places on its surface. Those shadows are large holes, or craters. How did the Moon get those features? Why are the craters different sizes?

What affects the size of craters on the Moon?

Form a Hypothesis

When rocks moving through space hit the Moon, they make holes called craters. Does a bigger rock make a larger crater? Write a hypothesis.

Test Your Hypothesis

1 **Make a Model** Place a large dish or tray on a sheet of newspaper. Cover the inside of the dish with wax paper. Pour in a layer of flour about 3 cm thick. This models the surface of the Moon. Do not touch it.

2 Press the clay into three balls. One ball should have a diameter of about 1 cm. The second should be about 3 cm. The third, 5 cm. These are your model space rocks.

3 **Measure** Drop a model rock onto the flour from a height of 25 cm. Measure the width of the hole it makes. Repeat three times. Record your data in a chart.

4 **Experiment** Repeat step 3 with the other models. Record these results in your chart.

Draw Conclusions

5 **Interpret Data** How does the size of the rock affect the size of the hole that it makes?

6 **Infer** How does this activity explain the Moon's appearance?

Explore More

What variable besides rock size affects the size of craters? Form a hypothesis. Make a plan to test it. Decide which variables will stay the same and which variable will change. Try it!

Materials

- large dish or tray
- newspaper
- wax paper
- flour
- metric ruler
- modeling clay

Step 2

Step 3

▶ **Essential Question**

What can we learn about the Moon?

▶ **Vocabulary**

crater, p. 371

phase, p. 373

lunar eclipse, p. 374

solar eclipse, p. 374

▶ **Reading Skill** ✓
Compare and Contrast

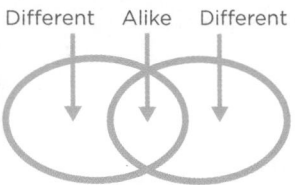

Different   Alike   Different

▶ **Technology** 🔵 LOG ON
e-Glossary and e-Review online
at www.macmillanmh.com

What is the Moon like?

On many nights, the Moon appears to be the largest, brightest object in the sky. Unlike stars, however, the Moon does not make its own light. Why does the Moon shine? It reflects the light of the Sun. Moonlight is reflected sunlight!

▲ The Moon reflects light from the Sun.

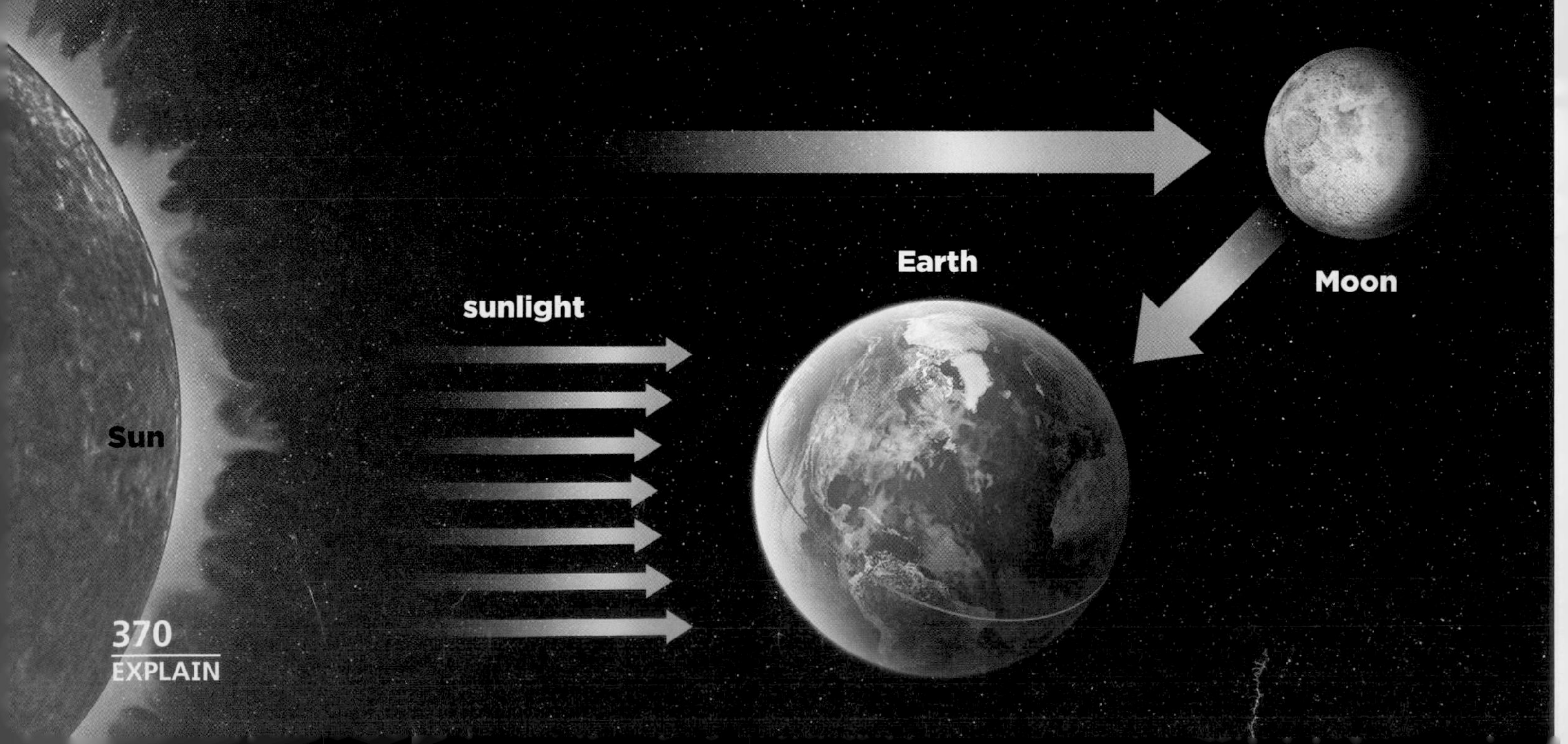

Sunlight strikes the surface of Earth as well as the Moon. The Moon reflects this light to Earth.

Earth

Moon

sunlight

Sun

The Moon and Earth

The Moon is Earth's closest neighbor in space. It is about 384,000 km (240,000 miles) from Earth. This is almost 400 times closer to Earth than the Sun.

Rocks on the Moon are similar to some Earth rocks. However, there are many differences between Earth and the Moon. For one, the Moon is much smaller than Earth. There is no air or atmosphere. It has almost no water either.

Because of these factors, the Moon has an extreme range of temperatures. In the daytime, it is hot enough to boil water. The nights are colder than any place on Earth. No wonder the Moon does not support life!

Surface Features

The Moon has a few tall mountains. It also has flat plains. But most of its surface is covered with craters (KRAY•turz). A **crater** is a hollow area or pit in the ground. Large rocks called *meteoroids* (MEE•tee•uh•roydz) made many of the Moon's craters. Meteoroids travel through space. They often crash into other space objects.

crater

Craters and Earth's Atmosphere

If meteoroids are always crashing into things, why isn't Earth covered in craters? Earth's atmosphere keeps them away. When meteoroids enter Earth's atmosphere, they become very hot. Most of them burn up before they hit Earth's surface.

✔ Quick Check

Compare and Contrast How are Earth and the Moon alike? How are they different?

Critical Thinking Why do people who go to the Moon need to wear space suits?

FACT The Moon does not make its own light.

third quarter moon
The Moon is three quarters of the way around Earth.

waning crescent moon
The left sliver of the Moon is the only part you can see.

waning gibbous moon
Slightly less of the lit side can be seen.

new moon
The lit side cannot be seen from Earth.

full moon
The entire lit side can be seen.

waxing crescent moon
Some of the lit side can be seen.

waxing gibbous moon
The Moon is almost full.

first quarter moon
The Moon is a quarter of the way around Earth.

Read a Diagram

You cannot see the Sun in this diagram, but you can infer its position. Where is the Sun?

Clue: Observe the small Moons along the blue circle.

What are the phases of the Moon?

Like the Sun, the Moon seems to rise and set. The Sun does not move around Earth, but the Moon does!

As Earth revolves around the Sun, the Moon revolves around Earth. It completes one orbit around Earth in just over 29 days. This is almost as long as an average month. In fact, some of the earliest calendars were based on the Moon's motion.

Apparent Shapes

As the Moon orbits Earth, its appearance seems to change. The apparent shapes of the Moon in the sky are called **phases** (FAYZ•ez). During one complete orbit, the Moon cycles through all of its phases. At the same time, the Moon completes about one rotation.

All this time, the Sun is shining. It lights one half of the Moon at a time. The other half is dark. During the Moon's orbits, we see different fractions of its lit half.

The Moon's Gravity

The Moon has gravity. It pulls slightly on Earth. On the side of Earth that faces the Moon, the water or land bulges slightly outward. The Moon pulls more than the Sun because it is closer to Earth.

≡ Quick Lab

Moon and Earth

1 Use a sticker to mark a spot on a small ball.

2 **Make a Model** Move the small ball in a revolution around a larger ball. Meanwhile, rotate the small ball in the same direction. Your rotation and revolution should finish at the same time.

3 How does this model the Moon and Earth?

4 **Infer** Will you ever see a different side of the Moon from Earth? Explain.

Earth's Tides

The Moon's gravity causes tides. *Tides* are the daily rise and fall of the ocean's surface. Most coasts on Earth have high tides and low tides.

✔ Quick Check

Compare and Contrast How is the Moon's first quarter phase like its third quarter phase?

Critical Thinking How much time passes between a full moon and a new moon?

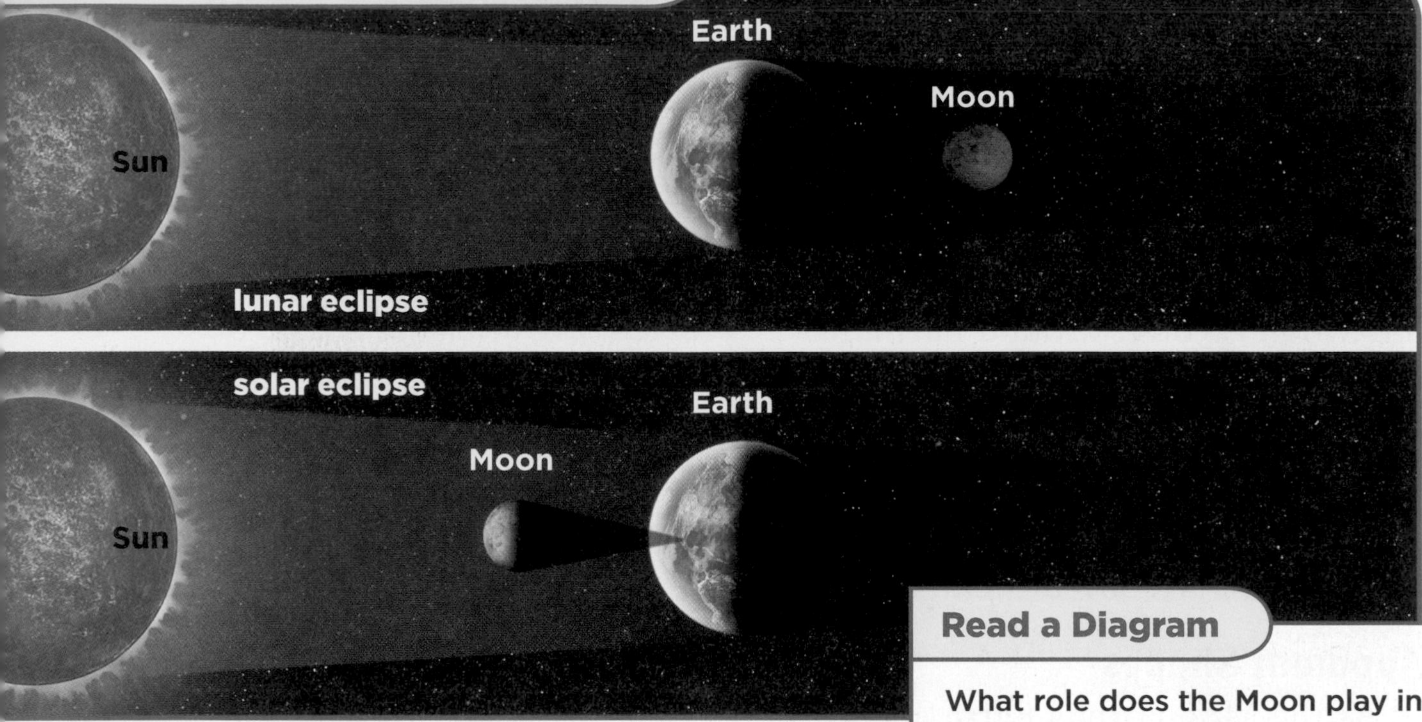

Earth

Moon

Sun

lunar eclipse

solar eclipse

Earth

Moon

Sun

What is an eclipse?

An *eclipse* (ih•KLIPS) occurs when a shadow is cast by Earth or the Moon. The diagram shows the two basic kinds.

Lunar Eclipses

In a **lunar eclipse**, Earth casts a shadow on the Moon. This happens when Earth is directly between the Sun and the Moon. The Moon passes through Earth's shadow.

Solar Eclipses

In a **solar eclipse**, the Moon casts a shadow on Earth. Solar eclipses happen only during the new moon. A *partial solar eclipse* is when part of the Sun is blocked. A *total solar eclipse* blocks all of the Sun.

Eclipse Safety

Only a lunar eclipse is safe to observe. Looking at a solar eclipse will damage your eyes or cause blindness. Sunglasses do not help. For this reason, you should never look directly at the Sun during an eclipse. Scientists use special tools to observe solar eclipses safely.

 Quick Check

Compare and Contrast How is a lunar eclipse like a solar eclipse? How is it different?

Critical Thinking Why is it safe to observe a lunar eclipse?

Lesson Review

Visual Summary

The Moon is Earth's nearest neighbor in space. It reflects the Sun's light. **Craters** cover its surface.

The Moon revolves around Earth about once every 29 days. As the Moon revolves, we see its different **phases.**

An **eclipse** occurs when a shadow is cast by Earth or the Moon. A solar eclipse is not safe to view.

Make a **FOLDABLES** Study Guide

Make a three-tab book. Use it to summarize what you read about Earth and Moon.

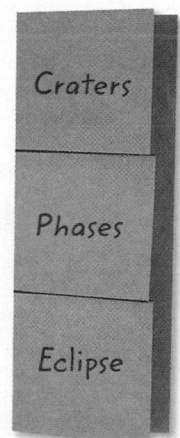

Craters

Phases

Eclipse

Think, Talk, and Write

1 **Vocabulary** During a(n) _____, the Moon's shadow is cast onto Earth.

2 **Compare and Contrast** Fill in the Venn diagram to show how Earth and the Moon are alike and different.

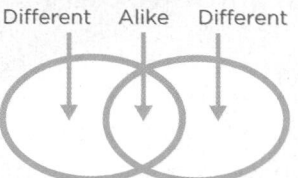

Different   Alike   Different

3 **Critical Thinking** You see a full moon in the night sky. Is there a new moon someplace else on Earth? Explain.

4 **Test Prep** What causes many of the craters on the Moon?

   A meteoroids striking the Moon

   B earthquakes on the Moon

   C landslides on the Moon

   D flooding on the Moon

5 **Essential Question** What can we learn about the Moon?

 **Math Link**

Diameter and Radius
The diameter of a circle is its width at the widest point. The radius of a circle is one half of its diameter. A picture of the full moon has a radius of 6 cm. What is its diameter?

 **Social Studies Link**

Learn about the Apollo Program
Research and report on NASA's Apollo program. Why was it important to Americans in the 1960s and 1970s? Try to interview an adult who remembers the events.

Focus on Skills

Inquiry Skill: Interpret Data

During any month, you can see different phases of the Moon. The changing positions of Earth and the Moon cause these phases. Scientists can predict when the Moon will be in any of its phases. To do so, they collect and **interpret data** about the Moon.

▶ Learn It

When you **interpret data,** you use information that has been gathered to answer questions or solve problems. Interpreting data from a written report can be difficult. It is better to organize your data into a table, chart, or graph. These tools help you see and understand your data at a glance. They help others understand your data too.

A calendar is a type of table. The one below shows data about moon phases during the month of May. Each drawing shows the phase that was observed on that day. The pattern in the calendar helps you predict other moon phases.

May

Sunday	Monday	Tuesday	Wednesday	Thursday	Friday	Saturday
						1
2	3	4	5	6	7	8
9	10	11	12	13	14	15
16	17	18	19	20	21	22
23	24	25	26	27	28	29
30	31					

▶ Try It

Interpret data in the moon phase calendar on the opposite page. Answer the following questions.

Materials moon phase calendar

1 On which day or days was there a new moon?

2 On which day or days was there a first quarter moon?

3 On which day or days was there a gibbous moon?

4 Is there a pattern of moon phases in this calendar? Describe it.

▶ Apply It

Interpret data by turning the information into a table like the one shown here.

1 Make a table with two columns. In one column, draw the phases of the Moon. You do not need to include the gibbous phase. In the other column, tally the number of times each phase appears in the calendar.

2 Find a new calendar at home or at school that shows moon phases. Look at the month of May. Make another table that shows the tally of moon phases.

3 Compare the two tables. Are your tallies the same in both cases? How are they different?

4 Look at the two moon phase calendars. Do the same phases of the Moon occur on the same days in May? Why or why not?

Lesson 3

The Solar System

Look and Wonder

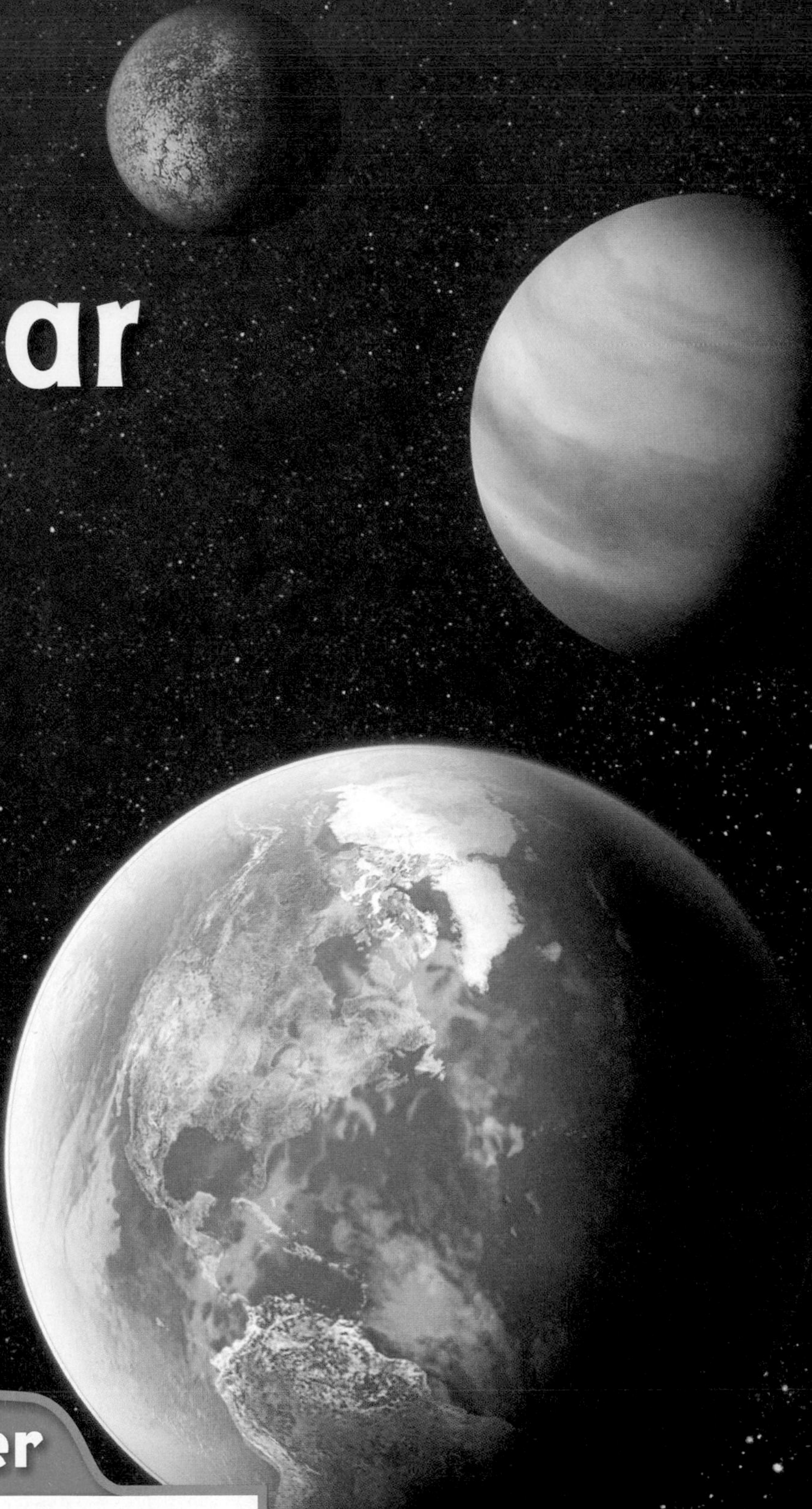

Is this a photograph taken from space? Look at the distances between the three objects. Are they really so close together in space?

How do sizes of objects in the solar system compare?

Purpose

Explore how Earth's size compares to the Moon's size and the sizes of other objects in the solar system.

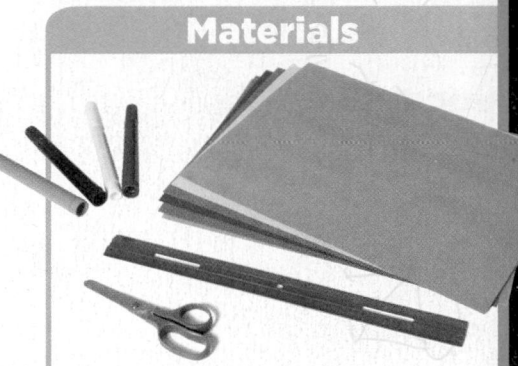

Procedure

⚠ **Be Careful!** Handle scissors carefully.

1. **Use Numbers** Study the table. Compare the diameters of the different objects.

- scissors
- construction paper
- metric ruler
- marker

2. **Measure** Cut a paper circle with a diameter of 16 cm. This circle models Earth. Measure and cut circles to model the other objects in the table. Label each object. For at least one of the models, you will need to tape two or more sheets of construction paper together.

Step 2

3. **Classify** Arrange the objects in a way that lets you compare their sizes.

Draw Conclusions

4. **Communicate** How do the sizes of the different objects compare?

5. **Infer** Why does the Moon appear larger than Mars in Earth's night sky? Why does the Sun seem larger and brighter than other stars?

Explore More

Research the sizes of other objects in the solar system. Make large and small circles to represent them. Find out how these objects are arranged in the solar system. Then arrange your models to represent those locations.

Comparing Diameters

Object	Size in Earth Diameters
Earth	1
Moon	$\frac{1}{4}$
Mars	$\frac{1}{2}$
Uranus	4

▶ **Essential Question**

How does Earth compare with other objects in the solar system?

▶ **Vocabulary**

solar system, p. 380

planet, p. 380

gravity, p. 381

telescope, p. 382

comet, p. 388

asteroid, p. 388

meteor, p. 388

meteorite, p. 388

▶ **Reading Skill** ✓

Main Idea and Details

Main Idea	Details

▶ **Technology**

e-Glossary and e-Review online at www.macmillanmh.com

What is the solar system?

You probably know that human-made satellites (SA•tuh•lites) orbit Earth. Did you know that the Moon is a satellite? A *satellite* is any object that moves in orbit around another body.

The Sun has many satellites. The Sun and all the objects in orbit around it make up our **solar system**. The solar system is millions of kilometers wide. At its center is the Sun.

Planets

On a clear night, you might see a planet or two in the sky. **Planets** are round objects in space that are satellites of the Sun. Scientists have identified eight planets in our solar system.

Planets are smaller and cooler than stars. Like the Moon, planets cannot make their own light. They reflect the light of the Sun.

The Solar System

Sun   Mercury   Venus   Earth   Mars

Orbiting the Sun

In the 1500s, a Polish scientist named Nicolaus Copernicus studied the planets. He found that they orbit the Sun. One hundred years later, a German scientist named Johannes Kepler showed that those orbits are ellipses (ih•LIPS•eez). An *ellipse* is a slightly flattened circle, or oval.

The English scientist Sir Isaac Newton lived in the late 1600s. He described how the planets stay in their orbits. Newton said it was a balance between gravity and inertia (ih•NUR•shuh). **Gravity** is a force of attraction between all objects. It pulls planets toward the Sun. *Inertia* is the tendency of a moving object to keep moving in a straight line.

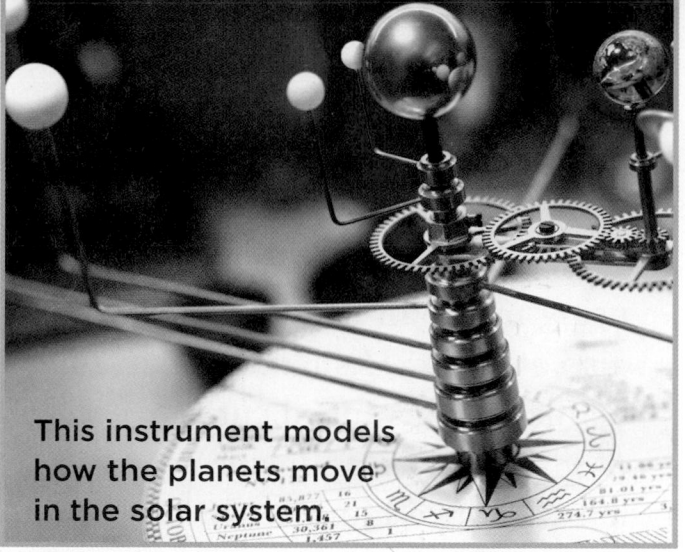

This instrument models how the planets move in the solar system.

✓ Quick Check

Main Idea and Details How do planets move in the solar system?

Critical Thinking Why is Earth's Moon a satellite?

Read a Diagram

Which planet has the shortest journey around the Sun?

Clue: Trace each line of orbit.

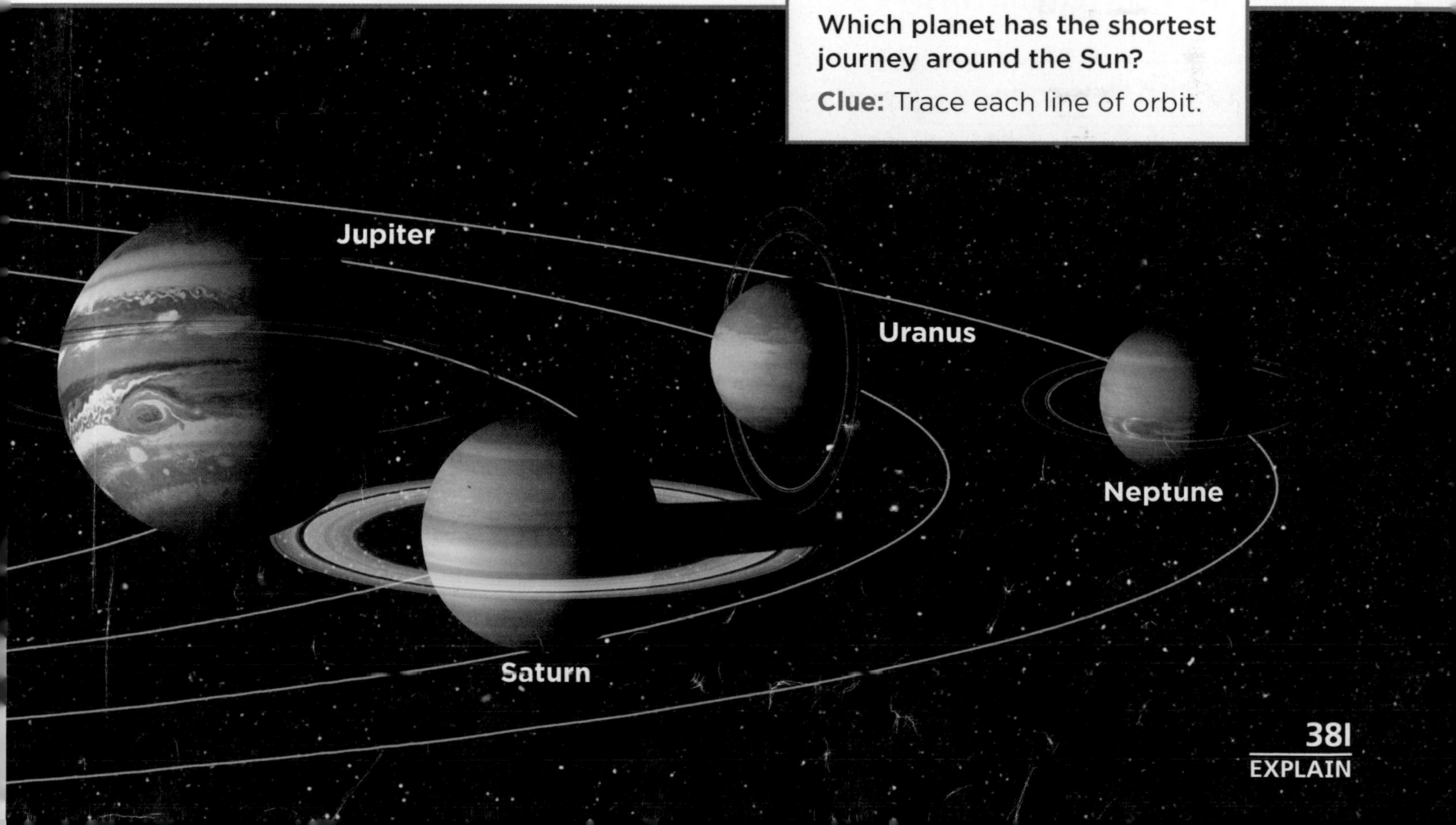

Jupiter

Uranus

Neptune

Saturn

How do we learn about the solar system?

While Kepler studied the planets using math, an Italian scientist was also at work. His name was Galileo Galilei. He arranged curved pieces of glass, or *lenses,* inside a tube. The lenses helped him peer into space.

Telescopes

Do you know what Galileo used to look into space? It was a telescope. **Telescopes** make faraway objects seem closer. Galileo found objects in space that no one had seen before. He was able to show that Copernicus's model of the solar system was correct. Some of today's telescopes work much like larger versions of the ones Galileo made. Others used curved mirrors.

Clouds and city lights make it hard to see through telescopes. For this reason, many telescopes are located in clear, deserted areas or on mountaintops. One of the best places for telescopes is in space.

Astronauts

Many countries have programs to explore space. The United States has *NASA*—the National Aeronautic (ayr•uh•NAW•tik) and Space Administration. In the 1960s, NASA launched rockets that took people into space. Those people were the first *astronauts* (AS•truh•nawts).

New and Old Telescopes

Read a Photo

How has technology that is used for learning about space changed over time?

Clue: Compare the telescopes in the photographs.

Shuttles and Space Stations

Space shuttles help astronauts conduct experiments and launch satellites. Many countries, including the United States, also use the International Space Station. Unlike shuttles, the space station remains in space for a long time.

Probes

Space probes are safer and less expensive than sending astronauts to space. A *probe* is an unmanned spacecraft that leaves Earth's orbit. NASA has launched probes to planets, moons, and other objects. The probes send pictures and other data from space to Earth.

In 2004 a space probe landed on Mars. Two robot explorers, called Mars rovers, studied the surface and recorded data. The names of these rovers are *Spirit* and *Opportunity*.

Because the solar system is so large, some probes need many years to reach their target. In 2006, NASA launched a probe. It will reach the edge of the solar system in 2015.

▲ The *Cassini* spacecraft is exploring the planet Saturn and its moons.

✔ Quick Check

Main Idea and Details How do scientists learn about space?

Critical Thinking Why is NASA exploring planets with space probes instead of astronauts?

This illustration shows a Mars rover on the surface of Mars.

383
EXPLAIN

Model the Solar System

1. As a class, discuss how to best model the solar system.

2. Have each class member pick an object to model.

3. **Make a Model** Carry out the plan in an open space. Observe the model in motion.

4. What did the class model show about the solar system? How could the model be improved?

1. Mercury 'Mars Sun
2. Venus Jupiter une
3. Earth 'aturn

What are the rocky planets?

The four planets closest to the Sun are called the *rocky planets*. They have much in common. Each is made up mostly of rock. They also seem to have solid cores made of iron. There are also important differences between these planets.

Mercury

Mercury is the closest planet to the Sun. That makes it very hot. It has almost no water and very little air. The surface has many craters like Earth's Moon. It is also the smallest rocky planet. At its equator, it is less than half the size of Earth. Mercury does not have a moon.

Mercury

Distance to the Sun: 58 million km
Diameter: 4,880 km
Rotation Time: 59 Earth days
Revolution Time: 88 Earth days
Fast Fact: Mercury's surface is covered with craters.

Venus

Distance to the Sun: 108 million km
Diameter: 12,100 km
Rotation Time: 243 Earth days
Revolution Time: 225 Earth days
Fast Fact: Temperatures on Venus can reach 500°C.

Venus

Venus is the second closest planet to the Sun. It has a thick atmosphere made mostly of carbon dioxide. The atmosphere does not allow heat to easily escape. This makes Venus the hottest planet. There are many volcanoes on Venus. Its surface is covered in lava flows. Venus also does not have a moon.

Earth

Earth is unique in our solar system. It has oxygen and liquid water. Earth's atmosphere keeps temperatures from getting too hot or too cold. These conditions are just right for life. Earth is the only planet known to support life.

Mars

Of all the planets, Mars is the most like Earth. It has two small moons and a thin atmosphere. Mars has volcanoes, but they are no longer active. The surface has many features that show evidence of erosion by floods and rivers. Today, Mars is much colder than Earth. Its water is frozen in ice caps near both poles. In addition to probes, NASA hopes to send astronauts to Mars.

 Quick Check

Main Idea and Details Name and describe the rocky planets.

Critical Thinking Why would Earth's living things be unable to live on the other rocky planets?

Earth

Distance to the Sun: 150 million km
Diameter: 12,756 km
Rotation Time: 1 Earth day
Revolution Time: 365 Earth days
Fast Fact: Earth's atmosphere makes it suitable for life.

Mars

Distance to the Sun: 228 million km
Diameter: 6,794 km
Rotation Time: About 1 Earth day
Revolution Time: 687 Earth days
Fast Fact: Iron oxide, or rust, gives Mars its reddish color.

What are the other planets?

The four planets beyond Mars are called *gas giants*. Can you guess why? They are huge in size and made mostly of gases. The nearest, Jupiter, is five times farther from the Sun than Earth is.

The gas giants do not have solid surfaces. They are mostly made up of hydrogen and helium. Scientists think that they might have some rock and ice at their core.

Each has a ring system, although most are difficult to see. Each also has many moons. Some are like the rocky planets and have atmospheres.

Jupiter

Jupiter is the largest planet in the solar system. Scientists have seen at least 63 moons in orbit around it. This planet's atmosphere is divided into bands. Each band has winds blowing in opposite directions. One band has a large red spot that is the size of Earth. It is a giant storm that has been raging for over 300 years!

Saturn

Saturn is the second-largest planet. It is famous for its large rings. The rings are made of pieces of ice and rock. Most of these pieces are less than a couple of meters in diameter. Saturn has at least 34 moons. The largest is named Titan.

Jupiter

Distance to the Sun: 778 million km
Diameter: 143,000 km
Rotation Time: 10 Earth hours
Revolution Time: 4,333 Earth days
Fast Fact: Jupiter's four largest moons were first observed by Galileo in 1610.

Saturn

Distance to the Sun: 1 billion, 429 million km
Diameter: 120,536 km
Rotation Time: 10 Earth hours
Revolution Time: 10,759 Earth days
Fast Fact: Winds on Saturn can blow at 500 meters per second.

Uranus

Have you ever heard of a "sideways" planet? The axis of Uranus is tilted so much that it rotates on its side! This means that one pole faces the Sun during parts of Uranus's orbit. The unusual color of this planet is caused by gases in its upper atmosphere. Uranus has at least 27 moons.

Neptune

Neptune is the farthest gas giant from the Sun. Winds on Neptune can blow at speeds of 2,000 km (1,200 mi) per hour! Scientists have observed 13 moons orbiting Neptune. Triton is the largest moon. It is known to have volcanoes.

Dwarf Planets

Scientists have been discovering smaller and smaller planets in the solar system. These are called *dwarf planets*. Most are round and made of rock and ice. Their orbits cross the orbits of other objects.

Pluto is the best known dwarf planet. For 76 years, it was considered the ninth planet. Scientists changed Pluto's classification in 2006.

 Quick Check

Main Idea and Details Name and describe the gas giants.

Critical Thinking Could humans live on the gas giants? Explain.

Uranus

Distance to the Sun: 2 billion, 871 million km
Diameter: 51,118 km
Rotation Time: 17 Earth hours
Revolution Time: 30,684 Earth days
Fast Fact: The axis of Uranus is tilted toward the Sun.

Neptune

Distance to the Sun: 4 billion, 504 million km
Diameter: 49,528 km
Rotation Time: 16 Earth hours
Revolution Time: 60,190 Earth days
Fast Fact: Neptune takes 165 Earth years to orbit the Sun.

Comet Hale-Bopp last approached the Sun in the 1990s.

What else is in our solar system?

Not all objects in the solar system are planets or moons. Smaller objects also revolve around the Sun.

Comets

A comet is mostly ice mixed with rocks and dust. It moves in a long, narrow orbit. When a comet nears the Sun, it heats up very quickly. This forms a tail of gas and dust pointing away from the Sun.

Asteroids

Asteroids (AS•tuh•roydz) are large chunks of rock or metal in space. The solar system has thousands of asteroids. Most of them lie in a belt between Mars and Jupiter.

Meteoroids

When comets or asteroids collide, pieces of rock or metal break off. These smaller pieces are meteoroids. There are millions of them in space!

If a meteoroid enters Earth's atmosphere, it is called a meteor. Small meteors burn up in the atmosphere, leaving streaks of light across the sky. We call them shooting stars, but they are not stars at all. If a meteor reaches Earth's surface, it is called a meteorite.

 Quick Check

Main Idea and Details Describe the smaller solar system objects.

Critical Thinking How do planets compare to asteroids and comets?

FACT Comets have a tail only when they are near the Sun.

Lesson Review

Visual Summary

 **The solar system** is made up of planets, moons, and other objects that orbit the Sun in space.

 **The planets** are round objects in space that are satellites of the Sun. They include gas giants and dwarf planets.

Smaller objects in the solar system include comets, asteroids, meteoroids, and meteors.

Make a **FOLDABLES** Study Guide

Make a three-tab book. Use it to summarize what you learned about the solar system.

The solar system

The planets

Smaller objects in the solar system

Think, Talk, and Write

1 **Vocabulary** The large rocks that are found in a belt between Mars and Jupiter are called _____.

2 **Main Idea and Details** Extend and fill in the graphic organizer to show the parts of the solar system.

Main Idea	Details

3 **Critical Thinking** Why might it be better for some experiments to be done in space or someplace away from Earth? Give an example of a variable that such an experiment might test.

4 **Test Prep** Which is the largest planet in the solar system?

A Mars

B Jupiter

C Saturn

D Earth

5 **Essential Question** How does Earth compare with other objects in the solar system?

 **Writing Link**

Write a Report
Research how the planets received their names. Present what you learn in a written report.

 **Social Studies Link**

Learn about a NASA Probe
In early 2006, NASA launched a probe to Pluto. Research the progress of this probe. What do scientists hope to learn from it?

TO THE MOON!

How have scientists explored our solar system? For thousands of years, people used their unaided eyes. Then about 400 years ago, scientists developed telescopes. In recent history, spacecraft have landed robots and people on the Moon. What we learn about the Moon may help us understand and explore other objects in the solar system.

1957 *Sputnik I* is the first spacecraft to travel into space.

Connect to

AMERICAN
MUSEUM ᴼⁿ
NATURAL
HISTORY

1959 *Luna 1* is the first spacecraft to approach the Moon closely. *Luna 2* lands on the Moon. *Luna 3* sends pictures of the Moon to Earth. This is the first time anyone can see what the far side of the Moon looks like.

NASA plans to send expeditions back to the Moon to learn more about it and what it takes to live in its extreme environment.

1972 *Apollo 17* is the last manned mission to the Moon. The crew spends 75 hours there. Astronauts Gene Cernan and Harrison Schmitt drive a lunar roving vehicle around the surface of the Moon to collect samples.

1969 *Apollo 11* mission is the first to land a person on the Moon. Neil Armstrong and Buzz Aldrin are the first astronauts to walk on the Moon and collect samples.

Main Idea and Details

▶ The main idea tells what the article is mostly about.

▶ Details, facts, and examples support the main idea.

Write About It

Main Idea and Details Reread the introduction and the captions on the time line. Then write a paragraph that explains the main idea and details of this article. Be sure to include facts and examples in your paragraph.

 **e-Journal** Research and write about it online at **www.macmillanmh.com**

Stars and Constellations

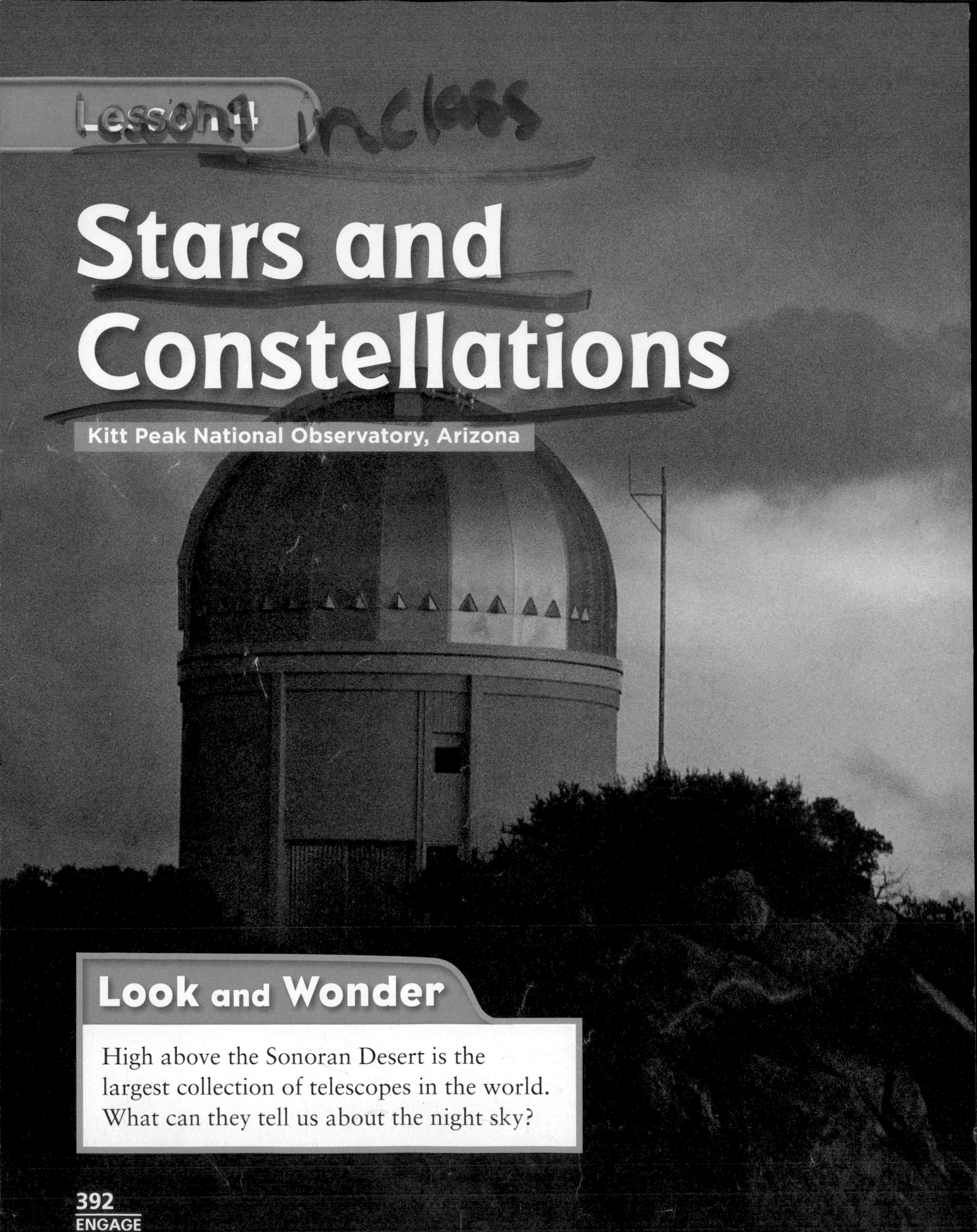

Kitt Peak National Observatory, Arizona

Look and Wonder

High above the Sonoran Desert is the
largest collection of telescopes in the world.
What can they tell us about the night sky?

Why do some stars seem brighter than others?

Form a Hypothesis

How does distance affect the apparent brightness of stars? Write a hypothesis.

Materials

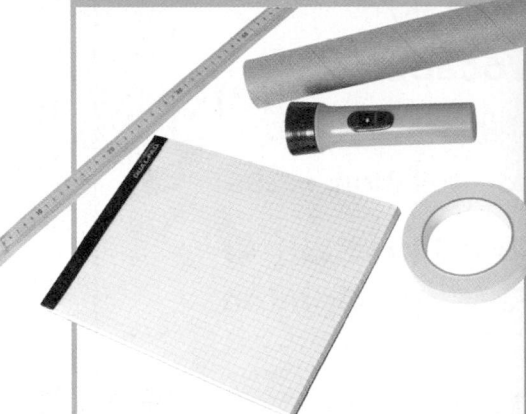

- masking tape
- cardboard tube
- flashlight
- graph paper
- meterstick

Test Your Hypothesis

1 Tape one end of a cardboard tube to the flashlight as shown.

2 **Measure** Hold the other end of the tube 10 cm above the center of the graph paper. Turn the flashlight on. Have a partner trace the circle of light on the paper. Label the circle *10 cm*.

3 Repeat step 2 at a distance of 20 cm. Repeat at 40 cm and at 80 cm. Label the circle each time. You might need to place the paper on the floor for the last tests.

Draw Conclusions

4 **Use Numbers** Count the number of squares filling each labeled circle on the graph paper.

5 **Interpret Data** How did the light change as the light moved farther from the paper?

6 **Infer** Why do you think some stars in the night sky seem brighter than others?

Explore More

Does the source of a light affect the apparent brightness? Form a hypothesis. Design a test to compare different sources of light. Predict how the number of lighted squares might change. Try it!

Step **1**

Step **2**

▶ **Essential Question**

How do stars appear in the sky?

▶ **Vocabulary**

star, p. 394

constellation, p. 396

▶ **Reading Skill** ✔

Fact and Opinion

Fact	Opinion

▶ **Technology** 🔵ON

e-Glossary and e-Review online at www.macmillanmh.com

The Andromeda Galaxy is wider than our own Milky Way Galaxy. ▼

What are stars?

For thousands of years, people have observed stars shining brightly in the night sky. A **star** is a sphere of hot gases that gives off light and heat.

The only star you can see in the daytime is the Sun. The Sun might look different from other stars, but it is rather ordinary.

Compared to other stars, the Sun has an average size. Its surface temperature is average too. Why does the Sun look bigger and brighter than any other star? The Sun is the closest star to Earth. Other stars are much farther away.

Colors and Temperature

Have you ever noticed different colors of stars? The colors are due to temperature. The Sun's temperature makes it look yellow. Cooler stars are red or orange. Warmer stars are white or blue.

A star glows for a very long time. Our Sun is about five billion years old. Scientists think it will glow for five billion more years!

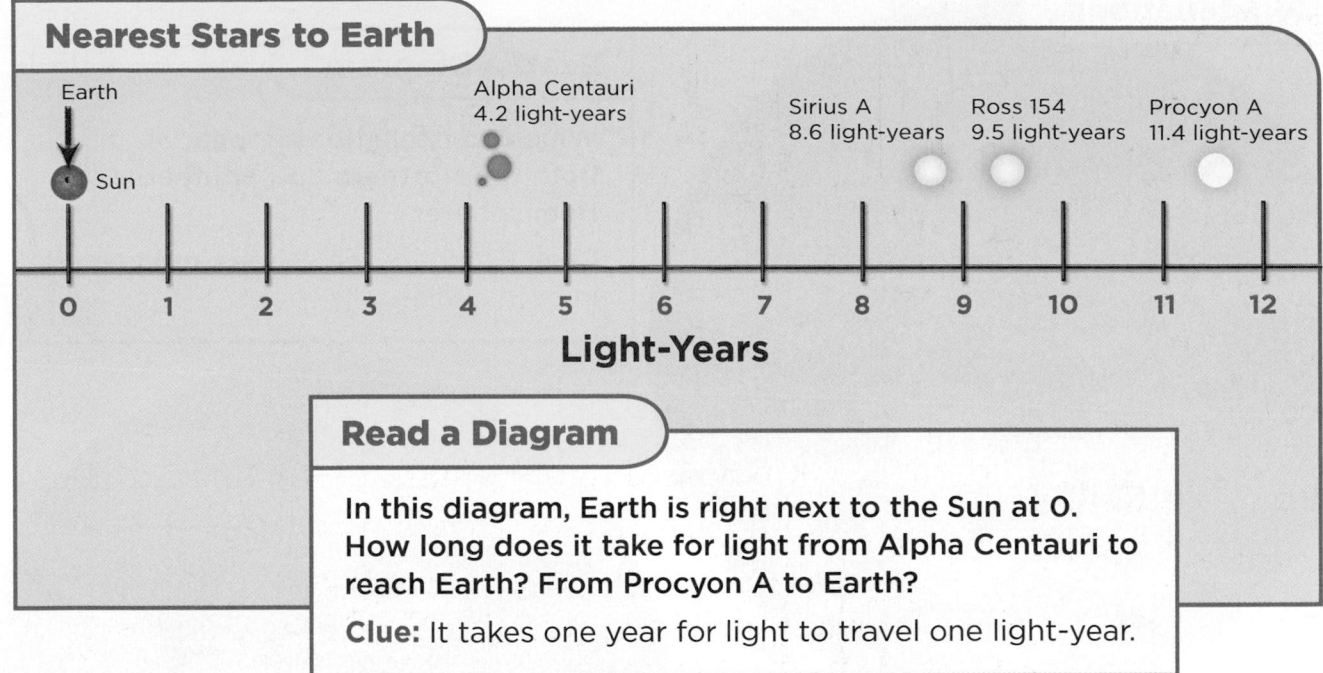

Nearest Stars to Earth

Earth

Sun

Alpha Centauri
4.2 light-years

Sirius A
8.6 light-years

Ross 154
9.5 light-years

Procyon A
11.4 light-years

0 1 2 3 4 5 6 7 8 9 10 11 12

Light-Years

Read a Diagram

In this diagram, Earth is right next to the Sun at 0. How long does it take for light from Alpha Centauri to reach Earth? From Procyon A to Earth?

Clue: It takes one year for light to travel one light-year.

Light-Years

When you observe the night sky, one star might seem brighter than another. Does that star give off more energy? Maybe not. It might simply be closer to Earth than others.

The Sun is about 150 million km from Earth. It takes about 8 minutes for its light to reach Earth. Most stars are much farther away. They are so far that scientists measure their distance in light-years. One *light-year* is the distance light travels in one year. That is nearly ten trillion km!

When you see a distant star, you are seeing what it looked like millions of years ago. A star you see today might have stopped glowing long ago. However, its light is still making its way to Earth.

Galaxies

Throughout the universe, stars are found in large groups called *galaxies* (GA•luk•seez). Our Sun is near the edge of a galaxy with billions of other stars. You know this galaxy as the Milky Way.

Our galaxy's nearest neighbor is the Andromeda (an•DRAH•muh•duh) Galaxy. It is shaped like a spiral. The universe might have many more galaxies, each with billions of stars. These are yet to be discovered.

✔ *Quick Check*

Fact and Opinion *Temperature determines a star's color.* Is this a fact or an opinion? Explain.

Critical Thinking How far away are stars? Use your own words to describe the distance.

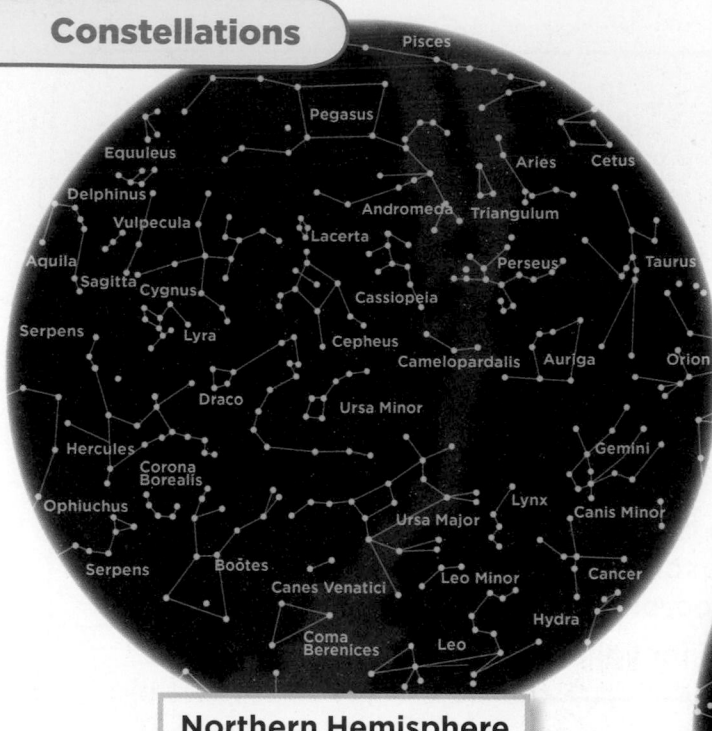

Northern Hemisphere

Read a Diagram

Which constellations appear in both the Northern and Southern Hemispheres?

Clue: Compare the shapes and names in both circles.

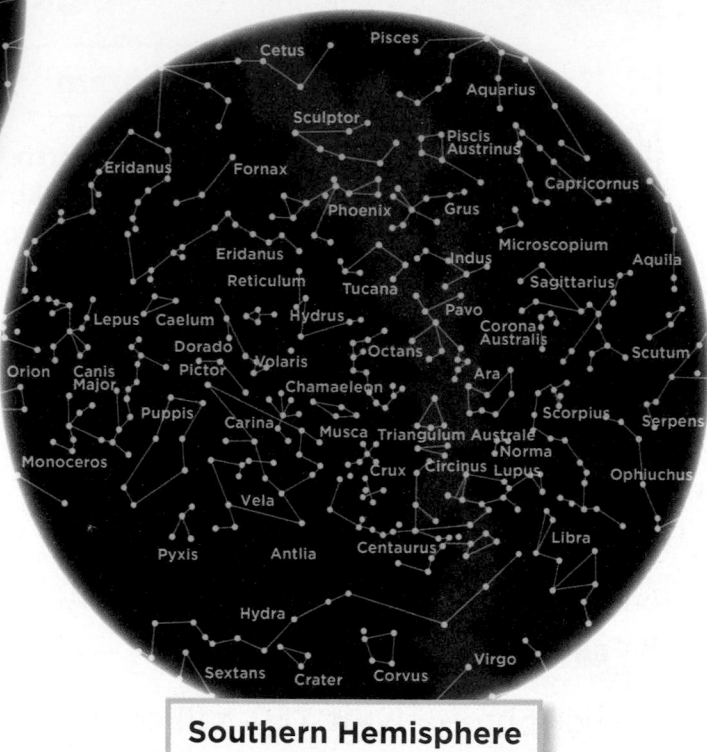

Southern Hemisphere

What are constellations?

There are billions of stars. How could you make sense of them all? One way is to group them into constellations (kahn•stuh•LAY•shunz). A **constellation** is a group of stars that make a pattern or picture in the sky.

Our constellations make sense only to an observer on Earth. Stars that seem close together are actually far apart. If you could move to a different part of the universe, those patterns would change.

Constellations also depend on the position of the observer on Earth. The night sky looks different in the Northern Hemisphere than it does in the Southern Hemisphere. Still, a few constellations appear in both.

Patterns of Stars

As Earth travels in its orbit around the Sun, we see different constellations. The constellations appear to move across the sky throughout the year. In fact, they always remain in the same patterns.

People named constellations after the pictures they saw in the sky. *Draco* is the Latin word for "dragon." The Draco constellation looks like a dragon to some people.

This ancient tool helped people tell time by the stars.

Marking Time and Seasons

Once there were no clocks to tell time. There were no satellites to help you find your position. Instead, people used constellations.

Farmers studied constellations to mark the seasons. The stars' positions helped them decide when to plant or harvest crops. Sailors used constellations to steer their ships at night. They knew which stars on the Big Dipper point to the North Star. The North Star is always in the northern sky.

Today scientists group the stars into 88 constellations. You can study constellations too. Star charts help you know where to look. Telescopes help you see each star. You can visit a local observatory to learn more.

Modeling Constellations

1. ⚠ **Be Careful!** Handle scissors with care. Remove one end of a cardboard shoe box. Cut a piece of black construction paper the same size as the cardboard piece.

2. **Make a Model** Choose one of the constellations. Draw it on the black paper. Use a pencil to punch out one hole for each star.

3. Cut a circle from the other end of the shoe box. Make it just big enough to fit a flashlight. Tape the paper from step 2 over the opposite end of the box.

4. **Observe** Dim the lights. Turn the flashlight on. Shine it through the hole in the box.

5. Share your observations with the class.

✔ Quick Check

Fact and Opinion *Draco is the best constellation.* Is this statement a fact or an opinion? Explain your answer.

Critical Thinking Why does a constellation appear to move across the sky every night?

What is the Sun like?

Like Earth, the Sun is made of layers. It has a core and three other layers. The layers are not distinct because the Sun is made of gas.

Unlike Earth, the Sun releases light into space. After all, the Sun is a star. The center, or core, of the Sun is the source of all its energy.

Light and Heat

Some of the Sun's energy is light that we can see. Much of the energy is released as heat. Earth receives just a fraction of the Sun's total energy. Yet that is enough to provide energy for nearly all living things. Producers turn the Sun's energy into food. Consumers take in the Sun's energy when they eat food.

Here you see parts of the Sun that you cannot see from Earth.

Power for the Water Cycle

The Sun's heat makes water evaporate. Evaporation is part of the water cycle that includes condensation and precipitation. The Sun also drives winds, ocean currents, storms, and other weather.

Sun Safety

Never look directly at the Sun. The energy the Sun releases could damage your eyes forever. Always wear sunscreen when you are outside. Even on a cloudy day, the Sun's energy can cause a sunburn.

✓ Quick Check

Fact and Opinion Is the Sun's energy good or bad for Earth? Support your answer with facts.

Critical Thinking How is the Sun like Earth? How is it different?

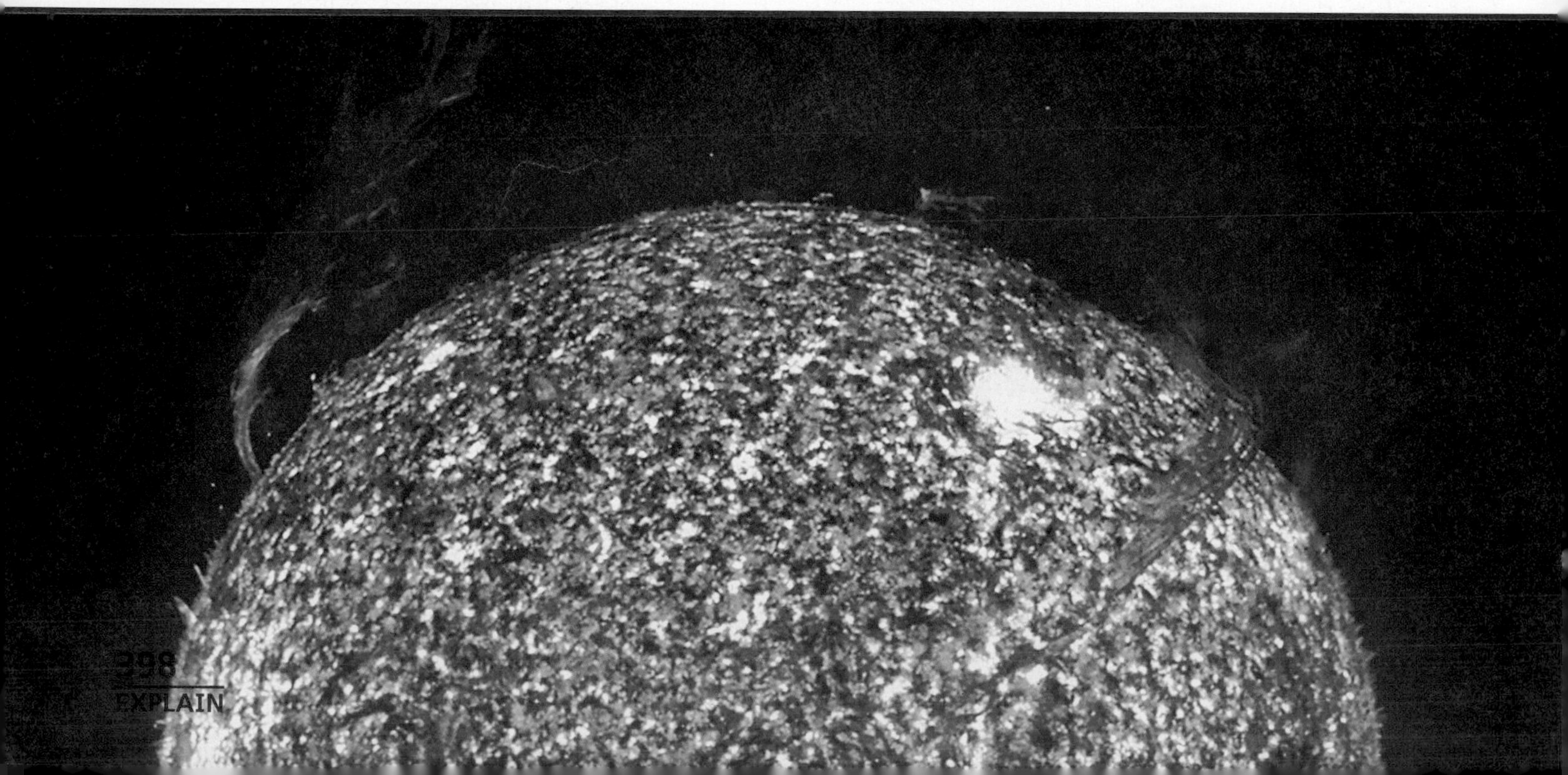

Lesson Review

Visual Summary

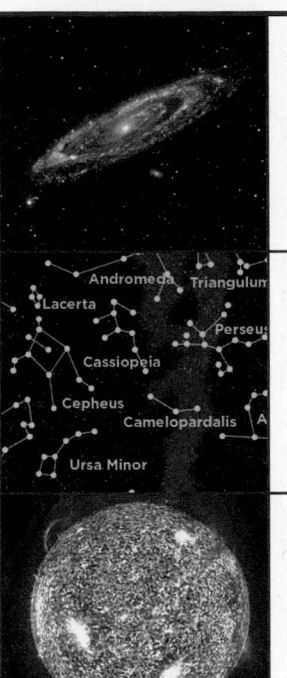

Stars are spheres of hot gases that give off light and heat. Most stars are light-years away from Earth.

Andromeda Triangulum
Lacerta
Perseus
Cassiopeia
Cepheus
Cameopardalis A
Ursa Minor

Stars can be grouped into **constellations.** Constellations help people tell time and position on Earth.

The Sun is the closest star to Earth. It provides energy for life, the water cycle, winds, currents, and weather.

Make a FOLDABLES® Study Guide

Make a trifold book. Use it to summarize what you read about stars and constellations.

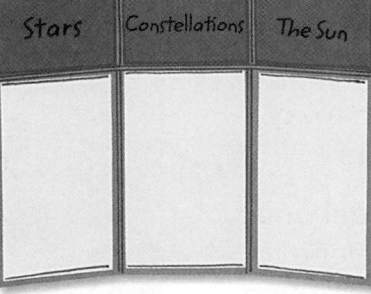

Stars | Constellations | The Sun

Think, Talk, and Write

1. **Vocabulary** What is a constellation?

2. **Fact and Opinion** Are constellations useful to people today? State your opinion. Support your opinion with at least one fact.

Fact	Opinion

3. **Critical Thinking** Why do some constellations appear only during certain seasons?

4. **Test Prep** How far away is the Sun from Earth?

   A 8 million km

   B 150 million km

   C 1 light-year

   D 71 million light-years

5. **Test Prep** Compared to other stars in the universe, the Sun is

   A much larger and hotter.

   B much smaller and colder.

   C much older and more massive.

   D about average.

6. **Essential Question** How do stars appear in the sky?

Writing Link

Write a Report
Write about a story, movie, or poem in which people travel among the stars. Discuss whether you think such travel is possible.

Math Link

Compare and Order Numbers
Write the following as numbers— four million, five trillion, two billion, eight thousand. Order them from smallest to largest.

Be a Scientist

aluminum foil

clear tape

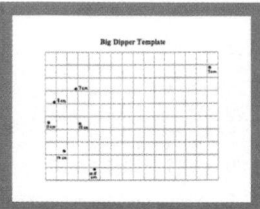

Big Dipper template

piece of cardboard

precut pieces of string

Structured Inquiry

Why do some distant stars appear to be close together?

Form a Hypothesis

Stars that are light-years apart can seem very close together. Does your viewing position affect how you see stars in the sky? Write your answer in the form "If I view a constellation from different positions, then I will observe..."

Test Your Hypothesis

1. **Make a Model** Make seven small balls of aluminum foil. These will model the stars in the Big Dipper.

2. Tape the Big Dipper template to the cardboard.

3. **Measure** Tape each length of string to the dot on the template marked with that length.

4. Open each foil ball partway. Insert the loose end of each string into a foil ball. Squeeze each ball tightly so that its string will stay in.

5. **Observe** Hold the cardboard up so the stars hang below it. Keep it one arm's length away. Observe the stars.

6. Rotate your model one turn to the left. Repeat step 5. Continue turning and observing until you have viewed the model from all four sides.

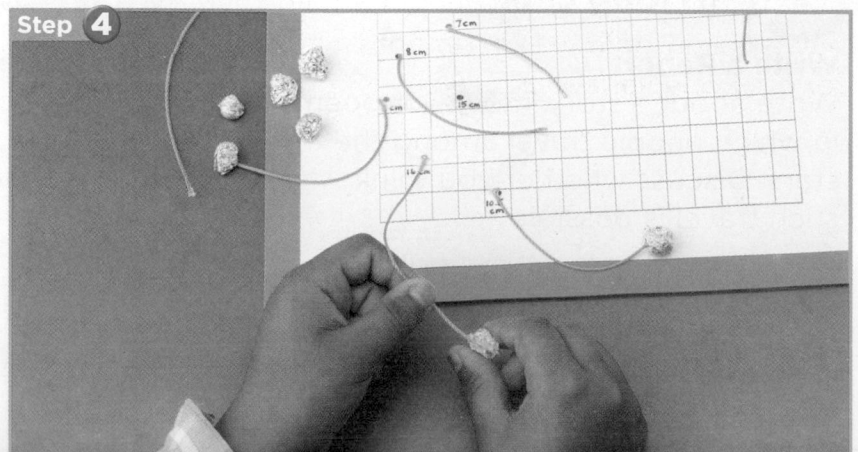

Draw Conclusions

7 What did your group of stars look like in step 5?

8 What changes did you observe each time you turned the model?

9 **Infer** When viewed from Earth, the stars in a constellation may seem close together. In space, those stars may be light-years apart. What can you infer about the stars in the Big Dipper?

Guided Inquiry

How does distance from Earth affect a star's apparent brightness?

Form a Hypothesis

How does a star's distance from Earth affect how bright it appears? Write a hypothesis.

Test Your Hypothesis

Make a plan to model how the distance from Earth affects the apparent brightness of a star. Write out the materials you need and the steps you will follow. Record your results and observations.

Draw Conclusions

Did your results support your hypothesis? Why or why not? Explain how you set up the experiment to test for only one variable.

Open Inquiry

What else can you learn about stars? For example, which constellations can you see during different seasons? Design an investigation to answer your question. Use reference materials to plan your activity. Write your procedure so that another group can complete the same activity by following your instructions.

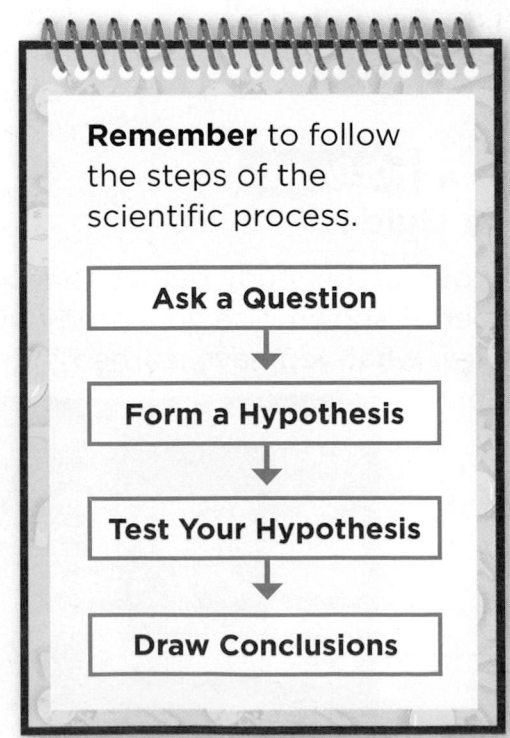

Remember to follow the steps of the scientific process.

Ask a Question
↓
Form a Hypothesis
↓
Test Your Hypothesis
↓
Draw Conclusions

Visual Summary

Lesson 1 Earth's movement in space causes day, night, and the seasons.

Lesson 2 As the Moon revolves around Earth, we observe its different phases.

Lesson 3 The Sun is at the center of the solar system. Planets, moons, and other objects orbit around the Sun.

Lesson 4 Stars are spheres of hot gases that give off light and heat.

Make a **FOLDABLES** Study Guide

Tape your lesson study guides to a piece of paper as shown. Use your study guide to review what you have learned in this chapter.

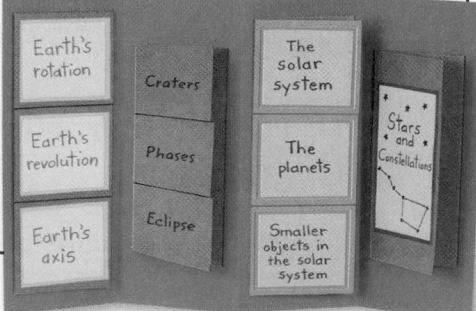

Fill each blank with the best term from the list.

comet, p. 388	**phase**, p. 373
crater, p. 371	**planet**, p. 380
gravity, p. 381	**revolution**, p. 362
lunar eclipse, p. 374	**rotation**, p. 360
meteor, p. 388	**star**, p. 394

1. Every 24 hours, Earth completes one __rotation__.

2. Each year, Earth completes one __revolution__ around the Sun.

3. Earth casts a shadow on the Moon during a __lunar eclipse__.

4. A chunk of ice, rocks, and dust that orbits the Sun is a _____.

5. If a meteorite enters Earth's atmosphere, it is called a _____.

6. A glowing sphere of gases that gives off heat and light energy is a _____.

7. A large, round object that orbits the Sun is called a _____.

8. When a meteor strikes the Moon, a _____ can form.

9. A full moon is a _____ of the Moon.

10. The attraction force between all objects is called _____.

Answer each of the following.

11. **Cause and Effect** What causes a solar eclipse?

12. **Interpret Data** Make a table showing how long it takes each planet to complete one rotation and one revolution. Does the planet with the shortest revolution also have the shortest rotation?

13. **Critical Thinking** Some people refer to comets as "dirty snowballs." Why do they use this term?

14. **Fictional Narrative** Suppose you moved to a new home near the South Pole. Write a story about the change of seasons there. Describe how those seasons differ from the ones where you live now.

15. **Infer** Explain why Mars will <u>most</u> <u>likely</u> be the easiest planet for people to visit someday.

16. **Critical Thinking** Explain why your shadow is longer in the morning than at midday.

17. **True or False** *People see the same stars all year long.* Is this statement true or false? Explain.

18. **Critical Thinking** Where on Earth does the Sun never set during the summer and never rise during the winter? Explain why.

19. We see the Sun rise and set because

   A Earth revolves around the Sun.

   B Earth rotates on its axis.

   C the Sun revolves around Earth.

   D the Moon revolves around Earth.

20. What objects are in the solar system and beyond?

Star Research

1. Choose and research a constellation. Explain what you find interesting about it.

2. Illustrate your constellation. Include labels for all of its stars.

3. Make a chart with details about your constellation. Include facts such as when it is visible in the sky and how it got its name. Also include the distance from Earth to the nearest star in the constellation.

4. Present your illustration and chart to the class.

1 Lorraine observed the Moon every other night for one week. Look at what she saw.

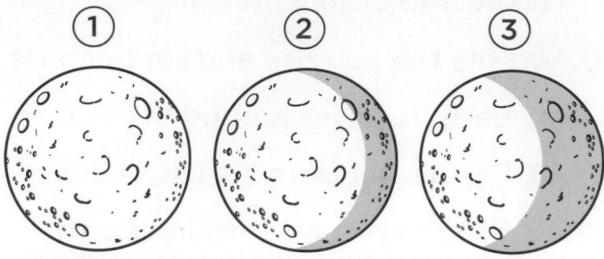

Which phase will she see next?

A

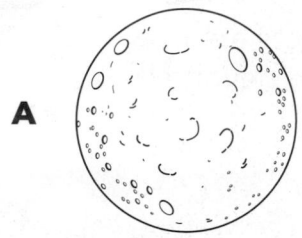

B

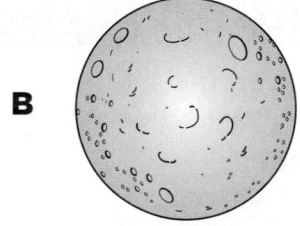

C

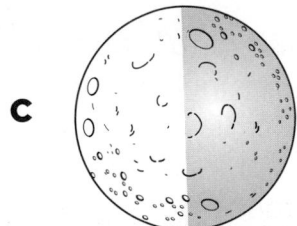

D

DOK 2

2 How is the Moon different from Earth?

A The Moon has no atmosphere.

B The Moon has no mountains.

C The Moon has different kinds of rock.

D The Moon has different kinds of living things.

DOK 1

3 How is the Sun different from all other stars?

A It is hotter than other stars.

B It is closer to Earth than other stars.

C It is bigger than other stars.

D It is brighter than other stars.

DOK 1

4 Your shadow is short when you go outside. What time of day is it?

A early morning

B late afternoon

C after sunset

D near noon

DOK 1

5 Meteoroids that strike Earth's surface are called

A comets.

B asteroids.

C meteors.

D meteorites.

DOK 1

6 A group of stars that form a pattern in the sky is called

A a constellation.

B an eclipse.

C a galaxy.

D a phase.
DOK 1

7 What causes Earth's changing seasons?

A Earth's rotation around the Sun

B the Sun's rotation around Earth

C Earth's tilted axis and revolution around the Sun

D Earth's rotation and the Moon's revolution around Earth
DOK 2

8 Which tool would <u>best</u> show the details of Saturn?

A telescope

B binoculars

C microscope

D rover
DOK 1

9 What do stars have in common with Jupiter, Saturn, Uranus, and Neptune?

A They give off their own light.

B They are beyond our solar system.

C They orbit around the Sun.

D They are made up of gases.
DOK 2

10 Which of these is a dwarf planet?

A Neptune

B the Sun

C Earth

D Pluto
DOK 1

Use the illustration below to answer questions 11–12.

11 What will the Moon look like in about two weeks?
DOK 1

12 What causes the phases of the Moon?
DOK 2

Check Your Understanding

Question	Review	Question	Review
1	pp. 372–373	7	pp. 362–363
2	pp. 370–371	8	pp. 382–383
3	pp. 394–398	9	pp. 386–387, 394–395
4	pp. 360–361	10	pp. 384–387
5	p. 388	11	pp. 372–373
6	pp. 396–397	12	pp. 372–373

Careers in Science

Planetarium Technician

Would you like to make star shows that are educational and fun? Think about being a planetarium technician. A planetarium is a place where people can watch representations of the solar system. These are usually light shows that are projected onto the ceiling and narrated.

As a planetarium technician, you would operate the audio and light equipment for the shows. You might work with teachers to help plan the programs. You would also get to see and hear the results of your work!

▲ A planetarium technician helps plan exciting star shows.

Air Traffic Controller

People depend on air traffic controllers to keep them safe during air travel. Some air traffic controllers direct planes on the runways. Others direct traffic between airports. All controllers make sure that airplanes keep a safe distance apart.

What does it take to become an air traffic controller? First, you need to be good at math. You should also have good speaking and computer skills. After college, you would train at the Federal Aviation Administration (FAA) Academy. Most graduates of this program have a lifelong career with the FAA.

▲ An air traffic controller keeps flight travel safe.

LOG ON e-Careers at www.macmillanmh.com

Matter

Heat can change sand
and minerals into glass!

worker making glass in a factory

Literature

Magazine Article

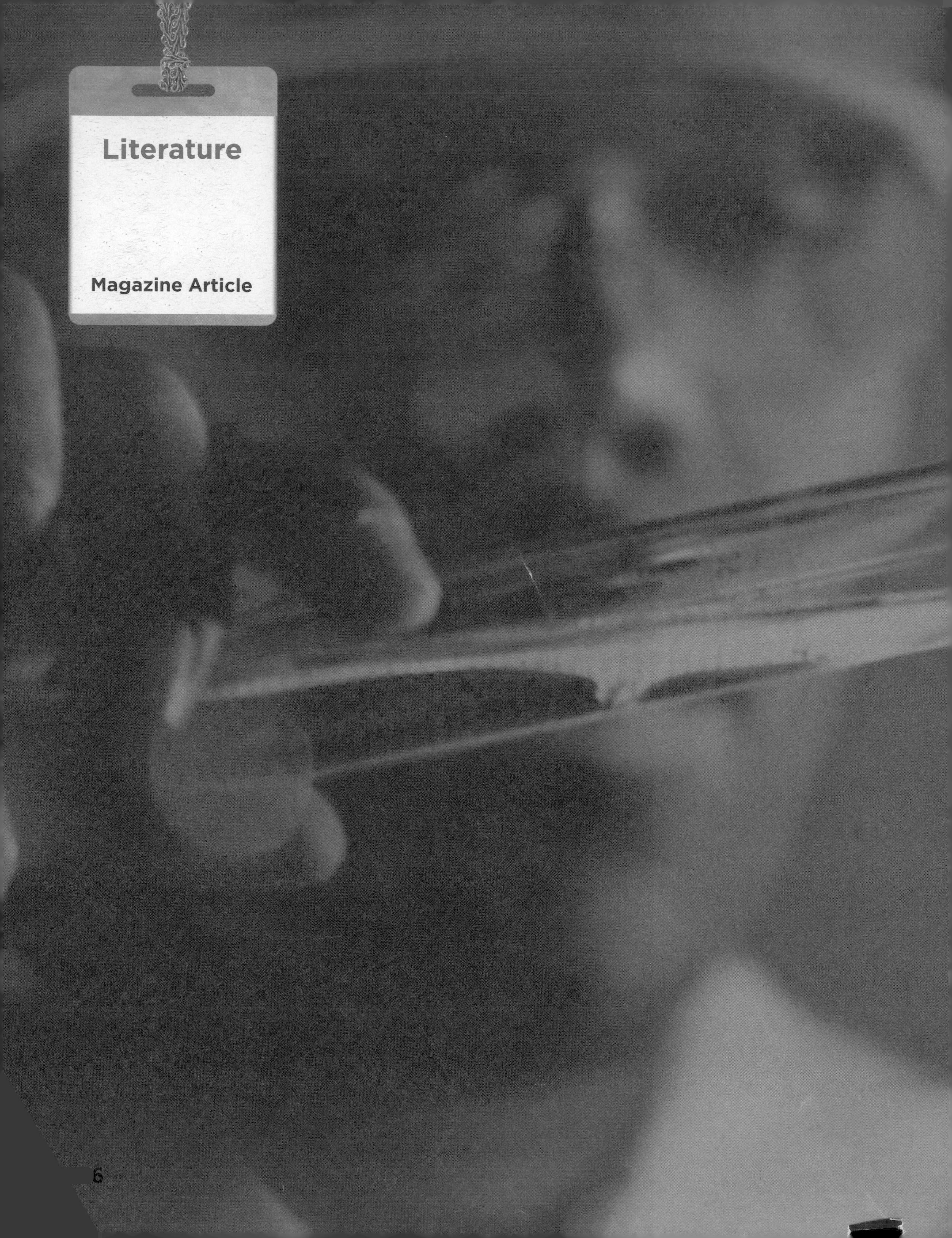

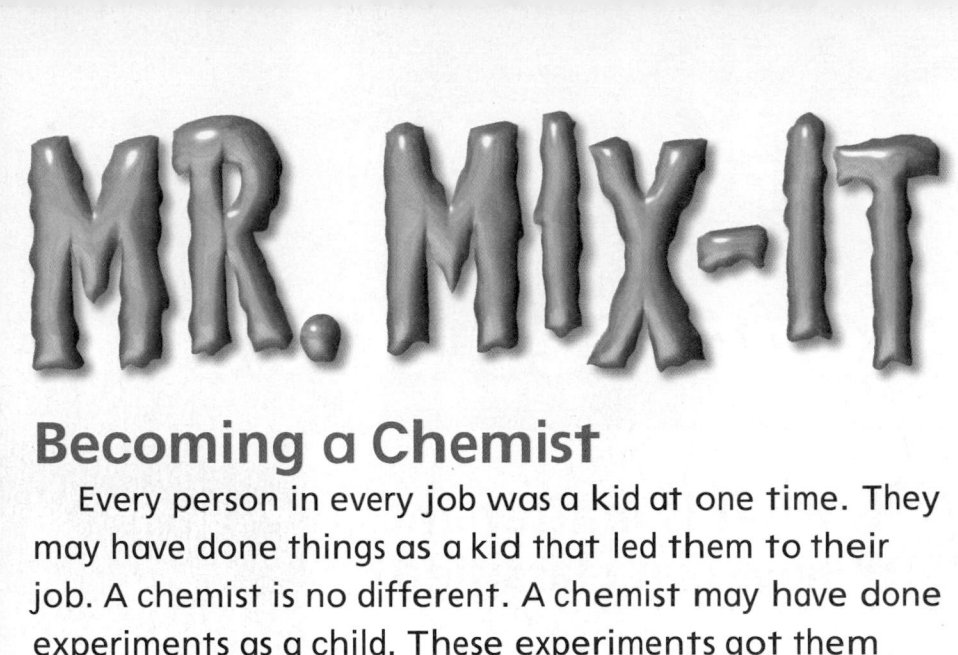

MR. MIX-IT

Becoming a Chemist

Every person in every job was a kid at one time. They may have done things as a kid that led them to their job. A chemist is no different. A chemist may have done experiments as a child. These experiments got them interested in chemistry.

You can study chemistry in school. You can learn how to mix different substances to make other substances. Sometimes the new substances look and act different from the old substances. This is how discoveries are made. If you like doing these kinds of experiments, maybe you should be a chemist.

A chemist can have many different types of jobs. Some chemists work in labs to produce new things you can buy in stores. Other chemists might work on making existing things better. Some chemists even teach other people about chemistry.

Look around your home or classroom. What substances do you see? Chances are, a chemist helped make that substance. So many things are made by mixing other things. There are always new discoveries to be made. Maybe you can be the chemist that makes a new invention.

 **Write About It**

Response to Literature What type of job would you like to have? What skills does it require? Write a paragraph about your plans.

 **-Journal** Write about it online at www.macmillanmh.com

CHAPTER 9

Properties of Matter

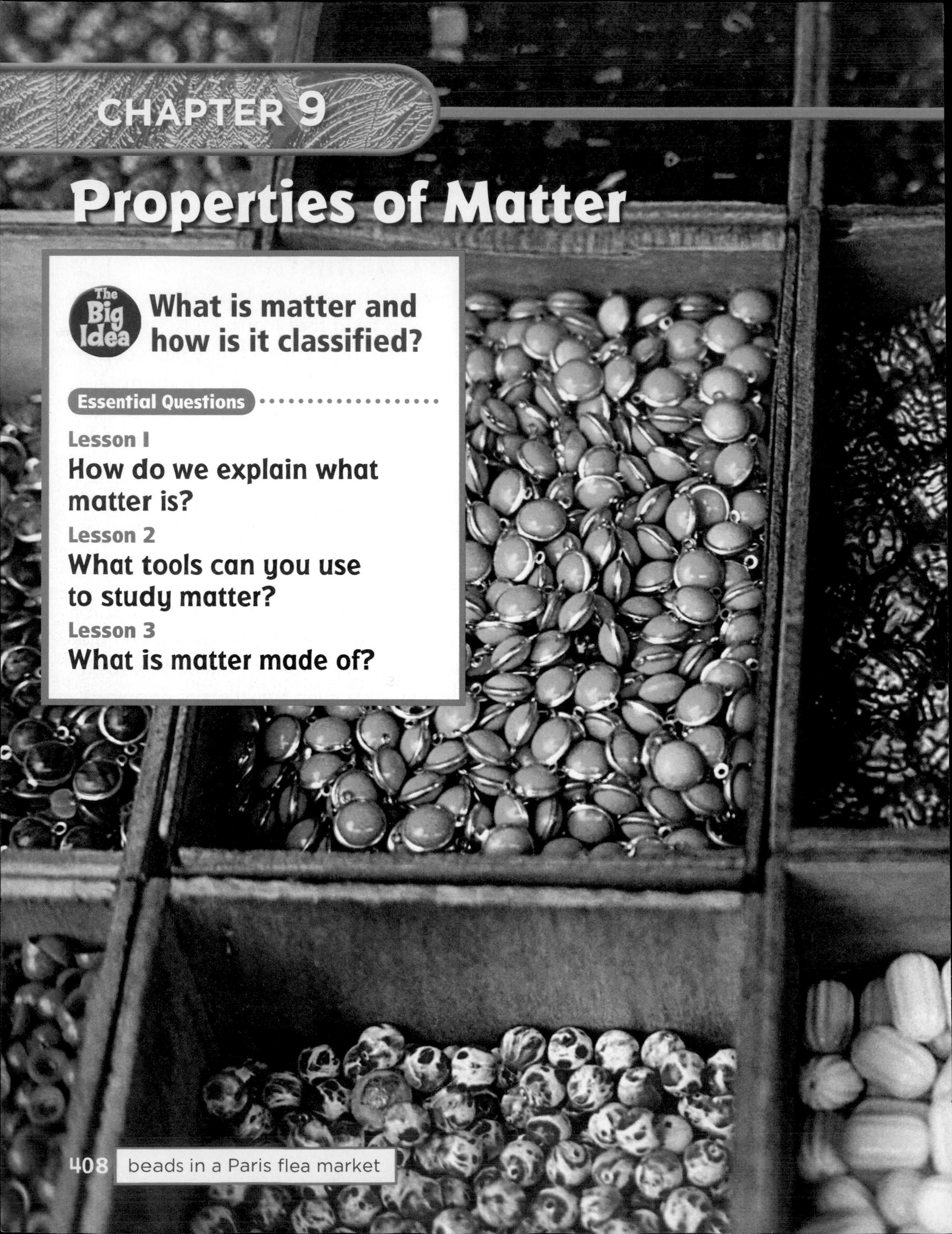

 **The Big Idea** What is matter and how is it classified?

Essential Questions ·····················

Lesson 1

How do we explain what matter is?

Lesson 2

What tools can you use to study matter?

Lesson 3

What is matter made of?

 # Big Idea Vocabulary

matter anything that has mass and takes up space (p. 412)

mass the amount of matter making up an object (p. 412)

length the number of units that fit along one edge of an object (p. 423)

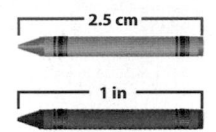

element a substance that is made up of only one type of matter (p. 432)

metal any of a group of elements found in the ground that conduct heat and electricity (p. 433)

periodic table a chart that shows the elements classified by properties (p. 434)

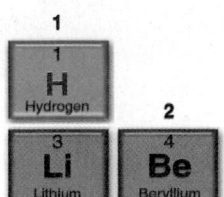

 Visit www.macmillanmh.com for online resources.

Describing Matter

Look and Wonder

In winter, rain can freeze to ice. The warmth of spring melts the ice. How can you tell the difference between rainwater and ice?

How can you tell if something is a solid or a liquid?

Make a Prediction

What is a solid? A liquid? Write a definition of each. If you mix cornstarch and water, will you have a solid or a liquid? Make a prediction.

Test Your Prediction

1. Pour the cornstarch and water into the bowl.

2. Use your fingers to mix the cornstarch and water together.

3. **Observe** Use your senses to observe the new substance. How does it feel? What does it look like? Record your description.

4. Tap the surface of the substance with your finger. Does it splash out of the bowl?

5. Place a small object such as a penny on the surface. Does it stay on top or sink?

Draw Conclusions

6. **Interpret Data** Compare your observations to your definitions. How is the new substance like a solid? How is it like a liquid?

7. **Infer** Is the mixture of cornstarch and water a solid or a liquid? Explain.

8. Do your results support your prediction? Why or why not?

Explore \ More

What would happen to this substance if you added more water? What if you let it dry out overnight? Make a prediction. Try it! Then report your results.

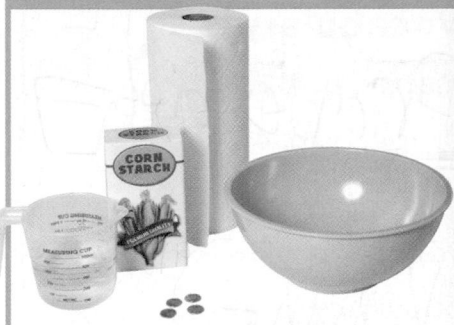

Materials

- **250 grams of cornstarch**
- **200 milliliters of water**
- **bowl**
- **penny**
- **paper towels**

Step 3

Read and Learn

▶ **Essential Question**

How do we explain what matter is?

▶ **Vocabulary**

matter, p. 412

property, p. 412

mass, p. 412

volume, p. 413

buoyancy, p. 413

solid, p. 414

liquid, p. 414

gas, p. 415

▶ **Reading Skill** ✓

Compare and Contrast

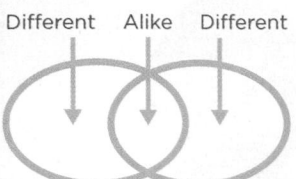

Different   Alike   Different

▶ **Technology**

e-Glossary and e-Review online at www.macmillanmh.com

What is matter?

When you mix cornstarch and water, you get a thick gooey substance. You can see and touch it. It takes up space in the container. Like many things, this substance is matter. **Matter** is anything that has mass and takes up space.

Most things are made of matter. The air you breathe and the book you are reading are made of matter. Light and heat are not matter, however. They do not take up space.

One way to describe matter is by its properties (PRAH•pur•teez). A **property** is a characteristic that you can observe. Color, shape, and size are some properties of matter.

Matter Has Mass

One very important property of matter is mass. **Mass** is the amount of matter making up an object. Mass is often measured in units called *grams* or *kilograms*. To measure mass, you use a tool called a *balance* (BA•luns).

Comparing Masses

Read a Photo

Which has more mass—the rock or the feather? How can you tell?

Clue: Which side of the balance is lower?

Matter Has Volume

Another property of matter is volume (VAHL•yum). **Volume** is how much space an object takes up. We measure volume by counting the number of cubic units in an object. We can also measure volume with tools like graduated cylinders.

Some Properties Are Unseen

Properties that cannot be seen can still be measured. Take magnetism, for example. This is the ability of matter to attract certain metal objects.

Another unseen property is the ability of matter to dissolve in a liquid. When a substance *dissolves,* it blends in and seems to disappear. Sugar and salt will dissolve in water. Sand will not.

Useful Properties

Properties help people choose the right kinds of matter for different jobs. When strength is needed, iron is a good choice. Wood is better when you need a light material that can easily be shaped.

Buoyancy (BOY•un•see) is a property that helps us build boats. **Buoyancy** is the upward force of a liquid or gas on an object. All objects are buoyant. Some objects are so buoyant that they float.

Magnetism is a property of matter. ▶

▲ Salt dissolves in water.

Sand does not dissolve in water. ▼

◀ Some objects can float in water. Other objects sink.

✔ Quick Check

Compare and Contrast How are mass and volume alike? How are they different?

Critical Thinking How do you know that your desk is matter?

What are the states of matter?

Matter is found in many forms. We call these forms *states*. Solid, liquid, and gas are the three common states of matter on Earth.

Solids

A **solid** has a definite shape and takes up a definite amount of space. The particles of matter in a solid are packed together tightly. Often the particles are packed in a regular pattern. This textbook and your desk are solids. What other solids can you name?

This musical instrument is a solid. Its particles are packed closely together.

Juice is a liquid. Its particles are less tightly packed as those in a solid. They can move past one another.

Liquids

Orange juice is a liquid. Unlike a solid, a **liquid** does not have a definite shape. It takes the shape of its container. However, a liquid does take up a definite amount of space.

Consider a glass of juice. The juice has the same volume whether it is in a glass or a graduated cylinder. If the juice spills, it will spread out. Its volume stays the same.

In a liquid, the particles of matter can move more than they do in a solid. The particles can change place and slide past one another. They are farther apart than in a solid.

Water, milk, and oil are all liquids. What other liquids can you name?

Gases

Helium (HEE•lee•um) is an example of a gas. A **gas** does not have a definite shape. In that way it is like a liquid.

Unlike a liquid, a gas does not take up a definite amount of space. It fills the shape and space of its container. The helium in a balloon takes the shape of the balloon. If the balloon bursts, the helium will spread out into the air.

In a gas, the particles of matter move about freely. The particles move farther apart from one another to fill the space around them. If there is less space to fill, the particles are closer together. A gas always spreads out to fill its container.

Quick Lab

States of Matter

1. Place several ice cubes in a pan. What state do they represent?

2. **Observe** Look at the ice cubes after 30 minutes. Now what states are represented?

3. Have your teacher heat the pan.

4. **Observe** What states do you see after the pan is heated?

Inside these balloons is a gas. Gas particles move about freely and spread far apart.

✔ Quick Check

Compare and Contrast How are solids, liquids, and gases the same? How are they different?

Critical Thinking A cornstarch and water mixture has both liquid and solid properties. How would you classify it?

FACT The particles that make up solids do move.

Objects Made by People

Objects in Nature

Read a Photo

How are these objects classified? How else could you sort them?

Clue: Think of the properties of each object.

What happens to the matter we use?

You use matter all the time. The food you eat is matter. Your chair is matter. You even breathe matter!

Some matter, like air, can be used again and again. Other forms of matter get thrown away. Too often, matter becomes trash. It goes into landfills or oceans.

Many people choose to *reuse* matter. This is when you use something again instead of throwing it away. An egg carton can be used to plant seeds. Are there other uses for things you throw away?

Matter can also be *recycled,* or made into something else. Cans, paper, plastic, and glass can all be recycled. What else can you recycle?

✔ Quick Check

Compare and Contrast What is the difference between using matter and reusing matter?

Critical Thinking How many uses can you think of for a milk carton? List them.

Lesson Review

Visual Summary

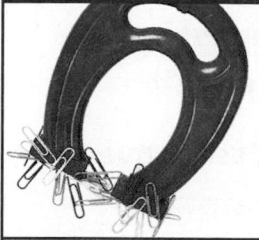
All **matter** has mass. Other properties of matter include volume, magnetism, and buoyancy.

The **three states** of matter are solid, liquid, and gas. Each has particles with different **physical properties.**

People use matter in many different ways. People can also reuse matter.

Make a **FOLDABLES** Study Guide

Make a three-tab book. Use it to summarize what you learned about describing matter.

Think, Talk, and Write

1 Vocabulary Solid, liquid, and gas are three _____ of matter.

2 Compare and Contrast Choose two states of matter. How are they alike? How are they different?

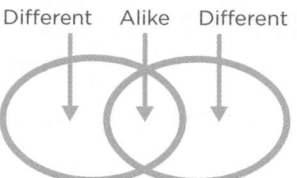

3 Critical Thinking Look around your school or classroom. List examples of solids, liquids, and gases.

4 Test Prep Which of the following is matter?

- **A** heat
- **B** sound
- **C** air
- **D** light

5 Essential Question How do we explain what matter is?

Writing Link

Write a Paragraph
Choose an object from around your classroom or home. Make a list of its properties. Write a short paragraph describing those properties. How do they make the object useful?

Math Link

Measure Water
Joel wants to empty a gallon of water into jars. He has jars that hold one quart and jars that hold two quarts. What combinations of jars can he use to empty all four quarts of the gallon?

Juggling Matter

You've seen jugglers throwing balls into the air and catching them. Usually the balls weigh about six ounces. They range in size from two to three inches in diameter. Most of the balls fit in thepalm of your hand. They can be soft so they do not cause any harm. Most juggling balls bounce. They are often brightly colored so the audience can see them.

Descriptive Writing

Good descriptive writing

▶ tells how things look, sound, smell, taste, and feel;

▶ uses details to compare and contrast.

Some jugglers use bowling balls. Bowling balls are larger than juggling balls. They are also a lot heavier. A bowling ball can weigh up to 16 pounds! Most are about eight or nine inches in diameter. Only professional jugglers should ever juggle bowling balls!

Write About It

Descriptive Writing

Choose three or four objects to describe. For example, you might choose a child's toy, your pet's toy, and a backpack. Write a paragraph describing them. Include the properties that make these objects useful to you.

 **e-Journal** Research and write about it online at **www.macmillanmh.com**

Math in Science

Taking Up Space

Volume is the amount of space that something takes up. Tools like measuring cups and beakers make it easy to find the volume of a liquid. You probably use measuring cups at home to add milk or water to a recipe. How can you find the volume of a solid?

To find the volume of a solid, you first take its measurements. Then, you make a calculation. For a rectangular solid, you measure its length, width, and height. Then, you multiply those numbers together.

Let's look at an example. A box measures 30 centimeters in length, 20 cm in width, and 10 cm in height. To find its volume, just multiply the numbers.

Calculating Volume

▶ The volume (V) of a rectangular object is the product of its length (l), width (w), and height (h). Another way of stating this relationship is: $V = l \times w \times h$

▶ In the example:
$V = 30 \text{ cm} \times 20 \text{ cm} \times 10 \text{ cm}$
$V = 6,000 \text{ cm}^3$

▶ What is a cm^3? It is a unit of volume called a cubic centimeter. One cm^3 is a cube with sides that are each 1 cm long. Six thousand of them would fit in a box with the measurements above.

Solve It

Calculate the volumes of the objects shown.

1. length = 6 cm, width = 4 cm, height = 2 cm

2. length = 31 cm, width = 18 cm, height = 11 cm

3. length = 5 cm, width = 25 cm, height = 38 cm

Measurement

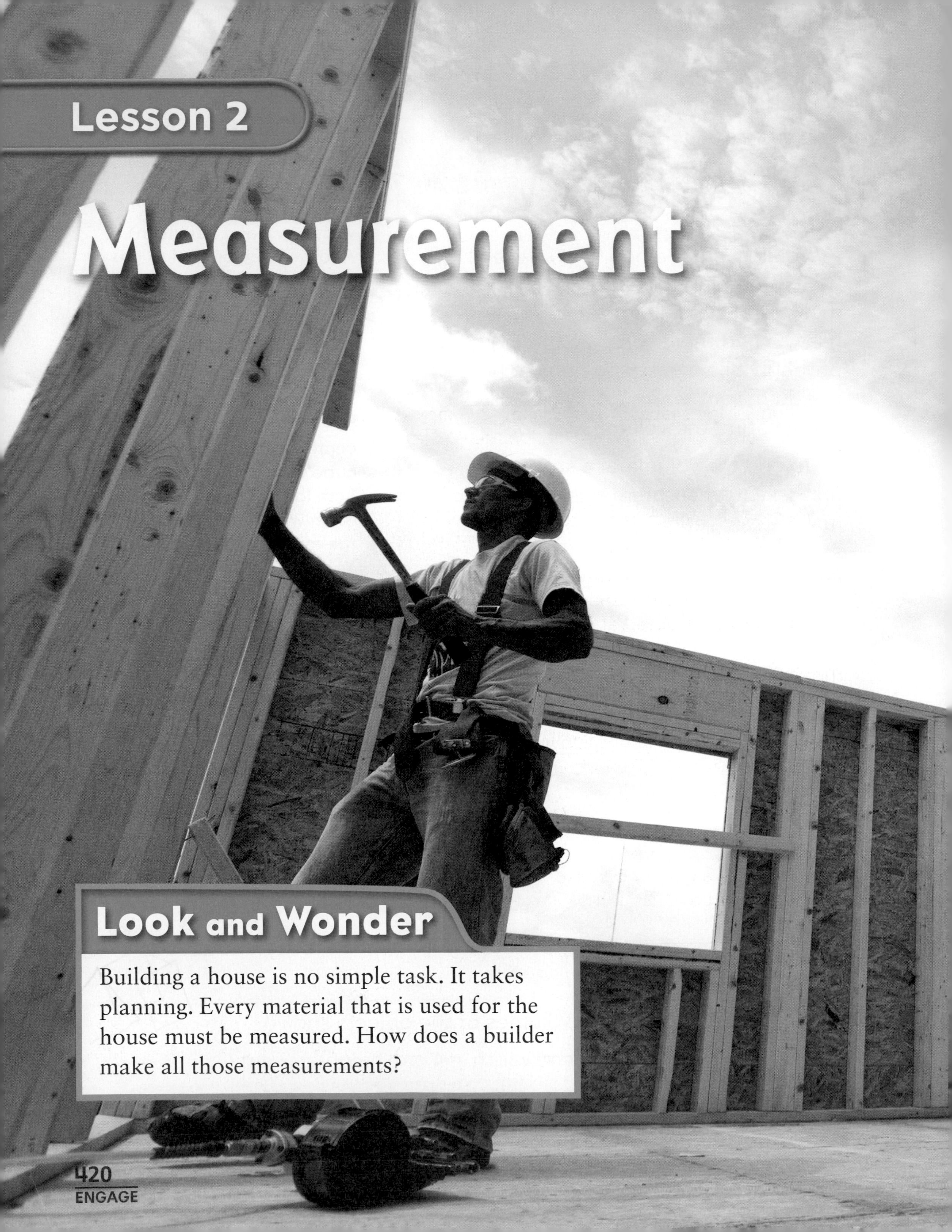

Look and Wonder

Building a house is no simple task. It takes planning. Every material that is used for the house must be measured. How does a builder make all those measurements?

How can you compare matter?

Make a Prediction

Look at shapes *A*, *B*, and *C*. Predict how you can use the ruler to determine the largest and smallest shapes. Make a prediction.

Materials

- 3 shapes labeled *A*, *B*, and *C*
- ruler
- pencil

Test Your Prediction

1 **Measure** Use the ruler to draw 1-inch squares on shapes *A* and *B*. Draw as many as you can fit. If you reach the edge, make a partial square.

2 **Use Numbers** Look at shapes *A* and *B*. How will you use the squares you drew to determine which shape is largest? Smallest?

3 **Observe** Repeat step 1 on shape *C*. Compare the three shapes again. Record your observations.

Draw Conclusions

4 Which shape is the largest? Smallest?

5 **Communicate** How did you use the 1-in. squares to compare the shapes?

6 Was your prediction correct? Explain.

Explore More

Can you use a different measuring tool to compare shapes *A*, *B*, and *C*? Make a prediction. Then try it.

▶ **Essential Question**

What tools can you use to study matter?

▶ **Vocabulary**

metric system, p. 422

length, p. 423

area, p. 423

density, p. 424

weight, p. 426

gravity, p. 426

▶ **Reading Skill** ✔
Problem and Solution

Problem
↓
Steps to Solution
↓
Solution

▶ **Technology** 🔵LOG ON
e-Glossary, e-Review, and animations online at www.macmillanmh.com

How do we measure matter?

Measuring and counting squares is one way to compare size. When we measure, we use *standard units*. A standard unit is a measurement on which people agree. Many people in the United States use standard units in the English system. The inch (in.), pound (lb), and ounce (oz) are standard units in that system.

Scientists use standard metric units. The **metric system** is based on units of ten. It uses prefixes such as *kilo-*, *centi-*, and *milli-* to define the size of measurements. For example, 1 meter is divided into 100 cm. There are 1,000 meters in 1 km.

Metric Units	Amount	Estimated Length
1 centimeter	$\frac{1}{100}$ of a meter	the width of your thumbnail
1 decimeter	10 cm $\frac{1}{10}$ of a meter	the length of a crayon
1 meter	10 dm 100 cm	the length of a baseball bat
1 kilometer	1,000 m 100,000 cm	the distance you walk in 10 to 15 minutes

Read a Table

How many centimeters are in a meter? In a kilometer?

Clue: Find each unit in the first column, then look across the row.

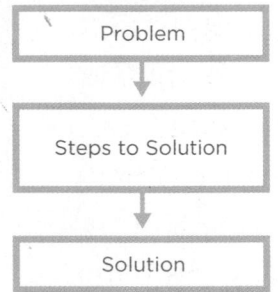

— 10 cm —

— 4 in. —

You can measure length in centimeters or inches.

Length and Width

An object's **length** is the number of units that fit from one end to the other. *Width* is the number of units that fit across. How wide is this page? How long is it?

Area

Area (AYR•ee•uh) describes the number of unit squares that cover a surface. An easy way to find the area of a rectangular shape is to multiply its length by its width. The area of this page, for example, is 27 cm × 20 cm, or 540 square cm (cm²).

What if a shape is not rectangular? Divide it into smaller squares. Find the area of each smaller shape. You might need to estimate parts of some shapes. Then add the area of each smaller shape to find the total area.

Kitchen tools measure volume in teaspoons or tablespoons.

Volume

Volume describes the number of cubes that fit inside an object. To find the volume of a rectangular solid, multiply its length by its width and height.

If a solid is not rectangular, you can use water. First, measure the amount of water in a container. Then, submerge the entire object below the water. Subtract the original water level from the new water level. The result is the volume of the object.

To find the volume of a liquid, pour it into a measuring cup, spoon, beaker, or graduated cylinder. Then read the markings on the container.

A baker can measure volume in cups or pints.

✔ Quick Check

Problem and Solution How can you measure the area and volume of your room?

Critical Thinking How can you find the area of a triangle?

What is density?

A plastic ball floats on water. If you fill the ball with sand, it will sink. Why? The volume of the ball is the same, but you increased its mass.

Mass Divided by Volume

The relationship between mass and volume is called density (DEN•suh•tee). **Density** is the mass of the matter in a given space. Scientists define density as the amount of mass in a unit of volume.

To find the density of an object, divide its mass by its volume. If the mass is in grams and the volume in cubic centimeters, then the result will have units in grams per cubic centimeter (g/cm^3). One cubic centimeter of steel is denser than one cubic centimeter of plastic.

The density of cork is 0.24 g/cm^3. The particles are loosely packed.

The density of marble is between 2.4 and 2.7 g/cm^3.

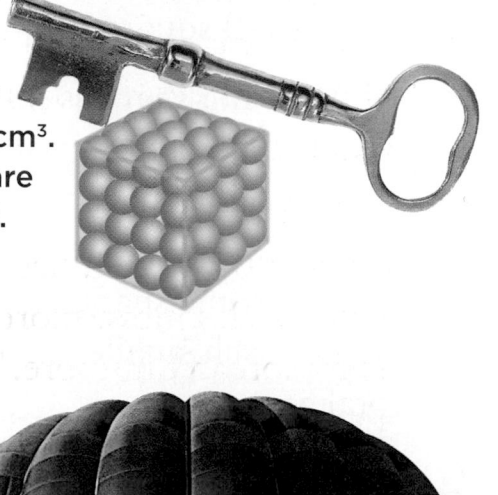

The density of brass is 8.5 g/cm^3. The particles are tightly packed.

Real-World Density

◄ air particles outside balloon

air particles inside balloon ►

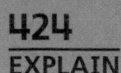

Density and Buoyancy

The density of an object also affects its buoyancy. Remember, buoyancy is the upward force of a liquid or gas on another object.

Float or Sink?

Consider cork and water. The density of water is 1 g/cm³. The density of cork is 0.24 g/cm³. Does cork float or sink?

An object floats when its density is less than the density of the liquid or gas in which it is placed. The density of cork is less than the density of water. So cork floats on water. Liquids can float on top of water too.

Can you change the density of matter? If you add heat to air, the air particles move more quickly. They spread out more. The heated air is less dense. It rises as cooler, denser air forces it upward.

Read a Diagram

Why does a hot-air balloon float?

Clue: Compare the density of the air inside the balloon with the density of the air outside.

LOG ON *Science in Motion* Watch density affect balloons at www.macmillanmh.com

☰ Quick Lab

Comparing Densities

❶ **Predict** Water, oil, and syrup have different densities. What will happen if you pour them into the same container?

❷ **Measure** Pour 100 mL of water into a cup. Then pour 100 mL of oil into the cup. Last, slowly add 100 mL of syrup.

❸ What happened when you added each liquid to the cup? Was your prediction correct?

❹ Drop a craft stick, a piece of pasta, and a crayon into the cup. Where does each float? Why? What can you say about the density of the liquids and solids?

✔ Quick Check

Problem and Solution What is the density of a cube with a mass of 8 g and volume of 1 cm³?

Critical Thinking What should a hot-air balloonist do to go higher? Explain.

What is weight?

Do you know your weight (WAYT)? Weight is another way to measure matter. Weight and mass may seem similar, but they are not the same.

Mass is the amount of matter in an object. **Weight** measures the amount of gravity between an object and a planet, such as Earth. **Gravity** is a force, or pull, between all objects.

How are weight and mass related? The force of gravity depends, in part, on an object's mass. The more mass, the stronger the pull of gravity. The stronger the pull of gravity, the more an object weighs.

Unlike mass, an object's weight is different on other planets and on the Moon. The pull of gravity on the Moon is about $\frac{1}{6}$ as strong as on Earth. So an object's weight on the Moon is only $\frac{1}{6}$ of its weight on Earth.

Do you weigh yourself with a scale? Mass is measured with a balance. Weight is measured with a scale. Ounces and pounds are the English units for weight. The metric unit for weight is the *newton* (N).

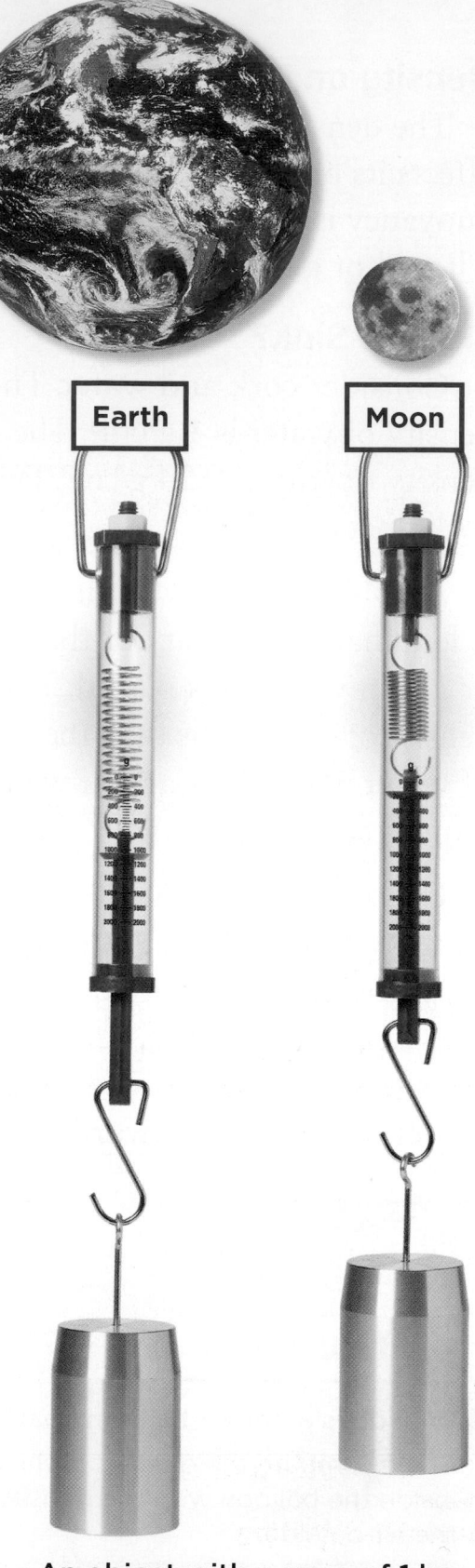

Earth

Moon

An object with a mass of 1 kg weighs 9.8 N on Earth. On the Moon, the same object weighs just 1.6 N.

✔ Quick Check

Problem and Solution How would you measure a rock's mass on the Moon?

Critical Thinking What is the difference between a balance and a scale?

Lesson Review

Visual Summary

 We use **standard units** to measure the length, width, area, and volume of an object.

 We calculate **density** by dividing the mass of an object by its volume.

 **Weight** is a measure of the pull of gravity. We measure weight with an instrument called a scale.

Make a FOLDABLES® Study Guide

Make a trifold book. Use it to summarize what you learned about measurement.

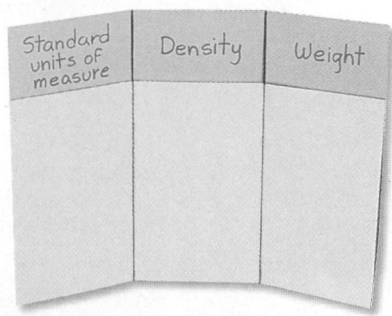

Think, Talk, and Write

❶ **Vocabulary** The number of unit squares that cover a surface describes its _____.

❷ **Problem and Solution** Describe how to find the volume of air in your classroom.

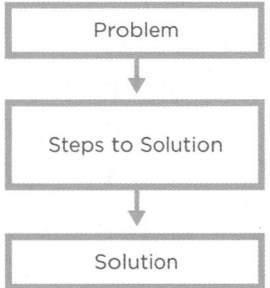

❸ **Critical Thinking** Why does 1 kg of foam take up more space than 1 kg of rock?

❹ **Test Prep** This property of matter changes depending on the pull of gravity.

   A density
   B length
   C mass
   D weight

❺ **Essential Question** What tools can you use to study matter?

 **Writing Link**

Scientific Writing
You are classifying several different objects. Write a report telling how you will determine the properties of each object.

 **Math Link**

Calculate Area and Volume
Measure the length, width, and height of your desk. What is its area? What is its volume?

Focus on Skills

Inquiry Skill: Measure

You know that there are many kinds of rocks and minerals. Scientists can describe a particular rock by its properties. Two properties that you can use to describe rock are mass and length. You **measure** to find an object's mass and length.

▶ Learn It

When you **measure,** you find the length, volume, area, mass, or temperature of an object. You can use tools to measure these properties. When you measure, it is a good idea to record your measurements on a table or chart. This helps you stay organized.

▶ Try It

Estimate and **measure** the mass and length of a rock.

Materials 3 rocks, gram masses, balance, metric ruler

1. Get a rock. Hold it in your hand. Estimate the mass of the rock. Compare the rock to gram masses that you hold in your other hand. Record your estimate, in grams, in a table like the one shown.

2. Measure the mass of the rock using a balance and gram masses. Place the rock on one side of the balance. One by one, place gram masses on the other side of the balance. When the two sides are even, stop. Add the gram masses to find the actual mass of the rock. Record it.

3. About how long do you estimate the rock is? Use the longest side of the rock. Record your estimate, in millimeters or centimeters, in the table.

4. Measure the length of the rock with a metric ruler. Record the actual length.

▶ Apply It

Estimate and **measure** the mass and length of two more rocks. Record this data in your table.

1 Look at your data. Did you closely estimate the mass of each rock? Did you closely estimate the lengths? Which was easier for you to estimate—mass or length? Why?

2 With practice, you can become better at estimating mass and length. Repeat the activity using different rocks. Record your estimates and actual measurements again in a table.

3 Were your estimates closer to your actual measurements this time?

4 Do you think you can now estimate the mass of a rock before you pick it up? Try it for several rocks. Then use the balance to measure the actual mass. What property or properties do some rocks have that might throw off your estimate?

Rocks	1	2	3
Estimated Mass			
Actual Mass			
Estimated Length			
Actual Length			

Classifying Matter

Look and Wonder

Everything in this picture is made of matter. Just as the basic unit of all living things is the cell, matter has a basic unit too. What is that unit? How does it differ among matter?

How can you identify a metal?

Purpose

Test the properties of some materials to find out which are metals.

Materials

Procedure

1. Obtain three unknown materials from your teacher. Make a table like the one shown.

- **3 unknown materials**
- **beaker**
- **ice water**

Property	Test
luster	Is the material shiny?
can bend	Can the material be bent into a shape?
carries heat	Does the material carry heat well?

2. **Observe** Examine each material. Is it shiny? Can it be bent? Answer yes or no.

3. Fill a beaker with ice water. For each material, place one end in the water. Does the end you are holding get cold? Record your answers.

Step 3

Draw Conclusions

4. **Interpret Data** How do the three materials compare? Is one more like another?

5. **Classify** Metals are shiny, can bend, and carry heat well. Which of the materials you tested are metals?

Explore More

Collect some materials from around your classroom or home. Test their properties using the same procedure. Record your findings in a table. Then classify the materials as metals or nonmetals.

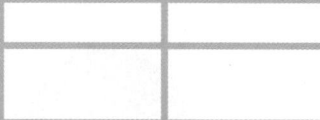
What are elements?

People once thought that all matter was made up of earth, air, fire, and water. We now know that all matter is made of elements (E•luh•munts). An **element** is a substance that is made up of only one type of matter.

Scientists call elements "the building blocks of matter." That is because an element cannot be broken down into a simpler form. Hydrogen and oxygen are elements. So are gold and silver.

Atoms

Elements are made up of atoms (A•tumz). An **atom** is the smallest part of an element. You can think of atoms as tiny particles—so tiny that you cannot see them!

All atoms in an element are alike. For instance, all the atoms in copper are copper atoms. They have different properties from the atoms of any other element.

Neon is a gas. In a tube, neon glows if electricity is added.

FACT ▶ Fire is not an element.

Metals and Nonmetals

How do we classify elements? One way is to decide wether they are metal (ME•tul) or nonmetal. A **metal** is shiny. It can be bent or hammered into a shape. Some metals are iron, aluminum (uh•LEW•muh•num), and copper.

Metals let heat and electricity pass through them easily. A metal pan over a flame or a heated burner gets hot very quickly.

Metalloids (ME•tul•oydz), such as silicon, have some but not all of the properties of metals. Oxygen and nitrogen are *nonmetals*. They have none of the properties of metals. They are gases.

Symbols for Elements

Scientists use symbols to stand for each element. Often, a symbol is the first letter of the element's name. For example, *C* is the symbol for carbon. Some elements take their symbols from their Latin names. The Latin word for gold is *aurum* (AW•rum). Gold's symbol is Au. The first letter in a chemical symbol is always capitalized.

▲ Aluminum is strong and light in weight.

▼ pure copper

▲ Artists use copper to make jewelry.

✔ Quick Check

Classify Name two elements that are gases. Name two elements that are metals.

Critical Thinking Table salt is made of two elements—sodium and chlorine. Is table salt an element? Why or why not?

Key

11 Na Sodium	— Atomic number — Element symbol — Element name	▨ Metals
		▨ Metalloids (semimetals)
		▨ Nonmetals

State at room temperature:
Solid   Liquid   Gas

1 H Hydrogen											
3 Li Lithium	4 Be Beryllium										
11 Na Sodium	12 Mg Magnesium	3	4	5	6	7	8	9	10	11	12
19 K Potassium	20 Ca Calcium	21 Sc Scandium	22 Ti Titanium	23 V Vanadium	24 Cr Chromium	25 Mn Manganese	26 Fe Iron	27 Co Cobalt	28 Ni Nickel	29 Cu Copper	30 Zn Zinc
37 Rb Rubidium	38 Sr Strontium	39 Y Yttrium	40 Zr Zirconium	41 Nb Niobium	42 Mo Molybdenum	43 Tc Technetium	44 Ru Ruthenium	45 Rh Rhodium	46 Pd Palladium	47 Ag Silver	48 Cd Cadmium
55 Cs Cesium	56 Ba Barium	57 La Lanthanum	72 Hf Hafnium	73 Ta Tantalum	74 W Tungsten	75 Re Rhenium	76 Os Osmium	77 Ir Iridium	78 Pt Platinum	79 Au Gold	80 Hg Mercury
87 Fr Francium	88 Ra Radium	89 Ac Actinium	104 Rf Rutherfordium	105 Db Dubnium	106 Sg Seaborgium	107 Bh Bohrium	108 Hs Hassium	109 Mt Meitnerium	110 Ds Darmstadtium	111 Rg Roentgenium	112 Uub Ununbium

58 Ce Cerium	59 Pr Praseodymium	60 Nd Neodymium	61 Pm Promethium	62 Sm Samarium	63 Eu Europium	64 Gd Gadolinium	65 Tb Terbium
90 Th Thorium	91 Pa Protactinium	92 U Uranium	93 Np Neptunium	94 Pu Plutonium	95 Am Americium	96 Cm Curium	97 Bk Berkelium

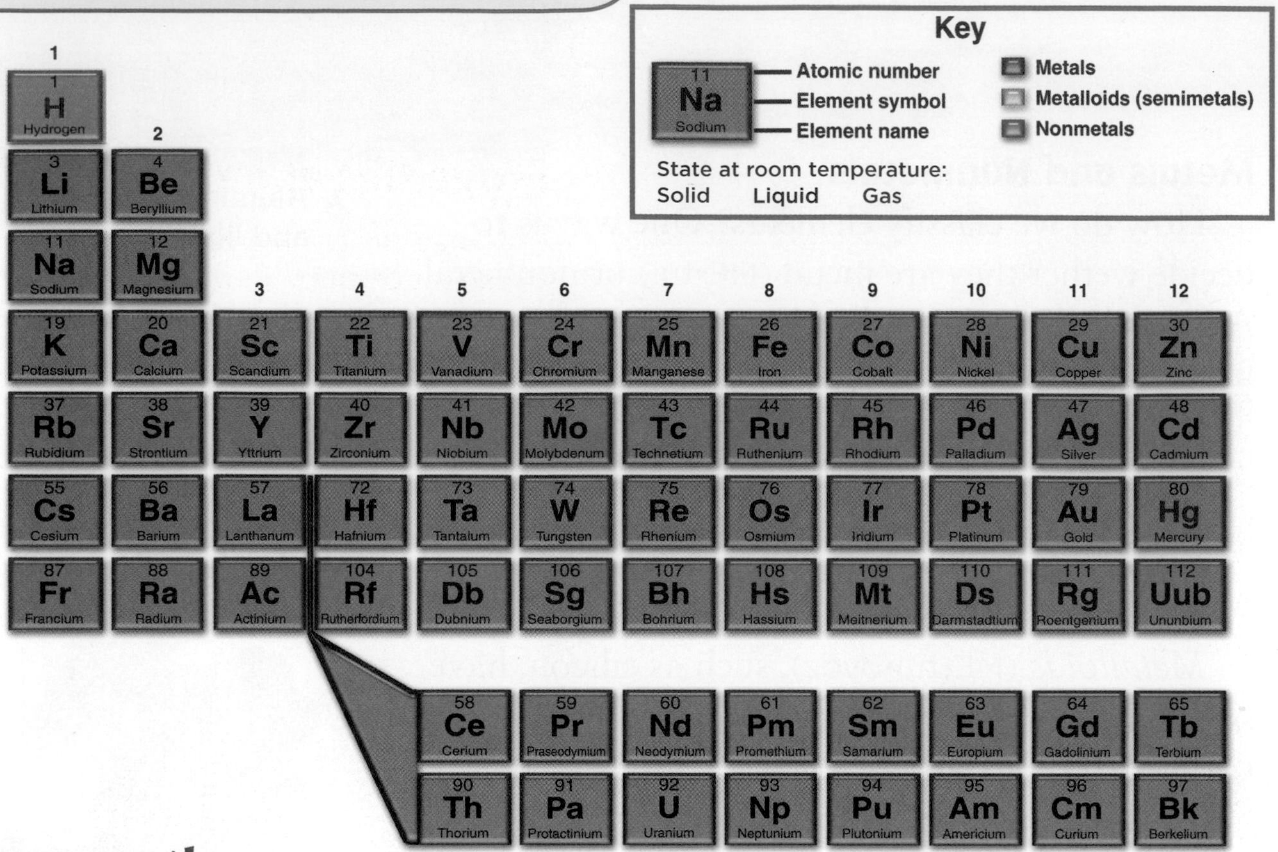

How are the elements organized?

Nearly 150 years ago, a Russian scientist named Dimitri Mendeleev (men•duh•LAY•uf) made a table of the elements. He based his table on the known properties of elements. It is the periodic (peer•ee•AH•dihk) table.

Atomic Number

As new elements were found, they were added to the table. The modern table is shown above. The three main groups are metals, metalloids, and nonmetals. You can see that the elements are arranged by their atomic numbers. The *atomic number* is related to the mass of each element.

Columns and Rows

So far, scientists have named 118 elements. The periodic table shows how they are similar and how they are different. Elements in the same column have similar properties. For example, all the elements in column 17 combine easily with other elements. They often form new substances. The elements in column 18 hardly ever combine with other elements.

The rows of the table are called *periods*. Iron, cobalt, and nickel are magnetic. They are in the same row in the table.

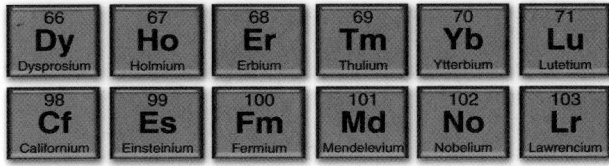

13	14	15	16	17	18
					2 **He** Helium
5 **B** Boron	6 **C** Carbon	7 **N** Nitrogen	8 **O** Oxygen	9 **F** Fluorine	10 **Ne** Neon
13 **Al** Aluminum	14 **Si** Silicon	15 **P** Phosphorus	16 **S** Sulfur	17 **Cl** Chlorine	18 **Ar** Argon
31 **Ga** Gallium	32 **Ge** Germanium	33 **As** Arsenic	34 **Se** Selenium	35 **Br** Bromine	36 **Kr** Krypton
49 **In** Indium	50 **Sn** Tin	51 **Sb** Antimony	52 **Te** Tellurium	53 **I** Iodine	54 **Xe** Xenon
81 **Tl** Thallium	82 **Pb** Lead	83 **Bi** Bismuth	84 **Po** Polonium	85 **At** Astatine	86 **Rn** Radon
113 **Uut** Ununtrium	114 **Uuq** Ununquadium	115 **Uup** Ununpentium	116 **Uuh** Ununhexium	117 **Uus** Ununseptium	118 **Uuo** Ununoctium

66 **Dy** Dysprosium	67 **Ho** Holmium	68 **Er** Erbium	69 **Tm** Thulium	70 **Yb** Ytterbium	71 **Lu** Lutetium
98 **Cf** Californium	99 **Es** Einsteinium	100 **Fm** Fermium	101 **Md** Mendelevium	102 **No** Nobelium	103 **Lr** Lawrencium

Read a Table

What is the atomic number for fluorine (F)? Is it a metal? What state is it at room temperature?

Clue: Look at the key at the top of the previous page.

Diamonds, coal, and the graphite in pencils are all made of carbon.

Properties of an Element

① **Observe** Look carefully at a sheet of aluminum foil. List the properties you can see and touch.

② Tear the foil in half, and then tear it in half again. Tear each piece in half once more. You will have eight small pieces of foil.

③ What properties do the pieces of foil have? Make a new list. Compare it to the properties you listed for the whole sheet of foil.

④ **Interpret Data** Are the pieces of aluminum foil still aluminum? How are the pieces of foil similar to the atoms of an element?

✓ Quick Check

Classify Describe how the periodic table is organized.

Critical Thinking How are coal and diamonds similar? Different?

Elements in column 1, such as potassium (K), react strongly with water and other substances.

Many nails contain iron (Fe), a magnetic metal. Cobalt (Co) and nickel (Ni) are magnetic metals too.

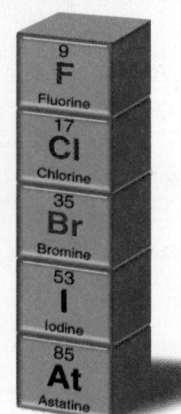

This mineral contains fluorine (F), a halogen. Elements in this column react with elements in column 1.

Read a Diagram

Which elements are magnetic? Which elements form salts?

Clue: Notice the symbols in each box.

How do scientists use the periodic table?

Scientists use the periodic table to predict how an element will react. For example, elements in the same column have similar properties. They react in similar ways.

Elements in column 18 are called noble gases. *Noble gases* rarely react with other elements. The neon (Ne) in neon signs is a noble gas.

Compare neon with hydrogen (H). Hydrogen is a gas in column 1. Unlike noble gases, hydrogen reacts and forms many new substances. The other elements in column 1 are metals called alkali (AL•kuh•lie) metals. Like hydrogen, *alkali metals* react to form other substances.

Potassium (K) is an alkali metal. It reacts easily. It even burns when placed in water! Most alkali metals react with elements in column 17, called *halogens* (HA•luh•juns). Alkali metals and halogens react to form many new substances.

 Quick Check

Classify How do scientists use properties to classify elements?

Critical Thinking Why might a scientist want to make a prediction about an element?

Lesson Review

Visual Summary

An atom is the building block of matter. Elements are made of atoms.

Matter can be organized by its properties.

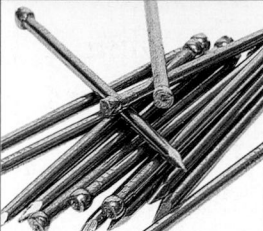

Elements in the periodic table are grouped by similar properties.

Make a FOLDABLES® Study Guide

Make a trifold book. Use it to summarize what you learned about classifying matter.

An atom is... | Matter can be organized... | Elements in the periodic table...

Think, Talk, and Write

1 Vocabulary A substance that is made up of only one type of matter is called a(n) _____.

2 Classify List all the elements mentioned in this lesson. Use the periodic table to classify each one as metal, nonmetal, or metalloid. Write the classification in your chart.

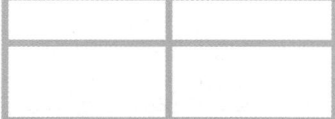

3 Critical Thinking Why are elements called "the building blocks of matter"?

4 Test Prep Nitrogen is listed in the periodic table. Therefore, nitrogen is a(n)

A atom.

B element.

C metal.

D liquid.

5 Essential Question What is matter made of?

 **Writing Link**

Explanatory Writing
Write a brief paragraph explaining some of the ways you use elements in daily life.

 **Math Link**

Compare and Order Numbers
Compare the number of solids, liquids, and gases in the periodic table. Place them in order from greatest to fewest in number.

Meet
Sisir Mondal

Every year, for about a month, Sisir Mondal travels across the globe to places like India and South Africa. Sisir travels to those places to study rocks.

In the field, Sisir studies large layers of igneous rock. Sisir collects rock samples. He studies them closely to figure out their textures and what kinds of minerals the rocks contain. Based on his observations, he makes a geological map of the area.

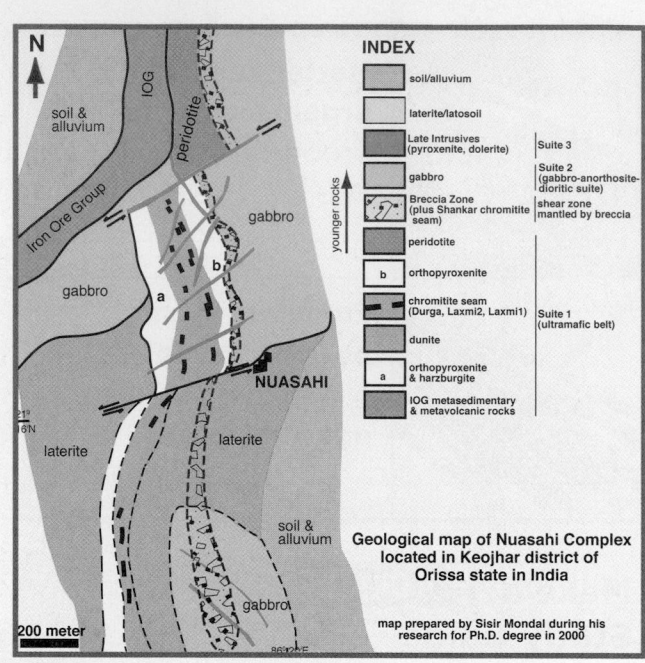

INDEX

soil/alluvium	
laterite/latosoil	
Late Intrusives (pyroxenite, dolerite)	Suite 3
gabbro	Suite 2 (gabbro-anorthosite-dioritic suite)
Breccia Zone (plus Shankar chromitite seam)	shear zone mantled by breccia
peridotite	
b	orthopyroxenite
chromitite seam (Durga, Laxmi2, Laxmi1)	Suite 1 (ultramafic belt)
dunite	
a	orthopyroxenite & harzburgite
IOG metasedimentary & metavolcanic rocks	

Geological map of Nuasahi Complex located in Keojhar district of Orissa state in India

map prepared by Sisir Mondal during his research for Ph.D. degree in 2000

200 meter

▲ Sisir made this geological map from data he collected in India.

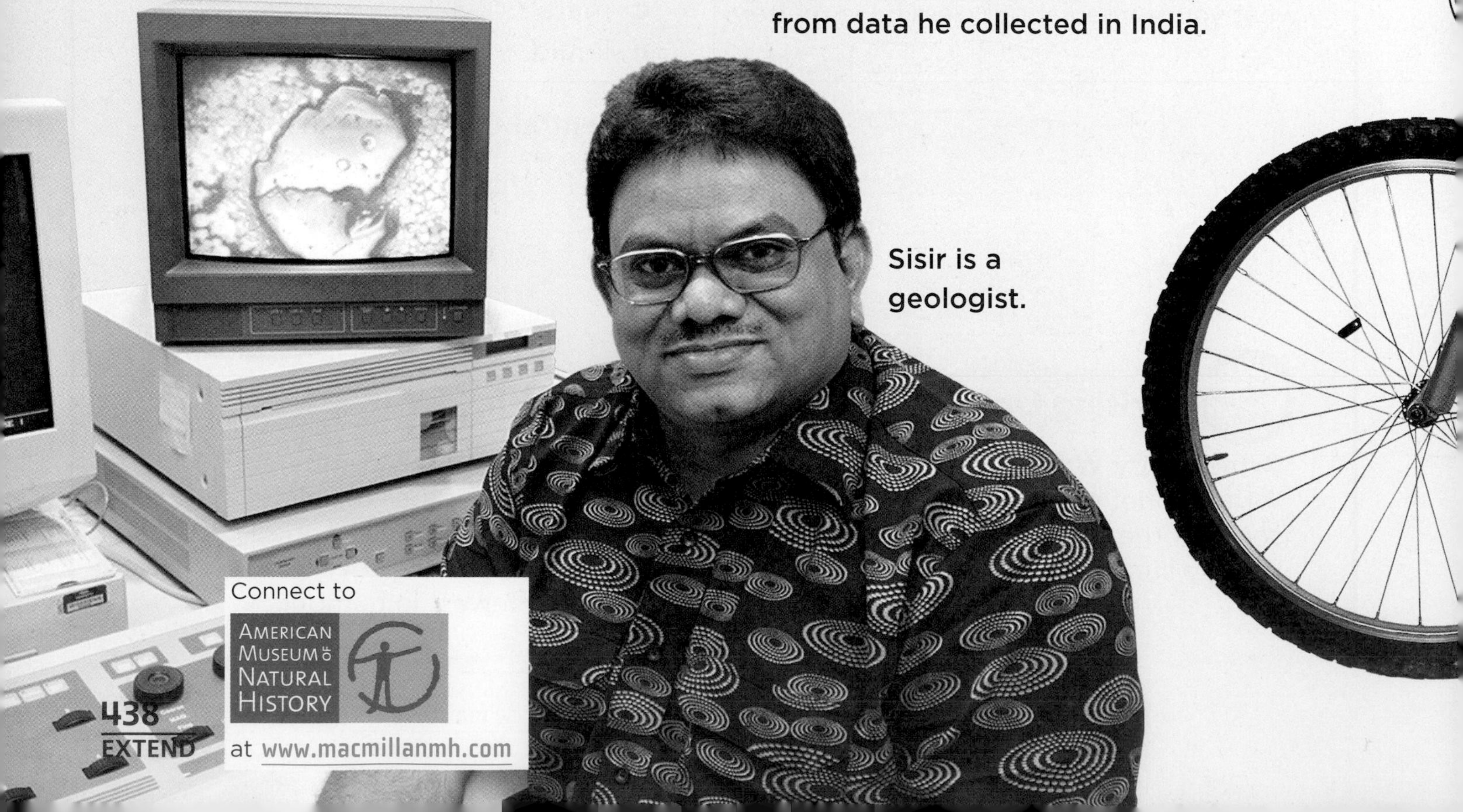

Sisir is a geologist.

Connect to

AMERICAN MUSEUM OF NATURAL HISTORY

at www.macmillanmh.com

Sisir looks for rock samples.

Back in the museum, Sisir takes a much closer look at the rock samples he collected. He uses microscopes and other tools to see what stories the rocks tell. Sisir wants to know why certain minerals are found in the rocks. He's particularly interested in finding rocks that contain metallic elements like chromium and platinum. Why are those metals important? People use them every day. Chromium is used to make many things, including steel. Platinum is a precious metal, used in everything from jewelry to catalytic converters in cars.

◄ Chromium is used in making bicycle frames.

▲ These rings are made of platinum.

Classify

▶ Arrange ideas or objects into groups that have something in common.

▶ List the properties that objects in the group share.

Write About It

Classify Read the article again. What does Sisir look for in the rocks he studies? How do you think Sisir classifies the rocks?

LOG ON **e-Journal** Research and write about it online at www.macmillanmh.com

Visual Summary

Lesson 1 Matter can be described by its properties, such as mass, volume, and state.

Lesson 2 Matter can be measured using standard units of length, area, volume, mass, density, and weight.

Lesson 3 All matter is made up of elements that can be classified by their properties.

Make a FOLDABLES Study Guide

Tape your lesson study guides to a piece of paper as shown. Use your study guide to review what you have learned in this chapter.

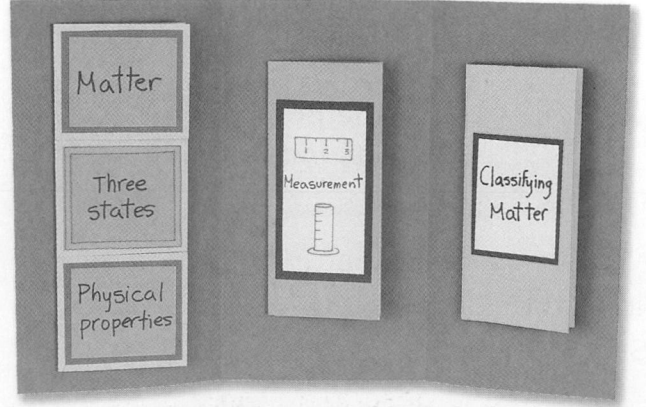

Fill each blank with the best term from the list.

atom, p. 432	**matter,** p. 412
density, p. 424	**metal,** p. 433
element, p. 432	**periodic table,** p. 434
gravity, p. 426	**property,** p. 412
mass, p. 412	**weight,** p. 426

1. Anything that has mass and takes up space is _____.

2. The measure of gravity's pull between an object and a planet is _____.

3. A substance that cannot be broken down into a simpler form is a(n) _____.

4. Color is an example of a(n) _____ of matter.

5. To calculate an object's _____, you divide its mass by its volume.

6. Gold, copper, and iron are examples of _____.

7. The pull between objects is called _____.

8. Elements are classified by their properties in the _____.

9. The _____ is the amount of matter in an object.

10. Every _____ in an element is alike.

Answer each of the following.

11. Problem and Solution How can you prevent the matter you use from becoming trash?

12. Measure You want to know the area of a sheet of paper. What would you measure? How would you calculate the area?

13. Critical Thinking How is hydrogen an example of an element?

14. Descriptive Writing Describe the properties of copper.

15. True or False *A sailboat is less dense than water.* Is this statement true or false? Explain.

16. True or False *Your mass would change if you traveled to the Moon.* Is this statement true or false? Explain.

17. What does density measure?

A the amount of matter in a given space

B the weight of matter in a given space

C the height of matter in a given space

D the force of gravity in a given space

18. Which two measuring tools could you use to find the density of a large plastic building block?

A a ruler and a thermometer

B a beaker and a ruler

C a ruler and a balance

D a scale and a balance

19. How can you measure the volume of air inside a balloon?

A Submerge the balloon in water. Subtract the original water level from the new water level.

B Measure the length and width of the balloon. Multiply the two numbers.

C Empty the contents of the balloon into a beaker. Record the volume.

D Weigh the balloon using a scale.

20. What is matter and how is it classified?

Performance Assessment

DOK 2

Elements Wanted

1. Choose an element from the periodic table.

2. Research properties of the element from available resources.

3. Create a "wanted" poster for the element, describing three of its properties. Include illustrations. Be creative!

Test Preparation

1 The pictures show one way to find the volume of a cube.

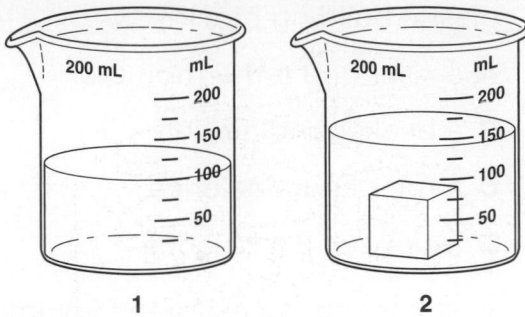

Which is the <u>best</u> description of the cube's volume?

A almost 50 mL

B almost 150 mL

C more than 50 mL

D less than 150 mL

DOK 2

2 The periodic table shows

A the names of elements that react to form new substances.

B the names of elements that change from a liquid to a gas.

C the amount of space taken up by certain objects.

D the elements with their names and symbols.

DOK I

3 Which tool can you use to measure an object's mass?

A thermometer

B scale

C pan balance

D tape measure

DOK I

4 The chart below shows the sizes of four boxes. All four boxes have the same mass.

Box	Length on Each Side
A	4
B	1
C	3
D	2

Which box has the greatest density?

A Box A

B Box B

C Box C

D Box D

DOK 2

5 Look at the column from the periodic table of elements.

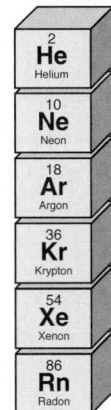

Which of these elements would <u>most likely</u> react with hydrogen?

A helium

B neon

C all of them

D none of them

DOK 2

441A

6 Look at the picture of the graduated cylinder shown below.

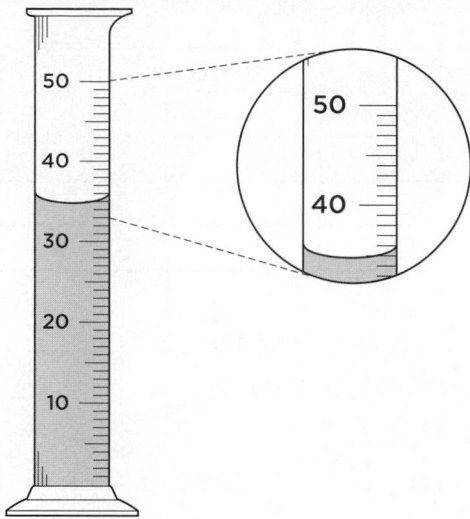

What is the volume of the liquid?

A 30 mL

B 35 mL

C 40 mL

D 50 mL
DOK 2

7 A substance that has no definite volume and no definite shape is a

A solid.

B liquid.

C gas.

D metal.
DOK 1

8 Which of these has mass?

A carbon dioxide

B electricity

C heat

D sound
DOK 1

9 Nails are made of metal.

Name three properties of a metal.
DOK 1

10 How does weight differ from mass?
DOK 2

Check Your Understanding			
Question	Review	Question	Review
1	pp. 413, 422–423	6	pp. 422-423, R5
2	pp. 432–436	7	pp. 414–415
3	pp. 412, R5	8	p. 412
4	pp. 423–425	9	pp. 413, 433
5	pp. 434–436	10	pp. 412, 426

CHAPTER 10

Matter and Its Changes

 **The Big Idea** How can matter change?

Spanish pottery

 # Big Idea Vocabulary

change of state a physical change in which one state of matter changes to another (p. 448)

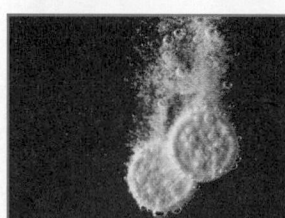

chemical change a change in which a new kind of matter is produced (p. 450)

mixture two or more types of matter that are combined but keep their original properties (p. 458)

solution a mixture in which one or more types of matter are mixed evenly in another kind of matter (p. 458)

compound a new substance made when two or more elements combine through a chemical change (p. 468)

acid a substance that turns blue litmus paper red (p. 470)

 Visit www.macmillanmh.com for online resources.

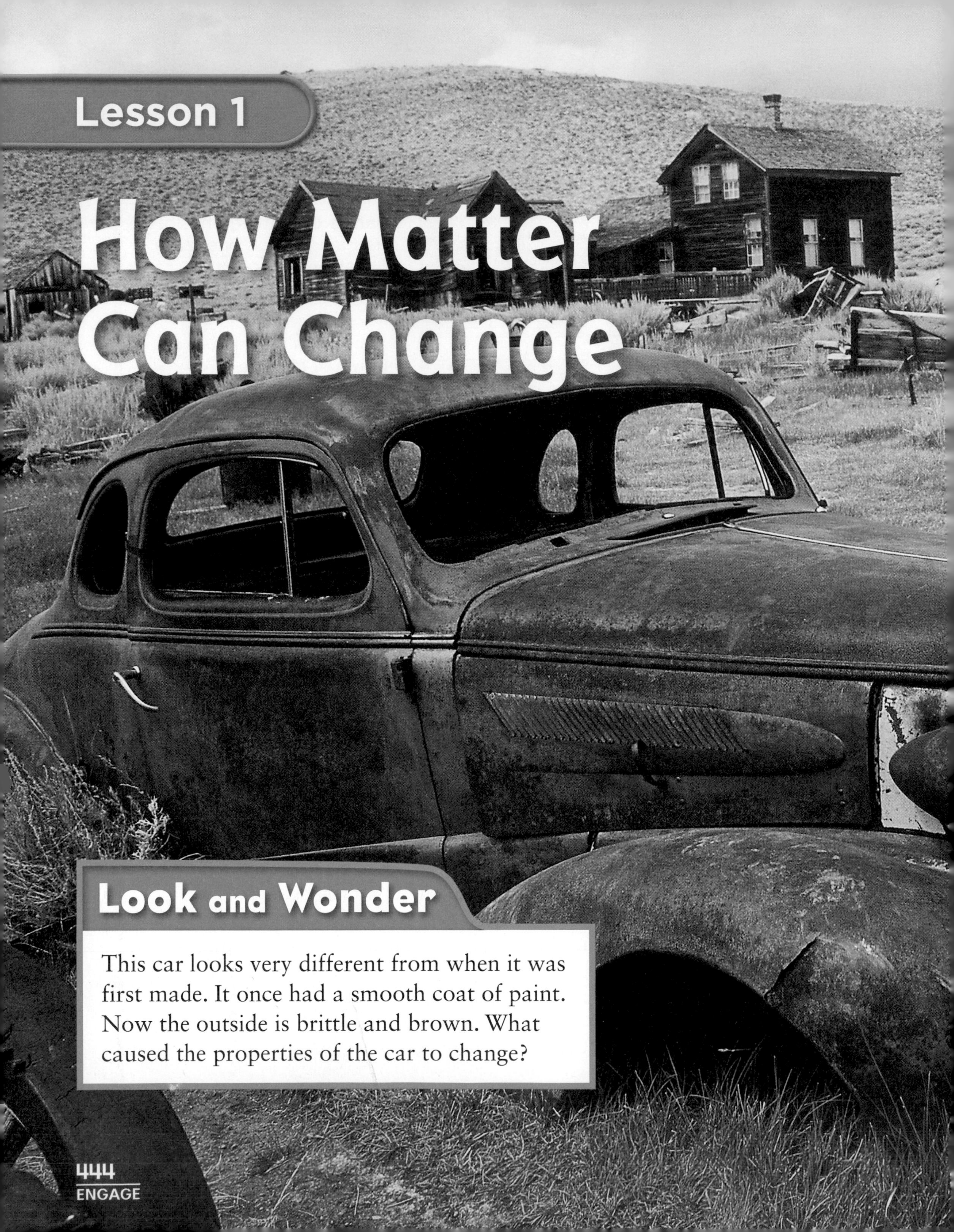

How Matter Can Change

Look and Wonder

This car looks very different from when it was first made. It once had a smooth coat of paint. Now the outside is brittle and brown. What caused the properties of the car to change?

Can you change the properties of a solid?

Make a Prediction

Will a piece of clay keep its original properties if you change its shape? What will happen to its mass and volume? Make a prediction.

Test Your Prediction

1 Measure Using a balance, measure the mass of a piece of clay. Measure its volume using a graduated cylinder and water. Record your observations in a table like the one shown.

2 Change the shape of the clay. For example, you might flatten it or cut it into pieces.

3 Measure With the balance and graduated cylinder, measure the mass and volume of the changed clay. Record your findings.

4 Change the shape of the clay in two or three other ways. Repeat step 3 each time.

Draw Conclusions

5 Interpret Data Did the mass change after you changed the shape of the clay? Did the volume change?

6 Infer What can you conclude about changing the properties of a solid?

Explore More

Will the mass or volume of the clay change if you let it dry out? Make a prediction. Try it!

Materials

- modeling clay
- balance
- graduated cylinder
- water
- plastic knife

Step **2**

What I Observed

Mass before change	Volume before change	Shape change	Mass after change	Volume after change

▶ **Essential Question**

How can you change matter?

▶ **Vocabulary**

physical change, p. 446

change of state, p. 448

evaporation, p. 449

rust, p. 450

chemical change, p. 450

tarnish, p. 450

▶ **Reading Skill** ✓

Sequence

First
Next
Last

▶ **Technology** 🔵 LOG ON

e-Glossary, e-Review, and animations online at www.macmillanmh.com

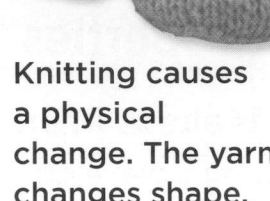

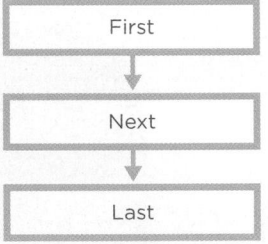

What are physical changes?

When you flatten a piece of clay or cut it in half, you cause a physical (FIH•zih•kul) change. A **physical change** begins and ends with the same type of matter. Even though the clay has a different shape, it is still clay. You can cut, crush, tear, bend, or stretch matter. Each action will cause a physical change.

Knitting causes a physical change. The yarn changes shape.

After a physical change, the physical properties of the matter change. However, the matter is still the same substance. For example, an ice cube is solid water. If the ice cube warms, it melts. Now it is liquid water. If you pour that water into a kettle and heat it, you get steam. Steam is water in the form of a gas. Physical changes can be caused by wind, rain, freezing, and heating.

People use physical changes to create useful products. For example, plastic, metal, and glass can be cut and molded into useful shapes. Rock can be sculpted. You will learn about many other ways matter can change throughout this chapter.

Steam shows a physical change.

Folding paper is a physical change.

Real-World Changes

How can you tell whether a physical change has taken place? Look for a change in size, shape, or position.

Physical changes happen around you all the time. The sidewalks in your town or city are made of concrete. When the concrete is new, it is one solid piece. In time, chips and cracks form. Small pieces of concrete break off. Wind and rain carry the pieces away. Cracking and breaking are physical changes in the sidewalks.

Water's physical changes allow fish and other organisms to survive in a pond during winter. In winter the surface of lakes and ponds can become solid ice. Beneath the ice the water is still a liquid. How is this possible?

Not all liquids have this property. Unlike most other liquids, water expands when it freezes. That means ice is less dense than liquid water. This is why ice floats in liquid water.

 Quick Check

Sequence What happens when ice turns to liquid water?

Critical Thinking Describe and explain some types of physical changes you can see every day.

Moving water can break apart even the hardest rocks.

How does matter change state?

You know that matter can exist in three states—solid, liquid, and gas. A pair of scissors is a solid. The air you breathe is a gas. Juice and milk are liquids.

Most types of matter can exist in more than one state. For example, water can exist as a solid, a liquid, or a gas. Matter can change its state. A **change of state** is a physical change in which one state of matter changes to another. After a change of state, the volume of a substance may change. Its mass stays the same.

Adding Energy

If you add energy to a solid by heating it, its particles move faster. If the particles gain enough energy, the solid changes to a liquid. *Melting* is a change of state from a solid to a liquid.

If you add energy to a liquid, it can change to a gas. *Boiling* is a rapid change of state from a liquid to a gas. Boiling is not the only way a liquid can become a gas.

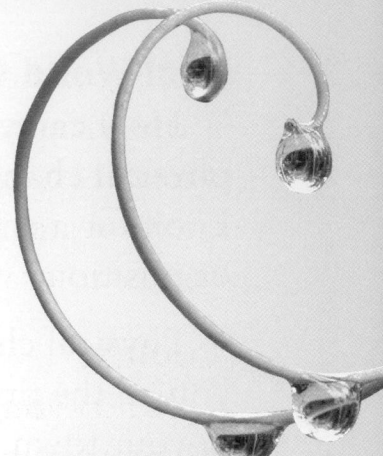

Dew is water that changed state from a gas to a liquid.

How Water Changes State

solid

Ice melts when energy is added. The particles move faster.

liquid

As energy is added to liquid water, the particles move faster. Some turn to gas.

Evaporation (ih•va•puh•RAY•shun) is the change of a liquid to a gas without boiling. The Sun's energy causes water in lakes, rivers, and oceans to evaporate.

Cooling

When you take away energy from any substance, its particles move more slowly. The substance cools and its particles move closer together. This is how a gas changes, or *condenses,* to a liquid.

If enough energy is taken away, a liquid *freezes* into a solid. Water particles move closer together until about 4°C. Then they start to move apart. Particles of other kinds of matter continue to move closer together as they freeze.

≡ *Quick Lab*

Heat and Evaporation

1 Pour an equal amount of water into two petri dishes.

2 **Predict** Place one petri dish under a lamp or in a sunny place. Put the other in a cooler or darker place. Predict which will evaporate first.

3 **Infer** Which dish of water evaporated first? Why?

gas

Water vapor is a gas. Its particles move very fast.

Read a Diagram

What happens when energy is added to ice? To liquid water?

Clue: Compare the particles in each state.

LOG ON *Science in Motion* Watch matter change state at www.macmillanmh.com

✓ *Quick Check*

Sequence What happens as water changes from a liquid to a gas? From a liquid to a solid?

Critical Thinking On a summer day, a puddle quickly disappears. What happened to the water?

What are chemical changes?

Most bicycle frames contain iron. If you leave a bicycle outside for too long, the frame might rust. **Rust** is a solid, brown substance. It forms when iron gets wet and reacts with oxygen in the air. Rust is neither iron nor oxygen, though it contains both of these elements.

The change from iron to rust is a chemical (KEM•ih•kul) change. A **chemical change** begins with one kind of matter and ends with another. A chemical change is also known as a chemical reaction.

All chemical reactions either give off energy or use energy. Chemical reactions may give off heat, light, or electricity.

Gas bubbles are evidence of a chemical change. ▼

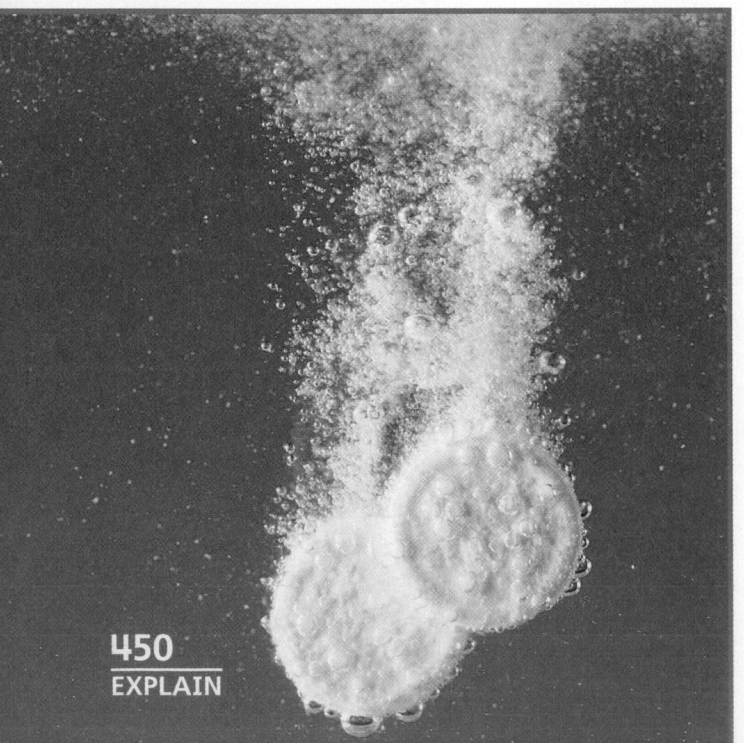

▲ Fireworks release so much energy that they can light up the sky!

Examples of Chemical Changes

When we cook food, we change the color, taste, and texture of food. Cooking and baking cause chemical changes in food.

Some chemical changes produce gases. When vinegar is mixed with baking soda, carbon dioxide gas is released. You can tell by the bubbles that form.

Have you ever noticed a black substance on silver jewelry? That substance is **tarnish**. It forms when silver reacts with sulfur in the air. If you use polish on the areas of tarnish, you cause another chemical change. The silver gets shiny again!

Signs of a Chemical Change

If you know what to look for, you will find evidence of chemical changes all around you. The most visible sign is a change in color. Rust and tarnish are good examples.

A change in scent is another sign of a chemical change. Have you ever toasted marshmallows over a campfire? While your marshmallow is toasting, it gives off a pleasant smell. If it burns, the odor is less pleasant.

If you see bubbling or hear fizzing, a chemical change may have occurred. These signs indicate the release of a gas. For example, dropping antacid tablets into water will release lots of fizzy bubbles.

Many reactions cause the matter involved to become warm or hot. Others cause the matter to become cold. Some reactions even give off light. Fire is a chemical reaction that gives off light and heat.

✔ Quick Check

Sequence Explain how tarnish forms on silver. How is it removed?

Critical Thinking Over time, a copper statue will turn green. Is this a chemical change? Explain.

Reaction of Iron and Sulfur

① Iron and sulfur are mixed together. Iron is a gray metal. It is also magnetic. Sulfur is a yellow powder.

② A metal rod is heated to a high temperature.

③ The heated rod causes a chemical change. Light and heat are released.

④ The result is iron sulfide—a black, nonmagnetic material.

Read a Diagram

How is iron sulfide different from the iron and sulfur that formed it?

Clue: Compare the top picture with the bottom one. Read both captions.

FACT ▶ Air is not the same as oxygen.

Physical Changes

As water vapor cools, it condenses into clouds. When enough drops condense in the cloud, they fall as rain.

outside ▲

The particles of sweat carry away energy as they evaporate, leaving you cooler.

in your body ▲

Moist dough is easy to stretch. It can be pulled apart, braided, or shaped into rolls.

in the kitchen ▲

Chemical Changes

Some rain falls as acid rain. Acid rain reacts with limestone. It eats away at rock, buildings, and statues.

outside ▲

Blood carries oxygen to your cells. In the cell, the oxygen reacts with sugars. This releases energy for your body to use.

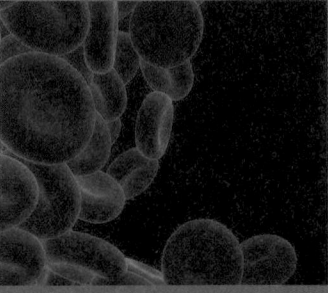

in your body ▲

In the presence of heat, the moist dough hardens. Baking changes it into crusty bread.

in the kitchen ▲

What are other real-world changes?

Changes take place around you all the time. What kinds of physical and chemical changes affect you and your environment? The chart shows some examples. Can you think of other cases of physical and chemical changes?

✔ Quick Check

Sequence Describe the physical and chemical changes that happen when making bread.

Critical Thinking The flesh of a cut apple turns brown when left out in the air. Is this change physical or chemical? Explain.

Lesson Review

Visual Summary

A **physical change** begins and ends with the same type of matter. An example is folding paper.

A **change of state** is a physical change from one state of matter to another.

A **chemical change** forms a new substance with different properties from the original matter.

Make a FOLDABLES Study Guide

Make a trifold book. Use it to summarize what you read about physical and chemical changes.

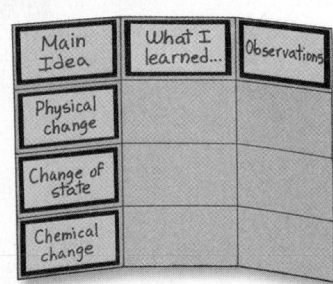

Think, Talk, and Write

1 **Vocabulary** The change of a liquid to a gas without boiling is called _____.

2 **Sequence** To build a campfire, wood must be gathered, dried, and cut into small pieces. Which changes are physical? Which are chemical?

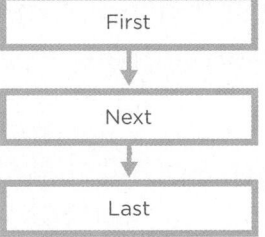

First

Next

Last

3 **Critical Thinking** What change could you make to a sheet of paper to illustrate a physical change? A chemical change?

4 **Test Prep** Which of the following is a chemical change?

  A tarnished spoon

  B popped corn

  C cloud formation

  D a change in state

5 **Essential Question** How can you change matter?

Writing Link

Speechwriter
You have been asked to speak to a third-grade class. Your goal is to explain physical and chemical changes. Write a speech for the class. Use examples.

Health Link

Digestion Chart
When you eat, physical and chemical changes take place. Research how food changes in the digestive system. Make an illustrated chart of your findings.

Lady Liberty

Did you know that the Statue of Liberty wasn't always green? When it was built, the statue was the color of a shiny new penny.

The Statue of Liberty is made of the metal copper. New copper has a reddish color. Twenty years after it was built, Lady Liberty was completely green. What caused this color change? A chemical reaction occurred!

Oxygen in the air reacts chemically with copper. The elements form a compound called copper oxide. This kind of change is called oxidation.

The Statue of Liberty was built in France in 1884. Two years later it came to America in 350 individual pieces.

Connect to

AMERICAN MUSEUM oᶠ NATURAL HISTORY

at **www.macmillanmh.com**

▲ copper

▲ copper oxide

▲ copper hydroxide

The Statue of Liberty changed from red to dark brown—the color of copper oxide. That color didn't last long. Over time, rainwater and carbon dioxide in the air reacted with the copper oxide. This reaction formed a different compound called copper hydroxide. What color is copper hydroxide? It's green, of course!

The green layer on Lady Liberty is only as thick as a postcard. Still, it protects the copper underneath. If the layer were removed, the statue would shine like a new penny again—but only for a while.

The Statue of Liberty was originally the color of a shiny new penny.

Sequence

▶ Give events in order.

▶ Use time-order words such as *first*, *then*, and *next*.

Write About It

Sequence

1. Make a sequence chart showing how the color of the Statue of Liberty has changed over time.
2. Use your chart to write a summary of those changes.

 **LOG ON** **e-Journal** Research and write about it online at www.macmillanmh.com

Mixtures

Look and Wonder

There are many different solids in this pond. Can you count them all? What happens when you mix solids with liquids?

How do solids and water mix?

Make a Prediction

What will happen when you mix salt into water? What about sand and water? Sugar and water? Gelatin and water? Make your predictions.

Test Your Prediction

1. Label one cup *salt* and a second cup *sand*.

2. **Measure** Pour 100 milliliters of water into each cup. Add one spoonful of salt to the cup marked *salt*. Stir well. Add one spoonful of sand to the cup marked *sand*. Stir well.

3. **Observe** Study the contents of both cups carefully. What happened to the salt? The sand? Record your observations.

4. Label a third cup *sugar* and the last cup *gelatin*. Repeat step 2 with both substances. After stirring, leave each cup alone for 20 minutes. What happened this time?

Draw Conclusions

5. **Communicate** Describe the similarities and differences you observed after the four solids were mixed with water. Were your predictions correct?

Explore More

Would you get the same results if the temperature of the water were higher or lower? Write a prediction that you could test.

Materials

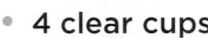

- 4 clear cups
- marker
- measuring cup
- water
- 4 plastic spoons
- salt
- sand
- sugar
- gelatin

Step 2

Read and Learn

Essential Question

How can mixtures be separated?

Vocabulary

mixture, p. 458

solution, p. 458

alloy, p. 459

filter, p. 461

filtration, p. 461

distillation, p. 462

Reading Skill ✓

Classify

Technology

e-Glossary and e-Review online at www.macmillanmh.com

Lemonade is a solution of water, lemon juice, and sugar.

Kinds of Mixtures

solids and solids

solids and gases

What is a mixture?

Did you ever make a salad? If you did, then you know how to make a mixture. A **mixture** is a physical combination of two or more kinds of matter. Mixtures are physical changes. They can be separated by physical means.

Everyday Mixtures

A salad is usually a mixture of lettuce, tomatoes, and other foods. The foods that go into a salad might be chopped up. However, they are still the same kinds of food.

You probably see mixtures every day. Some breakfast cereals are mixtures of solids. If you add milk, you get a mixture of solids and a liquid. Many products, such as food, drinks, and clothing, are made from mixing different kinds of matter.

Solutions Are Mixtures

Some solids mix easily with liquids. If you mix salt into water, the salt will break up. Salt water is a solution. A **solution** is a mixture in which two or more substances are blended completely.

liquids and liquids

solids and liquids

Read a Photo

What are four different ways to make a mixture?

Clue: Read the labels and identify the contents of each photograph.

Chemical Properties

In a mixture, matter might lose its physical properties. But the matter still keeps its original chemical properties. *Chemical properties* only change during chemical reactions.

Substances in a solution might have physical properties that the original substances did not have. Salt and water by themselves are both poor conductors of electricity. Salt water is a very good conductor.

Alloys Are Solutions

Thousands of years ago, people discovered how to make bronze by mixing melted copper and tin. Bronze is a kind of solution called an alloy. An **alloy** is a mixture of two or more elements. At least one of the elements is a metal. Alloys can be stronger or more flexible than the substances that formed them.

 **Quick Check**

Classify How are solutions, alloys, and mixtures related?

Critical Thinking A cook combines peas and carrots in a bowl. Is this a mixture or a solution? Explain.

How can you separate the parts of a mixture?

We can use physical properties to separate a mixture. For example, you can separate a mixture of beads and coins by picking out different shapes and colors. You can also use properties like volume, size, state, and density.

Settling

One way to separate the parts of a mixture is by settling. *Settling* is when differences in densities cause the parts of a mixture to separate. For instance, mud is a mixture of sediment and water. Over time, the sediment in mud will sink to the bottom. The sediment settles because it is denser than water.

Natural Separation

Read a Photo

How is this picture an example of settling?

Clue: Think about the kind of mixture that might be near the car.

A colander filters noodles from water.

Filtration

A **filter** separates things by size. Usually, the filter is a mesh, screen, net, or sieve (SIHV). Materials that are smaller than the holes in a filter can pass through it. Larger materials stay behind.

Have you ever cooked noodles? You probably used a colander to drain the water out. People often use filters to separate a solid from a liquid. This process is called **filtration**.

Magnets

You can use a magnet to separate the parts of some mixtures. Magnets are often used to separate scrap metal in junkyards. A magnet pulls, or attracts, the elements iron, nickel, and cobalt. This property is called *magnetic attraction*.

How can you separate the parts of a solution?

You have learned several ways to separate mixtures. How can you separate the parts of a solution like salt water? The tiny particles of salt can pass through most filters.

By Distillation

One way to separate a solution of liquid water and solid salt is distillation. In **distillation**, a solution is heated until the liquid becomes a gas. The solid is left behind.

The gas then passes through a *condenser*. This device cools the gas and collects it as a liquid. Distillation is used to make fuel. It separates gasoline from crude oil.

By Evaporation

Another way to separate the parts of a solution is by evaporation. Recall that evaporation is the change of a liquid to a gas without boiling. When salt water evaporates, the water becomes a gas—water vapor. The solid salt is separated from the mixture and left behind. The liquid water is separated from the mixture too. However, it is lost to the air.

 Quick Check

Classify List the methods you would use to separate the parts of a solution.

Critical Thinking To separate pure water from salt water, would you use evaporation or distillation? Explain.

In this Vietnamese salt factory, salt is collected as seawater evaporates.

Lesson Review

Visual Summary

Mixtures are combinations of two or more types of matter. Solutions and alloys are types of mixtures.

Mixtures can be separated by their physical properties.

Solutions can be separated using evaporation and distillation.

Make a **FOLDABLES®** Study Guide

Make a trifold book. Use it to summarize what you read about mixtures.

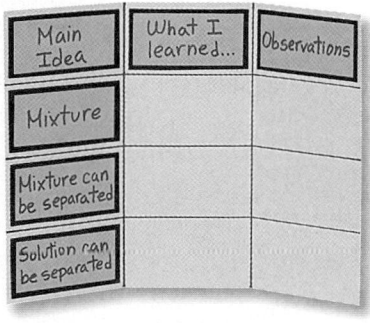

Think, Talk, and Write

1 **Vocabulary** To collect the evaporated water from a solution of salt water, you would use _____.

2 **Classify** Classify the following as mixtures or solutions—vegetable soup, salt water, bronze, smoke, apple juice and water, oil and water, trail mix.

3 **Critical Thinking** Blood is made of water, solids, and gases. Is blood an element or a mixture? How could you separate the solids from blood?

4 **Test Prep** How would you separate salt from a saltwater solution?

- A filtration
- B magnetism
- C evaporation
- D settling

5 **Essential Question** How can mixtures be separated?

 **Math Link**

Gold Standards

Copper and gold form a hard alloy. The amount of gold in the alloy is measured in karats. Pure gold is 24 karats. Gold with half copper is 12 karats. How much copper is there in 6-karat gold alloy?

 **Art Link**

Mix Colors

Red, blue, and yellow paint can be mixed to make other colors. What color does red and blue make? Red and yellow? Blue and yellow? What if you mix all three? Try it. Report your findings.

Focus on Skills

Inquiry Skill: Use Variables

You know that water evaporates all the time. How would you find out if heat affects evaporation? When scientists plan an experiment to answer questions like this, they **use variables.** Variables are the factors in an experiment that are changed or unchanged. The factor you test is the *independent variable*. The factors you measure or count are *dependent variables*. The factors you keep the same are the *controlled variables*. By controlling variables, you know that only one thing affected your results—the independent variable.

▶ Learn It

When you **use variables** in an experiment, you identify what you are testing and what you are not testing. The best experiments test only one independent variable at a time. It is good practice to decide in advance how you will change the independent variable. It is important to keep careful records of those changes. Then you can easily see how the change affects the dependent variables.

▶ Try It

Use variables in your experiment to find out how heat affects evaporation.

> **Materials** 3 air thermometers, graduated cylinder, water, 3 clear cups, 3 paper towels, 3 rubber bands, stopwatch

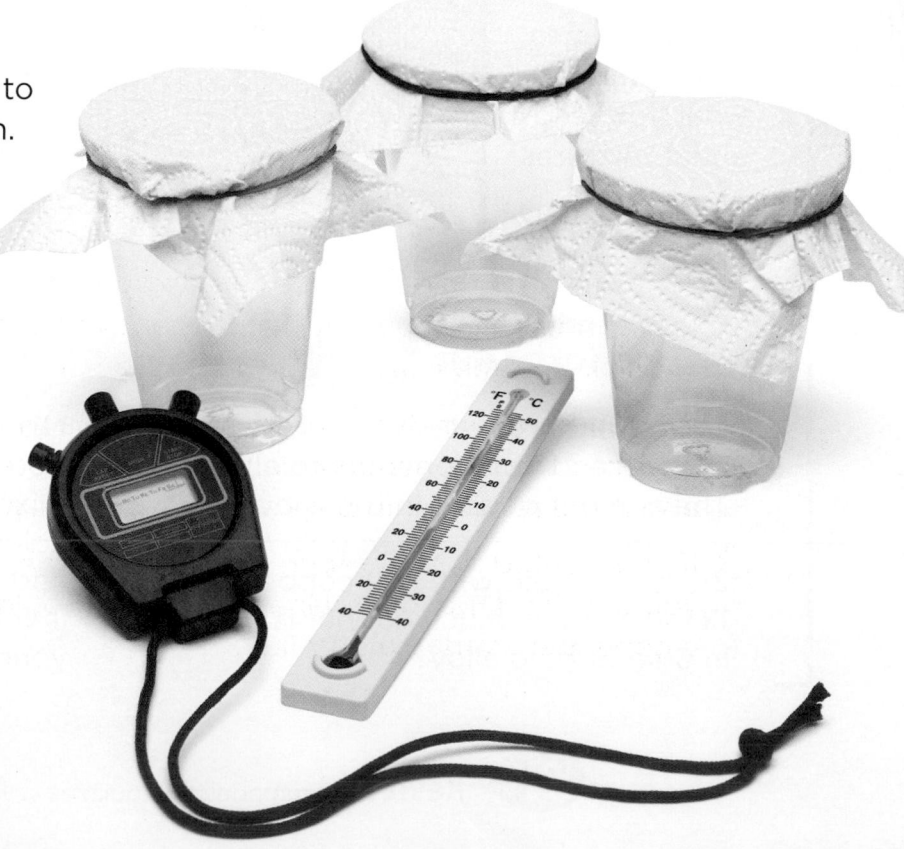

1 As a class, select three locations that you think will have different air temperatures. Place a thermometer at each location.

2 Fill a graduated cylinder with 25 mL of water. Then prepare your cups. Stretch a paper towel across the top of each cup. Secure each towel with a rubber band. Label the cups *1*, *2*, and *3*.

3 Slowly pour 5 mL of water onto the center of each towel.

4 Place one cup at each of the three locations you chose in step 1. When you set each cup down, record the temperature and the time. Use a table like the one shown.

5 Check the paper towels on the cups every minute. Record the time when each paper towel is dry.

	Area 1	Area 2	Area 3
Temperature			
Start Time			
End Time			

▶ **Apply It**

1 How did you **use variables** in this experiment? List your independent variable, your dependent variable, and your controlled variables.

2 Explain how your dependent variables changed as you changed the independent variable. What does this tell you about the relationship between heat and evaporation?

3 If you wanted to show your results in a graph, where would you put the dependent variable? Where would you put the independent variable? Try it.

Compounds

Look and Wonder

Each July 4 we celebrate Independence Day.
We gather with our families and our friends.
At night colorful fireworks light up the sky.
What gives these fireworks their bright colors?

How does iron react with air and moisture?

Make a Prediction

Steel wool has iron in it. What happens when you expose steel wool to air and moisture? Make a prediction.

Test Your Prediction

1. ⚠ **Be Careful.** Wear safety goggles. Take a small piece of steel wool. Soak it in vinegar for about one minute. Vinegar exposes the iron in the steel.

2. Fill a beaker slightly more than halfway with water. Using a pencil, push the steel wool into the bottom of an empty test tube. Place the tube upside down in the beaker of water.

3. **Observe** Put the beaker in a safe place. Observe it each day for four to five days. Record your observations. Add water to the beaker if the level gets low.

Draw Conclusions

4. **Communicate** Was your prediction correct? What happened to the steel wool? Describe any changes.

5. **Infer** Why do you think steel wool changes when it is exposed to moist air?

Explore More

Would you get the same results if the steel wool were completely underwater? Write a prediction that you can test. Design an experiment. Try it!

Materials

- safety goggles
- steel wool
- vinegar
- beaker
- water
- pencil
- test tube

Step 2

► **Essential Question**

What happens when matter goes through a chemical change?

► **Vocabulary**

compound, p. 468

acid, p. 470

base, p. 470

► **Reading Skill** ✔

Problem and Solution

Problem
↓
Steps to Solution
↓
Solution

► **Technology** 🔵 LOG ON!

e-Glossary and e-Review online at www.macmillanmh.com

What are compounds?

Most matter is made of more than one element. However, unlike mixtures, certain combinations of elements cannot be separated physically. No amount of crushing or filtering can separate the elements in table salt, for example. That is because table salt is a compound. A **compound** forms when two or more elements combine chemically. This happens during a chemical reaction.

When you form a compound, you form a new substance with new chemical properties. Elements in a compound can only be separated chemically—by going through another chemical reaction.

Rust is a compound formed from iron and oxygen. When iron and oxygen react chemically, the substance that results—rust—has new chemical properties. Water is a compound formed from hydrogen and oxygen. Like rust, water reacts differently than the elements it formed from.

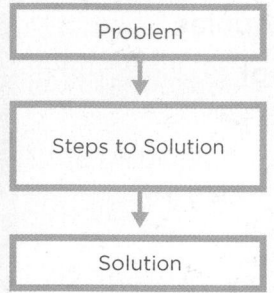

Comparing Compounds and Mixtures

	Compound	Mixture
How are the parts combined?	two or more elements are combined chemically	two or more types of matter are mixed together
Do the parts keep their own properties?	no	yes
How can it be separated?	by chemical means	by physical means

1 **H** Hydrogen	+	8 **O** Oxygen	→	Water is a liquid.
Hydrogen is a gas.		Oxygen is a gas.		

11 **Na** Sodium	+	17 **Cl** Chlorine	→	Table salt is a safe, nonmetal solid.
Sodium is a metal.		Chlorine is a dangerous gas.		

26 **Fe** Iron	+	8 **O** Oxygen	→	Rust is a weak, brown solid.
Iron is a strong, gray metal.		Oxygen is a gas.		

6 **C** Carbon	+	1 **H** Hydrogen	+	8 **O** Oxygen	→	Natural sugar is a brown solid.
Carbon is a black solid.		Hydrogen is a gas.		Oxygen is a gas.		

14 **Si** Silicon	+	8 **O** Oxygen	→	Quartz is a hard mineral.
Silicon is a dark metalloid.		Oxygen is a gas.		

Read a Diagram

1. What do the plus signs stand for?
2. What do the arrows stand for?

Clue: Think of each combination as a math equation.

 Quick Check

Problem and Solution How can you separate the elements in a compound?

Critical Thinking How are compounds different from mixtures?

469

Acids and Bases

1. **Measure** In a large cup, mix 1 teaspoon of baking soda in 50 mL of water. Test this substance with both red and blue litmus paper. Record your observations in a chart.

2. ⚠ **Be Careful.** Wear goggles. Pour about 70 mL of vinegar in a different cup. Test it with litmus paper. Record your observations.

3. ⚠ **Be Careful.** Slowly pour the vinegar into the baking soda solution. Test the new solution with litmus paper. Record your observations.

4. **Interpret Data** Which were acids? Bases? How do you know?

Acids and Bases

lemons   water   soap

most acidic   most basic

Read a Diagram

Why does the water have both blue and red litmus paper below it?

Clue: What do the colors represent?

What are acids and bases?

Acids and bases are compounds that react easily with other substances. You can identify them using *litmus* (LIT•mus) *paper*. Litmus is a dye that changes color when it touches acids or bases.

An **acid** is a substance that turns blue litmus paper red. A weak acid is what makes lemons sour. Strong acids can burn skin and dissolve metals. A **base** is a substance that turns red litmus paper blue. In foods, bases taste bitter. Strong bases can be found in some drain cleaners.

Strong acids and strong bases are dangerous chemicals. Never taste a substance to find out whether it is an acid or a base.

When an acid and a base combine, they react chemically. They form new compounds—a salt and water. A saltwater solution is *neutral*. It is neither an acid nor a base. Water is a neutral substance. It does not turn litmus paper blue or red.

✔ Quick Check

Problem and Solution How could you make a salt?

Critical Thinking Why should you never taste an acid or base?

Lesson Review

Visual Summary

 **Compounds** form when two or more elements combine chemically. Rust is a common example.

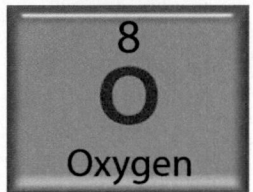 The properties of a compound are different from the properties of its original **elements**.

 You can use litmus paper to test for **acids and bases.**

Make a FOLDABLES Study Guide

Make a layered-look book. Use it to summarize what you read about compounds.

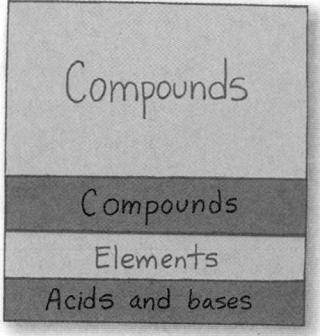

Compounds
Compounds
Elements
Acids and bases

Think, Talk, and Write

1 **Vocabulary** Red litmus paper turns blue when you place it in a(n) _____.

2 **Problem and Solution** You are looking for evidence of acid rain. What test or tests could you conduct?

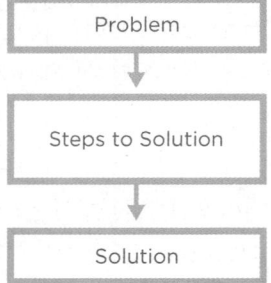

Problem

↓

Steps to Solution

↓

Solution

3 **Critical Thinking** Oxygen is found in air. Oxygen is also found in rust. You can breathe oxygen from air but not from rust. Why not?

4 **Test Prep** **Which of the following is a compound?**

A oxygen
B sodium
C water
D iron

5 **Essential Question** What happens when matter goes through a chemical change?

 **Writing Link**

Explanatory Writing
How do you use compounds in your daily life? Write a brief paragraph explaining what they are and how you use them.

 **Art Link**

Acid and Base Chart
Find pictures of different liquids you use every day. Establish whether they are acids or bases. Arrange the pictures into a chart that shows how to classify the liquids.

Be a Scientist

Materials

3 paper plates

markers

3 apple slices

toothpicks

lemon juice

water

Structured Inquiry

How can you change a chemical reaction?

Form a Hypothesis

Soon after an apple is cut, a chemical reaction takes place. Oxygen in the air turns the apple brown, just like iron turns to rust. Can you prevent this reaction? Use the materials in the list. Write your answer in the form "If you want a cut apple to not turn brown, then..."

Test Your Hypothesis

1. Label three plates *A*, *B*, and *C*. Place one slice of apple on each plate. Put one toothpick upright in each slice.

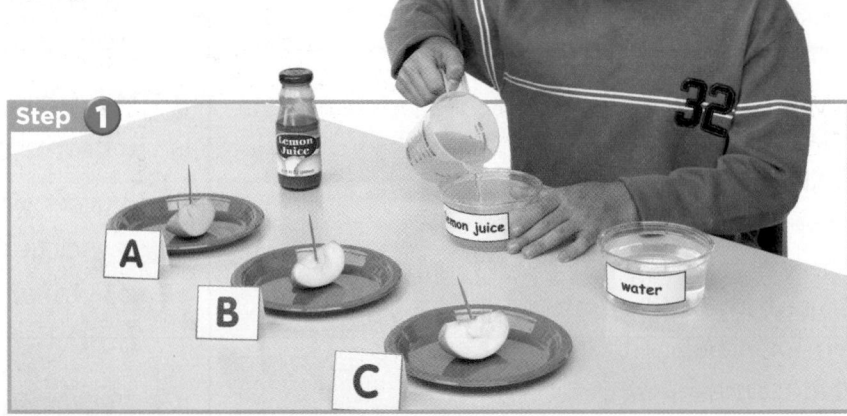

Step 1

2. **Communicate** In your journal, draw your setup. Describe the apple slices.

3. ⚠ **Be Careful.** Always wear safety goggles when using acids. Holding the toothpick, dip the entire apple slice from plate *A* in lemon juice. Then place the slice back on the plate. Dip the slice from plate *B* in water. Put it back on the plate. Leave the slice on plate *C* alone.

Step 3

lemon juice

④ Observe After ten minutes, observe each apple slice. Record your observations.

Draw Conclusions

⑤ Was your hypothesis correct? Explain your answer.

⑥ Interpret Data How can you prevent a cut apple from turning brown? Why do you think this is so?

Guided Inquiry

How else can you stop oxygen from reacting chemically?

Form a Hypothesis

What other ways could you prevent fruit from turning brown? Write a hypothesis.

Test Your Hypothesis

Design an investigation to find out whether other liquids stop fruit from turning brown. Write out the steps you will follow. Remember your safety tips. Record your results and observations in your journal.

Draw Conclusions

Did your results support your hypothesis? Why or why not? If you were making a fruit salad, what could you add to keep the fruit looking fresh?

Open Inquiry

What else would you like to learn about reactions between fruit and oxygen? For example, which fruits turn brown fastest? Design an investigation to answer your question. Another group should be able to complete the study by following your instructions.

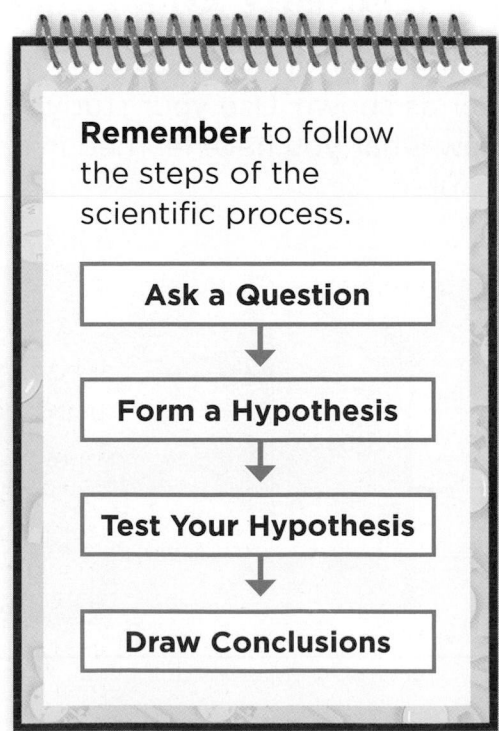

Remember to follow the steps of the scientific process.

> **Ask a Question**
> ↓
> **Form a Hypothesis**
> ↓
> **Test Your Hypothesis**
> ↓
> **Draw Conclusions**

Visual Summary

Lesson 1 Physical changes start and end with the same kind of matter. Chemical changes form new kinds of matter.

Lesson 2 Matter can be combined to form mixtures. Mixtures can be separated by their physical properties.

Lesson 3 Compounds are formed by chemical reactions between two or more elements.

Make a **FOLDABLES**® Study Guide

Tape your lesson study guides to a piece of paper as shown. Use your study guide to review what you have learned in this chapter.

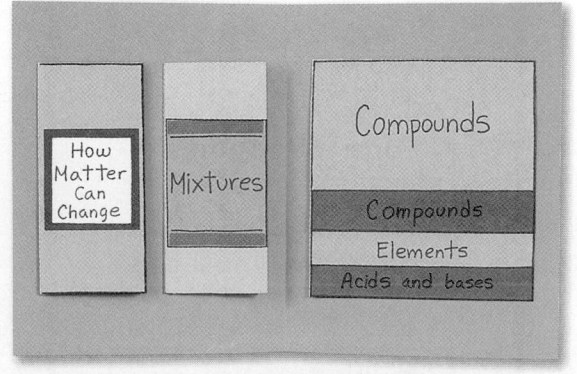

Fill each blank with the best term from the list.

alloy, p. 459   **distillation,** p. 462

base, p. 470   **evaporation,** p. 449

change of state, p. 448   **mixture,** p. 458

chemical change, p. 450   **solution,** p. 458

compound, p. 468   **tarnish,** p. 450

1. A combination of two or more types of matter is a(n) _____.

2. Rusting is a(n) _____.

3. Two or more elements that are chemically combined make up a(n) _____.

4. If energy is added to a solid, a(n) _____ can occur.

5. Bronze is a(n) _____ that is a mixture of several metals.

6. Baking soda is a(n) _____, which will react with an acid to form a salt and water.

7. A mixture in which two or more substances are blended together completely is called a(n) _____.

8. A liquid changes to a gas during _____.

9. The black substance that forms on silver, called _____, is an example of a chemical change.

10. Separating a solution by heating a liquid and capturing the gas that forms is called _____.

Answer each of the following.

11. Sequence How does water change from a solid to a liquid? Explain the sequence of events.

12. Use Variables You want to find out whether light affects how fast a nail rusts. You plan an experiment to test and compare two nails. Which variable will you change? Which variables will you keep the same?

13. Critical Thinking When carbon and oxygen combine, carbon dioxide forms. Is carbon dioxide a mixture, a solution, or a compound? Explain.

14. Descriptive Writing Describe the properties of a base.

15. Expository Writing What kinds of changes occur as you blend pancake mix, milk, and an egg and then heat the batter to make pancakes?

16. Which of the following foods is formed by a chemical change?

A celery stick

B toasted marshmallow

C sliced apple

D fruit salad

17. Charles put a tablespoon of salt in a glass of water. He placed the glass on a sunny windowsill. After two weeks, he noticed a solid deposit on the bottom of the glass. What <u>most likely</u> happened to the water in the glass?

A It dripped out of the glass.

B It changed into a solid.

C It evaporated from the glass.

D It combined with the salt.

18. Which method of separating a mixture involves a change of state?

A all physical changes

B magnetic attraction

C using a sieve

D evaporation

19. True or False *Steel is a mixture of iron and carbon.* Is this statement true or false? Explain.

20. How can matter change?

Mixture or Solution?

Your goal is to make one mixture and one solution.

1. Gather water, cooking oil, sugar, salt, and gravel or small rocks.

2. Choose two substances to make a mixture. Combine them. Tell how you know it is a mixture. Name the parts of your mixture.

3. Choose two substances to make a solution. Combine them. Tell how you know it is a solution. Name the parts of your solution.

Analyze Your Results

Would your mixture change if it were heated or cooled? Would your solution change? Explain.

1 On a sunny day, a puddle on the sidewalk disappears. What caused this change?

A melting

B condensation

C evaporation

D distillation
DOK 1

2 All of the following cause a change of state <u>except</u>

A boiling.

B freezing.

C melting.

D rusting.
DOK 1

3 Which method of separating a mixture involves a change of state?

A filtration

B magnetism

C distillation

D settling
DOK 2

4 How can you tell that lemon juice is an acid?

A It tastes bitter.

B It reacts with water.

C It turns red litmus paper blue.

D It turns blue litmus paper red.
DOK 1

5 The pictures below show different containers of heated water.

In which container will the water change from liquid to gas fastest?

A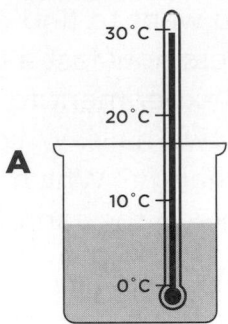

B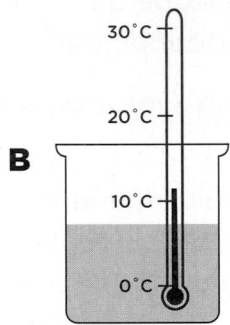

C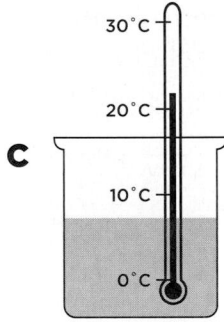

D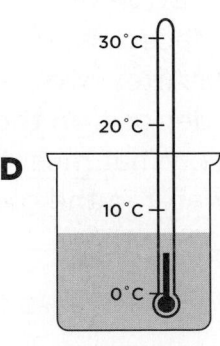

DOK 2

6 Look at the picture of carbon powder and iron filings shown below.

carbon powder   iron filings

If the carbon and iron were mixed together, which tool could separate them?

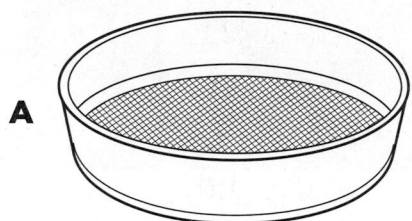

A

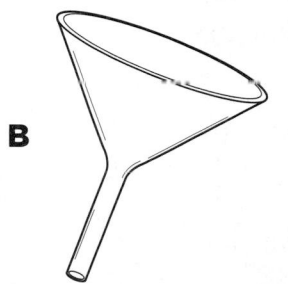

B

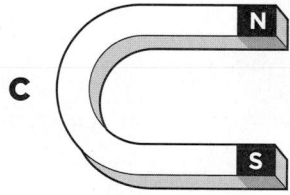

C

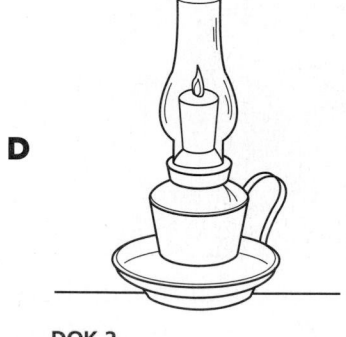

D

DOK 2

7 Look at the picture below.

Is this a physical change or a chemical change? Explain.
DOK I

8 Look at the picture below.

Is this a physical change or a chemical change? Explain.
DOK I

Check Your Understanding

Question	Review	Question	Review
1	pp. 446–449	5	pp. 448–449
2	pp. 446–449	6	pp. 460–461
3	pp. 448, 460–462	7	pp. 446–452
4	p. 470	8	pp. 446–452

Careers in Science

Pharmacy Technician

Do you look forward to doing science activities? Do you also like working with people? If so, you might enjoy a career in health care. A pharmacy technician works with pharmacists, or people who fill prescriptions. This person may work in a pharmacy in a drugstore, grocery store, hospital, or nursing home.

To qualify for this career, you would train on the job. You might take classes to earn a certificate. Then you could work with a pharmacist. You would help prepare medicines, counsel patients, and work with insurance companies. Best of all, you would help people recover from illnesses.

▲ A pharmacy technician helps people understand more about their medicine.

Pharmaceutical Researcher

Have you ever wondered where your medicine comes from? Some medicines, like aspirin, were first made from plants! Today, most are made in laboratories by pharmaceutical researchers.

If you are curious about how the body works and you want to make a difference, this career might be for you. To become a pharmaceutical researcher, you would study science in college. Then you would study medicine in graduate school.

▲ Pharmaceutical researchers work to develop new medicines.

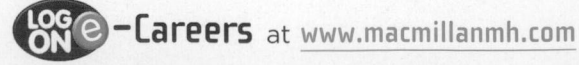

LOG ON e-Careers at www.macmillanmh.com

Forces and Energy

Windmills turn moving
air into energy.

Magnetic Migration

What would happen if you didn't know where to go? How would you find your way? You might rely on a map or signs, or even use a special tool called a compass. A compass tells you whether you are headed north, south, east, or west.

Birds make a long journey every year. Their journey can be thousands of miles long! Some birds travel alone while others travel in groups with birds that have already made the journey. If they travel by day, they seem to use the Sun's position or landmarks to find their way. If they travel at night, they can use the stars to help them navigate. They can even travel if the sun is hiding the Sun and stars!

People didn't know how birds were able to make this amazing journey for a very long time. Scientists discovered something fascinating! Some birds have a mineral called magnetite in their heads. The most magnetic of all minerals is magnetite. It can be compared to a compass needle. Just like we could use a compass to keep track of where we are going, birds have their own build-in compass!

 **Write About It**

Response to Literature Have you ever traveled to a different place? Where did you go? How did you get there? Write about a trip you have taken. Tell how you figured out the directions.

 **-Journal** Write about it online at www.macmillanmh.com

Forces

 **Why do things move?**

 The Big Idea

Big Idea Vocabulary

speed the distance an object moves in an amount of time (p. 485)

force a push or a pull (p. 486)

gravity a force of attraction, or pull, between all objects (p. 488)

work the use of force to move an object a certain distance (p. 504)

energy the ability to do work, either to make an object move or to change matter (p. 504)

simple machine something that has only a few parts and makes it easier to do work (p. 514)

· ·

 LOG ON Visit www.macmillanmh.com for online resources.

Motion and Forces

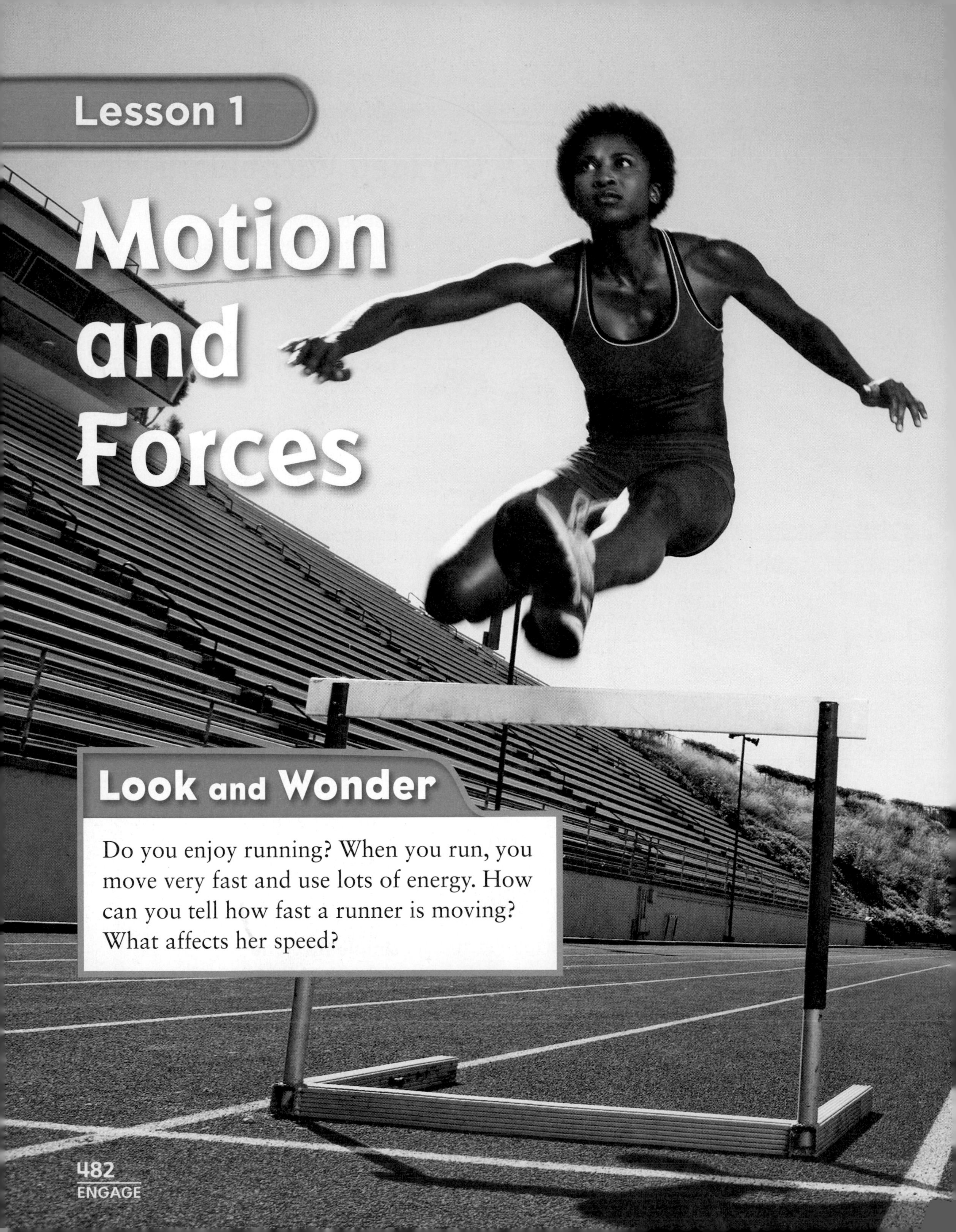

Look and Wonder

Do you enjoy running? When you run, you move very fast and use lots of energy. How can you tell how fast a runner is moving? What affects her speed?

How fast does it move?

Make a Prediction

How long will it take a marble to roll down a sloping path? Does the height of the slope affect the marble's motion? How? Make a prediction.

Test Your Prediction

1 **Make a Model** Stack three books on top of each other. Place one end of a cardboard tube on top of the stack. Let the other end of the tube touch a fourth book, which is on the table. Tape the tube in place on the outside.

2 Roll a marble down the tube. Start the stopwatch at the same moment. When you hear the marble hit the book, record the time. Repeat this step three times.

3 **Use Variables** Repeat step 2 with two books stacked. Then repeat with only one book.

Draw Conclusions

4 **Interpret Data** Make a bar graph comparing results for three books, two books, and one book. In which setup was the motion fastest?

5 Did your results match your prediction? Explain.

6 **Infer** Why was it important to repeat each test three times?

Materials

- 4 books
- cardboard tube
- tape
- marble
- stopwatch

Step **2**

Explore More

Would your results change if you used a longer tube? What if you had a higher stack of books? Make a prediction. Try it!

▶ **Essential Question**

How do objects move?

▶ **Vocabulary**

speed, p. 485

velocity, p. 485

force, p. 486

acceleration, p. 486

inertia, p. 486

friction, p. 487

gravity, p. 488

▶ **Reading Skill** ✔

Infer

Clues	What I Know	What I Infer

▶ **Technology**

e-Glossary and e-Review online at www.macmillanmh.com

What is motion?

When a marble rolls down a tube, it changes its location. Its starting place is at the top of the tube. Its ending place is at the bottom. An object is in *motion* if its location is constantly changing.

Position

How can you tell whether something is in motion? You look at its position. *Position* is the location of an object. You know something has moved when its position has changed.

Words such as *left* and *right*, *above* and *below*, and *east* and *west* give clues about position. When we describe an object's position, we compare it to surrounding objects. The objects used for comparing are called the *frame of reference*.

Another way to talk about position is to describe distance. *Distance* means how far apart two points or places are. We can give distance as a measurement. For instance, New York City is about 370 kilometers (230 miles) from Washington, D.C. We could add that New York is north and east of Washington, D.C.

▼ A horse is a fast runner, but a cheetah is faster! What are their speeds?

Speed

All moving objects have speed. **Speed** is the distance an object moves in an amount of time. A cheetah can run at nearly 112 kilometers per hour (70 miles per hour). You can write these values as 112 km/h and 70 mph. The speed of a racing horse might reach 76 km/h (47.5 mph).

How can you find the speed of an object? First, find out the distance the object moved. Next, find out how long it took to go that distance. Then, divide the distance traveled by the time spent moving. Suppose that in one hour, you pedal your bike 12 km (7.5 mi). Your speed is 12 km/h (7.5 mph).

Velocity

People sometimes confuse velocity (vuh•LAH•suh•tee) with speed. Speed tells you how fast an object is moving. **Velocity** describes the object's speed and direction of motion. A bike racer's speed may be 50 km/h (31 mph). If she goes 50 km/h *to the west*, however, that is her velocity.

A pendulum (PEN•juh•lum) is a mass on the end of a rod. After an initial push, the pendulum swings back and forth. It changes velocity on each swing.

▲ On each swing, the pendulum of a clock changes direction. That means its velocity changes too.

If this train's speed is 300 km/h, its velocity is 300 km/h to the east.

east

Quick Check

Infer An athlete running west crosses the finish line of a race. How can you tell she has moved?

Critical Thinking A girl runs 50 meters north. Then she runs 50 m west. Her speed does not change. Does her velocity change? Explain.

How do forces change motion?

How many times each day do you push or pull something to make it move? When you throw a ball, your muscles push on the ball. The ball moves away from you. Each push or pull is a **force**.

Forces can be big or small. The force a crane uses to lift a truck is huge. The force your hand uses to lift a feather is tiny. Forces can cause objects to start or stop moving. Forces can also change the speed or direction of a moving object.

Acceleration

As speed skaters race around a track, they go faster and slower. They turn left or right or skate in a circle. Any change over time in the speed or direction of a moving object is called **acceleration** (ak•se•luh•RAY•shun).

Inertia

A skater will not start to move unless a force acts on her. When moving, she cannot change her speed or direction unless a force acts on her. **Inertia** (ih•NUR•shuh) is the tendency of an object in motion to stay in motion or of an object at rest to stay at rest.

Acceleration

Read a Photo

How are these skaters accelerating?

Clue: The arrow shows their direction of movement.

A table-tennis ball changes its motion when a force acts on it.

Friction

All objects with mass have inertia. What is it, then, that brings a rolling marble to a stop? Why doesn't it keep rolling with the same velocity forever? The answer is that other forces stop it.

Friction (FRIK•shun) is a force that works against motion. Friction acts between the surfaces of objects that touch. The surfaces rub against each other, slowing the object or stopping it completely.

There is friction between a rolling marble and the surface on which it is rolling. That friction causes the marble to stop moving. There is very little friction between ice and the steel blades of ice skates.

≡Quick Lab

Inertia and Friction

1 Place a piece of paper flat on your desk. Put a plastic bowl on top of it.

2 **Predict** What do you think will happen if you pull the paper out from under the bowl as quickly as you can? Try it.

3 Was your prediction correct? What happened to the bowl when you pulled on the paper?

4 **Infer** Why was it necessary to pull the paper out so quickly?

5 What force would cause your results to change? Explain.

✔ Quick Check

Infer If there were no friction, would a moving object stop moving? Explain.

Critical Thinking You are riding in a car. The driver brakes suddenly. What happens to you? Why?

What is gravity?

A force is acting on you right now. This force acts on you all the time. It is pulling you and Earth together. Do you know the name of this force? It is gravity (GRA•vuh•tee).

Gravity is a force that acts over a distance and pulls all objects together. The pull of gravity depends on two things. One is the amount of matter in the objects. The other is the distance between the objects.

Objects with more mass have a stronger pull. For instance, the mass of Earth is huge. Its gravity pulls strongly on all objects, keeping your feet on the ground. The Moon is less massive than Earth, so its pull is weaker. Gravity is also stronger when objects are closer together. As objects move apart, the pull of gravity is weaker.

✓ Quick Check

Infer Mars is a planet that is smaller than Earth. How would the pull of gravity be different on Mars? Why?

Critical Thinking The Sun is much more massive than Earth. Do we feel the pull of the Sun's gravity? Explain.

Read a Diagram

How does gravity affect the motion of an apple falling from a tree?

Clue: What is the drawing trying to show?

Lesson Review

Visual Summary

Motion is a change in the position of an object. Speed and velocity are two ways to describe motion.

Acceleration is a change over time in the speed or direction of a moving object. Friction causes acceleration.

Gravity is a force that pulls all objects together. It depends on mass and distance.

Make a FOLDABLES Study Guide

Make a three-tab book. Use it to summarize what you learned about forces and motion.

Motion is...

Acceleration is...

Gravity is...

Think, Talk, and Write

1 **Vocabulary** What is the difference between speed and velocity?

2 **Infer** A bicyclist has been riding for 30 minutes at 20 km/h. He is to the east of where he began. What can you infer about his motion?

Clues	What I Know	What I Infer

3 **Critical Thinking** A leaf falls off a tree. It moves through the air before landing on the ground. What two forces affect the motion of the leaf?

4 **Test Prep** **This force causes an object to accelerate as it falls toward Earth.**

   A gravity
   B friction
   C inertia
   D velocity

5 **Essential Question** How do objects move?

 **Math Link**

Calculate Speed, Time, and Distance
What is the speed of a car that travels 400 km in 5 hours? If a bus goes 25 km/h, how long does it take to travel 50 km? If you can walk 4 km in 1 hour, how far can you walk in 3 hours?

 **Social Studies Link**

Changing Transportation
People's lives can change greatly when a faster means of transportation is invented. How would it affect your life if there were no transportation?

Focus on Skills

Inquiry Skill: Use Numbers

You know that gravity affects objects on Earth and elsewhere. Scientists can measure the motion of an object to learn how gravity affects its acceleration. To interpret the data, you may need to do some math or make a graph. You **use numbers** to measure, record, and interpret data.

▶ Learn It

When you **use numbers,** you order, count, add, subtract, multiply, and divide them. This is an important skill for a scientist to have. It is easier to use numbers if you organize them in a table, chart, or graph. That way, you can interpret your results more easily.

▶ Try It

When an object rolls downhill, gravity makes it accelerate. **Use numbers** to learn how quickly gravity makes objects accelerate.

> **Materials** long table, ruler, masking tape, four books, soup can, stopwatch, graph paper

1. Using the ruler and tape, divide the table into sections. Make each section 25 centimeters in length.

2. Place two books under each leg at one end of the table.

Distance	First Test	Second Test
	Time (in seconds)	Time (in seconds)
Start	0	0
Line 1		
Line 2		
Line 3		
Line 4		
Line 5		

3 Make a data table like the one shown. Add enough rows for each line of tape in step 1. The number of rows on your data table may be different from the one shown.

4 Place the soup can on its side at the raised end of the table. Start the stopwatch as you release the can. As the can rolls past each line of tape, record the time. Use the column labeled *First Test*. Have your partner catch the can before it rolls off the table!

5 Repeat step 4. Record the times under *Second Test*.

▶ **Apply It**

Now **use numbers** to make a line graph on graph paper.

1 Label the bottom line *Time (in seconds)* and the left side *Distance (in cm)*. Title your graph *Acceleration of a Soup Can*.

2 Mark off equal spaces along the left side in intervals of 25 (0, 25, 50, and so on). End this scale with the distance of the last line of tape on the table. Mark off the bottom in intervals of 1.

3 Using the data from your first test, write ordered pairs in the form (25, 1) and so on. For each ordered pair, place a point on the graph. You might need to estimate for fractions of seconds. Connect the points with straight lines. Do the same for your second test, but use a different color for the points and lines.

4 What does each ordered pair tell you? Where is the can moving slowest? Fastest? Is the can accelerating? How can you tell?

Changing Motion

Look and Wonder

A pitcher throws a fastball toward home plate. The batter prepares to swing. Then—contact! The baseball flies off into left field. What caused the motion of the baseball to change?

How do forces change motion?

Make a Prediction

If you roll a steel ball down a ramp, it should move in a straight line. How will a magnet affect the motion of the steel ball? Make a prediction.

Test Your Prediction

1 Stack three books. Lean the cardboard over the top book to form a ramp. Place the fourth book at the base of the ramp to stop the ball.

- 4 books
- cardboard
- steel ball
- marker
- magnet

2 **Observe** Roll a steel ball down the ramp. With a marker, draw the path the ball took as it rolled down the ramp.

3 **Observe** Point the magnet at one side of the ramp as shown. While holding the magnet, roll the ball again. Trace this new path.

4 **Use Variables** Move the magnet closer to the ramp. Repeat step 3.

Draw Conclusions

5 **Interpret Data** What happened to the path of the ball in step 3? How did the magnet affect its velocity? Did the ball accelerate? In what ways?

6 **Infer** Look at your tracings of the ball's path. Where was the pull of the magnet strongest? Where was it weakest?

Step 3

Explore More

What do you predict will happen if you use a stronger or weaker magnet? What if the magnet were underneath the cardboard? Test your predictions.

▶ **Essential Question**
How can pushes and pulls affect the way objects move?

▶ **Vocabulary**
balanced forces, p. 494
unbalanced forces, p. 495
newton, p. 495

▶ **Reading Skill** ✔
Predict

My Prediction	What Happens

▶ **Technology**
e-Glossary, e-Review, and animations online at www.macmillanmh.com

How do forces affect motion?

When a batter hits a baseball, he or she applies a force. Other forces are at work too. How do all the forces acting on an object affect its motion?

Balanced Forces

When you put a heavy backpack on your desk, the backpack does not move. Gravity pulls the backpack toward Earth, but your desk is in the way. The desk pushes up on the backpack with a force. The strength of that force is exactly equal to the pull of gravity. We describe these two forces as balanced.

Balanced forces are forces that cancel each other out when acting together on a single object. Each force is equal in size and opposite in direction. Balanced forces do not cause a change in motion. When an object is sitting still, all of the forces acting on it are balanced.

If the puppies pull with equal force, the shoe does not move. The forces are balanced. If one puppy pulls harder, the shoe moves toward him.

15 N

5 N

1 N

10 N

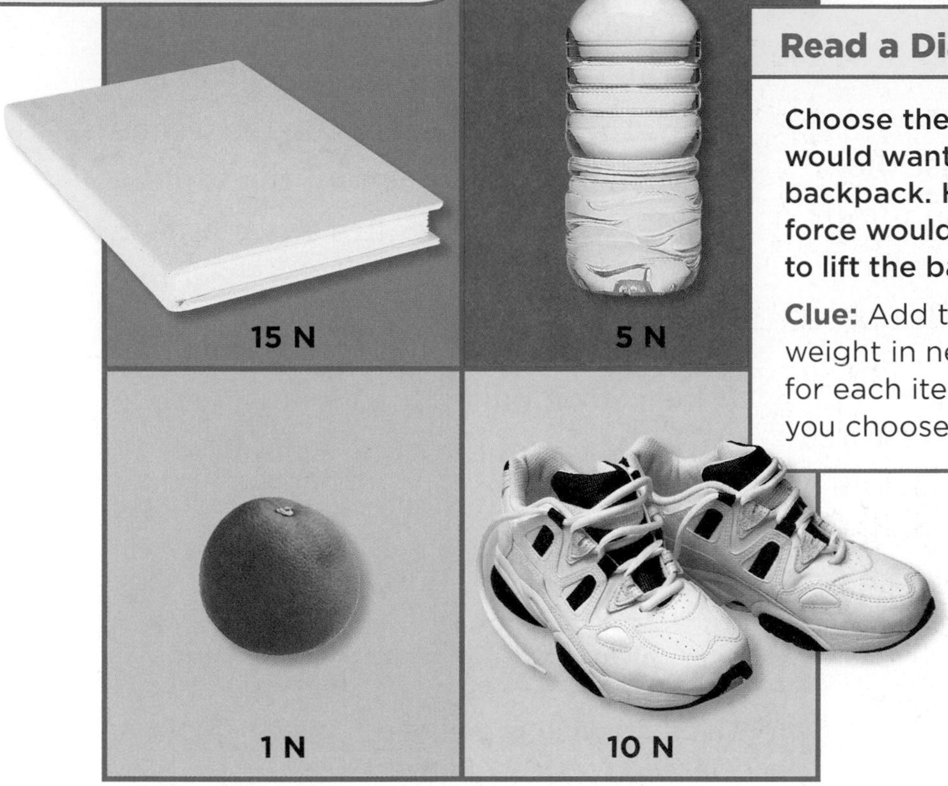

Read a Diagram

Choose the items you would want in your backpack. How much force would you need to lift the backpack?

Clue: Add the weight in newtons for each item you choose.

This empty backpack weighs 5 N.

Unbalanced Forces

Suppose you push that heavy backpack across your desk. There is friction between your backpack and the desk. The force of friction is weaker than your push. How can you tell? The backpack moves!

Forces that are not equal to each other are called **unbalanced forces**. Unbalanced forces cause a change in motion. The greater force determines the direction of motion. Unbalanced forces can affect an object's direction, speed, or both.

Over 300 years ago, the English scientist Sir Isaac Newton told how force and motion are related. Today we measure force in metric units called **newtons** (N).

Weight and Force

You may recall that the newton is also a metric unit for weight. How are weight and force related? An object has weight because the force of gravity pulls down on the object. Therefore, weight is a force. Like all forces, it is measured in newtons.

 Quick Check

Predict If the brown puppy pulls the ring twice as hard as the black puppy, what will happen? Why?

Critical Thinking A steel ball is placed between two magnets of equal strength. Does the ball move? Why?

How do forces affect acceleration?

Do you swim? If so, you know you must push harder against the water to swim faster. Do you jump? To jump higher, you push harder against the ground.

The size of a force affects an object's acceleration. A greater force gives the object more acceleration. The mass of the object matters too. If you apply the same force to an object with more mass, that object accelerates more slowly.

Forces Add Up

Look at the diagram below. In the first drawing, one person pulls the load. The load accelerates. In the second drawing, two people pull the same load. Now it accelerates twice as fast. Why? Two people exert twice the force.

What happens in the third drawing? One person pulls a load that is twice as large. She uses the same amount of force as in the first drawing. The load accelerates half as fast.

Read a Diagram

In which drawing is the load accelerating fastest? Slowest?

Clue: Compare the size of the arrows. What do they show?

LOG ON *Science in Motion* Watch accelerating masses at **www.macmillanmh.com**

Force and Acceleration

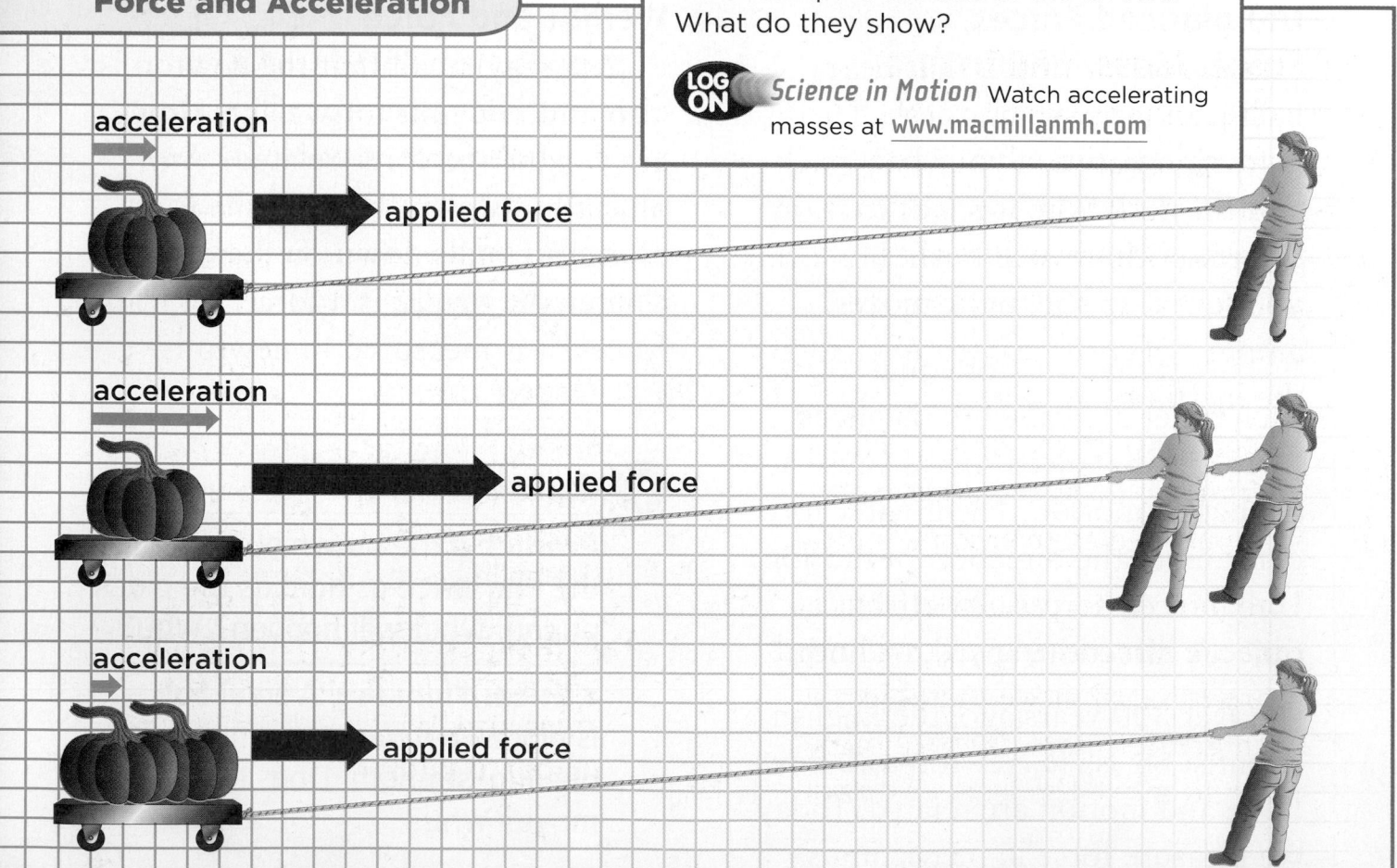

acceleration
applied force

acceleration
applied force

acceleration
applied force

If all the racers apply equal forces, who will win? The racer with the least amount of mass will win. She accelerates fastest.

Force, Mass, and Inertia

Recall that inertia is the tendency of a moving object to stay in motion. It is also the tendency for an object with no motion to stay at rest. The amount of force needed to change the inertia of an object depends on the mass of the object. The greater the mass the greater the force needed to overcome inertia.

If you try to push two objects that are not moving you will need more force to move the object with greater mass. It takes more force to move the heavier object.

FACT ▶ A force is needed to stop an object that is in motion.

Suppose you try to stop two different balls that are rolling down a hill. More force is needed to stop the ball with more mass. In this case, too, greater force is needed to change the inertia of heavier objects.

Quick Check

Predict Peg kicks a soccer ball with a force of 5 N. If she uses 10 N of force in a second kick, will the ball accelerate faster? Explain.

Critical Thinking Bowling balls and soccer balls are about the same size. Why is a bowling ball harder to throw?

Friction and Motion

1. Tie a piece of string through a book. Place the book on a smooth surface. Attach the string to a spring scale. Stack a second book on top of the first.

2. **Measure** Gently pull on the scale to measure the force of your pull just before the books move. Record your result.

3. Using the scale, drag the books quickly along the surface. Measure and record the force.

4. **Infer** Does an object at rest have more friction than a sliding object? Base your answer on your results from step 3.

The steel skates of an ice hockey player reduce friction on the icy surface.

How does friction affect motion?

Think of a hockey player skating on ice. Now think of a student wearing sneakers and standing at a bus stop. Will he slide if he pushes on the sidewalk? Why not?

Remember, friction is a force that works against motion. Friction acts between surfaces that touch. The amount of friction depends on the surfaces involved. A rough surface produces more friction than a smooth surface. There is little friction between steel blades and ice. There is more friction between sneakers and the sidewalk.

Why do you need to oil the moving parts of a bicycle? The oil reduces friction. It helps the parts work smoothly together.

 Quick Check

Predict Are you more likely to slip walking on ice or grass? Why?

Critical Thinking Have you ever seen a sign that says "Slippery When Wet"? Why would water make a surface slippery?

Lesson Review

Visual Summary

Balanced forces are forces that cancel each other when acting together on the same object.

A greater force will accelerate an object faster than a lesser force will.

Friction is a force that works against motion. It depends on the kinds of surfaces that are touching.

Make a FOLDABLES® Study Guide

Make a trifold book. Use it to summarize what you learned about changing motion.

Think, Talk, and Write

1 Vocabulary Scientists measure the amount of a force in units called _____.

2 Predict A student has two magnets of equal strength. She places one on each side of a ramp. She releases a steel ball down the ramp. The student repeats her procedure with both magnets on the same side of the ramp. Predict the path of the ball in each trial.

My Prediction	What Happens

3 Critical Thinking When moving down a giant slide, some people sit on slippery mats. Explain why.

4 Test Prep What force will stop a moving skateboard?

- **A** friction
- **B** gravity
- **C** newtons
- **D** balanced forces

5 Essential Question How can pushes and pulls affect the way objects move?

Writing Link

A Day without Friction

What would happen if you awoke one day and there were no friction? What would you do? How would you move? Write a short story titled "A Day without Friction."

Math Link

Tugboat Mathematics

Two tugboats are moving a barge. One tugboat pulls with a force of 7,000 N. The other tugboat pushes in the same direction with an equal force. What is the sum of the forces on the barge?

WHEELS IN MOTION

For thousands of years, people have been using the wheel to move things around. You might think that wheels help us move things by reducing friction. Did you know that without friction, wheels would not work?

You use wheels when you ride a bicycle. Tires cover the rims of the wheels. The bicycle pedals are attached to the rear wheel. A gear and a chain connect them. When you push the pedals, the wheels turn.

What would happen if there were no friction between the tires and the road? The wheels would spin, but the bike would not move. Friction causes the tires and the ground to push against each other. Friction is the force necessary to make the bike move.

Explanatory Writing

Good explanatory writing

▶ describes how to complete a task or how something works;

▶ gives clear details that are easy to follow.

Write About It

Explanatory Writing
Research how the brakes on a bike work. Write a description that explains how friction helps the bike stop moving.

LOG ON **e-Journal** Research and write about it online at www.macmillanmh.com

Math in Science

The Force of Friction

People have been bobsledding since the 1880s. The first bobsleds had waxed wood runners on the bottom. Later, people used steel runners. Steel reduces friction better than waxed wood does.

The amount of friction depends on the weight of the objects that are touching. It also depends on the materials from which those objects are made. This two-man bobsled experiences friction of 115 N. Rounding this value to the nearest ten, the force is about 120 N.

Rounding Numbers

► Look at the digit in the ones place.

► Compare the digit to 5. If it is less than 5, round down. If it is 5 or greater, round up.

Force of Friction on Bobsleds

Type of Runners	Four-Man Bobsled	Two-Man Bobsled	Two-Woman Bobsled
steel	185 N	115 N	100 N
waxed wood	617 N	382 N	333 N
rubber	2,470 N	1,529 N	1,333 N

Solve It

Make a new table that rounds the above values to the nearest tens.

Work and Energy

Look and Wonder

This skater looks like he is having fun! Would you say he is doing work? How did he get his skateboard off the ground?

How are position and force related?

Make a Prediction

Will the slope of a ramp affect how far a toy car travels? Make a prediction.

Test Your Prediction

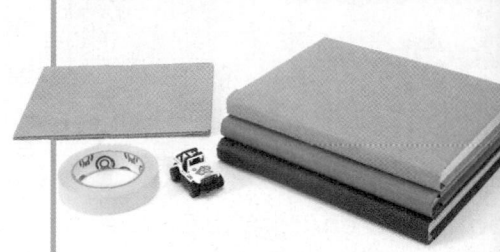

Materials

- **6 books**
- **cardboard**
- **masking tape**
- **toy car**
- **meterstick**

1. Stack three books on top of one another. Tape one edge of the cardboard to the edge of the top book. Tape the bottom edge of the cardboard to the floor or a tabletop. The cardboard and books should make a ramp.

2. Place the car at the top of the cardboard. Then let it go. Wait until the car comes to a full stop.

3. **Measure** Using a meterstick, measure the distance from the bottom of the ramp to the rear of the car. Record your measurement.

4. **Use Variables** Add another book to the stack. Repeat steps 2 and 3. Now repeat the test with five or six books. For each test, record the distance the car travels.

Draw Conclusions

5. **Interpret Data** How did the height of the ramp affect how far the car traveled? Was your prediction correct?

6. **Infer** What can you infer about position and force?

Explore More

If you use a toy car with more mass, will your results change? Make a prediction. Try it!

Step 2

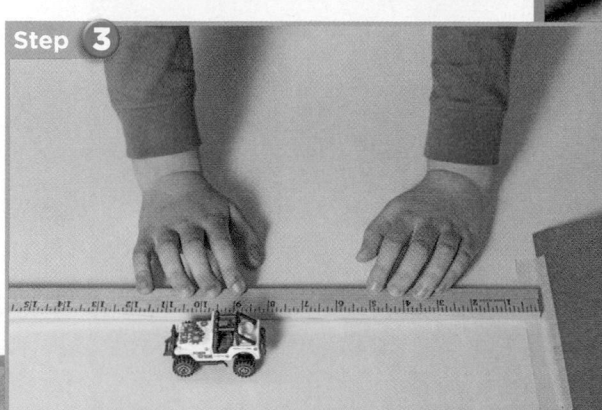

Step 3

▶ **Essential Question**
How are energy and work related?

▶ **Vocabulary**

work, p. 504

energy, p. 504

potential energy, p. 504

kinetic energy, p. 505

▶ **Reading Skill** ✓
Summarize

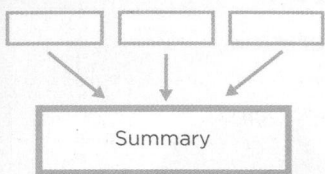

Summary

▶ **Technology** LOG ON
e-Glossary and e-Review online
at www.macmillanmh.com

504
EXPLAIN

What is work?

You ride a roller coaster up a steep ramp. Your car climbs higher and higher. It reaches the top of the ramp then—whoosh! It rushes down the other side. A roller coaster ride can be fun, but it also involves work. That means work can be fun!

Force and Distance

In science, work has a particular meaning. **Work** is done when a force is used to move an object a certain distance.

A roller coaster does work when it moves its cars up a ramp. Gravity does work when it pulls the cars down the other side. You do work when you run or skate. Anytime you push or pull to move an object some distance, you do work.

The weight lifter in the picture is using a force to keep the weights above her head. Is she doing work? Surprisingly, she is not. The force she is using is balanced by the force of gravity. These balanced forces keep the weights from moving. Of course, she did work to lift the weights up.

Potential Energy

You know that it takes energy to play sports. In science, **energy** is the ability to do work. That means energy is needed to apply a force that will move an object. Energy is also needed to make matter change.

Stored energy is called **potential energy** (puh•TEN•shul E•nur•jee). It has the potential, or future ability, to do work. A roller coaster at the top of a ramp has potential energy.

FACT Energy cannot be made or destroyed.

WELCOME AMUSEME

ROCKET RACER

1 The roller coaster pulls the cars to the top of the ramp. The cars gain potential energy.

2 Gravity pulls the cars down the other side of the ramp. Potential energy changes to kinetic energy.

3 The friction of the brakes brings the cars to a stop.

Read a Diagram

What forces and energies does this diagram show?

Clue: The numbers on the diagram relate to the captions.

Kinetic Energy

When an object is moving, it has the energy of motion. This is called **kinetic** (kuh•NE•tik) **energy**. A roller coaster has kinetic energy when it moves up or down a ramp.

A moving object can do work on any object it touches. As it travels, the roller coaster also moves the people inside the cars.

Potential energy can change to kinetic energy easily. Have you ever shot a bow and arrow? The stretched bow has potential energy. It works much like a spring. When you release the string on the bow, its potential energy becomes kinetic energy. The arrow moves.

Quick Check

Summarize Describe potential energy and kinetic energy.

Critical Thinking A student sits at his desk studying. Is he doing work?

What are some forms of energy?

Energy makes motion and change possible. Which forms do you use?

Chemical Energy

The energy your body uses to walk and lift things is chemical energy. Chemical energy is stored in the particles that make up food and other fuels. When you eat the food, chemical energy passes to you.

Electrical Energy

Electrical energy comes from the movement of charged particles. Some electrical energy comes from batteries. However, most of it comes from power plants that burn fuel to make electricity. The electrical energy is sent through wires and cables to homes and businesses.

Light Energy

The Sun is a major source of light energy on Earth. Plants use light energy for photosynthesis. When they make food, plants convert light energy into chemical energy. People can collect light energy with solar power cells. These cells change light energy into electrical energy.

Some scientists use light energy from lasers. A laser is a kind of light that can cut through some materials.

Mechanical Energy

Mechanical (mih•KA•nih•kul) energy is the sum of kinetic and potential energy. Moving objects have mechanical energy. Objects at rest have it only as potential energy.

light energy

chemical energy

electrical energy

Thermal Energy

A stove, a heater, and a match have thermal energy. They give off heat. Thermal energy depends on the motion of the tiny particles in matter. The faster these particles move, the warmer a substance gets. The warmer the substance, the more thermal energy it has.

Nuclear Energy

Nuclear (NEW•klee•ur) energy comes from the tiniest particles of matter. When these particles split apart or join together, huge amounts of nuclear energy are released.

✔ Quick Check

Summarize Name the different forms of energy. Give an example for each.

Critical Thinking What do all these forms of energy have in common?

Quick Lab

The Energy of a Pendulum

❶ Tie a washer to the end of a string. Hold the string in place on your desk with a heavy book. Pull the washer back. Then let go.

❷ **Observe** How does the washer move after you release it?

❸ When does the washer have the most kinetic energy? The most potential energy?

❹ **Infer** What form of energy is the pendulum showing?

mechanical energy

thermal energy

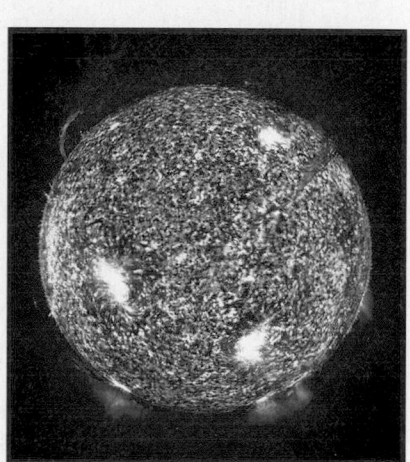

nuclear energy

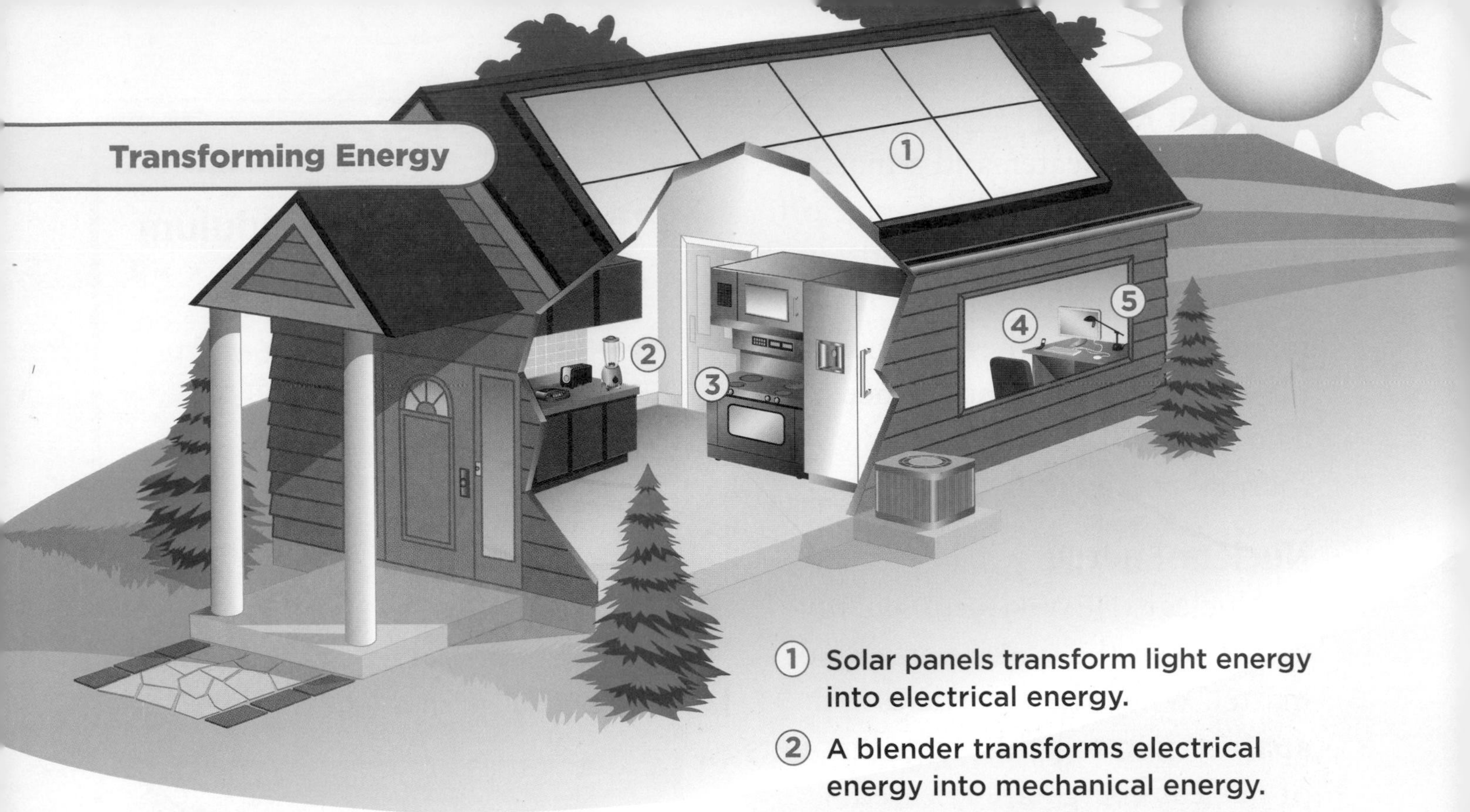

① Solar panels transform light energy into electrical energy.

② A blender transforms electrical energy into mechanical energy.

③ Electrical energy is transformed to thermal energy by a kitchen stove.

④ Electrical energy is transformed to chemical energy when a battery is charged.

⑤ A lamp transforms electrical energy into light energy.

How can energy change?

Energy does not always keep its form. It does not always stay in one place either.

Transforming Energy

Energy is *transformed* when it changes from one form to another. A lightbulb transforms electrical energy to light and thermal energy. A blender transforms electrical energy to mechanical energy.

Transferring Energy

Energy is *transferred* when it passes from one object to another. The kinetic energy of a marble is transferred to another marble when they collide. The first marble slows or stops. The second marble picks up the kinetic energy and moves.

Read a Diagram

In what ways is energy transformed inside a home?

Clue: Look at the numbers in the diagram. Then read the captions.

 Quick Check

Summarize What are some ways in which energy changes form?

Critical Thinking How is energy transformed when you rub your hands together?

Lesson Review

Visual Summary

 **Energy is** the ability to do work. This ability can be classified as potential energy or kinetic energy.

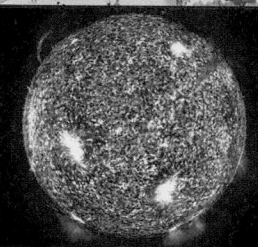 **Some forms of energy are** mechanical, chemical, electrical, light, thermal, and nuclear.

 **Energy can be** transformed from one form to another or transferred from one object to another.

Make a FOLDABLES Study Guide

Make a trifold book. Use it to summarize what you learned about work and energy.

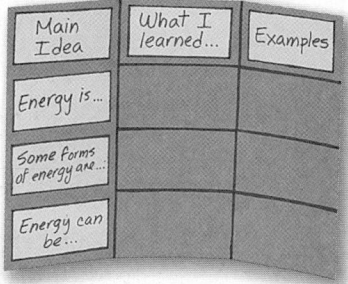

Think, Talk, and Write

1 **Vocabulary** A child sitting on the top of a slide has _____ energy.

2 **Summarize** List three examples of work as it is defined in the text. What do they all have in common?

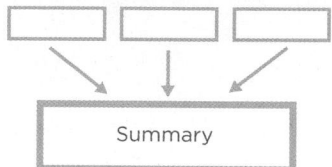

3 **Critical Thinking** Name five ways in which you observe energy changing form every day.

4 **Test Prep** Which is an example of kinetic energy?

- A a parked car
- B a rock at the top of a cliff
- C a train rolling down a track
- D a flashlight battery

5 **Essential Question** How are energy and work related?

 **Writing Link**

Explanatory Paragraph
Examples of potential energy and kinetic energy are all around you. Write a paragraph explaining the difference between them. Tell about the examples you have seen.

 **Social Studies Link**

Energy in Your Neighborhood
Take a walk around your school or home. Note how the different forms of energy are used. Share your observations with your classmates.

Hybrid Power

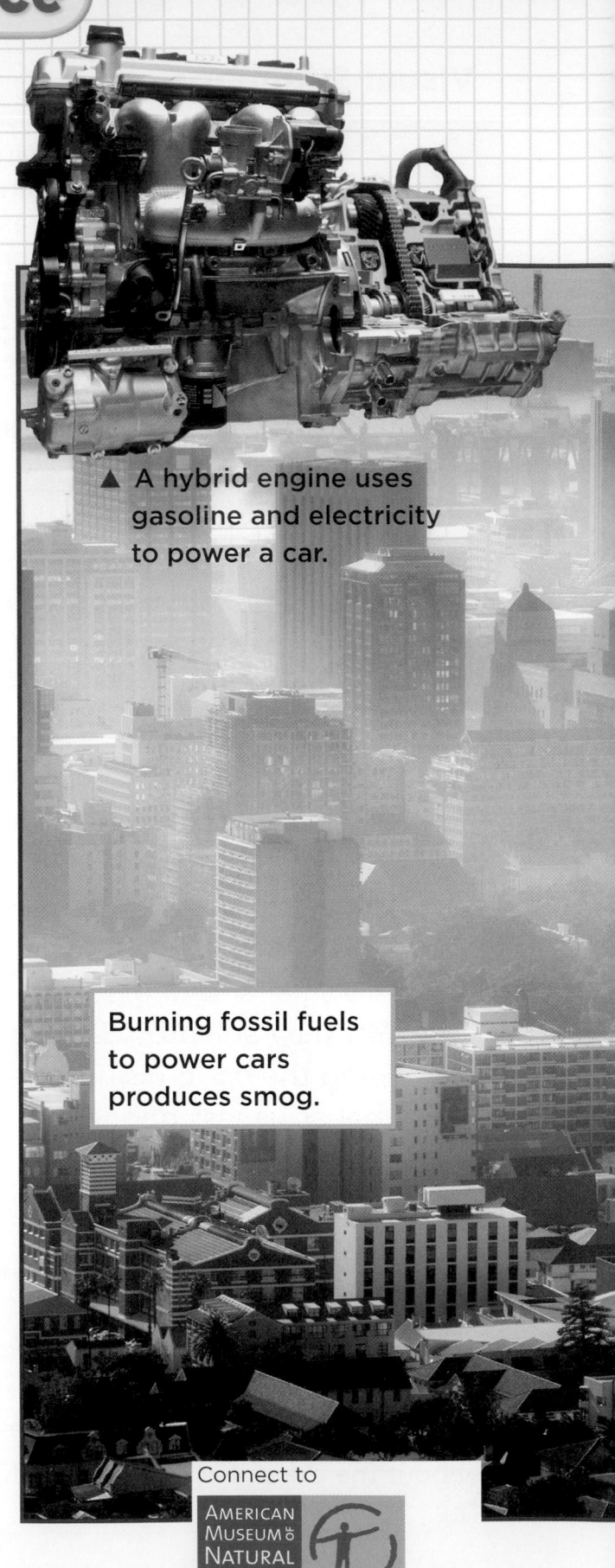

▲ A hybrid engine uses gasoline and electricity to power a car.

In cities like Los Angeles and Chicago, millions of people drive cars. Most of the cars run on gasoline, which is made from oil. There is a limited supply of oil in the world. Our cars make us very dependent on it. Also, the more gasoline the cars burn, the more they pollute the air. Pollution from cars contributes to clouds of smog that can cover a city like a blanket.

How can we become less dependent on gasoline and reduce air pollution? One way is to build cars that use less gasoline. Many car companies now make hybrid cars. *Hybrid* is a word that describes something that is a mix of two different things. Hybrid cars use two different power sources—gasoline fuel and electrical energy.

In a traditional car, the gasoline engine runs all the time. However, when the car is stopped at a light, sitting in traffic, or slowing down, it does not need the gas engine's power. At these times, the fuel used to keep the engine running is being wasted.

Burning fossil fuels to power cars produces smog.

Connect to

AMERICAN
MUSEUM OF
NATURAL
HISTORY

at www.macmillanmh.com

A hybrid car is designed to use much less fuel than a traditional car. It combines a gas-powered engine with an electric motor powered by batteries. When the car is stopped, slowing, or moving at slow speeds, the gas engine shuts off. The electric motor takes over to keep the lights, air conditioning, and radio working. The batteries that run the motor get recharged when the car slows to a stop. The car changes its energy of motion into electrical energy.

The gasoline engines in hybrid cars are smaller and more energy efficient. Yet they still provide enough power to keep the car cruising on a freeway. This makes us less dependent on gasoline—and causes less pollution from gasoline!

▲ Hybrid cars can help reduce air pollution.

Summarize

▶ State the main idea.

▶ Include the most important details.

▶ Use your own words.

 **Write About It**

Summarize Read the article again. How do hybrid cars work? How do hybrid cars help the environment?

 **e-Journal** Research and write about it online at www.macmillanmh.com

Simple Machines

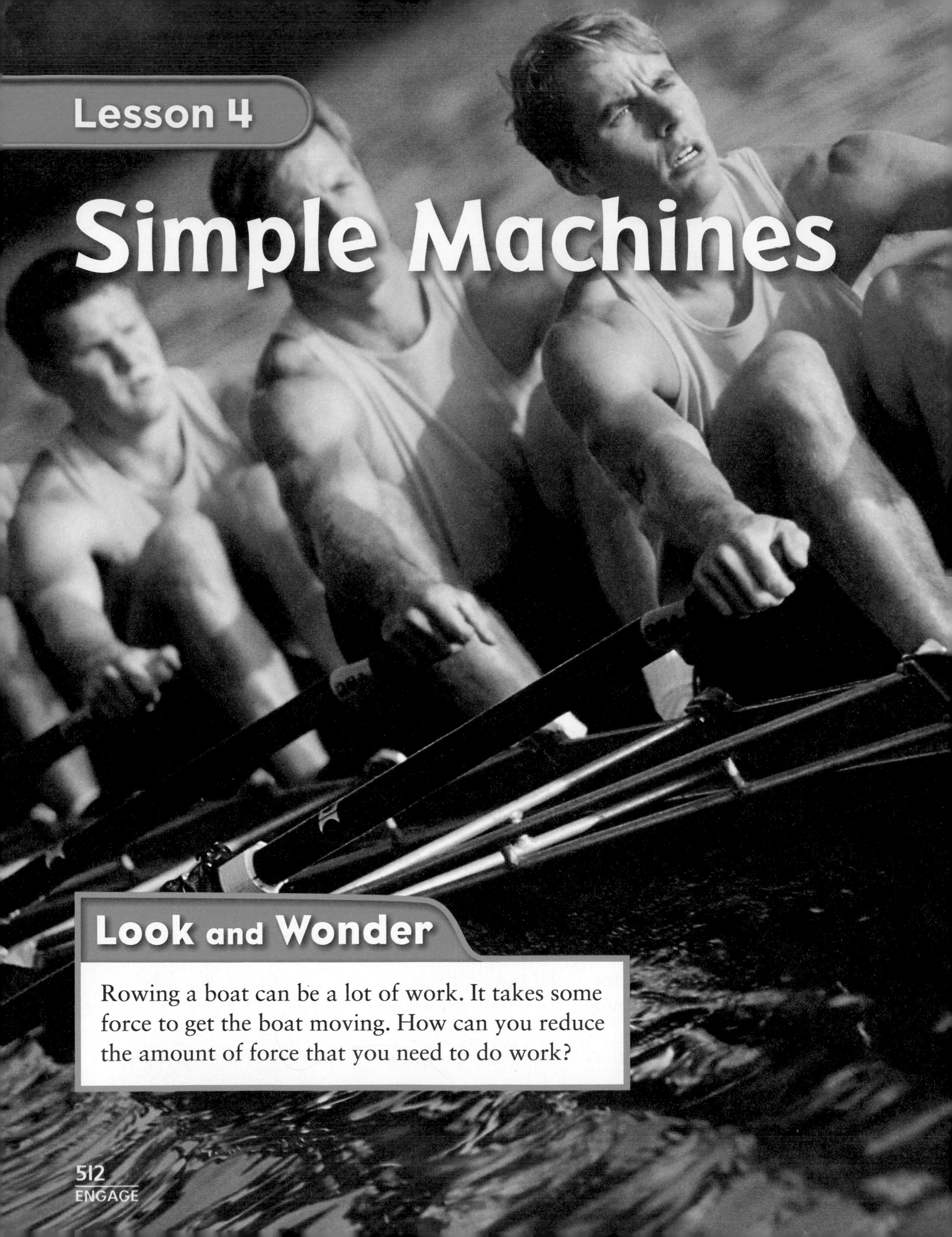

Look and Wonder

Rowing a boat can be a lot of work. It takes some force to get the boat moving. How can you reduce the amount of force that you need to do work?

How do pulleys reduce force?

Make a Prediction

A pulley is a machine that makes work easier. Look at the two pulleys shown here. Which will make it easier to lift a book? Make a prediction.

Test Your Prediction

1 Fasten a pulley to something that will not move. Tie the longer cord around a book. Rest the book on a surface. Thread the other end of the cord through the pulley's groove. Then attach the spring scale as shown

2 Measure Pull down on the spring scale until the book lifts off the surface. Measure and record how much force you are using.

3 Remove the cord from the pulley. Fasten the second cord to something that will not move. Thread the free end of the second cord through the groove. Attach the spring scale to the second cord. Attach the book to the pulley.

4 Observe Pull up on the spring scale. What happens? Measure and record the force.

Draw Conclusions

5 Interpret Data Which pulley system made it easier to lift the book? Did your results match your prediction?

6 Infer Why did the amount of force change when you changed the setup?

Explore More

Instead of measuring force, use a ruler to measure distance. Repeat the two setups. How far must you pull the cord to lift the book 10 cm? What do you notice in each case?

Materials

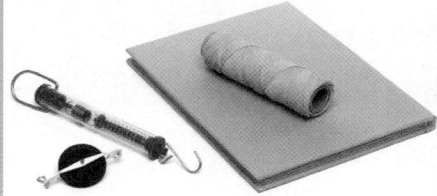

- pulley
- 2 pieces of cord, one longer than the other
- book
- spring scale

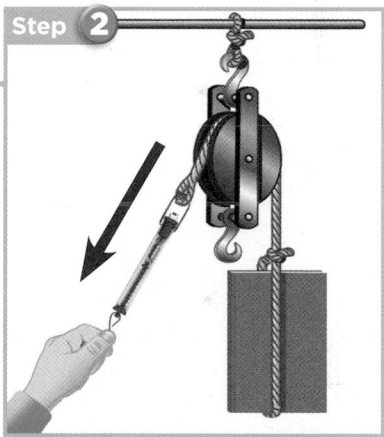

Step **2**

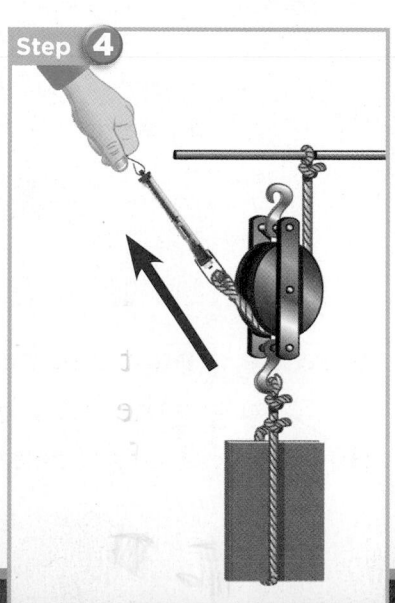

Step **4**

Read and Learn

▶ **Essential Question**

How do simple machines make work easier?

▶ **Vocabulary**

simple machine, p. 514

lever, p. 514

effort force, p. 515

output force, p. 515

inclined plane, p. 518

compound machine, p. 520

▶ **Reading Skill** ✔

Compare and Contrast

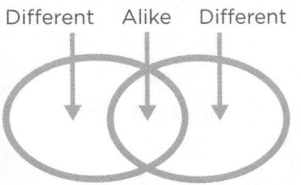

Different   Alike   Different

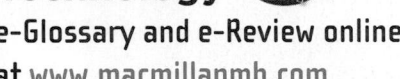

▶ **Technology** 🔵LOG ON🔵

e-Glossary and e-Review online at www.macmillanmh.com

What are simple machines?

Name some machines people use. You might think of a bulldozer, a washing machine, or a power saw. A machine uses force to do work.

Not all machines have a motor or lots of parts. A machine with only a few parts is a **simple machine**. Pulleys, levers, and wedges are simple machines. So are wheels and axles, screws, and inclined planes.

Levers

Have you ever used a screwdriver to open a can of paint? If so, you have used a lever. A **lever** (LE•vur) has two parts. It has a bar and a fixed point called a fulcrum. The *fulcrum* supports the bar and allows it to turn. The *load* is the object the lever moves. Levers can apply a small force over a long distance. They can also apply a large force over a small distance.

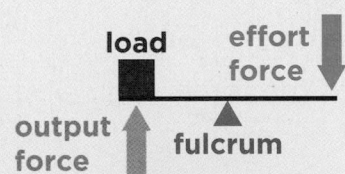

A *first-class lever* has its fulcrum between the load and the effort force.

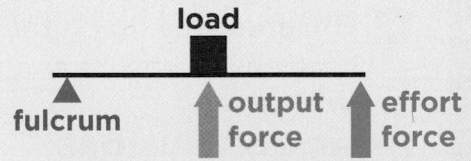

A *second-class lever* has its fulcrum at one end and the load in the middle.

Effort and Output Forces

To do work with a lever you apply force to a load. The force you apply is the **effort force**. The lever then applies a force to the load called the **output force**.

A lever does not make you stronger. A lever makes the output force greater than the effort force. This makes work easier.

The position of the fulcrum determines how a lever does work. When a load is closer to the fulcrum, less effort force is needed to move the load. When a load is farther from the fulcrum, more effort force is needed to move the load. In this case, the load moves farther or faster than the effort force.

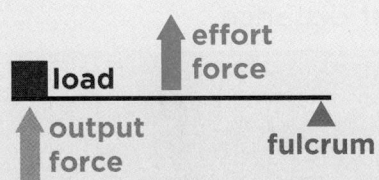

A *third-class lever* has its fulcrum at one end and its effort force is in the middle.

Quick Lab

Comparing Levers

1. Make a lever that changes the direction of a force. Place about an inch of a ruler under the edge of a book. Put a pencil under the ruler, close to the book. Push down on the other end of the ruler. What happens?

2. Now make a lever that does not change the direction of a force. Remove the pencil. Center the book on the ruler. Lift up on one end of the ruler. What happens?

3. **Interpret Data** Where was the fulcrum, load, and effort force in step 1? In step 2?

4. **Infer** Which kind of a lever did you make in step 1? In step 2? Explain.

✔ Quick Check

Compare and Contrast How are the three kinds of levers similar? How are they different?

Critical Thinking How does a tennis racquet make work easier?

What are two simple machines with wheels?

Pulleys and wheels and axles are simple machines with wheels.

Wheels and Axles

A *wheel and axle* is a two-part simple machine. It has a wheel fixed to a central axle. When the axle turns, so does the wheel. Likewise, a turn of the wheel moves the axle.

A faucet is a wheel and axle. The faucet handle is the wheel. You only need to use a small effort force on the handle to turn on the water. The axle turns it into a larger force. The larger force is needed to open the valve.

A Ferris wheel is also a wheel and axle. The effort force is applied to the axle. When the axle makes one turn, the cars attached to the wheel also make one turn. The wheel moves faster and farther than the axle when the axle is turned.

Fixed Pulleys

A *pulley* is a simple machine with a grooved wheel and a cord. A pulley attached to an unmoving object is a *fixed pulley*. The cord passes around the wheel and attaches to the load. Pulling the end of the cord in one direction moves the load in the other direction. A fixed pulley simply changes the direction of a force.

▼ A small effort force is applied to a faucet handle. A greater output force opens the valve.

▼ The axle of a Ferris wheel turns a small distance. The cars move a greater distance.

Fixed pulley	Movable pulley	Multiple fixed and movable pulleys

effort force | effort distance | output force | effort force | effort distance | output force | effort force | effort distance | output force

8 cm
6 cm
4 cm
2 cm
0 cm

surface

Read a Diagram

Compare and contrast the effort force and output distance of the three types of pulleys.

Clue: What do the arrows show?

Movable Pulleys

A pulley attached to the load is called a *movable pulley*. Unlike a fixed pulley, a movable pulley does not change the direction of a force. However, it does change the amount of force needed to do work. A movable pulley makes the output force twice as large as the effort force. For example, one movable pulley needs only 5 kilograms of effort force to lift a 10 kg load.

Although a single movable pulley doubles the output force, the effort distance also doubles. To lift a load 2 meters, for example, a person must pull the cord 4 meters.

Often, movable and fixed pulleys work together. The more movable pulleys attached to the load, the easier it is to move the load.

✓ Quick Check

Compare and Contrast How does a wheel and axle differ from a pulley?

Critical Thinking A wrench tightening a bolt is a wheel and axle. Which part is the wheel? The axle?

What are inclined planes?

Suppose you are moving a heavy desk onto a truck. Would you lift the desk to get it on the truck? That would take a lot of effort and probably a helper or two. A better plan would be to push the desk up a ramp.

A ramp is an inclined plane. An **inclined plane** is a simple machine with one part—a flat, slanted surface. This machine holds up, or supports, most of the load's weight.

Inclined planes make it easier for people with physical disabilities to get around. You find inclined planes at the corners of sidewalks. They are often at the entrances to buildings.

With an inclined plane, you need only a small effort force. However, you have to move the load farther. Look at the photo. The cable must pull the car up the entire length of the ramp. The ramp is longer than the distance needed to lift the car straight up.

A screw is an inclined plane wrapped around a cylinder.

An inclined plane is a flat, slanted surface.▼

inclined plane

screw

wedge

Wedges and Screws

What do you get if you put two inclined planes back to back? You get a simple machine called a *wedge* (WEJ). A wedge changes a downward or forward force into a sideways force.

A *screw* is an inclined plane twisted into a spiral. To see how, cut a piece of paper into a right triangle. Wrap the triangle around a pencil. Watch how the slanted edge of the triangle goes up and around the pencil.

Like all inclined planes, wedges and screws trade distance for effort force. Often we use them on a very small scale. We use knives to cut food. We use drills to make holes in wood and metal.

✓ Quick Check

Compare and Contrast How are a knife and a screw similar? How are they different?

Critical Thinking Which would be easier—walking up a gradual ramp or a steep ramp? Why?

Read a Photo

Choose a simple machine from the photos above. How is it making the task easier?

Clue: Identify the load and the effort force.

How do simple machines work together?

Two or more simple machines form a **compound machine**. Most machines are compound machines. Scissors, for example, are made of two wedges and a first-class lever. When you push the handles together, the blades cut paper.

Bicycles combine many simple machines. The handlebars and brakes are levers. The gears and chain are part of the axle that turns the rear wheel. Screws hold the parts in place.

Efficiency

Imagine trying to turn an old rusty doorknob. Compared to a new doorknob, it is very difficult to turn. Rust has increased friction on the doorknob's moving parts. You must work harder to turn it.

Efficiency (ih•FIH•shun•see) means how much work a machine produces compared to the amount of work applied. Friction reduces the efficiency of any machine.

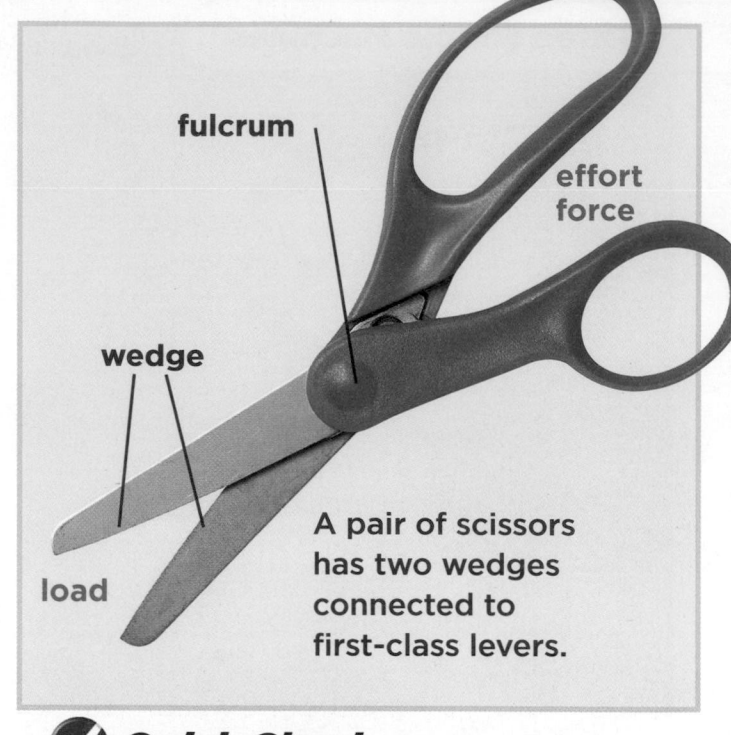

fulcrum

effort force

wedge

load

A pair of scissors has two wedges connected to first-class levers.

✔ Quick Check

Compare and Contrast How is a compound machine different from a simple machine?

Critical Thinking Which of a bicycle's wheels—front or rear—is a wheel and axle? Explain.

A bicycle is a compound machine. ▶

Velocity describes the objects speed and direction of motion.

Each push or pull is force.

Any change in speed or direction is called acceleration.

Friction is force that works against motion.

Ahmed
Elbasha

the sun and all of the objects around
it make up the solar system.

Planets are round objects in space
that are satilites of the sun.

Gravity is the force of attraction beetween
all objects.

Telescopes make for away objects came
closer.

A comet is mostly mixed with
rocks and ice.

Astroids are large chunks of metal or
rocks.

If a meteorid enters earth's
atmosphere it is called a meteor

If a meteor reaches Earths surface
it is called a meteorte

Speed is the distance and objects moves
in an amount of time.

Lesson Review

Visual Summary

Simple machines include levers, wheels and axles, pulleys, inclined planes, wedges, and screws.

Levers and inclined planes change the amount or direction of the effort force needed to move a load.

Compound machines are made of two or more simple machines.

Make a **FOLDABLES®** Study Guide

Make a three-tab book. Use it to summarize what you learned about simple machines.

Simple machines...

Levers and inclined planes...

Compound machines...

Think, Talk, and Write

1. **Vocabulary** A simple machine that has a bar pivoting on a fulcrum is called a(n) _____ .

2. **Compare and Contrast** You can use a lever, a pulley, or an inclined plane to lift an object. How are the machines alike? How are they different?

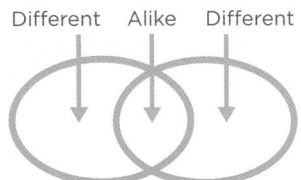

Different   Alike   Different

3. **Critical Thinking** Most simple machines are more efficient than compound machines. Why?

4. **Test Prep** Which of the following is a compound machine?

   A scissors
   B lever
   C pulley
   D wedge

5. **Essential Question** How do simple machines make work easier?

Writing Link

Write an Advertisement
Choose a simple machine. Write an advertisement that persuades someone to buy it. How does this machine make life easier? Explain all the benefits.

Math Link

Inclined Plane Problem
An inclined plane leads to a building. The plane is 255 cm long. The distance between the ground and the building is 160 cm. How much farther must you push an object up the inclined plane than if you lifted it straight up?

Be a Scientist

books

tape

short piece
of cardboard

string

wooden cube

spring scale

long piece
of cardboard

Structured Inquiry

How do ramps help move objects?

Form a Hypothesis

Ramps make it easier to do work. Does the length of a ramp affect the amount of force needed to move a load? Write your answer in the form "If the length of a ramp increases, then the amount of effort force..."

Test Your Hypothesis

1. Stack two books. Tape one side of the short piece of cardboard to the top of the stack. Let the other side rest on a flat surface, forming a ramp.

2. Tie one end of the string around a wooden cube. Tie a loop in the other end and attach it to the spring scale. Hold the spring scale in the air. Record the weight of the cube.

3. **Measure** Place the cube at the bottom of the ramp. Slowly pull on the spring scale to bring the cube up the ramp. Try to keep the force steady as you pull. Record the amount of force you used.

4. Remove the short cardboard ramp. Tape the long piece of cardboard in its place. Repeat step 3.

Step 3

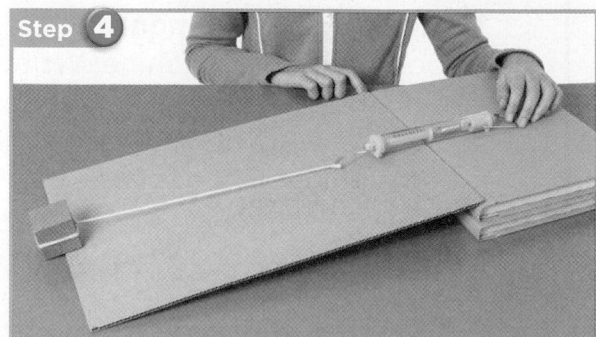

Step 4

Draw Conclusions

5 **Interpret Data** How much force was needed to pull the cube in step 3? In step 4? Describe any changes.

6 **Communicate** Do your results support your hypothesis? Explain.

7 **Infer** How do you think the force would change if you used an even longer ramp?

Does the length of a lever change the effort force?

Form a Hypothesis

Do levers work the same way as ramps do? Does the length of a lever change the amount of force needed to move a load? Write a hypothesis.

Test Your Hypothesis

Design an investigation to test whether the length of a lever changes the effort force. Write out the materials you will need and the steps you will follow. Record your results and observations.

Draw Conclusions

Did your results support your hypothesis? How did the amount of effort force change in your investigation? Discuss your conclusions with your classmates.

What else would you like to learn about simple machines? For instance, how does a wheel and axle reduce the force needed to move an object? Design an investigation to answer your question. Write your investigation so that another group can complete it by following your instructions.

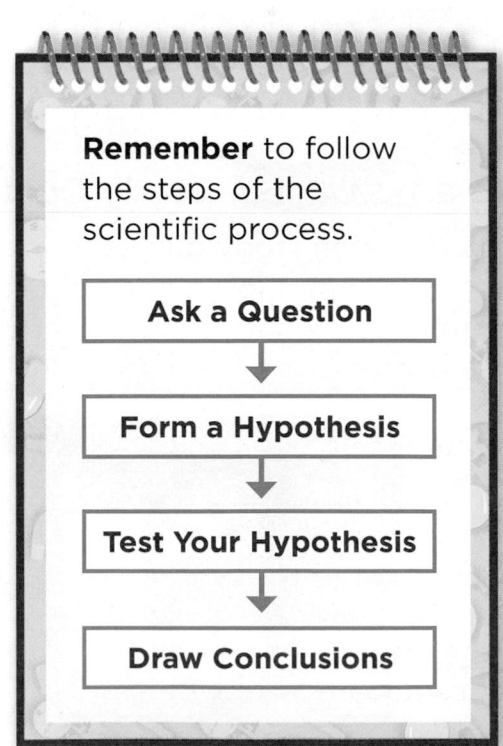

Remember to follow the steps of the scientific process.

Ask a Question
↓
Form a Hypothesis
↓
Test Your Hypothesis
↓
Draw Conclusions

Visual Summary

Lesson 1 Motion occurs when an object changes its position. Many forces act on motion.

Lesson 2 Unbalanced forces change the speed or direction of an object's motion.

Lesson 3 Work is done when a force is used to move an object a certain distance. Energy is the ability to do work.

Lesson 4 Simple machines make it easier to do work. There are six kinds of simple machines.

Make a FOLDABLES Study Guide

Tape your lesson study guides to a piece of paper as shown. Use your study guide to review what you have learned in this chapter.

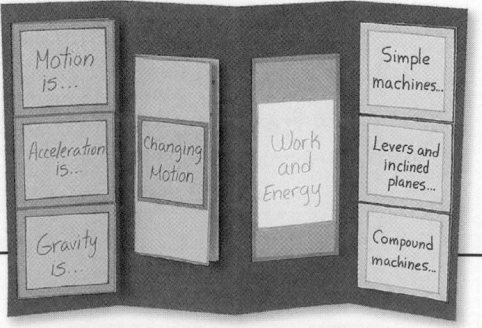

Vocabulary

DOK I

Fill each blank with the best term from the list.

balanced forces, p. 494	**potential energy**, p. 504
inclined plane, p. 518	**speed**, p. 485
kinetic energy, p. 505	**unbalanced forces**, p. 495
lever, p. 514	**velocity**, p. 485
load, p. 515	**work**, p. 504

1. An object's speed and direction of motion are described by its _____.

2. Forces are _____ if they are equal in size and opposite in direction.

3. The object being moved by a simple machine is called the _____.

4. Stored energy is also called _____.

5. The change in distance over time is called _____.

6. A ramp is an example of a(n) _____.

7. Stored energy becomes _____ when an object moves.

8. When forces are not equal or not opposite in direction, they are _____.

9. To move an object a certain distance is called _____.

10. A bar or plank that pivots, or turns, on a fulcrum is a(n) _____.

Skills and Concepts

DOK 2–3

Answer each of the following.

11. Summarize Name the six types of simple machines. Provide an example of each.

12. Use Numbers Alex's train travels 30 miles in 1 hour. Luisa's train travels 45 miles in 30 minutes. After one hour has passed, how much farther than Alex will Luisa have traveled?

13. Critical Thinking Two teams of equal strength are playing tug of war. Each team pulls equally on the rope. Are the forces balanced? Explain.

14. Explanatory Writing Write a paragraph describing how astronauts experience gravity in space and on the Moon.

15. Critical Thinking The oars on a rowboat are what class of lever? Explain your answer.

16. Infer Your soccer team practices on a wet, but not soggy, field. Does the ball roll faster or slower than normal? Explain your answer.

17. True or False *As a car slows down, it accelerates.* Is this statement true or false? Explain.

18. True or False *Holding a heavy object in place is hard work.* Is this statement true or false? Explain.

19. Which type of energy is in food?

A electrical **C** kinetic

B thermal **D** chemical

20. Why do things move?

Performance Assessment

DOK 3

Potential Energy

1. Build a ramp using a piece of cardboard and some books. Make a mark halfway down the ramp. Place a sponge at the bottom of the ramp.

2. Roll a toy car down the ramp toward the sponge from the halfway mark. Measure how far the sponge moves. Then, roll the car from the top of the ramp. Measure the sponge's movement again. Next, tape three pennies to the top of the toy car. Repeat the process. Record the distances on a chart.

Analyze Your Results

When did the sponge move the most? How does the movement of the sponge show the car's potential energy? What factors affected the car's potential energy? Explain your answers.

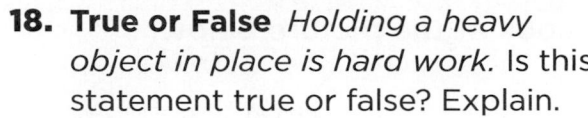

Test Preparation

1 Look at the picture below.

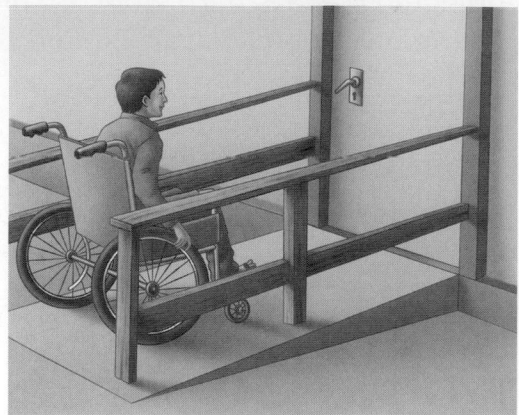

Which simple machine helps this person move toward the door?

A pulley

B lever

C wedge

D inclined plane
DOK 2

2 In the picture below, the apple is even, or level, with the gram masses.

The forces on each side are

A balanced.

B unbalanced.

C inertia and friction.

D friction and gravity.
DOK 2

3 Look at the picture below.

The change in speed created by kicking this ball is called

A gravity. **C** friction.

B inertia. **D** acceleration.
DOK I

4 Andrea tosses a baseball toward an empty field. Which two forces will <u>most likely</u> affect the ball's motion?

A balanced and unbalanced

B magnetism and gravity

C gravity and friction

D friction and magnetism
DOK 2

5 Which statement is true?

A Faster objects move farther in a certain period of time.

B Speed is a change in position of an object.

C Motion is how far an object moves in a certain period of time.

D Velocity and speed are equal.
DOK I

6 Mike and Tito have two different types of bicycles. They want to know which bicycle is faster.

Which experiment will <u>most likely</u> answer their question?

A Mike picks one bike and Tito picks another. Then they race. The winner has the faster bike.

B Mike and Tito take turns riding each bike on a track. They record their speeds on each bike. Then they compare notes.

C Mike rides one bike near his house. Tito rides the other bike near his house. Then they compare notes.

D Mike rides each bike in a circle. Tito rides each bike in a straight line. Then they compare notes.
DOK 2

7 Using a smoother surface on a playground slide reduces

A friction.

B speed.

C motion.

D weight.
DOK I

8 The picture below shows an example of work.

Define *work* and give your own example of work.
DOK 2

9 How do you know whether an object has moved?
DOK I

10 How do you transform energy when you switch on a flashlight?
DOK 2

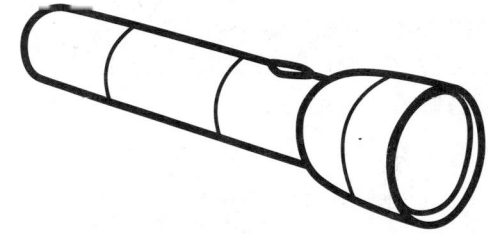

Check Your Understanding

Question	Review	Question	Review
1	pp. 514–519	6	pp. 4–7, 485
2	pp. 494–495	7	pp. 487, 498
3	pp. 496–498	8	p. 504
4	pp. 486–488, 498	9	p. 484
5	pp. 484–485	10	pp. 506–508

Energy

The Big Idea How do we use energy?

Big Idea Vocabulary

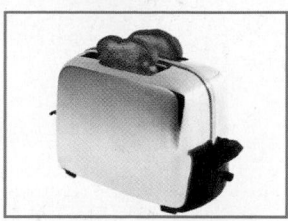

heat the flow of thermal energy from warmer to cooler objects (p. 530)

echo repetition of a sound produced by the reflection of a sound wave (p. 542)

pitch the highness or lowness of a sound (p. 545)

reflection the bouncing of light or sound waves off a surface (p. 556)

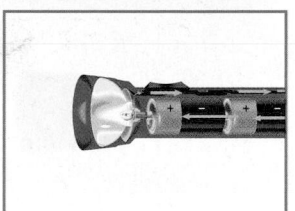

circuit a complete path through which electricity can flow (p. 567)

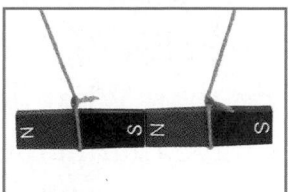

pole one of two ends of a magnet where a magnet's pull is strongest (p. 577)

 Visit www.macmillanmh.com for online resources

Heat

Look and Wonder

A lizard can warm itself by sitting in the sun. Animals that live in cold climates cannot always do that. What do animals in cold climates have that helps them keep warm?

What keeps mammals warm in places with little heat?

Purpose

Explore how certain mammals—such as whales and seals—stay warm in cold water.

Procedure

Materials

1. Put on a latex glove. Have a partner time how long you can comfortably keep your hand in ice water. Record the time. ⚠ **Be Careful.** Remove your hand as soon as it feels chilled!

2. **Make a Model** Dry your hand and let it warm. Move your gloved hand around in the shortening to coat it. Get a thick layer over your entire hand and between your fingers.

- latex gloves
- bucket of ice water
- stopwatch
- paper towels
- vegetable shortening

3. How long can you keep your hand in the ice water now? Have your partner time you. Record the results.

4. **Use Numbers** Trade places and let your partner repeat the procedure. Compute the average of both sets of results.

Draw Conclusions

5. **Interpret Data** How long on average could you keep your hand in ice water in step 1? In step 3?

6. **Infer** The shortening represents fat. How might an extra layer of fat help you survive in a cold climate?

Step ❶

Explore \ More

What other substances or materials can help mammals stay warm? List the ones you know. Then research some you do not know. Report your findings to the class.

Read and Learn

▶ **Essential Question**

What is heat?

▶ **Vocabulary**

heat, p. 530

conduction, p. 532

convection, p. 532

radiation, p. 533

insulator, p. 533

conductor, p. 533

▶ **Reading Skill** ✔

Cause and Effect

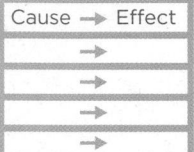

▶ **Technology** 🔵ᴸᴼᴳ ᴼᴺ

e-Glossary and e-Review online at www.macmillanmh.com

What is heat?

Energy is needed for animals to stay warm. Whether it's from the Sun or your body, thermal energy keeps you warm. *Thermal energy* is the energy of the moving particles of matter. The faster the movement of particles the greater the amount of thermal energy.

Heat is the flow of thermal energy from one object to another. Heat always moves from warmer objects to cooler objects. A warm object cools as it transfers heat.

Transferring Heat

What happens when you use a toaster? Not only do you heat the bread, but you also heat the air around it. Touch the warm toast, and that same thermal energy moves to your hand.

The hot particles of the toaster move quickly. The particles slow as they transfer their thermal energy. The cooler particles speed up. In time, all the particles move at the same speed.

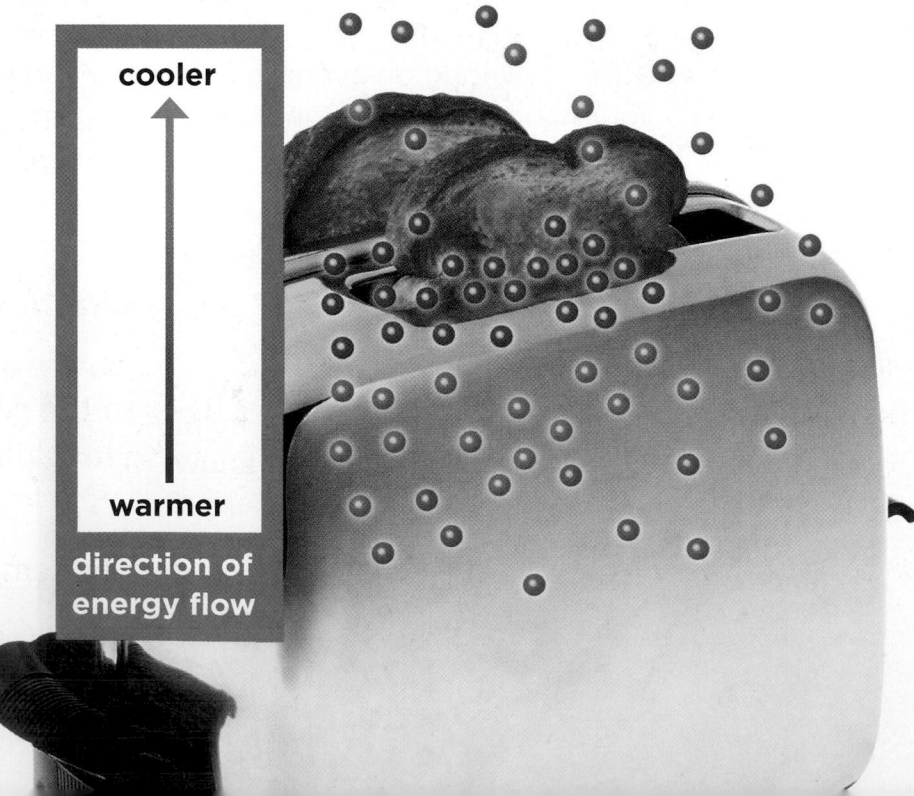

cooler

warmer

direction of energy flow

Heat moves from the warm toaster to the cooler air around it.

Friction between the match head and the surface creates heat.

Changing Temperature

Some sources of heat include burning wood and fossil fuels. Heating can change an object's temperature (TEM•puh•ruh•chur). *Temperature* measures the thermal energy of the particles in a substance.

We measure temperature with a *thermometer* (thur•MAH•muh•tur). Inside most thermometers is a liquid such as alcohol. As the thermometer warms, the particles of the liquid move faster and farther apart. This movement makes the liquid expand and rise inside the thermometer.

Did you ever have a fever? You probably measured your temperature in degrees Fahrenheit, or F. Most scientists use the Celsius, or C, scale to measure temperature.

The thermometer on this page shows the Fahrenheit and Celsius scales. Water freezes at 32° F. This is in the same place as 0° C. Water boils at 212° F. As you can see, that amount is 100° C.

boiling point of water

freezing point of water

Read a Photo

What is the temperature in Fahrenheit? In Celsius?

Clue: Find the marks near the top of the red line.

 Quick Check

Cause and Effect What happens to the particles of an ice cube when placed in a glass of juice?

Critical Thinking How are heat and temperature related?

How does heat travel?

You have learned what happens when thermal energy is transferred. How does heat transfer take place?

Heat Transfer

Heat is transferred through the water by convection.

Heat is transferred from the burner to the pot by conduction.

Read a Diagram

Describe how heat is flowing in this pot of water.

Clue: The red circles are hot particles. The blue circles are cooler particles.

Conduction

Solids are heated mainly by conduction (kun•DUK•shun). **Conduction** occurs between two objects that are touching. Conduction can also occur within an object, such as a metal pot.

What happens when you heat a pan on a stove? The fast moving particles of the burner or flame hit the cooler particles of the pan. The collision gives the cooler particles more thermal energy. The particles of the pan start to move faster. Soon, the entire pan gets hot.

Convection

Another way to transfer heat is by convection (kun•VEK•shun). **Convection** transfers heat through liquids or gases.

If you want to boil water, you can heat it in a pot. As the pot heats, it transfers energy to the water. The water particles at the bottom of the pot heat first. They move faster and farther apart. The hot water becomes less dense. The dense cooler water sinks, replacing the hot water. When all particles of water move at the same rate, the water boils.

FACT Heat is not the same as temperature.

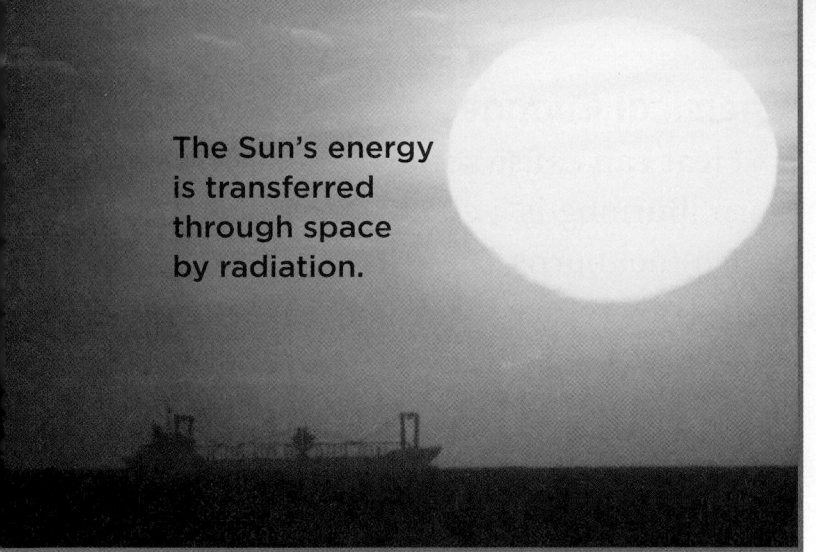

The Sun's energy is transferred through space by radiation.

Radiation

The third way to transfer heat is by radiation (ray•dee•AY•shun). **Radiation** does not need matter to transfer heat. It can travel through space. Without radiation, energy from the Sun would not reach Earth. Hot surfaces transfer thermal energy to the air by radiation.

Insulators and Conductors

In winter, you might wear a fleece jacket to stay warm. Fleece is an insulator (IN•suh•lay•tur). **Insulators** do not transfer heat very well. Fat is an insulator that mammals have in their bodies. It helps keep their body heat from escaping into the cold air.

The opposite of an insulator is a conductor (kun•DUK•tur). **Conductors** transfer heat easily. Metal, for instance, is a good conductor. That is why many pots and pans are made of metal.

✔ Quick Check

Cause and Effect A metal object feels cooler than a wood object at room temperature. Why?

Critical Thinking How is radiation different from conduction and convection?

≡ Quick Lab

Temperature and Air

1. **Predict** Place a deflated balloon over the mouth of an empty plastic bottle. What will happen if you put the bottle in hot water? In cold water?

2. **Observe** Place the bottle in a bucket of warm water. Wait five minutes. What happens to the balloon?

3. Now place the bottle in a bucket of ice water. What happens?

4. What do you think caused the balloon to inflate and deflate?

Wool mittens are good insulators for your hands.

A copper kettle is a good conductor for hot liquids. ▼

How does heat change matter?

The particles that make up matter are always moving. By adding energy to those particles or by taking energy away, you can change matter.

Physical Changes

If you increase thermal energy, the particles of matter move faster and farther apart. The matter expands, taking up more space. The opposite happens if you reduce thermal energy. When cooled, most matter contracts (kun•TRAKTS), or shrinks. The particles move closer together.

Chemical Changes

Heat can cause some matter to burn. Burning is a chemical change. When fuel burns, the energy stored inside it is released.

Changes of State

If enough heat is added, matter can change state. The welder below is using a torch to heat metal. The flame is hot enough to melt the metal. If more energy were added, the liquid metal would change to a gas.

 Quick Check

Cause and Effect How does heat cause matter to expand?

Critical Thinking Why do people burn coal and oil?

Heat can change solid metal to a liquid.

Lesson Review

Visual Summary

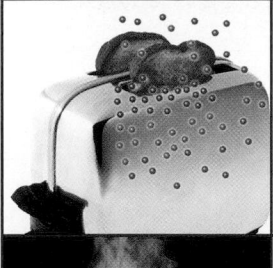

Heat is the flow of thermal energy from a warmer object to a cooler object.

Heat is transferred through conduction, convection, and radiation.

Heat can cause matter to expand, contract, change state, or burn.

Make a **FOLDABLES®** Study Guide

Make a three-tab book. Use it to summarize what you learned about heat.

Heat is

Heat is transferred through

Heat can cause matter to

Think, Talk, and Write

1 Vocabulary The transfer of heat through space is called _____.

2 Cause and Effect What happens when heat is added to ice? To liquid water? To a balloon filled with air?

Cause	→	Effect
	→	
	→	
	→	
	→	

3 Critical Thinking Explain why heat will not flow from an ice cube to a hot drink.

4 Test Prep Many pots and pans are made of metal because metal is a good

   A conductor.

   B insulator.

   C heat source.

   D radiator.

5 Essential Question What is heat?

Writing Link

Compare and Contrast
Write a paragraph comparing a metal cup and a foam cup. Which would you choose for a hot drink? A cold drink? Explain your choices.

Art Link

Heat Transfer Picture
Draw a picture that shows examples of the three ways that heat is transferred. Add labels and captions to your picture.

Focus on Skills

Inquiry Skill: Infer

You just read that insulators do not transfer heat very well. One way to keep ice cubes from melting is to insulate them. Scientists experiment to find out which materials prevent the most heat transfer. After the experiment, they can **infer** which material will make the best insulator.

▶ Learn It

When you **infer,** you form an idea from facts or observations. It is easier to form an idea about a result when the information is organized. You can use charts, tables, or graphs to organize your data. That way you can quickly see differences and form an opinion about the results.

▶ Try It

Use different materials to insulate ice cubes. **Infer** which material is best for slowing the melting.

 **Materials** **scissors, paper, aluminum foil, plastic wrap, 4 ice cubes, tape, shallow dish**

1. Make a chart like the one shown.

2. Cut a piece of paper just large enough to cover one ice cube. Do the same with the aluminum foil and plastic wrap.

3. Wrap one of the ice cubes in the paper. Seal the paper well with tape. Place the sealed ice cube in the dish. Record the time in your chart.

4 Repeat step 3 with the aluminum foil. Repeat with the plastic wrap. Leave one ice cube unwrapped. Record the time you place each ice cube in the dish.

5 Observe the ice cubes in the dishes. Record the time when each ice cube melts completely.

6 Calculate the time it took for each ice cube to melt. Enter the times in your chart.

	Foil	Paper	Plastic	Unwrapped
Start Time				
Melted				
Time to Melt				

▶ **Apply It**

Interpret your data to **infer** which wrapper best insulated the ice cube.

1 Compare your result for the unwrapped ice cube to each of your other results. Which material was the best insulator? What was the time difference between that one and the unwrapped cube?

2 Which material was the poorest insulator? Why do you think so?

3 Why was it a good idea to keep one ice cube unwrapped?

4 What type of heat transfer did you investigate? Explain your thinking.

Sound

Look and Wonder

Which of these are stringed instruments? How can you change the sound of a stringed instrument?

How can strings make music?

Make a Prediction

To make music on a guitar, you pluck its strings. What happens to the sound if you use a tighter string? A shorter string? A thicker string? Make your predictions.

Test Your Prediction

1 ⚠ **Be Careful.** Handle scissors carefully. Make a small hole in the top of the box.

2 Thread the string through the hole. Tie a large knot on the bottom end to fasten the string. Make sure it cannot come undone.

3 **Observe** Hold the box steady at the end near the hole. Pull the string over the top of the ruler as shown. Pluck the string. What do you hear? Record your observations.

4 Repeat step 3, but pull the string tighter. Do it again, but let the string hang loosely. Record your observations.

5 Remove the string. Cut several pieces with the same length. Twist them together. Repeat steps 2–4. Record your observations.

Draw Conclusions

6 **Communicate** How did the sound change in each case? Were your predictions correct?

7 **Classify** What type of instrument is your cardboard box device?

Explore More

Do different materials make different types of sounds? Repeat the experiment using a long rubber band. How does the sound change?

Materials

- **cardboard box**
- **scissors**
- **thin string**
- **wooden ruler**

Step **3**

Read and Learn

▶ **Essential Question**
How can you
make sounds?

▶ **Vocabulary**

vibration, p. 540

sound wave, p. 541

echo, p. 542

wavelength, p. 544

frequency, p. 544

pitch, p. 545

amplitude, p. 545

volume, p. 545

▶ **Reading Skill** ✔
Infer

Clues	What I Know	What I Infer

▶ **Technology** 🔵LOG ON
e-Glossary and e-Review online
at **www.macmillanmh.com**

What is sound?

Think of all the sounds musical instruments can make. The sound of a guitar can be soothing or piercing. A bass drum makes a deep, thumping sound. A marching drum makes a sharp, cracking sound. Have you ever wondered how different sounds are made?

Vibration

What happens when you pluck the string of a guitar? It moves back and forth very quickly. This back-and-forth motion is called a **vibration** (vi•BRAY•shun).

What can you notice if you place your fingers against your throat while you talk or hum? You can feel a vibration. You can feel the vibration of your vocal cords. Vocal cords in your throat vibrate when air moves past them. This allows you to speak.

All sounds begin with a vibration. Consider the bell of an alarm clock. When the alarm goes off, the bell vibrates. How does the sound reach your ear?

When the drummer strikes the drum, it vibrates. The vibration forms a sound wave.

A ringing bell sends sound waves in all directions.

Sound Waves

Think about what happens when an ocean wave rolls under a floating object. The object moves up and down. Overall, the object does not change position. Yet the wave's energy moves through the water.

In some ways an ocean wave is like a sound wave. A **sound wave** is a wave that transfers sound through matter. Sound waves spread outward from a vibration in all directions. Unlike an ocean wave, a sound wave does not move up and down.

Energy Transfer

Study the picture above. The blue dots show what happens to air particles when the bell rings. First, energy from a vibration causes air particles to move. Then, air particles bump into one another. Some air particles are crowded together. Some are spaced apart. The air particles move back and forth. However, they do not change their overall position as they transfer energy.

✔ Quick Check

Infer Can a sound wave move through water? Explain.

Critical Thinking When you pluck a string, it vibrates and makes a sound. How can you stop the sound?

FACT ▶ Sound cannot travel through a vacuum.

How does sound travel?

You know that sound travels through the air. Sound travels through other substances too, including solids, liquids, and gases.

Echoes

Sometimes sound waves bounce off a surface. The surface reflects the sound, causing it to repeat. An **echo** (E•koh) is a specific, reflected sound.

The dolphins in this picture use echoes to navigate and find prey. The sounds they make are reflected by underwater objects, such as fish.

▲ Dolphins use echoes to locate underwater objects.

Read a Diagram

Does sound travel faster in seawater or air? How much faster?

Clue: Find where each arrow is on the scale.

Sound Speeds

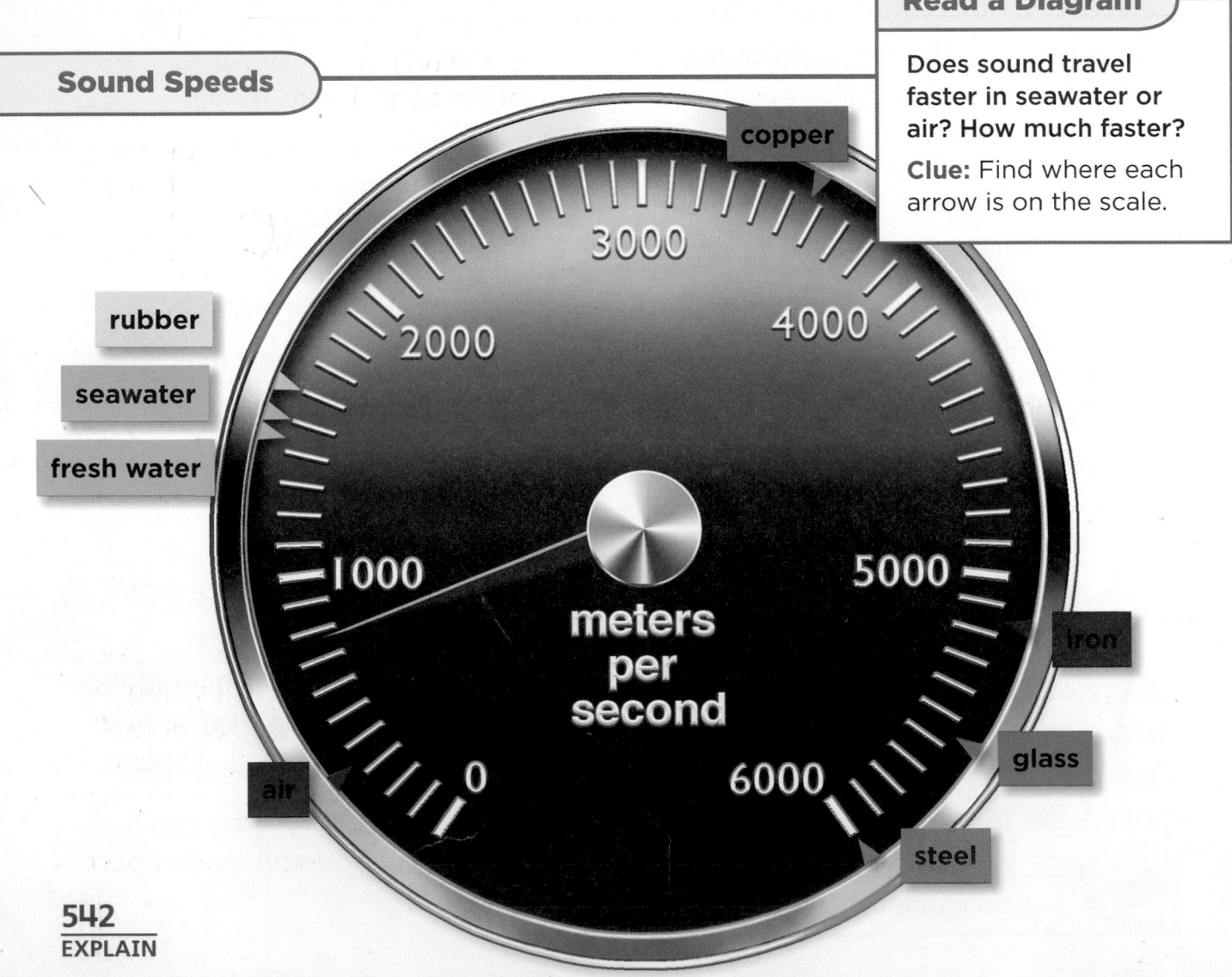

copper

rubber

seawater

fresh water

3000

2000

4000

1000

meters
per
second

5000

iron

air

0

6000

glass

steel

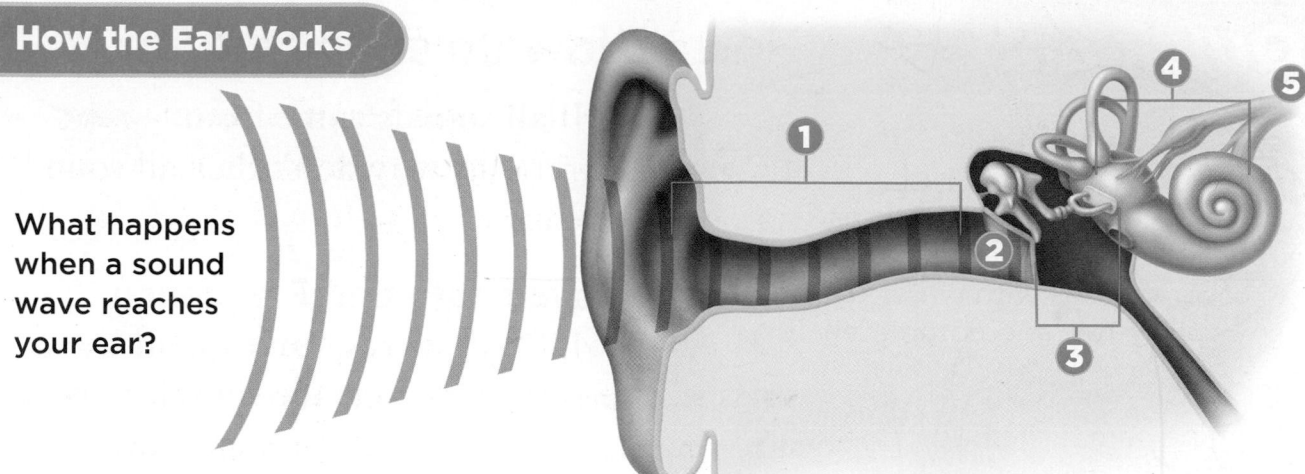

What happens when a sound wave reaches your ear?

1. **outer ear** The outer ear collects sound waves. Like a funnel, it directs the waves into the ear.

2. **eardrum** Sound waves make the eardrum vibrate like the head of a drum.

3. **middle ear** The vibrations are picked up by three tiny bones in the middle ear. The bones are the hammer, anvil, and stirrup.

4. **inner ear** The stirrup passes the vibrations to a coiled tube in the inner ear. The tube is filled with fluid and lined with tiny hair cells.

5. **nerve to brain** The moving hair cells signal a nerve in the ear. The nerve carries these signals to the brain. The brain interprets the signals as sound.

The Speed of Sound

Sound does not travel at the same speed through all materials. Sound travels slowest in a gas, such as air. Sound travels faster through a liquid, such as water. Sound travels fastest through a solid, such as glass or a metal.

Sound cannot travel through a vacuum. A vacuum does not contain matter. There can be no vibrations in a vacuum. Therefore, there can be no sound waves.

The Human Ear

When your friend speaks to you, sound waves travel through the air. What happens when those waves reach your ear?

First, the waves carry sound energy to tiny organs in the ear. The energy makes these organs vibrate. The diagram shows how the sounds move from the ear to the brain. All these steps happen in an instant!

✔ Quick Check

Infer A friend shouts from the far side of an empty gym. You hear the word several times. Explain.

Critical Thinking Why can't sound travel through outer space?

Pitch and Water

1 Set up five identical glasses. Leave one glass empty. Fill the other glasses with water to $\frac{1}{4}$ full, $\frac{1}{2}$ full, $\frac{3}{4}$ full, and completely full.

2 Observe Tap each glass lightly with a spoon. Which container makes the lowest sound? The highest sound?

3 Place the containers in order from lowest to highest pitch.

4 Infer What causes the pitch to vary? What vibrates? Explain.

How do sounds differ?

If all sounds come from vibrations, why don't they all sound the same?

Wavelength and Frequency

Like all waves, each sound wave has a wavelength and a frequency (FREE•kwun•see). In sound waves, **wavelength** is the distance from one area of crowded particles to the next. Look at the diagrams on page 545. The wavelength is the distance from the top of one sound wave to the top of the next wave.

Frequency is the number of vibrations a sound source makes in a given amount of time. When you strike a small bell, it vibrates quickly. The vibrations produce sound waves with a high frequency.

The size of each bell determines the pitch of the sound.

Pitch

The frequency of a sound wave determines its pitch. **Pitch** is the highness or lowness of a sound. High sounds, like the beat of a mosquito's wings, have high frequencies. Low sounds, like the croaks of a toad, have low frequencies.

Do you play a stringed instrument, such as a guitar? You can change its pitch by changing the strings. Shorter, thinner, or tighter strings vibrate more quickly. The sounds have a higher pitch.

Amplitude and Volume

The amount of energy in a sound wave is related to its **amplitude** (AM•pluh•tewd). Sound waves with high amplitude are made by objects that vibrate with a lot of energy.

Amplitude affects the **volume**, or loudness, of sound. As an airplane takes off, sound waves with high amplitude fill the air. The sounds are loud. When you whisper, your vocal cords vibrate just a little. The sound waves have low amplitude. The volume is low.

Quick Check

Infer You tune a guitar. One string makes a sound that is too low. How can you fix this?

Critical Thinking Why do flutes and tubas make different sounds?

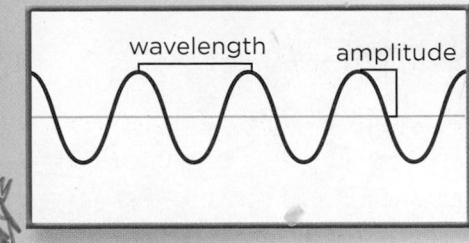

long wavelength
medium amplitude

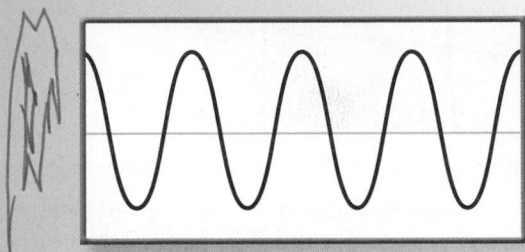

long wavelength
high amplitude

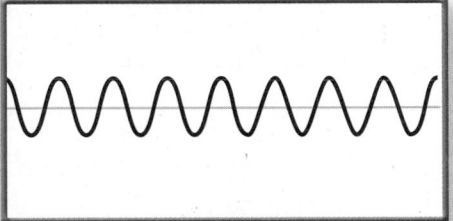

short wavelength
low amplitude

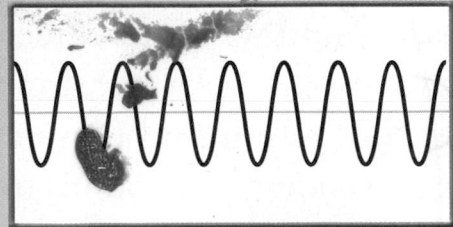

short wavelength
medium amplitude

Read a Diagram

Which sound has a loud, low pitch?

Clue: Compare amplitudes and wavelengths.

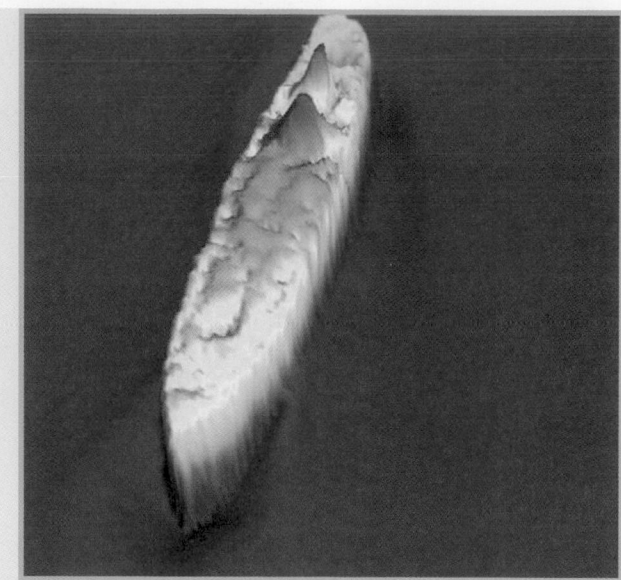

Sonar is used to find sunken ships on the ocean floor. These images show two different shipwrecks.

What is sonar?

Sound travels through water more quickly than it does through the air. It also travels farther. We can use these characteristics of sound to "see" below the water.

Sound Navigation and Ranging

Sonar (SOH•nahr) is a technology that uses sound waves to detect underwater objects. Sonar stands for <u>SO</u>und <u>N</u>avigation <u>A</u>nd <u>R</u>anging. It works by sending out sounds and receiving echoes.

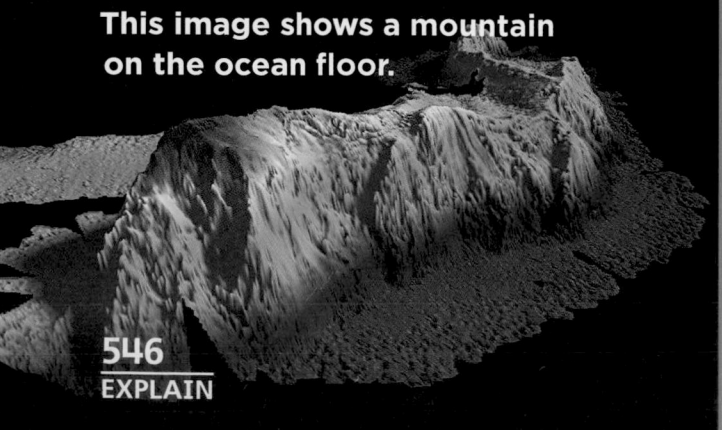

This image shows a mountain on the ocean floor.

Using Echoes

You know that when a sound hits a surface it is reflected. A sonar device measures the time difference between the sound and its echo. The device then makes an image of the object or surface.

How do we use sonar? Sailors use it to measure how deep the water is. Fishers use sonar to find schools of fish. Scientists use sonar to map the ocean floor. Others use it to search for shipwrecks or hazards.

 Quick Check

Infer Why might it be important to locate shipwrecks underwater?

Critical Thinking Why doesn't sonar work well in air?

Lesson Review

Visual Summary

Sound is produced when particles vibrate. Sound waves carry energy away from the source of the vibration.

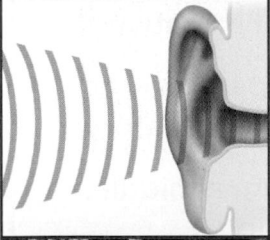

Sound travels through solids, liquids, and gases. Reflected sounds are called echoes.

The characteristics of sound include frequency, pitch, amplitude, and volume.

Make a FOLDABLES Study Guide

Make a trifold book. Use it to summarize what you learned about sound.

Sound is Produced	Sound Travels	Characteristics of Sound

Think, Talk, and Write

1. **Vocabulary** The highness or lowness of a sound is its _____.

2. **Infer** Trudy is at a rock concert. She must shout to be heard by her friend. What do you know about sounds and hearing? About loud noises? What can you infer about the sounds at the concert?

Clues	What I Know	What I Infer

3. **Critical Thinking** A piano tuner loosens one string and then tightens another. Why?

4. **Test Prep** A sound wave with high amplitude produces which kind of sound?
   - A high
   - B low
   - C loud
   - D fast

5. **Essential Question** How can you make sounds?

 **Math Link**

Write an Expression

An underwater object is 750 meters deep. It takes 1 second for an echo to return from it. How deep is the object if the echo takes 3 seconds? Write a math expression that you can use to find the answer.

 **Music Link**

Make an Instrument

Use the glasses of water from the Quick Lab on page 544 to play a simple tune. Experiment with your "instrument." How can you make different sounds? Be careful to tap gently on the glasses!

The Voice in the Well

One night my brother Tom told me about the well behind our house. "I've heard strange sounds coming from the woods," he said. "I also heard a voice coming from the old well." Gathering all my courage I said, "I don't believe you. Prove it!"

We slowly walked to the well. I shouted down, "Hello?" A moment later, I heard a voice say, "Hello? Hello?" I was shocked. I shouted, "Are you all right? Can I help?" The voice said, "Help! Help!"

My brother was laughing. "Calm down," he said. "It's an echo. The sound of your voice bounced off the water at the bottom and was reflected back to you!"

Personal Narrative

A good personal narrative

▶ uses the first-person pronoun *I* to tell the story;

▶ has an opening, a middle, and a conclusion.

 ## Write About It

Personal Narrative Have you ever heard an echo? What made the sound? Write a personal narrative about your experience.

 **e-Journal** Research and write about it online at www.macmillanmh.com

Hearing Echoes

How can you calculate your distance from the surface that reflects an echo? Time how long it takes between making the sound and hearing its echo. Multiply by the speed of sound. Then divide by 2. Why divide by 2? The sound makes a two-way trip before you hear its echo!

Suppose it takes 1 second for the narrator of "The Voice in the Well" to hear an echo. The speed of sound in air is 340 meters per second. How far is the narrator from the well? Write a multiplication sentence to solve the problem.

number of seconds	X	speed of sound	÷ 2 = distance
1 s	X	340 m/s	÷ 2 = 170 m

The narrator is 170 m from the bottom of the well.

Solve It

Write and solve a multiplication sentence for each problem.

1. You shout in a canyon. Your echo returns 2 seconds later. How far away is the canyon wall?

2. In the ocean, sound travels at 1,500 m/s. A ship's sonar signal returns in 3 seconds. How far away is the ocean floor?

Writing a Number Sentence

▶ Read the problem carefully.

What do you know? (The time was 1 second. The speed of sound is 340 m/s. The sound makes a two-way trip.)

What do you need to find out? (the distance)

▶ Decide whether to use addition, subtraction, multiplication, or division.

▶ Write a number sentence and solve it.

1 s x 340 m/s ÷ 2 = x
x = 170 m

▶ Check to see that your answer makes sense.

Light

Look and Wonder

Have you ever seen a rainbow in the sky? Rainbows are made of light. How do they form? Why do they have different colors?

What makes white light?

Purpose

Learn about white light using prisms.

Procedure

1 **Observe** Hold the long side of a prism up to the sunlight. Direct the light through the prism so it shines on the floor. Slowly rotate the prism. How does the light change? Record your observations.

2 Place the cardboard box on a table near a sunny window. Face the slit side toward the window. Place the prism inside the box about three inches from the slit. Stand the prism on one of its triangular sides.

3 Have a partner hold a mirror so it reflects the sunlight toward the slit as shown. Slowly rotate the prism. What happens to the light on the bottom of the box? Record your observations.

4 **Predict** What will happen if you place a second prism in the path of light coming from the first prism? Try it. Slowly rotate the second prism. Record your observations.

Draw Conclusions

5 What happened to the light in step 4?

6 **Infer** Review your observations. What can you conclude about white light?

Explore More

What would happen if you crossed the light beams from two different prisms? How would you design this investigation? Try it.

Materials

- 2 prisms
- large cardboard box with a precut slit
- mirror

Step **2**

Step **3**

Read and Learn

▶ **Essential Question**
How does light behave?

▶ **Vocabulary**

prism, p. 552

electromagnetic spectrum, p. 552

refraction, p. 554

reflection, p. 556

transparent, p. 558

translucent, p. 558

opaque, p. 558

▶ **Reading Skill** ✔
Main Idea and Details

Main Idea	Details

▶ **Technology**
e-Glossary and e-Review online at www.macmillanmh.com

What is light?

You live in a world full of color. Look around. All the colors you see are part of light. *Light* is a form of energy we detect with our eyes. It comes from the Sun, lightbulbs, fire, and other sources. It can even come from living things like fireflies!

Newton's Prism

In the mid-1660s, young Isaac Newton wanted to learn about light and colors. One sunny day, Newton darkened his room. He made a small hole in his window shutter. The hole was just big enough for a beam of sunlight to shine through.

Newton then held a glass prism (PRIH•zum) in the sunbeam. A **prism** is an object that separates white light into bands of colored light. With his prism, Newton saw all the colors of the rainbow!

The Visible Spectrum

Newton was the first to show that white light is made of all the colors we can see. These colors make up the *visible* (VI•zuh•bul) *spectrum*.

We now know that the visible spectrum is not the only part of light. Like sound, light travels in waves. The **electromagnetic spectrum** is the range of waves that make up light. Where is visible light in the electromagnetic spectrum? Look for it in the diagram below.

The Electromagnetic Spectrum

radio waves   microwaves   infrared waves

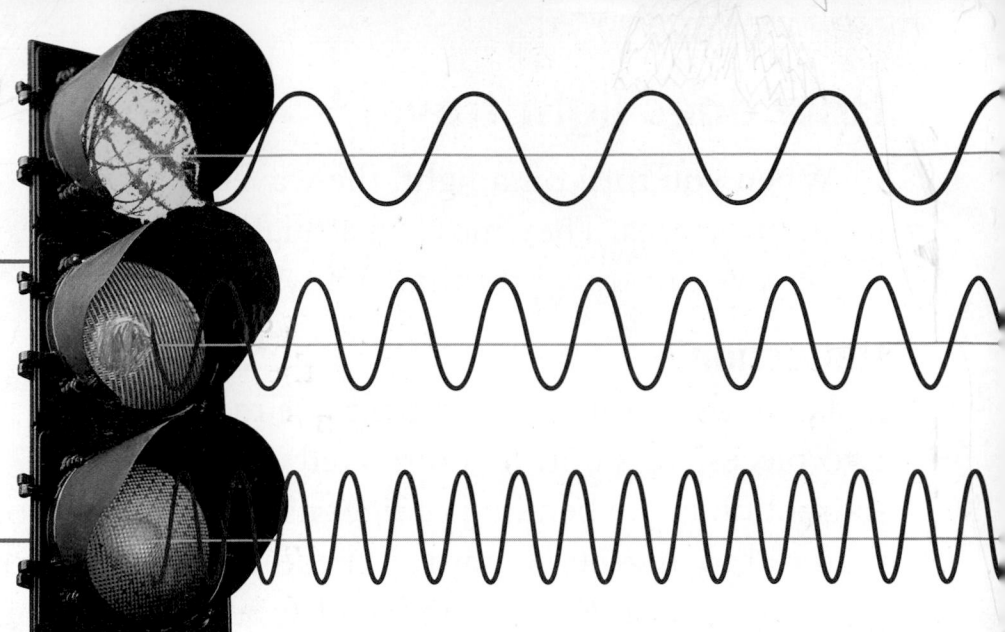

Read a Diagram

Which color has the longest wavelength?

Clue: Look at the distances between the tops of each wave.

Wavelengths and Energy

The light waves in the electromagnetic spectrum have different wavelengths. Each wavelength carries a different amount of energy. The longer the wavelength, the less energy it has.

The light waves with the longest wavelengths are radio waves. They have the least energy. On the other end of the spectrum are gamma waves. They have the shortest wavelengths and the most energy.

Light waves can be both helpful and harmful. Did you know that a microwave oven actually uses light waves? Or that heat is really infrared light waves? Ultraviolet, or UV, waves are dangerous and can burn your skin. X-ray waves help doctors see inside the body.

 Quick Check

Main Idea and Details What is the electromagnetic spectrum?

Critical Thinking What light sources can you name?

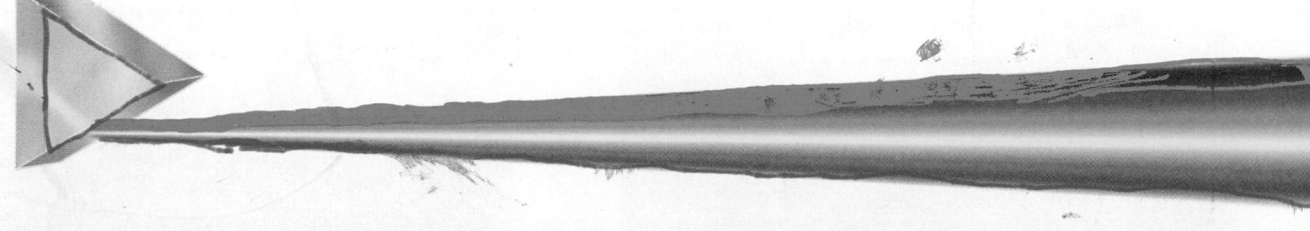

visible waves   ultraviolet waves   x-ray waves   gamma waves

How does light travel?

When you turn on a light, the waves spread out in all directions. They move in straight lines, or *rays*. Light rays can travel through air, water, and space.

Refraction

Is the thermometer in the picture really cut in two pieces? It is not. It is refracted (rih•FRAKT•ed). **Refraction** is the bending of light as it passes from one material to another. Light rays bend as they pass from glass into water. Refraction also happens when light moves from cold air into warm air.

Light travels at different speeds through different materials. Unlike sound, light travels more slowly through denser materials. At the point where light hits the denser material, it slows and bends. Water is denser than air, so light rays refract where water and air meet.

Refraction makes the thermometer appear to be in two pieces.

Lenses

A *lens* is a tool that refracts light. A *concave* lens curves inward. Light bends outward from the center of the lens. The rays spread apart. Glasses that help you see faraway objects are made with concave lenses.

A *convex* lens bulges outward. Light rays bend inward toward its center. This makes objects near the lens seem bigger. Reading glasses have convex lenses.

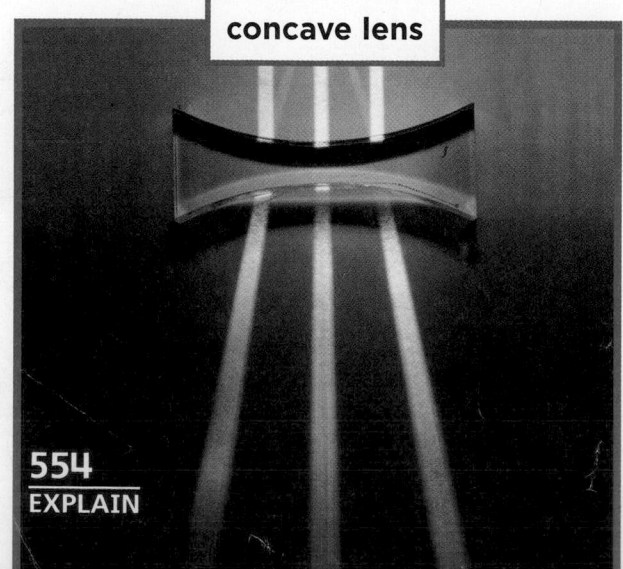

concave lens

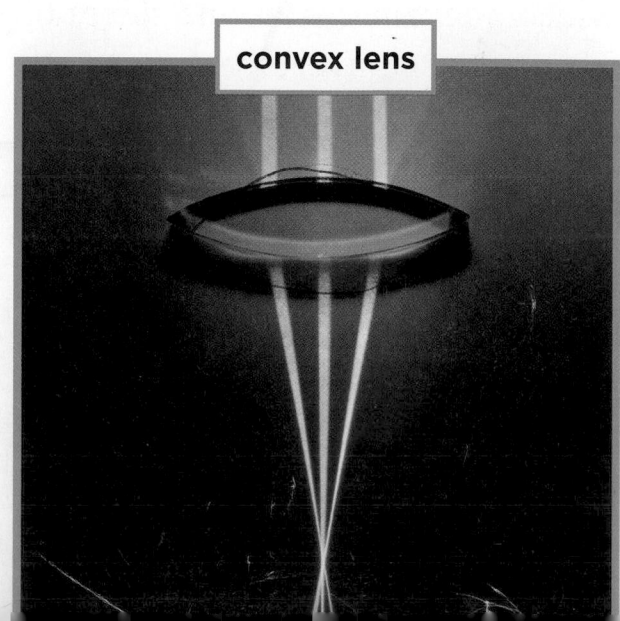

convex lens

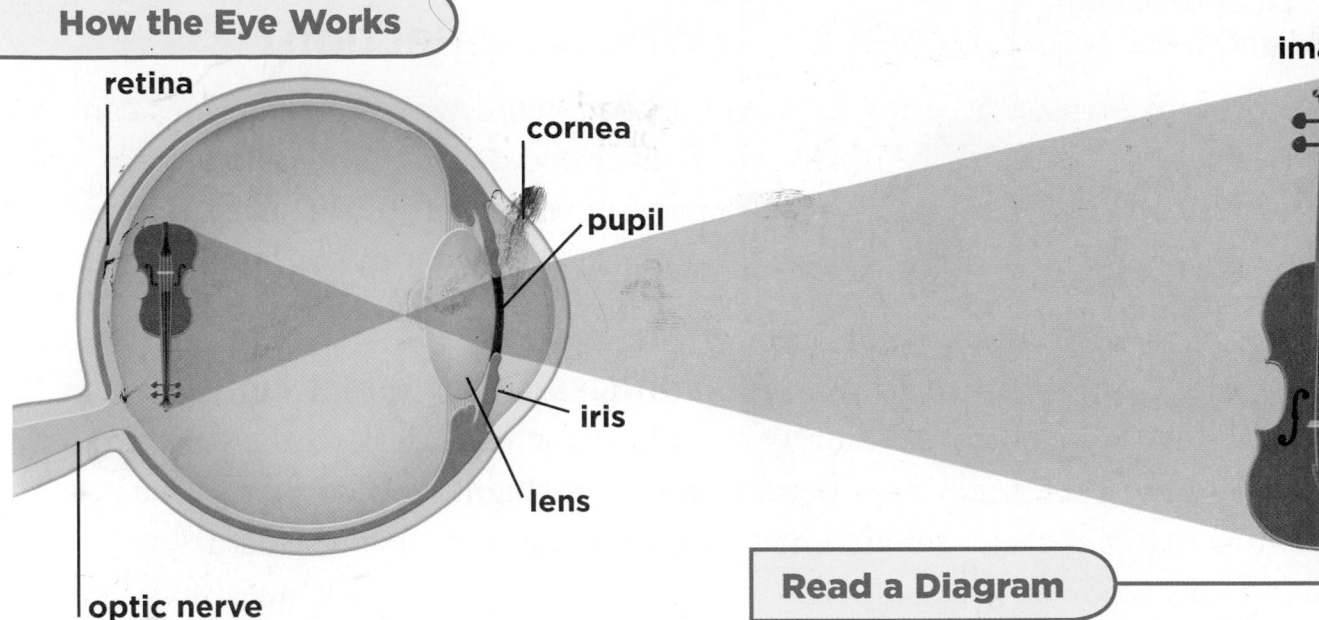

How the Eye Works

retina

cornea

pupil

iris

lens

optic nerve

image

Read a Diagram

In what order does light pass through the parts of the eye?

Clue: Trace the path of the light as it enters the eye.

The Human Eye

How do we see objects? Light bounces off them and enters the eye. First, the light passes through a thin, clear tissue covering each eye. This tissue is the *cornea* (KOR•nee•uh).

Next, the light passes through an opening in the eye called the *pupil* (PYEW•pul). The pupil is the black spot in the center of the eye.

The *iris* (I•ris) is the colored part of the eye. Eye muscles make the iris widen and narrow around the pupil. This action controls the amount of light entering the pupil.

From the pupil, light moves through a lens at the front of the eye. The lens refracts the light from the image. It focuses the image onto the back of the eye.

From Eye to Brain

Covering the back of the eye is a tissue called the *retina* (RET•nuh). The image that the lens focuses onto the retina is upside down. The retina changes that image into signals. The *optic nerve* brings these signals to the brain. The brain interprets them as right side up!

✓ Quick Check

Main Idea and Details What is refraction? Give an example.

Critical Thinking Which kind of lens would a telescope have? Why do you think so?

FACT The eye receives upside-down images.

The river near the Taj Mahal in India is smooth enough to reflect light.

What is reflection?

Like sound waves, light waves can bounce too. **Reflection** is the term for any wave that hits a surface and bounces off. Most of the light that reaches your eyes is reflected light.

Surfaces That Reflect Light

Look at your desk. If the desk did not reflect light, you could not see it! Most surfaces reflect at least some light. Smooth, shiny surfaces such as mirrors reflect almost all of the light falling on them. Dull, rough surfaces reflect the least amount of light. Surfaces do not have to be solid to reflect light. Liquid and gas surfaces can also reflect light.

Reflection and Color

Why do some leaves look green? The color of an object depends on the colors it reflects. When you look at a leaf, you are seeing reflected light. A green leaf reflects only the green wavelengths of the visible spectrum. It *absorbs,* or takes in, all the others.

The Law of Reflection

incoming angle   outgoing angle

The incoming angle of light is always equal to the outgoing angle of light.

How Mirrors Work

Like a lens, a mirror can be convex or concave. A convex mirror spreads out the reflected light rays. This gives a wide view of the reflected image. For this reason, convex mirrors are often used as rearview mirrors in vehicles.

Concave mirrors focus the reflected light rays together at a point. What you see depends on how close you are to the mirror. The toy penguin appears larger when it is close to the mirror. It appears upside down when it is far from the mirror.

The Law of Reflection

When light reflects off a surface, it changes direction. The light rays moving toward the surface are called *incoming rays*. The reflected light rays are called *outgoing rays*.

Incoming rays strike a surface at an angle called the *incoming angle*. Outgoing rays reflect at an angle called the *outgoing angle*. The incoming and outgoing angles are always equal. This relationship is called the *law of reflection*. It is shown in the diagram.

Quick Lab

Angle of Reflection

1. Tape a mirror to the wall at eye level. Tape a piece of paper to the wall to cover the mirror.

2. **Predict** Work with a partner. Predict where you each need to stand to see the other in the mirror. Mark your predictions on the floor with a piece of tape.

3. Uncover the mirror. Stand at the places you marked. Were your predictions correct? If not, cover the mirror and repeat step 2.

4. Place a strip of tape on the floor where you each stood. Run the tape from each spot to the wall under the center of the mirror.

5. **Infer** What is true about the angles that the two strips of tape make with the mirror?

✔ Quick Check

Main Idea and Details What is reflection? How does light reflect?

Critical Thinking Compare convex and concave mirrors. How are they different? Alike?

transparent

translucent

opaque

| Glass lets light pass through it. | Plastic scatters light in different directions. | Wood prevents light from getting through. |

What can light pass through?

When light shines on an object, it may or may not pass through.

Transparent Objects

Some materials are **transparent**. They allow light to pass through in a straight line. Light travels in a straight line through air, water and outer space. You can see through them clearly.

Translucent Objects

Translucent (trans•LEW•sunt) materials scatter light in different directions. It is hard to see through them clearly. Some shower doors are made of translucent plastic. This cloudy material offers some privacy.

Opaque Objects

For total privacy, people use opaque (oh•PAYK) materials. **Opaque** materials block light completely. Wood and metal are opaque. So is this textbook.

How do you know whether an object is opaque? Hold it in front of a light source. If no light passes through, then the object is opaque.

✓ Quick Check

Main Idea and Details Which materials can light pass through? Which materials block light?

Critical Thinking You are designing a window that protects people's privacy. What material would you use? Explain why.

Lesson Review

Visual Summary

Light is a form of energy that travels in waves. We can use the electromagnetic spectrum to classify it.

Refraction is the bending of light as it passes from one material into another.

Reflection occurs when light strikes a surface and bounces off it. Some surfaces block light completely.

Make a  Study Guide

Make a trifold book. Use it to summarize what you learned about light.

Main Idea	What I learned...	Sketches
Light is		
Refraction is		
Reflection occurs when		

Think, Talk, and Write

1 **Vocabulary** When light rays bounce off a surface, it is called _____.

2 **Main Idea and Details** How do lights help the eye see objects? Support your answer with details.

Main Idea	Details

3 **Critical Thinking** How is refraction similar to reflection?

4 **Test Prep** Light cannot pass through a(n) _____ object.

   A transparent

   B opaque

   C translucent

   D convex

5 **Test Prep** Which light has the most energy?

   A radio waves

   B x-rays

   C gamma waves

   D microwaves

6 **Essential Question** How does light behave?

 **Math Link**

Making Angles
Use a protractor to draw several light rays reflecting off a mirror at different angles. Remember to apply the law of reflection. Label the incoming and outgoing angles.

Art Link

Draw a Color Spectrum
Draw a diagram that shows how a prism separates white light. Include all the colors in the visible spectrum.

Be a Scientist

large sheet of
white paper

markers

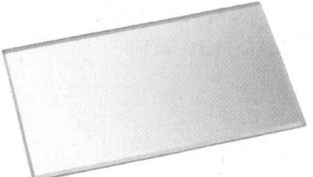

flat mirror

flashlight

Structured Inquiry

What happens to light when it is reflected?

Form a Hypothesis

When you look in a mirror, incoming rays from your body strike its surface. Those rays reflect off the mirror. What happens if you change the angle of the mirror? How will the outgoing rays change? Write your answer in the form "If I tilt the surface of a mirror, then the outgoing angle of light..."

Test Your Hypothesis

1. Work with a partner. Use the mirror as a straight edge. Draw a line across the center of a large sheet of paper. Hold the long end of the mirror upright along this line.

2. Your teacher will darken the room. Hold the flashlight directly in front of the mirror. Aim the light at the base of the mirror where it meets the paper. Hold the mirror straight up and down. You should see the reflected ray of light on the paper.

3. **Observe** Slowly move one end of the mirror's base away from the flashlight. What happens to the reflected light? Record your observations.

4. Continue to move the base of the mirror. Stop when the reflected light is parallel to the line on the paper. Trace a new line along the mirror base. Label it *mirror*.

Step 3

⑤ Have your partner draw a line along the incoming light ray. Label it *incoming ray*. Then draw a line along the reflected ray. Label it *outgoing ray*.

⑥ **Use Numbers** Identify the angle that the incoming ray makes with the mirror. Do the same for the outgoing ray and the mirror. Compare these two angles.

Draw Conclusions

⑦ **Communicate** How did the incoming angle and the outgoing angle compare?

⑧ **Infer** What is true about the relationship between these angles?

Guided Inquiry

How do curved mirrors reflect light?

Form a Hypothesis

The images reflected in concave and convex mirrors are different from those in flat mirrors. How does the shape of the mirror change the angle of reflection? Write a hypothesis.

Test Your Hypothesis

Design a way to investigate how the shape of a mirror changes the angle of reflection. Write out the materials you need and the steps you will follow. Record your results and observations.

Draw Conclusions

Did your results support your hypothesis? Why or why not? Explain how you set up the investigation to test for only one variable.

Open Inquiry

What else would you like to learn about mirrors? Design an investigation to answer your question. Your investigation must be written so that another group can complete it by following your instructions.

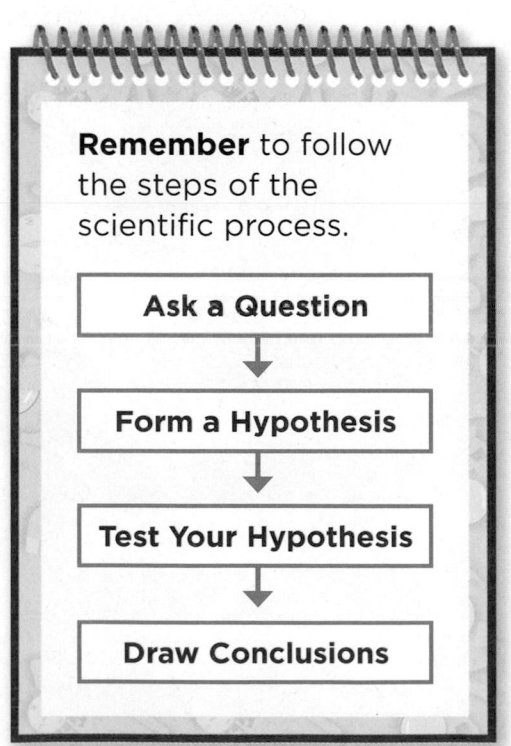

Remember to follow the steps of the scientific process.

> **Ask a Question**
> ↓
> **Form a Hypothesis**
> ↓
> **Test Your Hypothesis**
> ↓
> **Draw Conclusions**

Electricity

Look and Wonder

Lightbulbs have different parts inside them. It takes electricity to make these parts give off light. What is electricity? How does it work?

How do rubbed balloons interact?

Make a Prediction

How will two balloons interact if you rub one with a wool cloth? What if you rub both balloons with the cloth? Make your predictions.

Test Your Prediction

1 Tape a piece of string to each inflated balloon. Have a partner hold the balloons in the air about 1 meter apart.

2 **Observe** Rub one balloon ten times with a piece of wool cloth. What happens? Record your observations.

3 Rub the other balloon ten times with the cloth. Record your observations.

4 Hold the wool cloth between the two balloons. Observe and record what happens.

5 Put your hand between the two balloons. Observe and record what happens.

Draw Conclusions

6 **Communicate** Did your results match your predictions? Why or why not? How did the two balloons interact?

7 **Infer** What did the wool do to the balloons?

Explore More

Untie one balloon. Rub it with the wool. Then try sticking it to the wall. What happens? Why?

Materials

- **2 inflated balloons**
- **2 pieces of string, 50 centimeters each**
- **tape**
- **wool cloth**

Step **1**

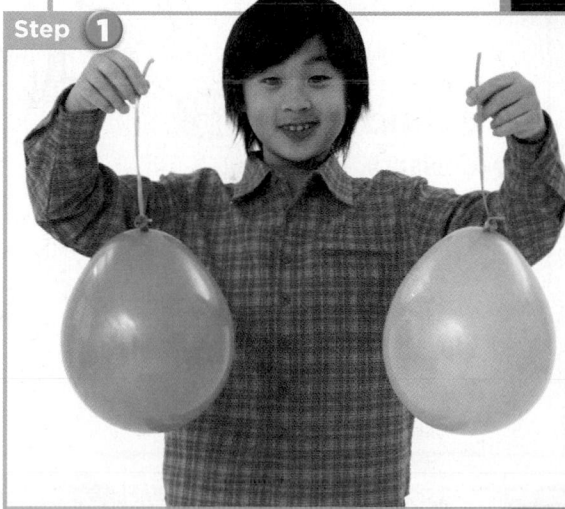

Step **2**

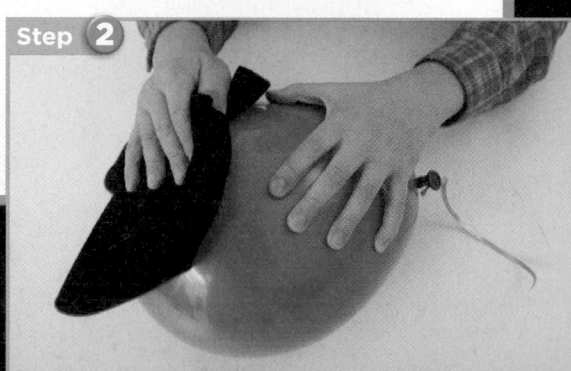

- ## Essential Question
How does electricity affect your life?

- ## Vocabulary
static electricity, p. 565

discharge, p. 566

circuit, p. 567

current electricity, p. 567

series circuit, p. 568

parallel circuit, p. 568

- ## Reading Skill ✔
Draw Conclusions

Text Clues	Conclusions

- ## Technology 🔵 LOG ON
e-Glossary, e-Review, and animations online at www.macmillanmh.com

What is electrical charge?

Have you ever watched a baseball game under bright ballpark lights? If so, then you have seen electricity in action. Electricity is the result of electrical charges. An electrical charge is not something you can see, smell, or weigh. Like color and hardness, it is a property of matter.

Positive and Negative Particles

You know that matter is made up of tiny particles called atoms. Inside atoms are even tinier particles! Some of them have a positive electrical charge. Others have a negative electrical charge. We can show a positive charge as a plus sign (+) and a negative charge as a minus sign (−). Positive and negative charges are opposites.

Overall Charge

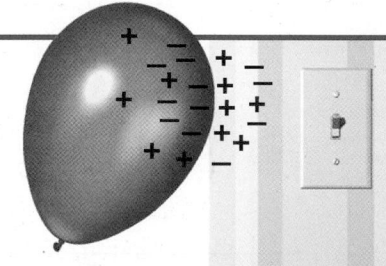

③ The negative charges on the balloon attract the positive charges on the wall. The balloon sticks to the wall.

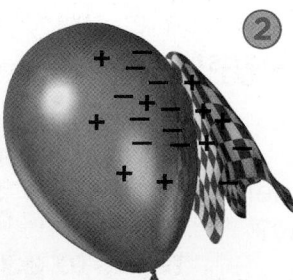

② By rubbing the balloon with the wool, negative charges build up on the balloon.

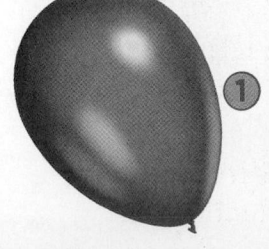

① A balloon and a wool cloth are neutral. Each has as many negative charges as positive charges.

Charges Interact

You cannot see or feel electrical charges the way you can see color or feel hardness. However, you can observe how charges interact.

A positive charge and a negative charge *attract*, or pull, each other. Like charges *repel*, or push away, each other. Positive repels positive. Negative repels negative.

Most matter has the same number of positive charges as negative charges. The charges cancel each other out. This means that the matter is *neutral* (NEW•trul)—it has no overall electrical charge.

④ In time, the charges move around. The balloon becomes neutral. It is no longer attracted to the wall, so it falls.

Charges Add Up

When two objects touch, charged particles can move from one object to the other. Negative charges move more easily than positive charges do.

Suppose you rub a balloon with a wool cloth. Negative charges move from the wool to the balloon. The balloon is left with a buildup of negative charges. A *buildup* means something has more of one kind of charge than the other. The wool has a buildup of positive charges.

Static Electricity

The buildup of electrical charges on an object is called **static electricity**. Rubbing objects together causes them to touch in more places. It produces more static electricity.

What happens if you hold a negatively charged balloon near a wall? It repels the wall's negative charges. It also attracts the positive charges in the wall. This pull makes the balloon stick to the wall. Over time the charges move around. The balloon becomes neutral and falls.

✔ Quick Check

Draw Conclusions Plastic wrap tends to accept negative charges. What would happen if you rubbed a balloon with plastic wrap?

Critical Thinking How have you experienced static electricity?

How do charges move?

Have you ever been shocked? You walk across a carpeted floor and then touch a metal doorknob. Zap! The shock you feel is the fast movement of charged particles.

Electrical Discharge

When you move across a carpet, negative charges rub off the carpet onto you. Your body gets a buildup of negative charges.

The charges keep building until you touch something. Then they move to whatever you touch. This fast movement of charge is called a **discharge** (DIS•charj). You might feel the discharge as a small shock. You might even see or hear it.

Lightning

Not all discharges result in small shocks. *Lightning* is the discharge of static electricity during a storm. Inside a storm cloud, ice and water droplets rub against one another. Some pick up positive charges and move to the top of the cloud. Negative charges move to the bottom. If the buildup gets large enough, the charges jump to the ground as lightning.

Electric Current

Now you know how electrical charges can be discharged. Charges can also flow through a material the way water flows in a river. A flow of electrical charges is known as an *electric current*.

How many uses of current electricity can you find at this carnival?

Circuits

To make an electric current, you need a path that will carry it. The path along which electric current flows is called a **circuit** (SUR•kut).

The simplest circuit has three parts. It has a power source, such as a battery. The source powers a *load,* such as a lightbulb or motor. Connectors, such as wires, carry electrical charges between the power source and the load.

The flow of electrical charges through a circuit is called **current electricity**. Most of the appliances you see run on current electricity.

To keep the current moving, a circuit cannot have any gaps or breaks. A complete, unbroken circuit is called a *closed circuit.* A circuit with gaps is called an *open circuit.* The path in an open circuit is incomplete. Current will not flow through it.

Switches

Many circuits have a switch. A *switch* turns current electricity on and off. The lights in your classroom are controlled by a switch. When the switch is in the closed position, the circuit is closed. Current flows through it. The lights are on. With an open switch, the circuit is open. No current flows. The lights are off.

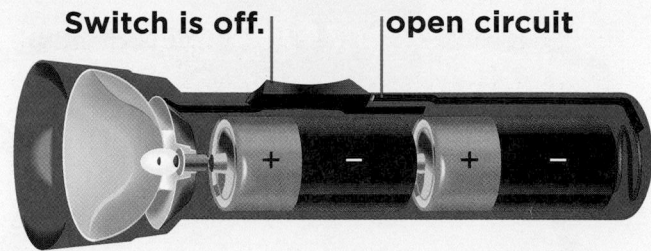

Open and Closed Circuits

Switch is off.   open circuit

Light is off.

Switch is on.   closed circuit

Light is on.

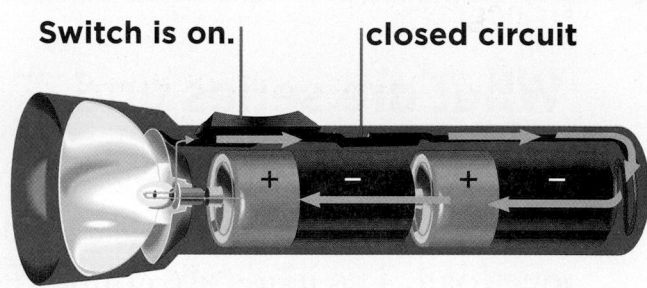

Read a Diagram

What happens to the circuit when the switch is on? What happens to the light?

Clue: Trace the path of the circuit in each flashlight.

 Quick Check

Draw Conclusions How is current electricity different from static electricity?

Critical Thinking You connect the ends of a wire to the terminals of a battery. What happens?

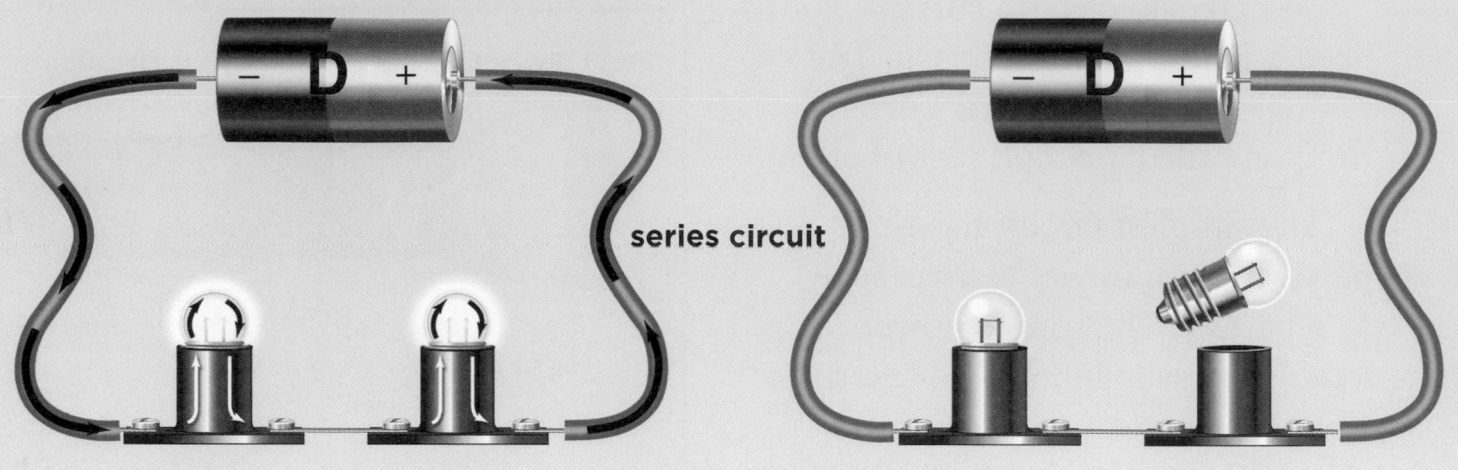

series circuit

What are series and parallel circuits?

Many circuits have more than one load. The loads are connected in a circuit in one of two ways.

Series Circuits

Picture a one-way circular road. All the cars on this road travel in the same direction. This is how a series circuit works. In a **series circuit**, electric current flows in the same direction along a single path.

The diagram above shows a series circuit. One loop of wire connects all the parts. When both lightbulbs are in place, the circuit is closed. With one lightbulb removed, the circuit is open. Charges do not flow through a series circuit when one part is removed. The parts must be connected one after another.

Parallel Circuits

A parallel circuit is like a group of roads that lead to the same place but take different paths. In a **parallel circuit**, electric current flows through more than one path. These different paths are often called *branches*.

In a series circuit, the same current goes through all the loads. In contrast, the branches of a parallel circuit divide the electric current among them. Some of the current flows through one branch. Some flows through another branch.

The diagram at the top of the next page shows a parallel circuit. Both lightbulbs connect to the power source through separate paths. However, if you remove one lightbulb, the other bulb will still light. Current still flows through the complete circuit in the other branch.

FACT Batteries do not have electricity inside them.

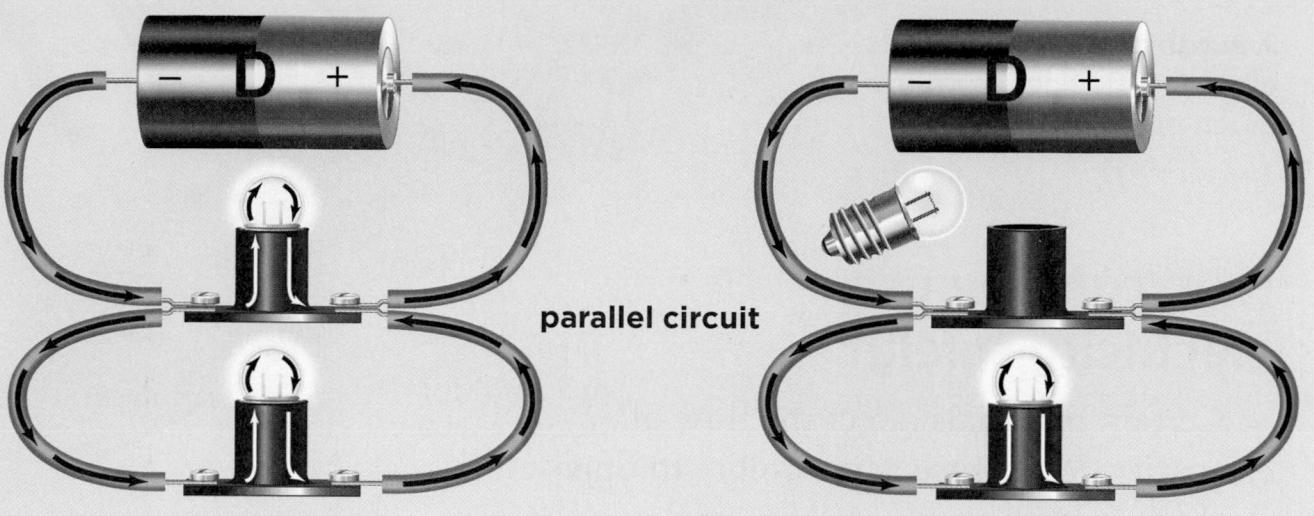

parallel circuit

Read a Diagram

How is a parallel circuit different from a series circuit?

Clue: The arrows show the flow of current electricity.

 Science in Motion Watch current flow through a parallel circuit at **www.macmillanmh.com**

Electrical outlets in most homes are connected in parallel circuits. You turn off one electrical device in a room, and others stay on. If the outlets were connected by a series circuit, all the electricity would turn off at once!

 Quick Check

Draw Conclusions A parallel circuit has two lightbulbs. One of them burns out. What happens to the other bulb?

Critical Thinking Which kind of circuit is best for a string of holiday lights? Why?

≡Quick Lab

Make a Parallel Circuit

1 Carefully screw two bulbs into two sockets.

2 Connect one socket to the other socket with two wires as shown.

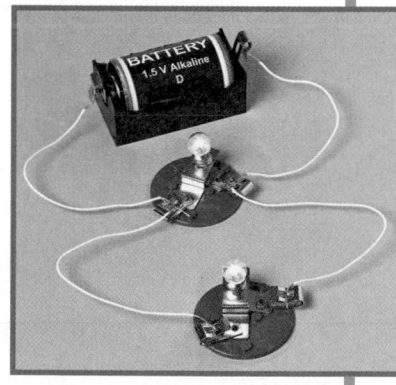

3 **Observe** Using two more wires, connect one socket to the terminals of a battery cell. What happens?

4 Remove one lightbulb from its socket. Now what happens? Why?

⚠ **Be Careful.** The lightbulbs might become hot.

A surge protector protects electrical devices from too much electricity. ▶

How can you use electricity safely?

Certain materials affect the flow of electricity. *Resistance* is the ability to oppose or slow electric current. If current flows through a path with little resistance, a *short circuit* can result.

Short circuits can be dangerous. The wire in the circuit can heat up and cause a fire. For this reason, you should never touch or use wires that are torn or frayed.

Fuses and Circuit Breakers

A *fuse* is a device that helps prevent short circuits. A fuse has a thin strip of metal in it. The strip has high resistance. If too much current flows through, it heats up and melts. The circuit opens. Current stops flowing.

Fuses can be used only once, but circuit breakers can be reset. A *circuit breaker* is a switch that protects circuits. When a dangerously high current flows through it, the switch opens. Current does not flow.

If a fuse breaks, it cannot be reused.

Most homes have circuit breakers.

 Quick Check

Draw Conclusions In new buildings, circuit breakers are used more often than fuses. Why?

Critical Thinking Should a circuit breaker be connected in series or in parallel? Why?

Visual Summary

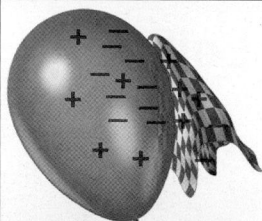

Static electricity is the buildup of charged particles on a surface.

Current electricity is the flow of electrical charges through a circuit.

The path along which electric current flows is an **electric circuit.** There are two basic kinds of circuits.

Make a FOLDABLES Study Guide

Make a three-tab book. Use it to summarize what you learned about electricity.

Static electricity

Current electricity

Electric circuits

Think, Talk, and Write

1 Vocabulary The path that electricity can flow through is called a(n) _____.

2 Draw Conclusions Peggy plugs in an electric heater. All the appliances in the room shut off. Why? What should she do?

Text Clues	Conclusions

3 Critical Thinking If you add lightbulbs to a series circuit, the circuit has more resistance. What happens to the electric current in the circuit?

4 Test Prep Which has separate paths connecting each load to its power source?

A a short circuit

B a circuit breaker

C a series circuit

D a parallel circuit

5 Essential Question How does electricity affect your life?

Writing Link

Lightning Safety Poem
Write a poem about how to stay safe during a thunderstorm. Be sure to include lightning safety.

Art Link

Design a Circuit
Draw a diagram of a series circuit and a parallel circuit. Each circuit should power two buzzers.

Be a Scientist

Materials

inflated balloon

wool

puffed-rice cereal

paper towels

water

Does the number of times a balloon is rubbed affect its charge?

Form a Hypothesis

When you rub wool across a balloon, the balloon builds up negative charges. If you continue to rub the balloon, what happens to its charge? Write your answer in the form "If I continue to rub a balloon with wool, then its charge..."

Test Your Hypothesis

1. Make a data table like the one shown. Spread one or two handfuls of puffed-rice cereal on a table.

Number of Rubs	Pieces of Cereal Attracted
1	
2	
3	
4	
5	

2. **Use Numbers** Rub the balloon once with the wool. Gently roll the balloon over the cereal. Count the pieces of cereal that stick to the balloon. Record the number in your table.

3. Remove all the cereal pieces from the balloon. Clean the balloon by wiping it gently with a damp paper towel.

4. Repeat steps 2 and 3 four more times. Each time, increase the number of rubs by one.

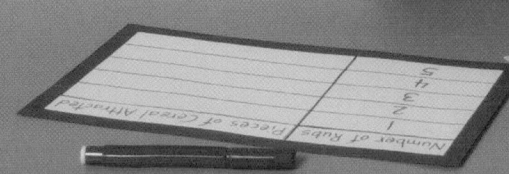

Draw Conclusions

5 **Interpret Data** Review your data table. How did the number of rubs affect the pieces of cereal on the balloon? Was your hypothesis correct?

6 **Infer** Why was it necessary to wipe the balloon with a wet paper towel after each test?

7 **Communicate** Make a graph of your results. Plot the number of cereal pieces on one axis. Plot the number of rubs on the other axis. Remember to give your graph a title.

Guided Inquiry

Does the type of material affect its charge?

Form a Hypothesis

What other materials produce static electricity? Does paper affect a balloon's charge the same way wool does? Write a hypothesis you can test.

Test Your Hypothesis

Design an experiment to test whether paper produces static electricity. Write out the steps you will follow. Then conduct your test. Record your results and observations.

Draw Conclusions

Add these results to the graph you made in step 7. Compare the two sets of data. What similarities and differences do you see? Does the type of material affect its charge? Did your classmates reach the same conclusion?

Open Inquiry

What else would you like to know about static electricity? For example, which materials are insulators or conductors? Design an investigation to answer your question. Your investigation must be written so that another group can complete it by following your instructions.

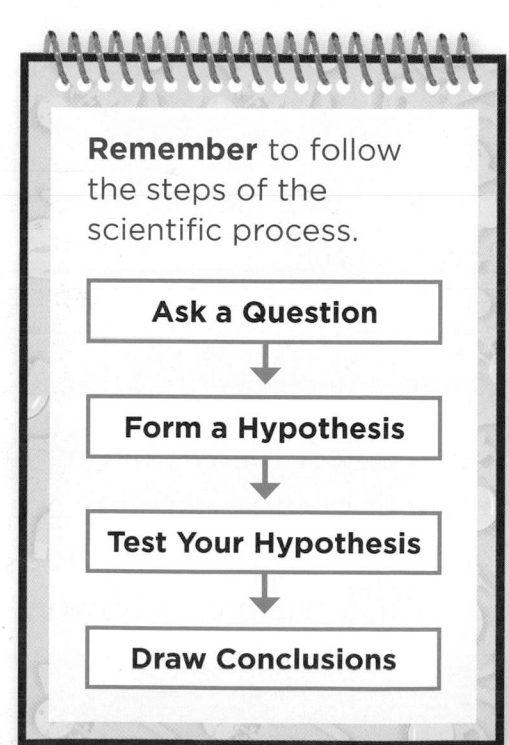

Remember to follow the steps of the scientific process.

Ask a Question
↓
Form a Hypothesis
↓
Test Your Hypothesis
↓
Draw Conclusions

Magnetism and Electricity

superconducting magnet

Look and Wonder

Is this object really floating in the air? Both magnetism and electricity are at work here. Magnetism is a force of attraction. How does this invisible force work?

How do magnets interact?

Make a Prediction

A magnet has two poles—north (N) and south (S). How can you make two magnets attract each other? How can you make them push apart from each other? Predict the positions of the poles in each case.

Test Your Prediction

1 **Observe** Bring the north pole of one magnet close to the north pole of the other magnet. What happens? Record your observations.

2 **Observe** What happens when you bring the south poles of the magnets near each other? Record your observations.

3 **Observe** Bring the north pole of one magnet close to the south pole of the other magnet. Record your observations.

Draw Conclusions

4 **Communicate** What happens when the like poles (S–S or N–N) of two magnets are brought together? What happens when opposite poles are brought together?

Explore More

Are certain places on a magnet stronger than other places? How could you find the strongest parts of a magnet? Make a plan and try it.

Materials

- **2 bar magnets with the poles marked**

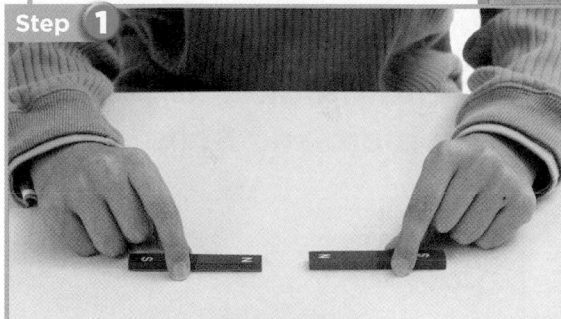

Step **1**

Step **3**

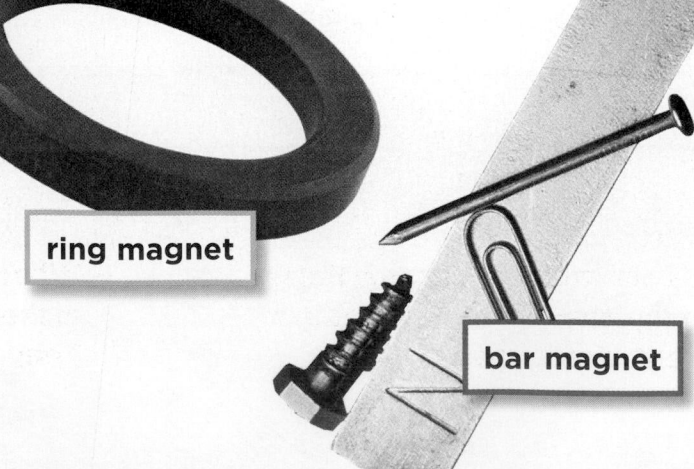
horseshoe magnet

Essential Question

How are electricity and magnetism related?

Vocabulary

attract, p. 576

repel, p. 576

pole, p. 577

magnetic field, p. 578

electromagnet, p. 581

motor, p. 581

generator, p. 582

Reading Skill ✓
Problem and Solution

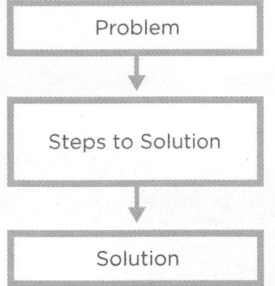

| Problem |
| Steps to Solution |
| Solution |

Technology
e-Glossary, e-Review,
and animations online
at www.macmillanmh.com

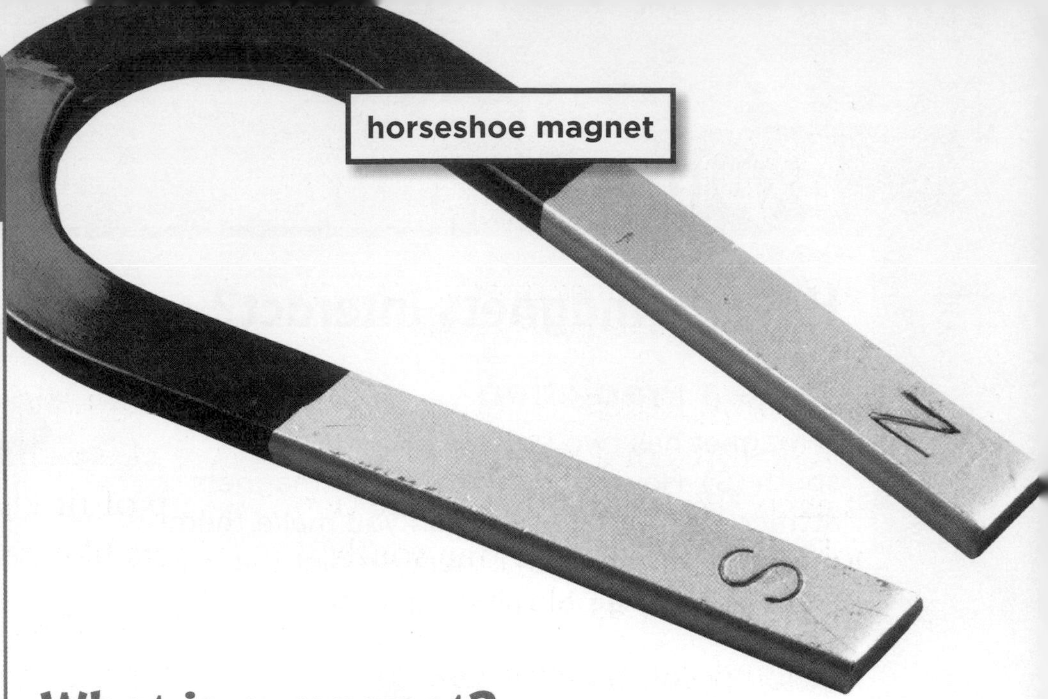

What is a magnet?

You have explored how two magnets can pull, or **attract**, each other. Two magnets can also push away, or **repel**, each other. These pushes and pulls are known as magnetic forces.

A *magnet* is something that can attract iron and certain other metals. These other metals include nickel and cobalt. A magnet also produces a magnetic field. You will read about magnetic fields later in this lesson.

Magnets come in all shapes and sizes. Some are simple bars. Others are shaped like horseshoes. Still other magnets are the shape of a ring.

ring magnet

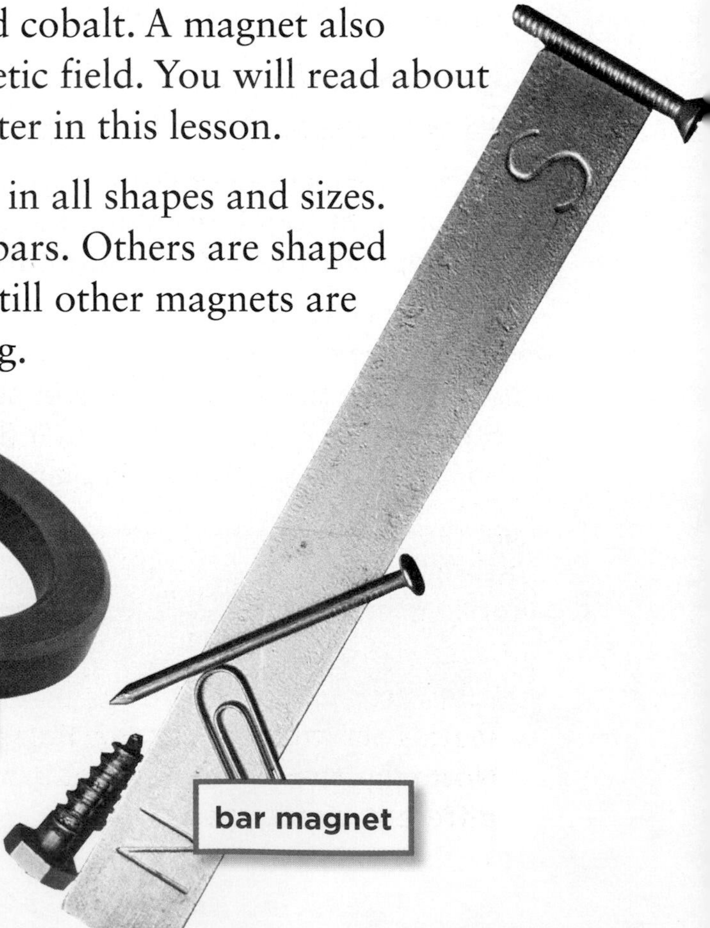

bar magnet

Like poles repel each other.

Opposite poles attract.

Magnetic Poles

The force of a magnet is strongest at each **pole**. All magnets have two poles—one north and one south. We label these as *N* or *S*.

Each pole acts a bit like an electrical charge. You know that negative charges attract positive charges. Similarly, the north pole of a magnet attracts the south pole of another magnet. How do like poles interact? They repel each other, just as like charges repel one another.

The attraction of two magnets is strongest when the magnets are closest together. Magnetic force gets weaker with distance.

Magnetic Particles

Like all matter, metals are made up of tiny particles. Each particle acts like a small magnet.

In objects made of iron, the tiny magnetic particles push and pull in different directions. If an iron object nears a magnet, these particles turn around and line up. North poles face one direction. The south poles face the other. The object becomes a temporary magnet.

✔ Quick Check

Problem and Solution How can two magnets repel each other?

Critical Thinking How is the pole of a magnet similar to an electrical charge?

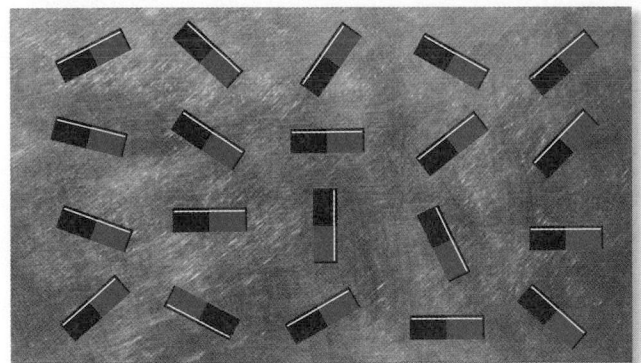

▲ Metals are made of tiny particles. Normally, the particles point in different directions.

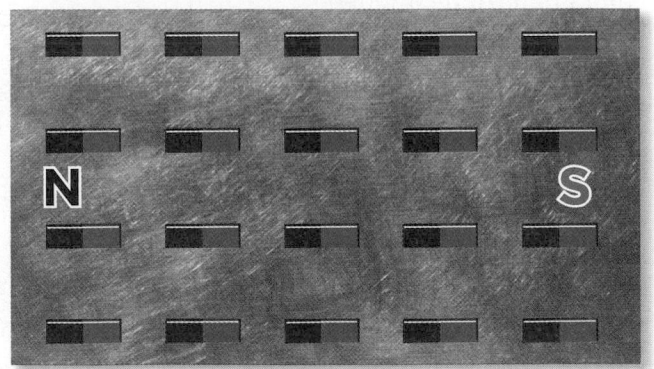

▲ When a magnet is brought near iron, nickel, or cobalt, the particles line up. They point in the same direction.

What are magnetic fields?

To pull a wagon or push a cart, you have to touch them. Magnets can pull or push objects without touching them at all!

A **magnetic field** is the area of magnetic force around a magnet. Every magnet has a magnetic field that wraps around it.

Objects can move when the magnetic fields of two magnets overlap. As you have seen, this can happen even if the magnets do not touch.

Earth's Magnetic Field

Did you know that our planet is actually a giant magnet? Some of the inside of Earth is made up of melted iron. This iron creates a magnetic field that surrounds Earth.

It might surprise you that Earth actually has two north poles. The geographic (jee•uh•GRA•fik) North Pole is located at one end of Earth's axis. Recall that an axis is a line that a spinning object turns around. Earth's magnetic north pole is near its geographic North Pole. However, the two locations are not the same. This is also true of the South Pole.

Earth's Magnetic Field

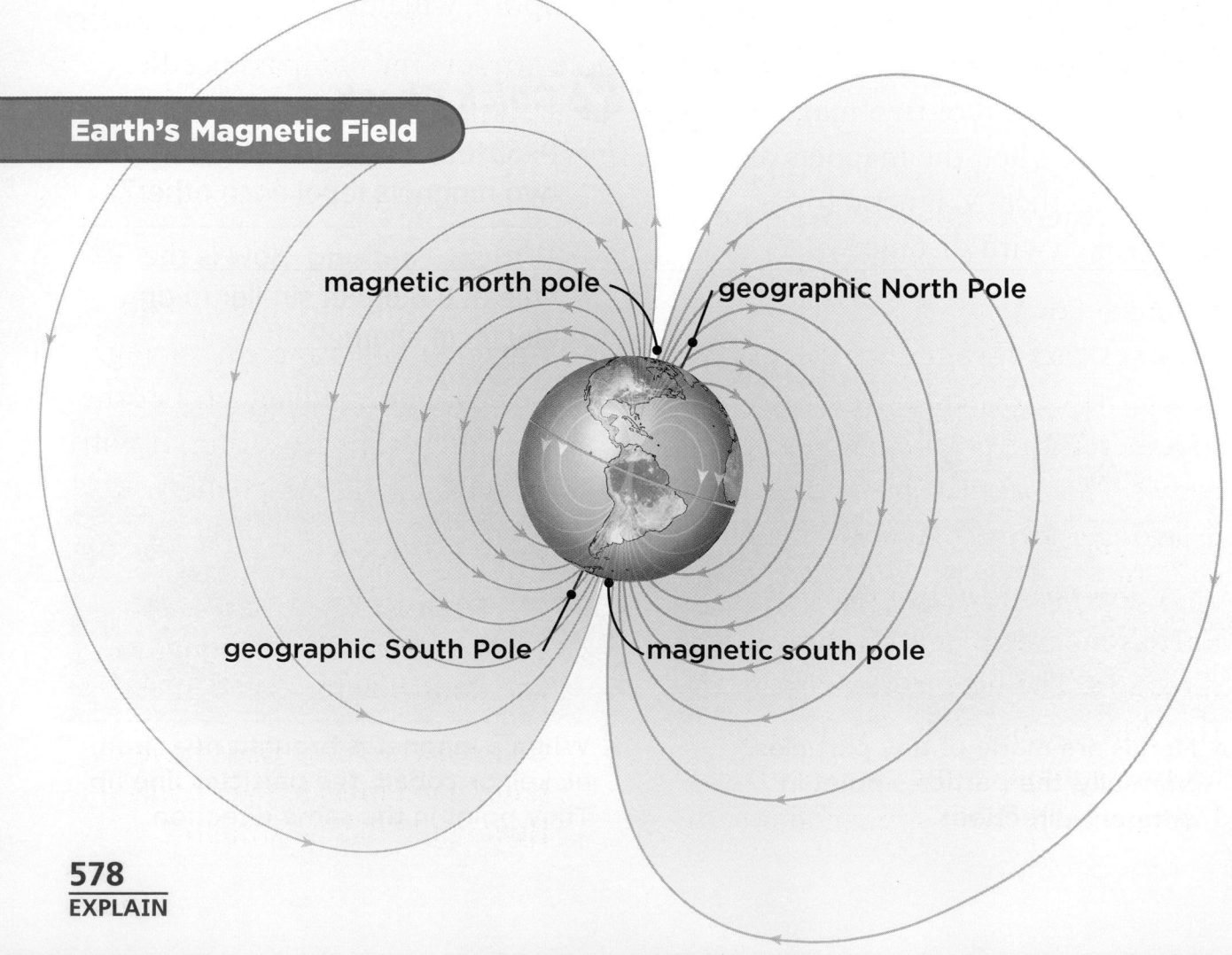

magnetic north pole

geographic North Pole

geographic South Pole

magnetic south pole

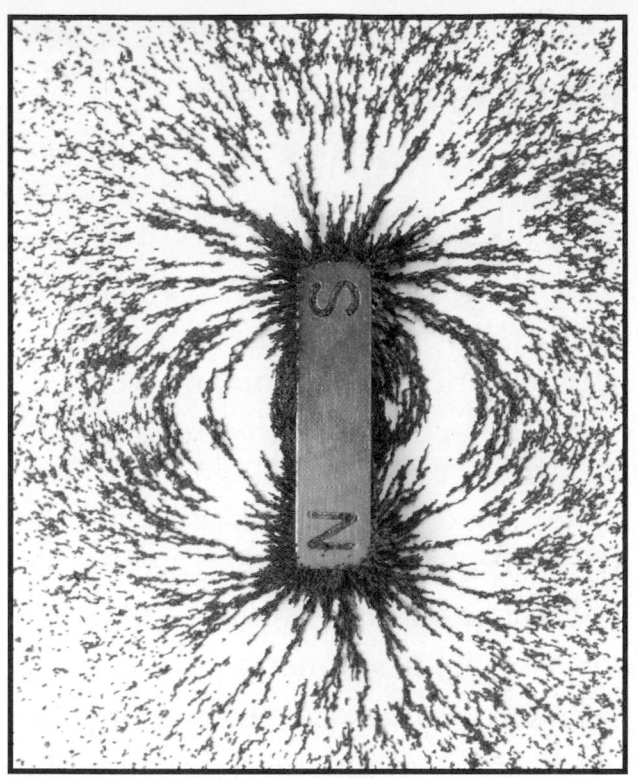

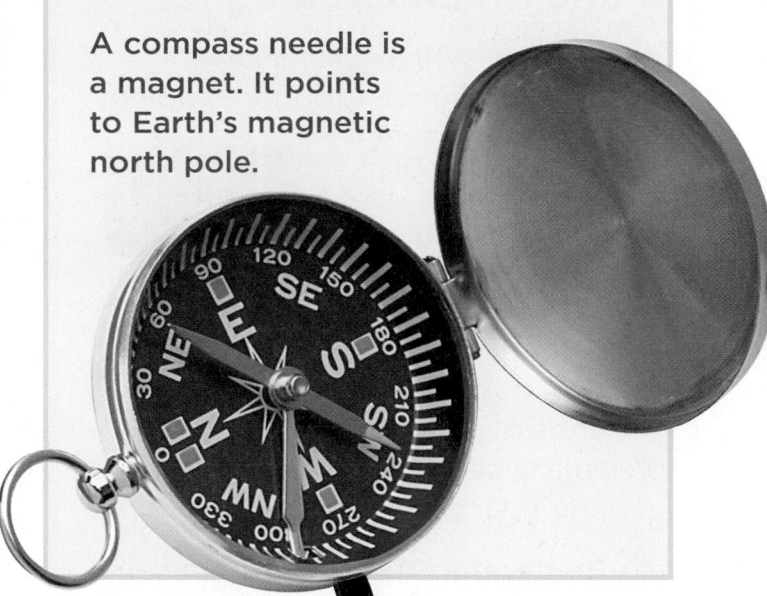

◀ Iron filings can show the magnetic field of a bar magnet.

A compass needle is a magnet. It points to Earth's magnetic north pole.

Seeing Magnetic Fields

You cannot actually see a magnetic field. It is invisible. However, you can use small pieces of iron to see what one looks like.

Iron filings were sprinkled on the bar magnet shown above. The filings lined up along the lines of the magnetic field. Observe how the lines curve from one pole to the other. Also, the field lines are closer together at the poles than near the center. This pattern shows the field is stronger at the magnet's poles.

A *compass* is a tool that uses Earth's magnetic field to show direction. The needle of a compass is a thin magnet. Like iron filings around a magnet, the compass needle lines up with Earth's magnetic field.

Using a Compass

A compass needle always points north. Why? Earth's magnetic north pole attracts the compass needle. This property is useful if you are lost. A compass helps you find east, south, west, and other directions in between.

You can make a compass with a bar magnet and a piece of string. Tie the string to the magnet. Let the magnet hang from the string. It will line up with Earth's magnetic field.

Quick Check

Problem and Solution How could you use a bar magnet to find your way through a forest?

Critical Thinking Birds have natural magnets in their bodies. How might this help them?

Make an Electromagnet

1 **Predict** Which will make an electromagnet stronger—adding electric current or adding coils?

2 Wind 40 cm of insulated wire around an iron nail 20 times. Attach the ends of the wire to a D cell. How many paper clips can your electromagnet pick up?

⚠ **Be Careful.** The wire may become warm.

3 Attach a 10-cm piece of wire to make a series circuit. Add another D cell. Now how many paper clips can your device pick up?

4 **Use Numbers** Remove the second D cell. Double the number of coils around the nail. Count the number of paper clips.

5 Was your prediction correct? Explain your results.

What is an electromagnet?

You know that electric current is a flow of charged particles. When charged particles move, they form magnetic fields. This means we can use electric current to make magnets.

The Effect of Current

Electric current moving through a wire sets up a magnetic field around that wire. The more current, the stronger the magnetic field. Turn the current off and the field goes away.

The Effect of Coils

Suppose you wind the wire into a long coil. With current flowing, the magnetic field around the coil is even stronger. Each loop in the coil is like a little magnet. All the loops pull and push in the same direction.

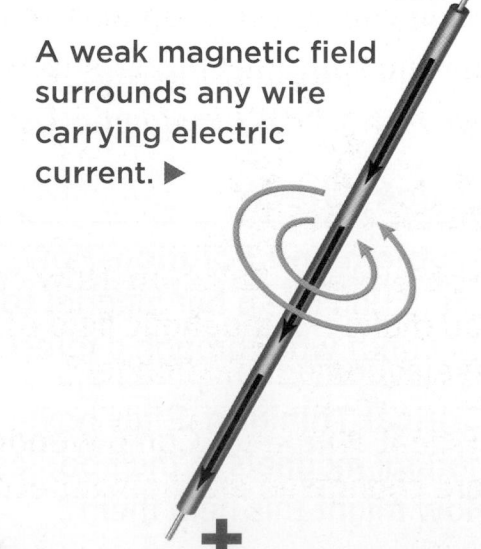

A weak magnetic field surrounds any wire carrying electric current. ▶

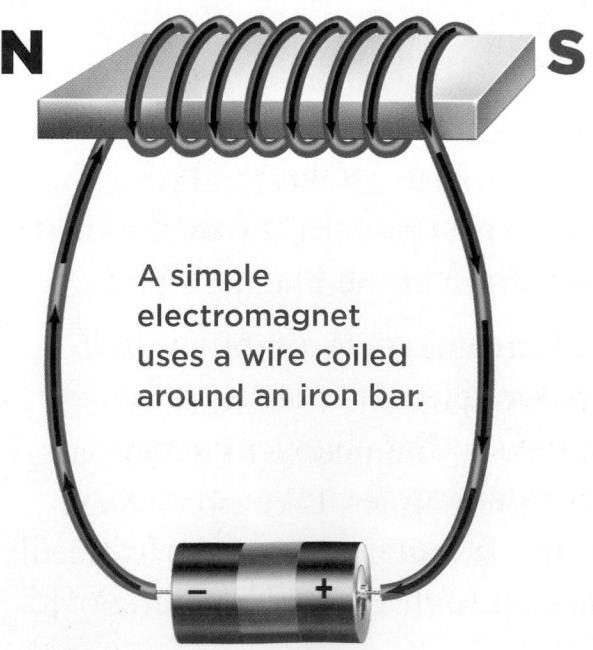

A simple electromagnet uses a wire coiled around an iron bar.

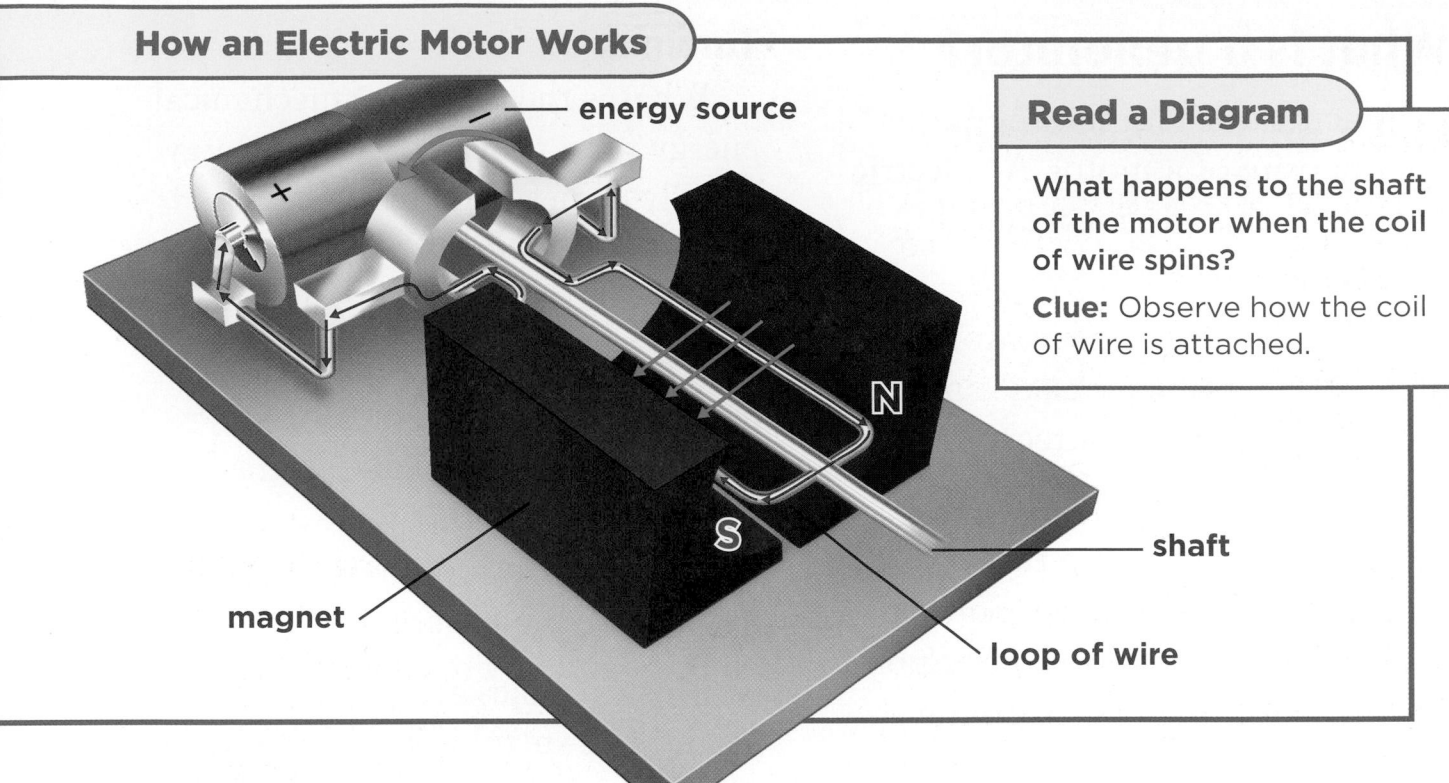

How an Electric Motor Works

energy source

+

−

N

S

magnet

loop of wire

shaft

Read a Diagram

What happens to the shaft of the motor when the coil of wire spins?

Clue: Observe how the coil of wire is attached.

The Effect of Iron

By adding a metal core, you can make the strongest magnetic field of all. An **electromagnet** is a coil of wire wrapped around a metal core, such as iron. Electric current flowing through the coil sets up a magnetic field. The particles inside the iron core line up, increasing the magnetic field around the coil.

An electromagnet can be turned on or off with a switch. This is a useful feature in electric devices such as headphones and telephones.

Electromagnets are often used to power electric motors. A **motor** is a device that changes energy into mechanical energy, or motion. Electric motors change electrical energy into mechanical energy.

Electric Motors

A simple electric motor has three parts. It has a power source, a magnet, and a wire loop attached to a shaft. The *shaft* is a rod that can spin.

The power source produces electric current. The current runs through the wire loop, making it an electromagnet. The magnet pushes and pulls on this electromagnet. The force then causes the loop and shaft to spin. The spinning shaft usually attaches to a wheel or a gear.

 Quick Check

Problem and Solution How can you make the magnetic field of an electromagnet stronger?

Critical Thinking Can a wooden core change an electromagnet?

What is a generator?

A generator (JE•nuh•ray•tur) is the opposite of a motor. An electric **generator** changes mechanical energy into electrical energy.

A simple electric generator has many of the same parts as a motor. It too has a power source, a magnet, and a wire loop attached to a shaft.

Motion is needed to turn the shaft and wire loop. The loop rotates between two magnetic poles. The magnetic field between the poles produces electric current in the wire loop. Each time the loop gets close to the poles, electrical charges are pushed through. These moving charges are electric current.

Turbines

What is the source of mechanical energy for a generator? This energy usually comes from a turbine (TUR•bine). A *turbine* is a set of angled blades attached to a shaft.

A simple turbine looks like an electric fan. Steam, water, or air is used to turn the blades of the turbine. The turning blades spin the shaft. The shaft then spins the wire loop inside the generator.

Most turbines are powered by steam. The steam is sent through a pipe pointed at the blades. Air turbines need strong, steady winds. Dams are usually built for turbines that use flowing water.

How A Generator Works

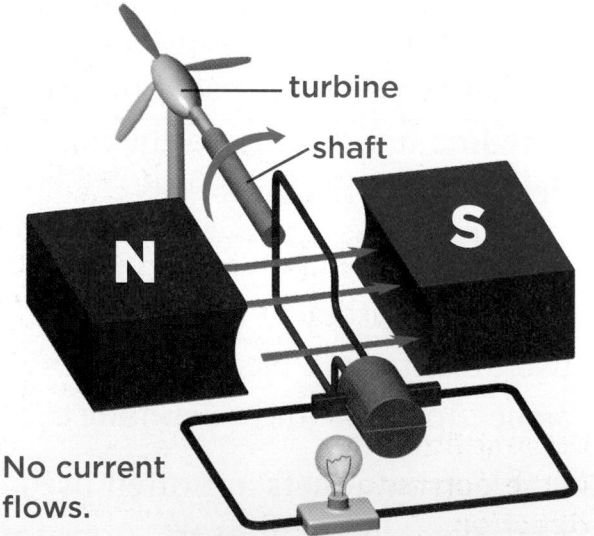

Current flows.

Mechanical energy turns the blades of the turbine. The blades turn the shaft. The shaft spins the wire loop through the magnetic field between the poles.

No current flows.

As the wire loop spins, it moves outside of the magnetic field. The circuit is open for less than one second. The loop turns so quickly you can't see the light flicker.

Alternating Current

We depend on generators to produce nearly all of our electricity. Most electric generators produce an alternating current, or AC.

Alternating current flows in one direction and then flows in the opposite direction. The electrical charges flow back and forth continuously. Most electrical wall outlets, such as those in your home or school, use AC.

Direct Current

When electric current flows in only one direction, it is called *direct current,* or DC. Like AC, the electrical charges in direct current flow continuously. However, the charges do not stop or reverse direction. A battery is an example of a DC power source. Some devices, like computers, change AC from wall outlets into DC.

Water spins the turbines of these generators. The turbines produce electricity.

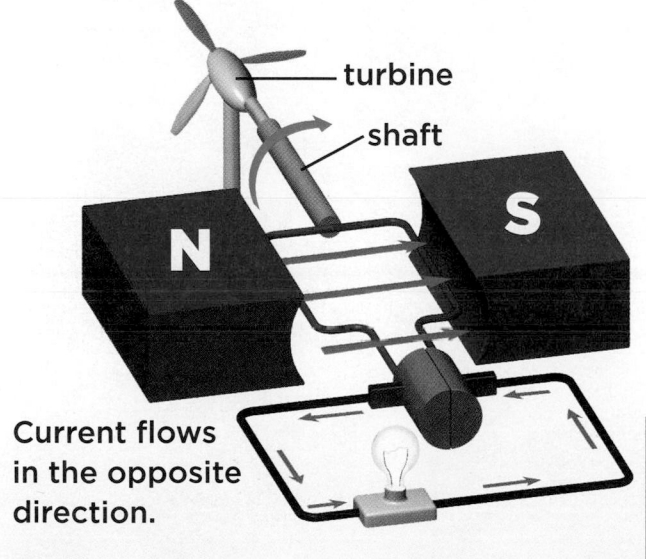

turbine

shaft

Current flows in the opposite direction.

As it continues to spin, the loop moves into the magnetic field again. The poles of the loop face opposite magnets. The current reverses direction.

✓ Quick Check

Problem and Solution
How do generators change motion into electricity?

Critical Thinking What do simple generators and simple electric motors have in common?

Read a Diagram

How is current produced in a generator?

Clue: Follow the moving parts in each step.

LOG ON *Science in Motion* Watch how a generator works at www.macmillanmh.com

How does electricity get to your home?

Power plants produce electrical energy. Electric current carries the energy to homes, schools, and businesses. The current moves in a circuit that connects to wall outlets.

Voltage

Voltage (VOHL•tij) is the strength of a power source. It is measured in *volts*. Power plants generally produce electric current at about 25,000 volts. To prevent loss of power over long distances, the voltage is increased. Increasing the voltage reduces the current. This reduces energy loss.

Transformers

Transformers change the voltage of electric current. A step-up transformer boosts the voltage. Current from a power plant goes through a step-up transformer. It leaves the transformer with a strength of about 400,000 volts.

Before entering your home, the current must be changed to a lower voltage. A step-down transformer decreases the voltage. Most homes use electric current at 120 or 240 volts.

✔ Quick Check

Problem and Solution What problem does a transformer solve?

Critical Thinking Why are power lines that move high voltages tall?

The Path of Electrical Energy

1. A power plant produces electrical energy.

2. A step-up transformer increases the voltage of the electric current.

3. The voltage decreases at a step-down transformer.

4. Another transformer makes the current safe for homes to use.

5. Power lines carry electric current back to the power plant.

Visual Summary

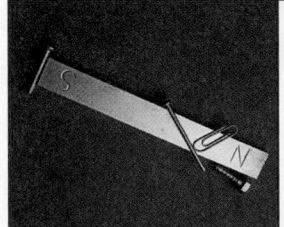

Magnets are objects that can attract certain metals and produce magnetic fields.

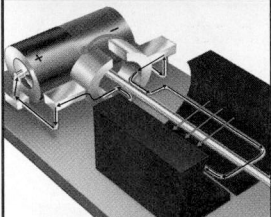

Electric motors change electrical energy into mechanical energy.

Electric generators change mechanical energy into electrical energy.

Make a FOLDABLES® Study Guide

Make a trifold book. Use it to summarize what you read about magnetism and electricity.

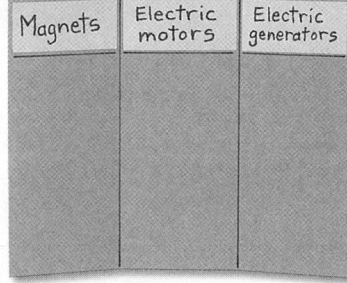

Think, Talk, and Write

① **Vocabulary** A magnet created by electric current is a(n) _____.

② **Problem and Solution** How can you make an electromagnet stronger?

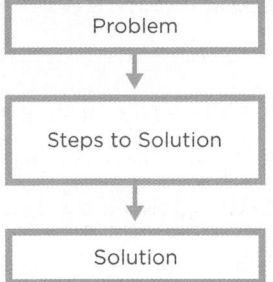

Problem

Steps to Solution

Solution

③ **Critical Thinking** When electric currents flow in the same direction through two wires, the wires attract each other. Why?

④ **Test Prep** Which changes electrical energy into motion?

A a power line

B a toaster

C an electric fan

D a lamp

⑤ **Essential Question** How are electricity and magnetism related?

Math Link

Calculate Voltage
A power plant uses a transformer to double the voltage of electric current. Current that enters at 100 volts leaves at 200 volts. How would this transformer change a current of 22,000 volts?

Health Link

Electromagnets in Medicine
Electromagnets have important uses in medicine. Research how electromagnets are used to diagnose or treat diseases. Write a report of your findings.

MOTORS AT WORK

Refrigerators, vacuum cleaners, hair dryers, and fans have one thing in common. They all have an electric motor. We use motors today because of people such as Joseph Henry and Michael Faraday. In 1831 these two scientists discovered how to use electromagnets. People could now turn electrical energy into motion.

Thomas Davenport was a blacksmith in Vermont. He learned about electromagnets. A few years after Henry's and Faraday's discovery, he built the first simple motor. It used electromagnets to separate iron from iron ore.

1882 Schuyler Wheeler makes the world a cooler place by inventing the electric fan.

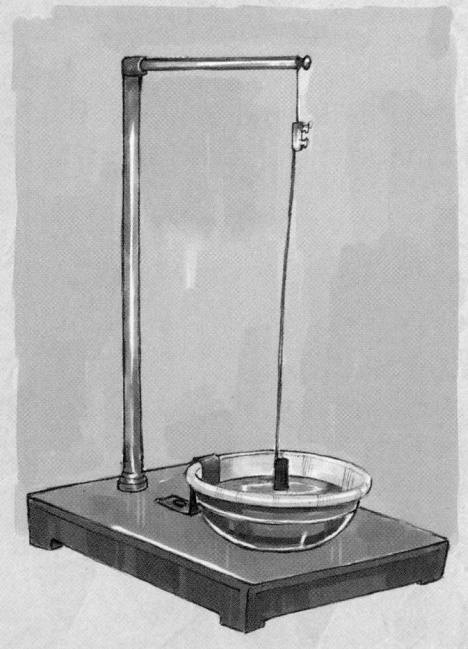

1831 Joseph Henry and Michael Faraday produce motion using electromagnets.

1837 Thomas Davenport gets the first patent for the electric motor.

1899 Baker Motor Vehicle Company builds electric hybrids. This is their 1902 "runabout."

1901 H. Cecil Booth patents the Red Trolley British Vacuum Cleaner.

It wasn't long before people started inventing new devices that used motors. Washing machines, invented in the early 1900s, use a motor to turn and wash clothes. Another motor in a washing machine turns the water on and off.

Some of the first automobiles ran on electrical energy. Today many new cars use electric motors in addition to gasoline engines. Motors are useful in many ways. Can you think of other machines that use electric motors?

1908 The Hurley Machine Company introduces electric washing machines.

Problem and Solution

▶ A problem is something that needs to be solved.

▶ A solution is a plan that helps you solve a problem.

Write About It

Problem and Solution What problem did Thomas Davenport solve with his motor? Write about a problem you have had, such as a messy room. How did an electric motor help you solve the problem?

 **e-Journal** Research and write about it online at www.macmillanmh.com

Connect to

  AMERICAN MUSEUM OF NATURAL HISTORY

at www.macmillanmh.com

Visual Summary

Lesson 1 Heat flows from warmer to cooler objects. There are three main ways heat is transferred.

Lesson 2 Sound is produced when energy causes particles to vibrate.

Lesson 3 Light is made up of waves with different wavelengths. Light travels in a straight line.

Lesson 4 Static electricity is a buildup of charges. Current electricity flows through a circuit.

Lesson 5 Magnets attract certain metal objects. Electromagnets use electric current.

Make a FOLDABLES® Study Guide

Tape your lesson study guides to a piece of paper as shown. Use them to review what you have learned in this chapter.

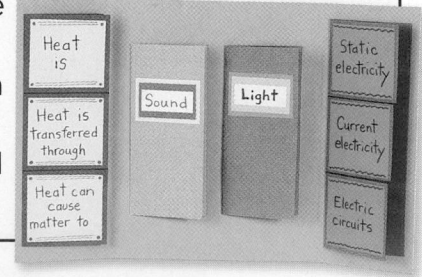

Fill each blank with the best term from the list.

amplitude, p. 545   generator, p. 582

circuit, p. 567   heat, p. 530

convection, p. 532   magnetic field, p. 578

discharge, p. 566   prism, p. 552

echo, p. 542   transparent, p. 558

1. When sound is reflected, it creates a(n) _____.

2. The movement of static electricity is called a(n) _____.

3. You can see through glass easily because it is _____.

4. The process that transfers heat through liquids or gases is _____.

5. Earth is surrounded by an invisible _____.

6. Motion can be turned into electrical energy by an electric _____.

7. When a sound has a great deal of energy, it also has a high _____.

8. White light can be separated into different colors by a(n) _____.

9. The flow of thermal energy between objects is _____.

10. Electric current flows through a(n) _____.

Answer each of the following.

11. **Main Idea and Details** A boy touches a metal doorknob. He feels a shock. How can you explain this?

12. **Infer** Your teacher gives you an unknown object and asks you to describe its properties. You test the object and find that it is attracted to a magnet. What can you infer about this object?

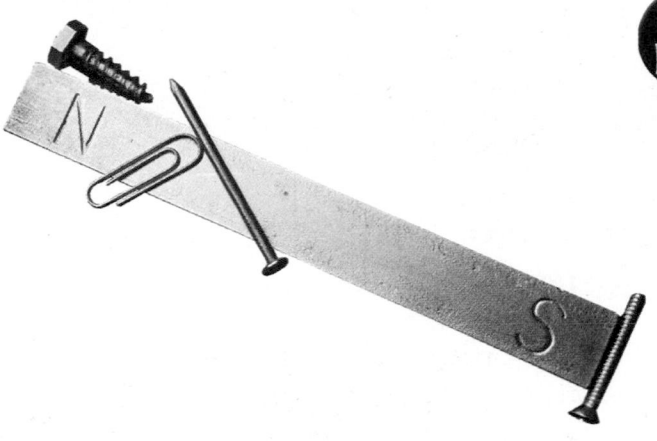

13. **Critical Thinking** What materials would make good insulators for a hot cup of soup?

14. **Personal Narrative** Research has shown that color can affect your mood. For instance, blue is a color that makes many people feel calm. Write a paragraph about how color affects you.

15. **Critical Thinking** Could sound travel through empty space?

16. **Descriptive Writing** Describe how a generator makes electricity.

17. **True or False** *Heat cannot travel through space.* Is this statement true or false? Explain.

18. **True or False** *A transparent object absorbs or reflects all light.* Is this statement true or false? Explain.

19. What happens to the beam of a flashlight when it hits a mirror?

   A It disappears.

   B It becomes a new form of energy.

   C It is reflected off the mirror.

   D It goes into the mirror.

The Big Idea

20. How do we use energy?

I'm All Ears!

Make a model that shows how the human ear allows us to hear sound.

1. Use the diagram for reference. Make sure you understand the steps involved in producing and hearing sound.

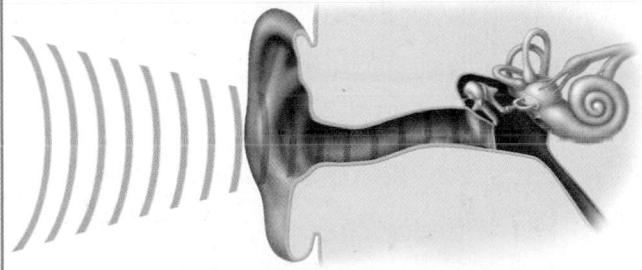

2. Use clay, papier-mâché, or other classroom materials to build your model. Include labels and a written explanation of each step.

1 The diagram below shows light rays passing through two kinds of lenses.

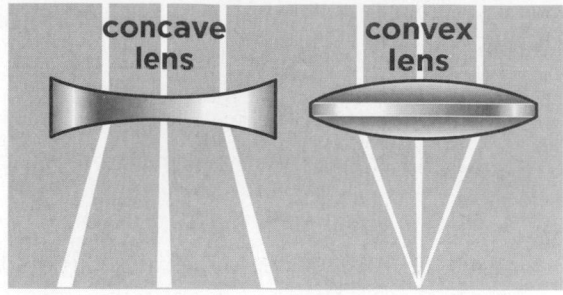

Which property of light does this diagram show?

A refraction

B absorption

C reflection

D translucence

DOK I

2 What is the <u>best</u> way for a drummer to increase the volume of her drumming?

A Hit a smaller drum.

B Hit a larger drum.

C Hit the drum with less energy.

D Hit the drum with more energy.

DOK I

3 Which statement is true about opposite magnetic poles?

A They will attract each other.

B They will repel each other.

C They will not affect each other.

D They will vibrate when brought together.

DOK I

4 A window curtain blocks light. The curtain is

A translucent.

B transparent.

C opaque.

D convex.

DOK I

5 Look at the diagram below.

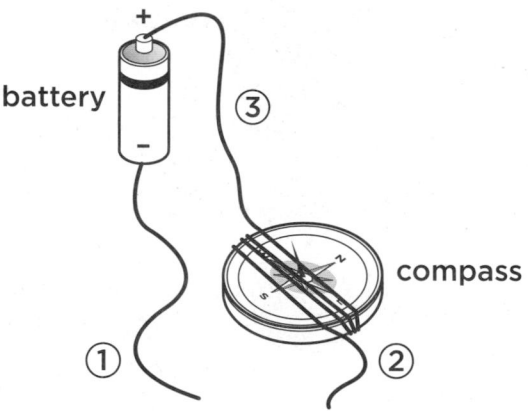

How could you make the compass's needle move?

A replace the wires

B replace the battery

C connect wires 1 and 2

D connect wires 2 and 3

DOK 2

6 In which of the diagrams below could <u>both</u> the lightbulb and the motor function?

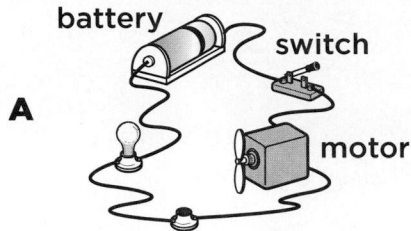

A

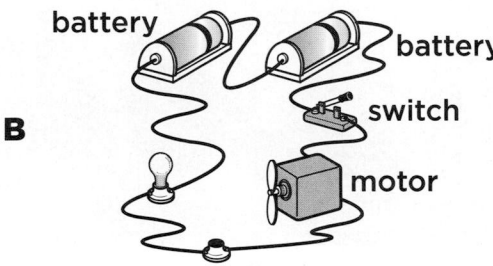

B

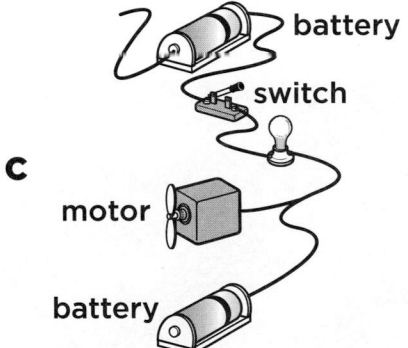

C

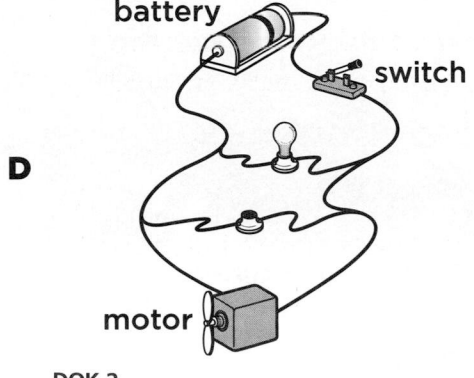

D

DOK 2

7 Look at the electromagnet shown below.

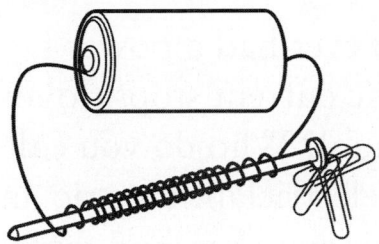

Suggest two ways you could make the electromagnet stronger.
DOK 2

8 Describe how heat is transferred by conduction, convection, and radiation. Give an example of each.
DOK 2

9 What is the difference between heat and temperature?
DOK 2

10 A guitar string makes a high-pitched sound. How could you lower its pitch?
DOK 1

Check Your Understanding

Question	Review	Question	Review
1	p. 554	6	pp. 568–569
2	pp. 544–545	7	pp. 580–581
3	pp. 575–577	8	pp. 532–533
4	p. 558	9	pp. 530–531
5	pp. 568, 578–581	10	p. 545

Careers in Science

Electrician

Have you ever had a power failure in which electric current stops flowing to your home or school? Who do you call to fix it? You call an electrician! Electricians install alarms, repair switches, and replace wiring. They know how to handle anything that runs on current.

Electricians know all about electricity. They are skilled in the use of tools. You can learn those skills in training programs in high school, college, and the military. You will need to be an apprentice, or helper, first. Then you can get your license.

▲ Electricians install wires, switches, and outlets.

Electrical Engineer

Suppose a new stadium is being built. The lights, roof, and scoreboard run on electricity. So do the heating, cooling, and security systems. These features were designed by electrical engineers. Electrical engineers plan and build many types of electrical systems.

To become an electrical engineer, you need to have good math and science grades. Many electrical engineers go through college in four years. Some work for industries. Others may teach college and do research. It can be an exciting career!

▲ Electrical engineers design electrical systems and keep them working.

Reference

You can use a ruler to find the length of an object. ▶

Science Handbook

Measurements

Units of Measurement

Temperature

▶ The temperature on this thermometer reads 74 degrees Fahrenheit. That is the same as 26 degrees Celsius.

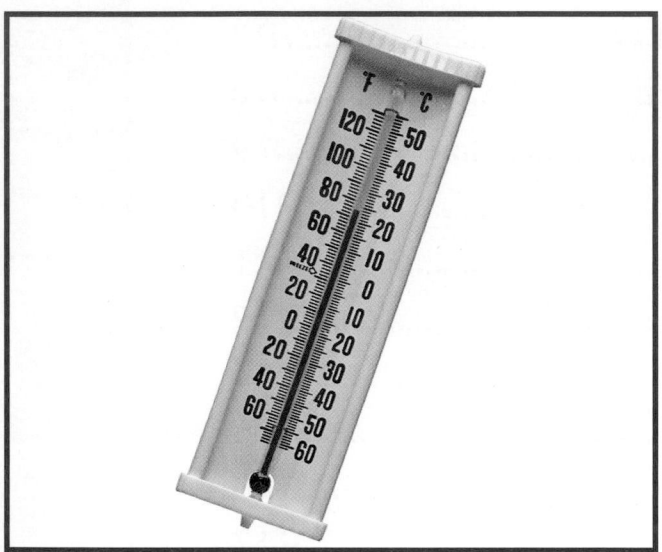

Length

▶ This student is 3 feet plus 9 inches tall. That is the same as 1 meter plus 14 centimeters.

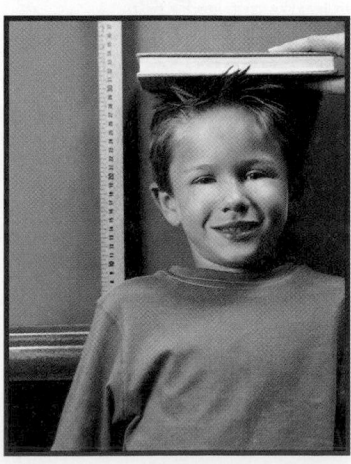

Mass

▶ You can measure the mass of these rocks in grams.

Volume of Fluids

▶ This bottle of water has a volume of 2 liters. That is a little more than 2 quarts.

Weight/Force

▶ This pumpkin weighs about 7 pounds. That means the force of gravity is 31.5 newtons.

Speed

► This student can ride her bike 100 meters in 50 seconds. That means her speed is 2 meters per second.

Table of Measures	
SI International Units/Metric Units	**Customary Units**

Temperature
    Water freezes at 0 degrees Celsius (°C) and boils at 100°C.
	Temperature
	    Water freezes at 32 degrees Fahrenheit (°F) and boils at 212°F.

Length and Distance
    10 millimeters (mm) = 1 centimeter (cm)
    100 centimeters = 1 meter (m)
    1,000 meters = 1 kilometer (km)
	Length and Distance
	    12 inches (in.) = 1 foot (ft)
	    3 feet = 1 yard (yd)
	    5,280 feet = 1 mile (mi)

Volume
    1 cubic centimeter (cm³) = 1 milliliter (mL)
    1,000 milliliters = 1 liter (L)
	Volume of Fluids
	    8 fluid ounces (fl oz) = 1 cup (c)
	    2 cups = 1 pint (pt)
	    2 pints = 1 quart (qt)
	    4 quarts = 1 gallon (gal)

Mass
    1,000 milligrams (mg) = 1 gram (g)
    1,000 grams = 1 kilogram (kg)
	Area
	    1 square foot (ft²) = 1 ft × 1 ft
	    43,560 square feet (ft²) = 1 acre

Area
    1 square meter (m²) = 1 m × 1 m
    10,000 square meters (m²) = 1 hectare
	Speed
	    miles per hour (mph)

Speed
    meters per second (m/s)
    kilometers per hour (km/h)
	Weight/Force
	    16 ounces (oz) = 1 pound (lb)
	    2,000 pounds = 1 ton (T)

Weight/Force 1 newton (N) = 1 kg × 1m/s²	

Measurements

Measure Time

You measure time to find out how long something takes to happen. Stopwatches and clocks are tools you can use to measure time. Seconds, minutes, hours, days, and years are some units of time.

Try it **Use a Stopwatch to Measure Time**

1. Get a cup of water and an antacid tablet from your teacher.

2. Tell your partner to place the tablet in the cup of water. Start the stopwatch when the tablet touches the water.

3. Stop the stopwatch when the tablet completely dissolves. Record the time shown on the stopwatch.

0 minutes   25 seconds
              75 hundredths
              (0.75) of a second

▲ Push the button on the top right of the stopwatch to start timing. Push the button again to stop timing.

Measure Length

You measure length to find out how long or how far away something is. Rulers, tape measures, and metersticks are some tools you can use to measure length. You can measure length using units called meters. Smaller units are made from parts of meters. Larger units are made of many meters.

Look at the ruler below. Each number represents 1 centimeter (cm). There are 100 centimeters in 1 meter. In between each number are 10 lines. The distance between each line is equal to 1 millimeter (mm). There are 10 millimeters in 1 centimeter.

Try it **Find Length with a Ruler**

Place a ruler on your desk. Line up a pencil with the "0" mark on the ruler. Record the length of the pencil in centimeters.

◀ The length of this caterpillar is about 3 cm.

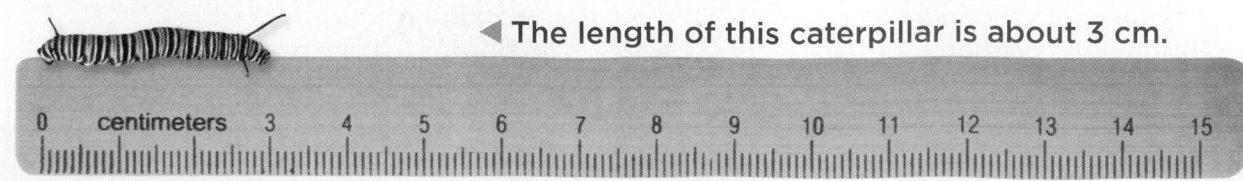

Measure Liquid Volume

Volume is the amount of space something takes up. Beakers, measuring cups, and graduated cylinders are tools you can use to measure liquid volume. These containers are marked in units called milliliters (mL).

Try it **Measure Liquid Volume**

1. Gather a few empty plastic containers of different shapes and sizes.

2. Use a graduated cylinder to find the volume of water each container can hold. To start, fill the graduated cylinder with water, then pour the water into the container. Continue pouring this until the container is full. Keep track of the number of milliliters you add.

▲ **This graduated cylinder can measure volumes up to 100 mL. The distance between each number on the cylinder represents 10 mL.**

Measure Mass

Mass is the amount of matter an object has. You use a balance to measure mass. To find the mass of an object, you compare it with objects whose masses you know. Grams are units people use to measure mass.

Try it **Measure the Mass of a Box of Crayons**

1. Place a box of crayons on one side of a pan balance.

2. Place gram masses to the other side until the two sides of the balance are level.

3. Add together the numbers on the gram masses. This total equals the mass of the box of crayons.

Measurements

Measure Force/Weight

You measure force to find the strength of a push or pull. Force can be measured in units called newtons (N). A spring scale is a tool used to measure force.

Weight is a measure of the force of gravity pulling down on an object. A spring scale measures the pull of gravity. One pound is equal to about 4.5 N.

Try it **Measure the Weight of an Object**

1. Hold a spring scale by the top loop. Put a small object on the bottom hook.
2. Slowly let go of the object. Wait for the spring to stop moving.
3. Read the number of newtons next to the tab. This is the object's weight.

Measure Temperature

Temperature (TEM•puh•ruh•chur) is how hot or cold something is. You use a tool called a thermometer (thur•MAH•muh•tur) to measure temperature. In the United States, temperature is often measured in degrees Fahrenheit (°F). However, you can also measure temperature in degrees Celsius (°C).

Try it **Read a Thermometer**

1. Fill a beaker with ice water. Then put a thermometer in the water.
2. Wait several minutes. Read the number next to the top of the red liquid inside the thermometer. This is the temperature.
3. Repeat with warm water.

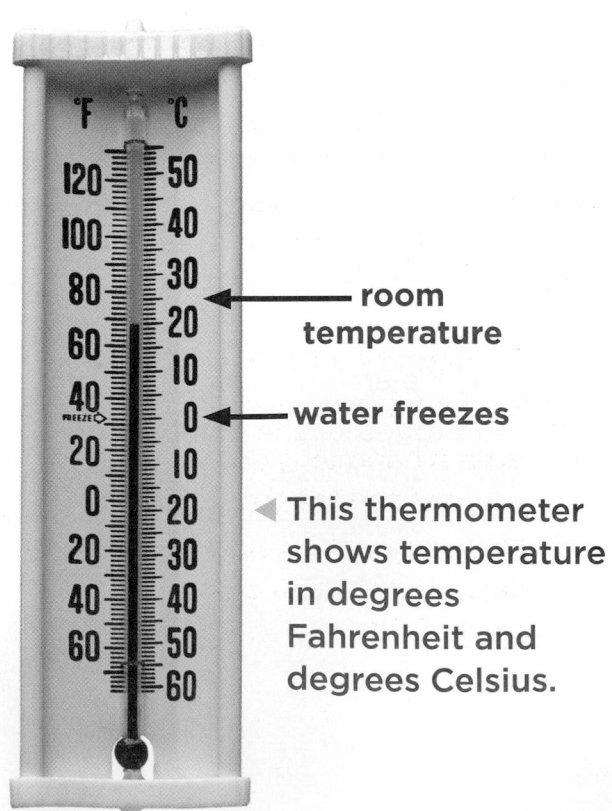

◄ room temperature

◄ water freezes

◄ This thermometer shows temperature in degrees Fahrenheit and degrees Celsius.

Tools of Science

Use a Microscope

A microscope (MI•kruh•skop) is a tool that magnifies objects, or makes them look larger. A microscope can make an object look hundreds or thousands of times larger. Look at the photo to learn the different parts of a microscope.

Try it Examine Salt Grains

1 Move the mirror so it reflects light upward toward the stage. ⚠ **Be Careful.** Never point the mirror at bright lights or the Sun. This can cause permanent eye damage.

2 Place a few grains of salt on a slide. Put the slide under the stage clips on the stage. Be sure that the salt grains are over the hole in the stage.

3 Look through the eyepiece. Turn the focusing knob slowly until the salt grains come into focus. Draw a picture of what you see.

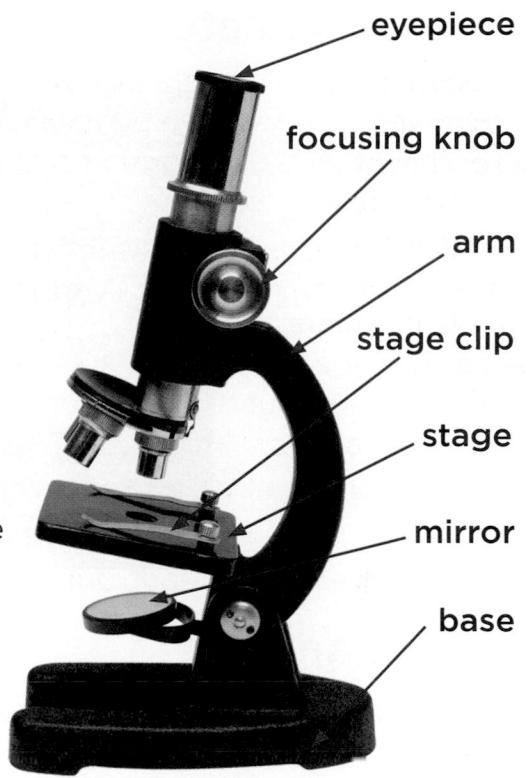

eyepiece

focusing knob

arm

stage clip

stage

mirror

base

Use a Hand Lens

A hand lens is another tool that magnifies objects. It is not as powerful as a microscope. However, a hand lens still allows you to see details of an object that you cannot see with your eyes alone. As you move a hand lens away from an object, you can see more details. If you move a hand lens too far away, the object will look blurry.

Try it Magnify a Rock

1 Look at a rock carefully. Draw a picture of it.

2 Hold a hand lens above the rock so you can see the rock clearly.

3 Fill in any details on your original drawing that you did not see before.

Tools of Science

Use a Calculator

Sometimes during an experiment, you have to add, subtract, multiply, or divide numbers. A calculator can help you carry out these operations.

Try it Convert from °F to °C

Water boils at 212°F. Use a calculator to convert 212°F into degrees Celsius.

1. Press the ON key. Then, enter the number 212 by pressing 2 1 2.
2. Subtract 32 by pressing - 3 2.
3. Multiply by 5 by pressing x 5.
4. Finally, divide by 9 by pressing ÷ 9. Press =. This is the temperature in degrees Celsius.

Now, convert 100°F into degrees Celsius.

Use a Camera

During an experiment or nature study, it helps to observe and record changes that happen over time. Sometimes it can be difficult to see these changes if they happen very quickly or very slowly. A camera can help you keep track of visible changes. Studying photos can help you understand what happens over the course of time.

Try it Gather Data from a Photo

The photos below show a panda eight days after birth and then several months later. What differences do you notice? How has the panda changed over those months? Now think of something else that changes over time. With the help of an adult, use a camera to take photos at different times. Compare your photos.

Computers

Use a Computer

A computer has many uses. You can use a computer to get information from compact discs (CDs) and digital video discs (DVDs). You can also use a computer to write reports and to show information.

The Internet connects your computer with computers around the world, so you can collect all kinds of information. When using the Internet, visit only Web sites that are safe and reliable. Your teacher can help you find safe and reliable sites to use. Whenever you are online, never give any information about yourself to others.

Try it Use a Computer for a Project

1. Choose an environment to research. Then use various resources to find out about this environment. Where is the environment located in the world? What is the climate like in the environment? What kinds of plants and animals live there?

2. Use DVDs or other sources from the library to find out more about your chosen environment.

3. Use the computer to write a report about the information you gathered. Then share your report with others.

Organizing Data

Make Maps

Locate Places

A map is a drawing that shows an area from above. Many maps have numbers and letters along the top and side. The letters and numbers help you find locations. The Buffalo Zoological Garden, for example, is located at D4 below. To find it, place a finger on the letter D along the left side of the map and another finger on the number 4 at the top. Move your fingers straight across and down the map until they meet. Now find B1. What is there?

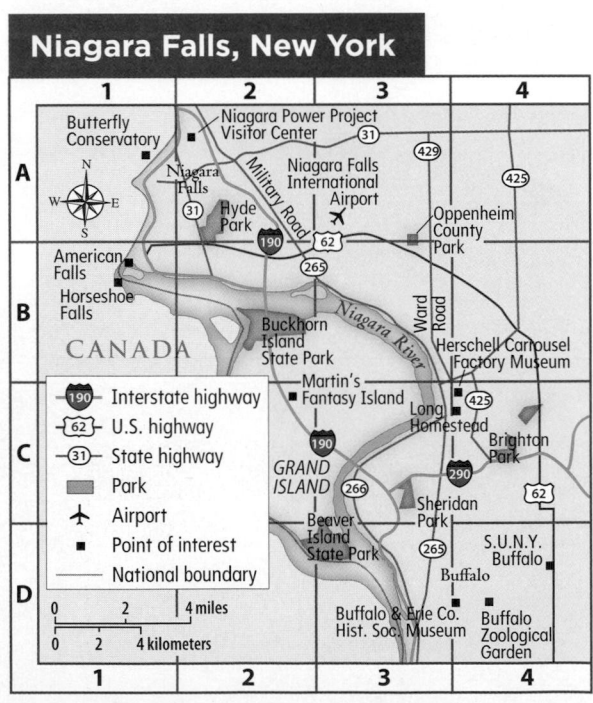

Niagara Falls, New York

Try it Make a Map

Make a map of an area in your community. It might be a park or the area between your home and school. Include numbers and letters along the top and side. Use a compass to find north, and mark north on your map.

Idea Maps

The Niagara Falls map shows how places are connected to each other. Idea maps, on the other hand, show how ideas are connected to each other. Idea maps help you organize information about a topic.

Look at the idea map below. It connects ideas about water. This map shows that Earth's water can be freshwater or salt water. The map also shows three sources of freshwater. You can see that there is no connection between rivers and salt water on the map. This can remind you that salt water does not flow in rivers.

Try it Make an Idea Map

Make an idea map about a topic you are learning in science. Your map can include words, phrases, or sentences. Arrange your map in a way that makes sense to you and helps you understand the connection between ideas.

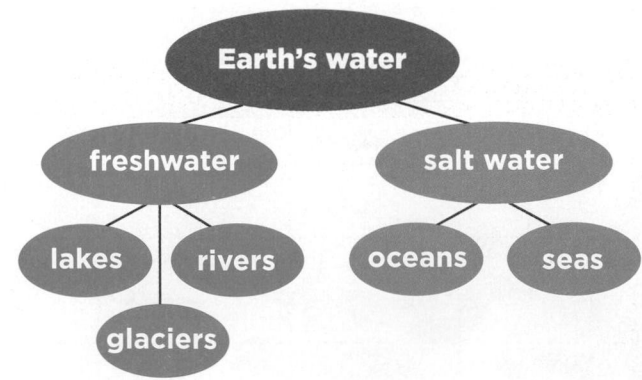

Make Charts

Charts are useful for recording information during an experiment and for communicating information. In a chart, only the column or the row has meaning but not both. In this chart, one column lists living things. A second column lists nonliving things.

Living	Nonliving
tree	rock
chipmunk	puddle
bird	cloud

Try it **Organize Data in a Chart**

Take a survey of your class. Find out each student's favorite kind of pet. Make a chart to show this information. Remember to show your information in columns or in rows.

Make Tables

Tables can also help organize data, or information. Tables have columns that run up and down and rows that run across. Headings tell you what kind of data each row or column contains.

The table below shows the properties of some minerals. Which mineral in the table has a white streak? Which mineral is yellow in color?

Try it **Organize Data in a Table**

Collect a few minerals from your teacher. Observe the properties of each. Make a table like the one shown. Use the same column headings. Record the properties of each mineral.

Mineral Identification Table					
	Hardness	Luster	Streak	Color	Other
pyrite	6–6.5	metallic	greenish-black	brassy yellow	called "fool's gold"
quartz	7	nonmetallic	none	colorless, white, rose, smoky, purple, brown	
mica	2–2.5	nonmetallic	none	dark brown, black, or silver-white	flakes when peeled
feldspar	6	nonmetallic	none	colorless, beige, pink	
calcite	3	nonmetallic	white	colorless, white	bubbles when acid is placed on it

Make Graphs

Graphs also help organize data. Graphs make it easy to notice trends and patterns. There are many kinds of graphs.

Bar Graphs

A bar graph uses bars to show data. What if you want to find the warmest and coldest months for your city? Every month you find the average temperature in the newspaper. You can organize the temperatures in a bar graph so you can easily compare them.

Month	Temperature (°C)
January	6
February	8
March	10
April	13
May	16
June	19
July	22
August	20
September	19
October	14
November	9
December	7

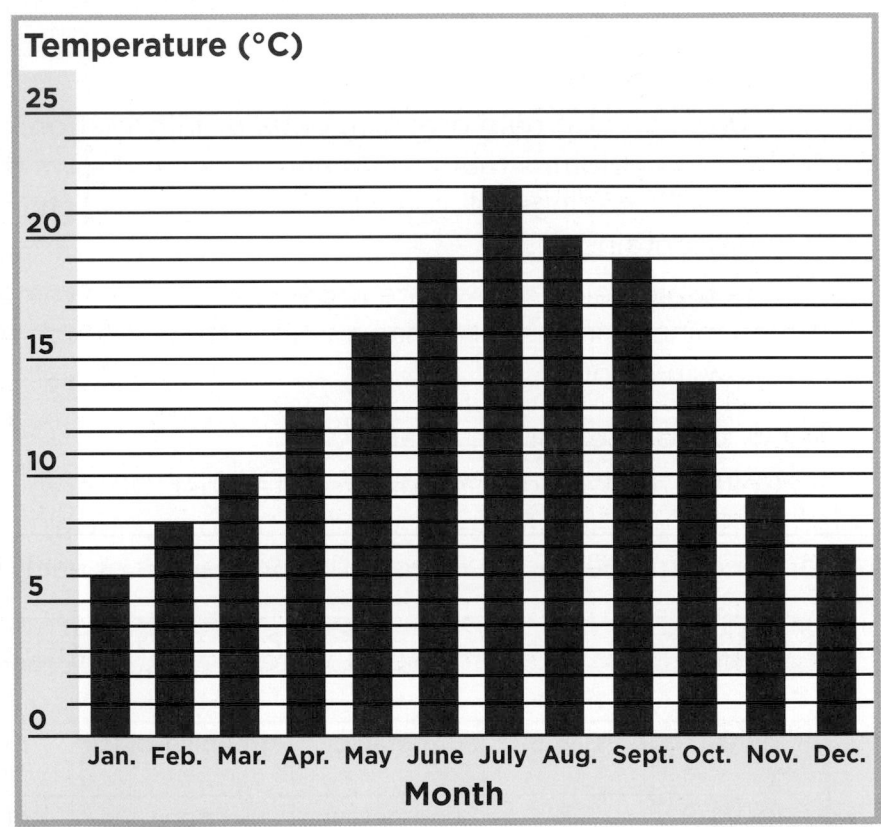

1. Look at the bar for the month of April. Put your finger at the top of the bar. Move your finger straight to the left to find the average temperature for that month.

2. Find the highest bar on the bar graph. This bar represents the month with the highest average temperature. Which month is it? What is the average temperature for that month?

3. Look at the bars on the graph. What pattern do you notice in the temperatures from January to December?

Pictographs

A pictograph uses symbols, or pictures, to show information. What if you collect information about how much water your family uses each day?

Water Used Daily (L)	
drinking	10
showering	100
bathing	120
brushing teeth	40
washing dishes	80
washing hands	30
washing clothes	160
flushing toilet	50

You can organize this information into a pictograph. In the pictograph below, each bucket means 20 liters (L) of water. A half bucket means half of 20 L, or 10, liters of water.

1 Which activity uses the most water?

2 Which activity uses the least water?

Water Used Daily	
drinking	🪣
showering	🪣🪣🪣🪣🪣
bathing	🪣🪣🪣🪣🪣🪣
brushing teeth	🪣🪣
washing dishes	🪣🪣🪣🪣
washing hands	🪣🪣
washing clothes	🪣🪣🪣🪣🪣🪣🪣🪣
flushing toilet	🪣🪣🪣

🪣 = 20 liters of water

Line Graphs

A line graph can show how information changes over time. What if you measure the temperature outdoors every hour starting at 6 A.M.?

Time	Temperature (°C)
6 A.M.	10
7 A.M.	12
8 A.M.	14
9 A.M.	16
10 A.M.	18
11 A.M.	20

Now organize your data by making a line graph. Follow these steps.

1 Make a scale along the bottom and side of the graph. Label the scales.

2 Draw a point on the graph for each temperature measured each hour.

3 Connect the points.

4 How do the temperatures and times relate to each other?

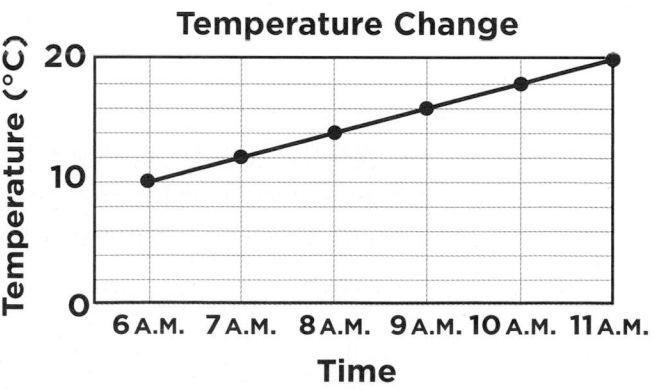

Temperature Change

Human Body Systems

The Skeletal System

Feel your elbows, wrists, and fingers. What are those hard parts? Bones! Bones make up the skeletal system. The skeletal system is one of many body systems. A body system is a group of organs that work together to perform a specific job.

The skeletal system is made up of 206 bones. Each bone has a particular job. The long, strong leg bones support the body's weight. The skull protects the brain. The hip bones help you move. Together, bones do important jobs to keep the body active and healthy.

▶ Bones support the body and give the body its shape.

▶ Bones protect organs in the body.

▶ Bones work with muscles to move the body.

▶ Bones store minerals and produce blood for the body.

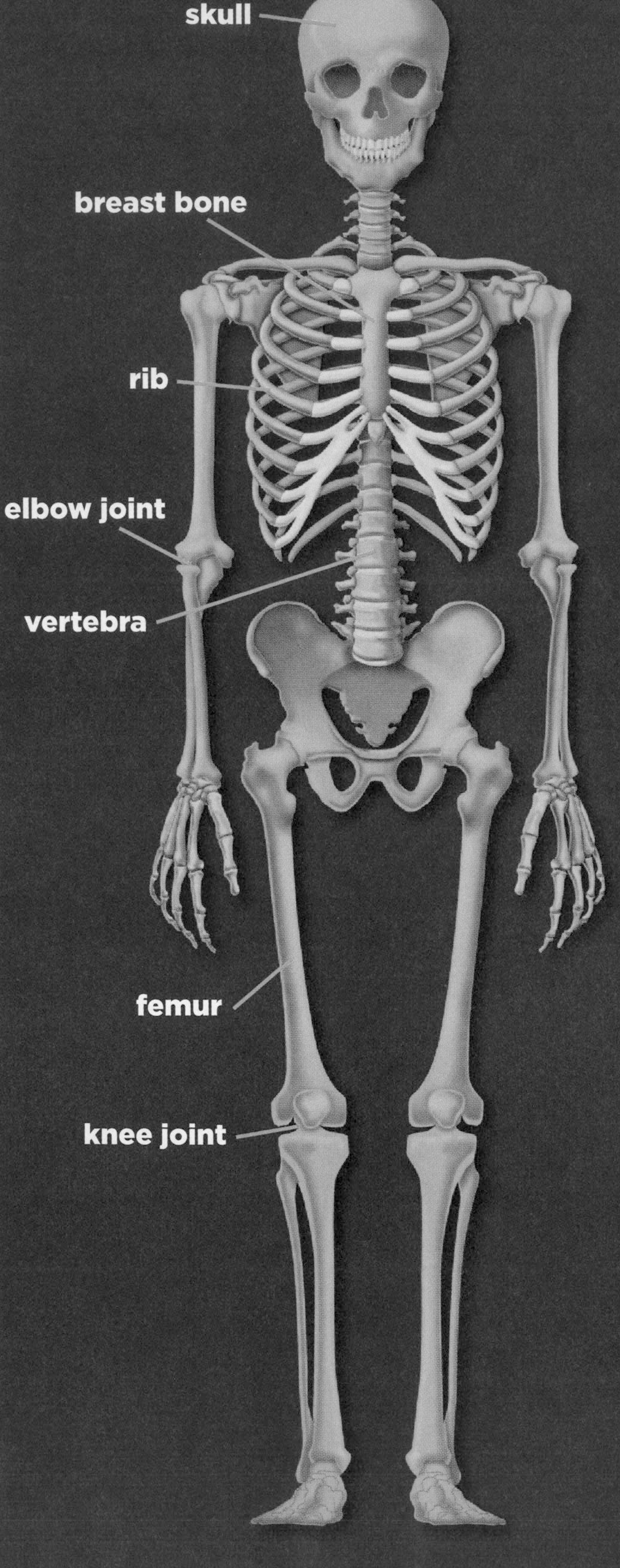

Joints

A joint is a place where two or more bones meet. There are three main types of joints.

Immovable joints form where bones fit together too tightly to move. The 29 bones of your skull meet at immovable joints. Partly movable joints are places where bones can move a little. Ribs are connected to the breastbone with these joints. Movable joints, like the knee, are places where bones can move easily. The knee lets the bones of your leg move.

skull

breast bone

rib

elbow joint

vertebra

femur

knee joint

The Muscular System

Together, all the muscles in the body form the muscular system. Muscles allow the body to move. Without muscles, you would not be able to run, smile, breathe, or even blink.

Most muscles are attached to bones and skin. These are called skeletal muscles. To move bones back and forth, skeletal muscles usually work in pairs. Each pulls on a bone in a different direction. When you want to move, your brain sends a message to a pair of skeletal muscles. One muscle contracts, or gets shorter. It pulls on the bone and skin. The other muscle relaxes to let the bone move.

▲ There are 53 muscles in your face. You use 12 of them whenever you smile.

◀ To bend his arm, this boy's biceps contract while his triceps relax.

biceps

triceps

thigh muscles

calf muscles

Some muscles work without you even thinking about it. The heart is made of muscle. It pumps blood throughout the body even while you sleep. A muscle in the chest helps you breathe. Muscles in the stomach help you digest food.

Human Body Systems

The Circulatory System

The body's cells need a constant supply of oxygen and nutrients. The circulatory (SUR•kyuh•luh•tawr•ee) system is responsible for sending these things throughout the body. The circulatory system is made up of the heart, blood vessels, and blood.

Blood rich in oxygen travels from the lungs to the heart. The heart is an organ about the size of a fist. It beats about 70 to 90 times each minute, pumping blood through the blood vessels.

Blood vessels are tubes that carry blood. There are two main types of blood vessels. Arteries are blood vessels that carry blood away from the heart. Veins carry blood back to it.

Blood contains plasma, red blood cells, white blood cells, and platelets. Plasma is the liquid part of blood. It carries nutrients and other things the body needs. Red blood cells carry oxygen to all the cells of your body. Red blood cells and plasma also carry wastes, such as carbon dioxide, away from cells. White blood cells work to fight disease. Platelets keep you from bleeding too much when you get a cut.

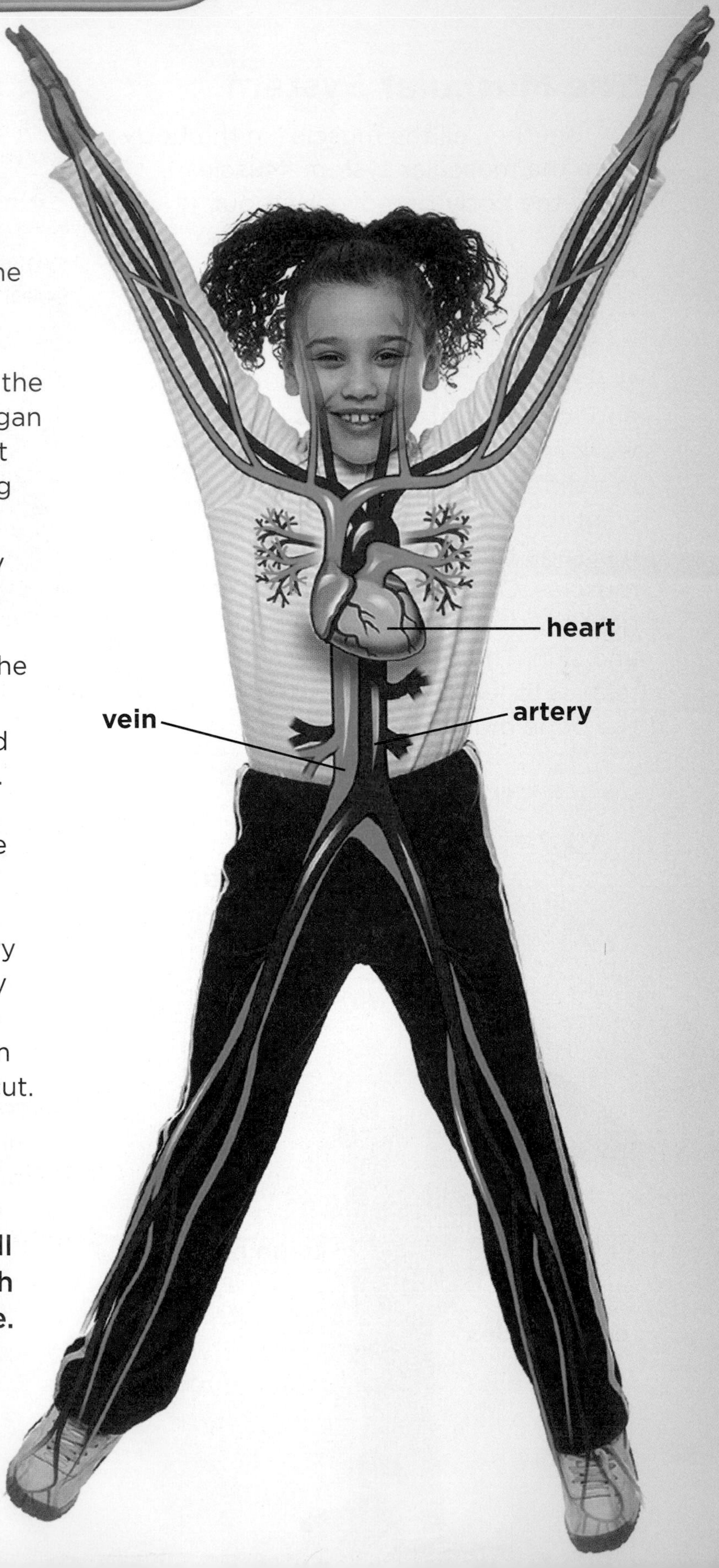

heart

vein

artery

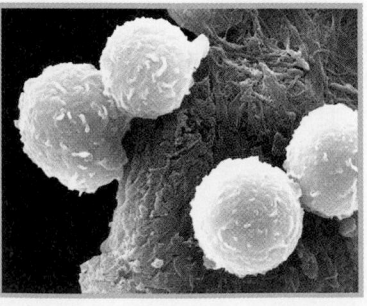

◄ **This is how a red blood cell looks through a microscope.**

The Respiratory System

The respiratory (RES•pruh•tor•ee) system helps the body take in oxygen and give off carbon dioxide and other waste gases. All the cells in your body require oxygen to work properly. You take in oxygen from the air when you breathe.

Every time you inhale, a muscle called the diaphragm (DI•uh•fram) contracts. This makes room in your lungs for air. Air is taken in through the nose or mouth. This air travels down the throat into the trachea (TRAY•kee•uh).

In the chest, the trachea splits into two bronchial (BRONG•kee•ul) tubes. Each tube leads to a lung. Inside each lung, the bronchial tube branches off into smaller tubes called bronchioles (BRONG•kee•olz). At the end of each bronchiole are millions of tiny air sacs. Here, red blood cells release carbon dioxide, a waste gas, and absorb oxygen. When you breathe out, the diaphragm relaxes. This causes the lungs to deflate and push carbon dioxide out of your body through the nose and mouth.

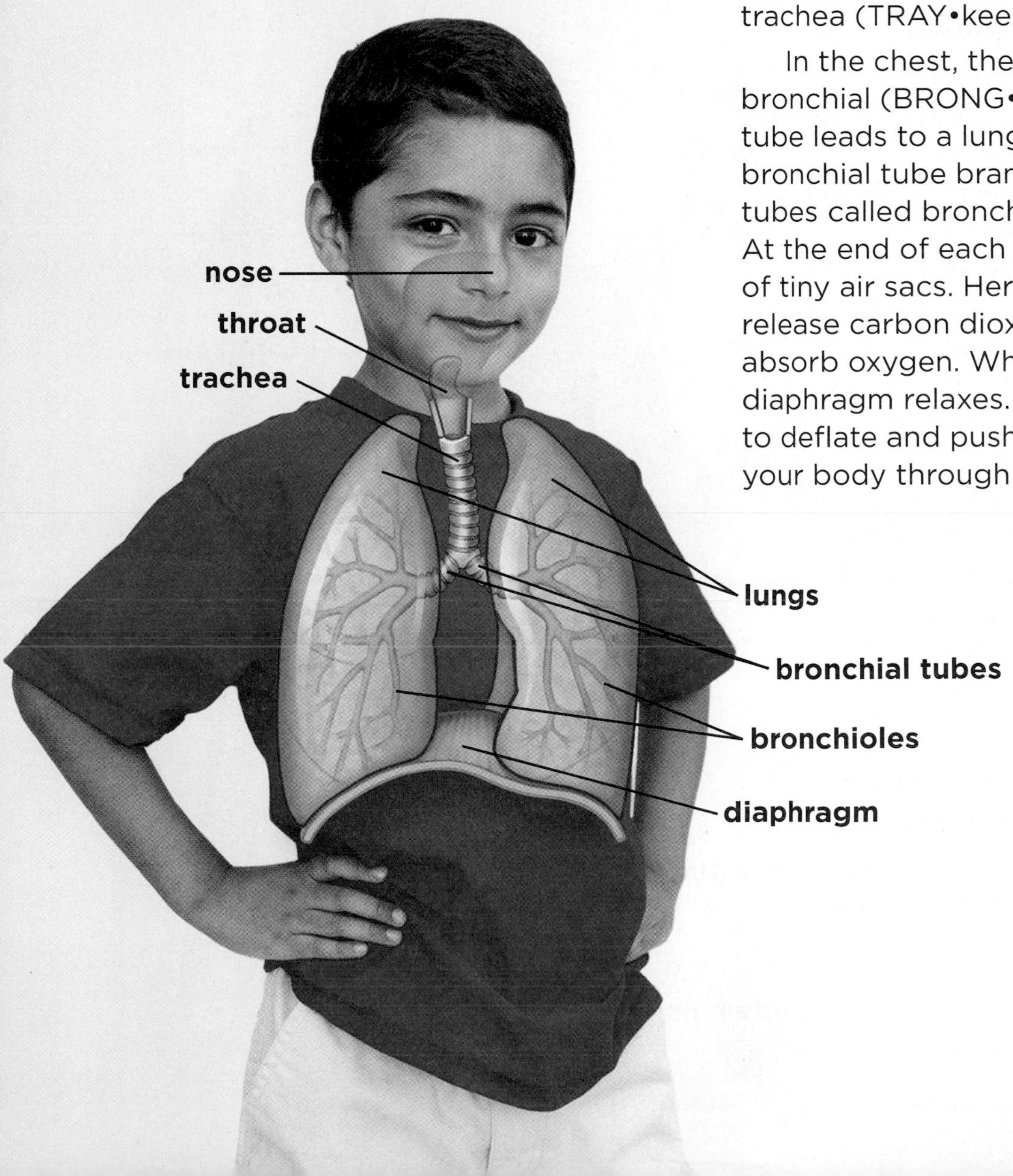

nose

throat

trachea

lungs

bronchial tubes

bronchioles

diaphragm

The Digestive System

The digestive (di•JES•tiv) system is responsible for breaking down food into nutrients the body can use. Digestion begins when you chew food. Chewing breaks food into smaller pieces and moistens it with saliva. Saliva helps food travel smoothly when you swallow. The food travels down your esophagus (ih•SAH•fuh•gus) and into your stomach.

Inside the stomach food is mixed with strong, acidic juices. This causes the food to break down further, making it easier for your body to absorb nutrients from the food.

After passing through the stomach, food moves into the small intestine (in•TES•tun). This is where most nutrients are absorbed. The small intestine is a narrow tube about 6 meters (20 feet) long. It is coiled tightly so it fits inside the body. As food passes through the small intestine, digested nutrients are absorbed into the blood. The blood then carries these nutrients to other parts of the body.

After food has passed through the small intestine, it enters the large intestine. The large intestine removes water from the unused food that is left. Then the unused food is removed from the body as waste.

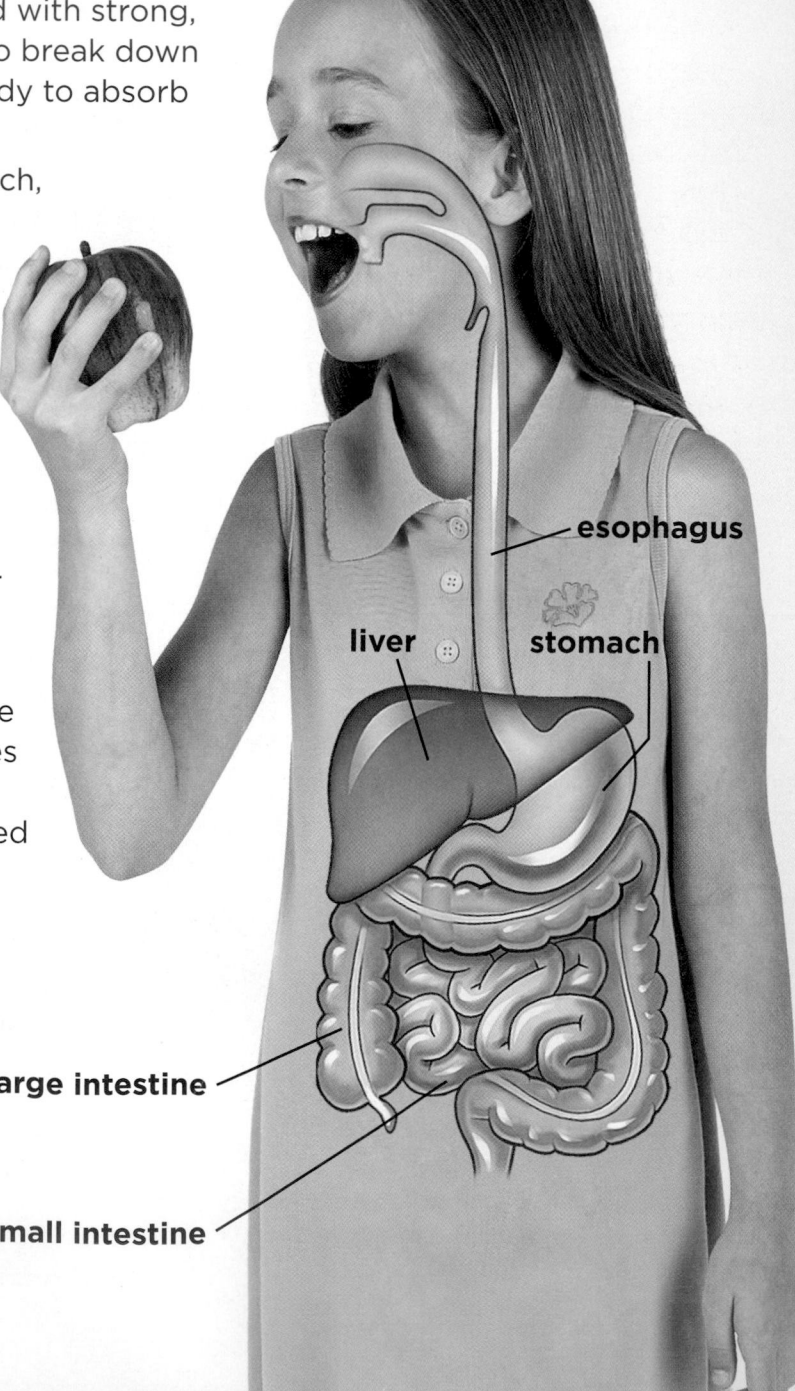

esophagus

liver

stomach

large intestine

small intestine

The Excretory System

The excretory (EK•skruh•tor•ee) system gets rid of waste products from your cells. Waste products are materials that the body does not need, such as extra water and salts. The liver, kidneys, bladder, and skin are some organs of the excretory system.

Liver, Kidneys, and Bladder

The liver filters wastes from the blood. It changes wastes into a chemical call urea and sends the urea to the kidneys. Kidneys turn urea into urine. Urine flows from the kidneys to the bladder. It is stored in the bladder until it is pushed out of the body through the urethra.

Skin

The skin takes part in excretion when a person sweats. Sweat glands in the inner layer of skin produce sweat. Sweat is made of water and minerals that the body does not need. Sweat is released through the outer layer of the skin. Sweating cools the body and helps it maintain an internal temperature of about 98°F (37°C).

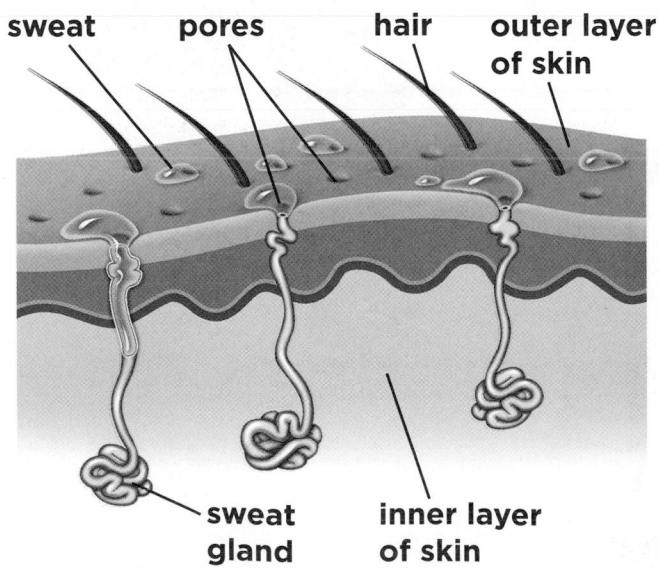

sweat   pores   hair   outer layer of skin

sweat gland   inner layer of skin

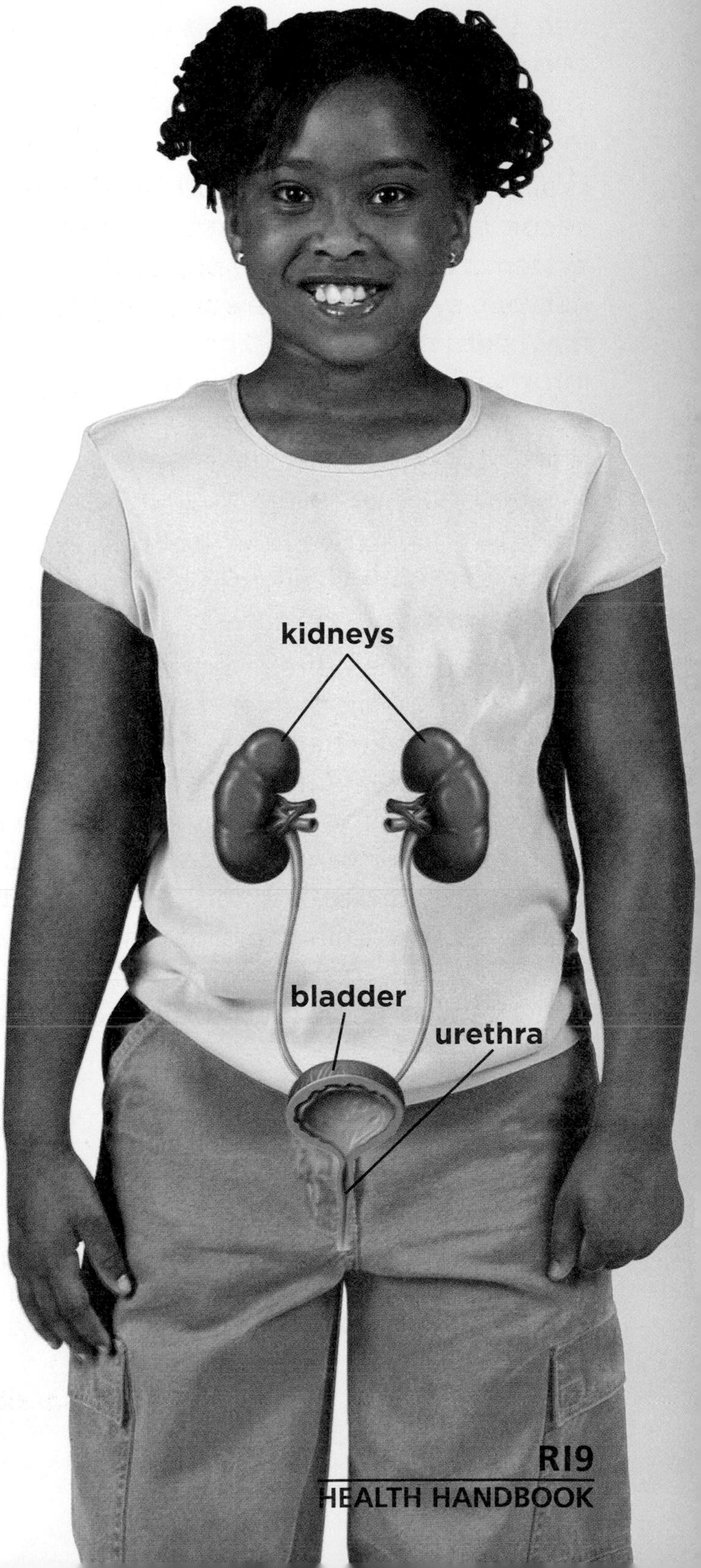

kidneys

bladder

urethra

Human Body Systems

The Nervous System

The nervous system is responsible for taking in and responding to information. It controls muscles and helps the body balance. It allows a person to think, feel, and even dream.

The nervous system is made of two main parts. The first part, the central nervous system, is made of the brain and spinal cord. All other nerves make up the second part, the peripheral (puh•RIH•frul) nervous system. Nerves from the peripheral nervous system receive sensory information from cells in the body. They pass this information on to the brain through the spinal cord. When the brain receives this information, it makes decisions about how the body should respond. Then it passes this new information back through the spinal cord to the nerves, and the body responds.

The Brain

The brain has three main parts, the cerebrum (suh•REE•brum), the cerebellum (ser•uh•BE•lum), and the brain stem. The cerebrum is the largest part of the brain. It stores memories and helps control information received by the senses. The cerebellum helps the body keep its balance and directs the skeletal muscles. The brain stem connects to the spinal cord. It controls heartbeat, breathing, and blood pressure.

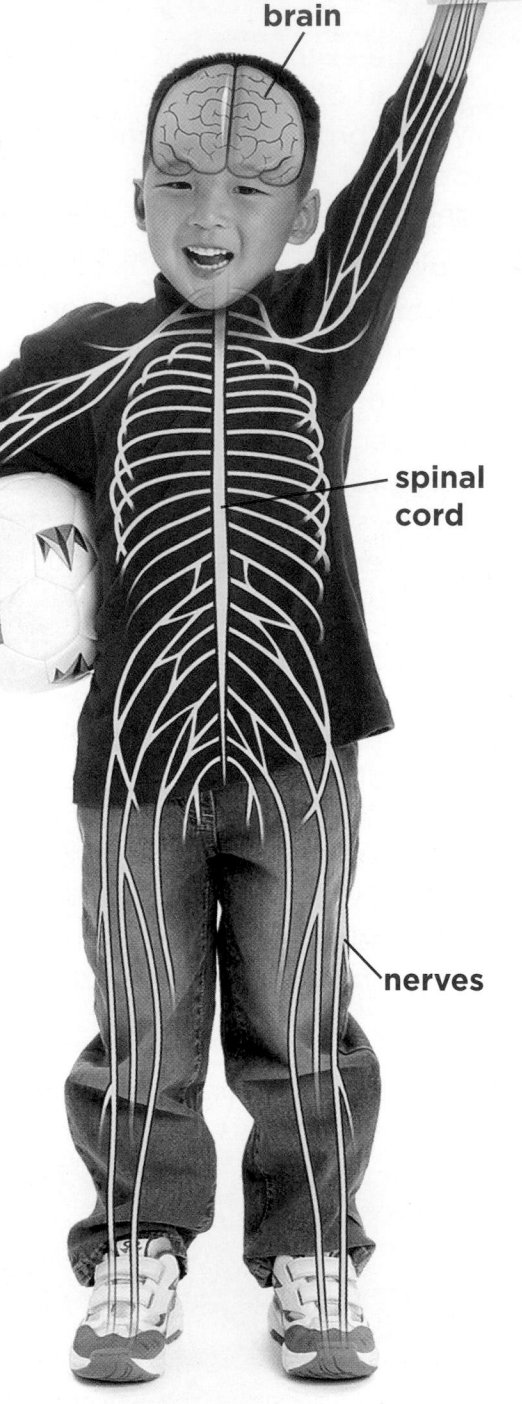

brain

spinal cord

nerves

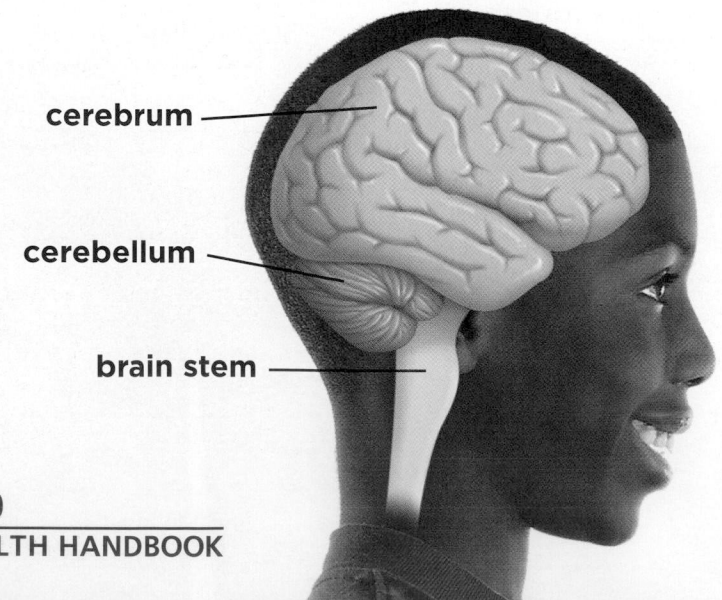

cerebrum

cerebellum

brain stem

The Senses

Different nerves in the body take in information from the environment. These nerves are responsible for the body's sense of sight, hearing, smell, taste, and touch.

Sight

Light reflects off an object, such as a leaf, and into the eye. The reflected light passes through the pupil in the iris. Cells in the eye change light into electrical signals. The signals travel through the optic nerve to the brain.

Hearing

Sound waves enter the outer ear. They reach the eardrum and cause it to vibrate. Cells in the ear change the sound waves into electrical signals. The signals travel along the auditory nerve to the brain.

Smell

As a person breathes, chemicals in the air mix with mucus in the upper part of the nose. When they reach certain cells in the nose, those cells send information along the olfactory nerve to the brain.

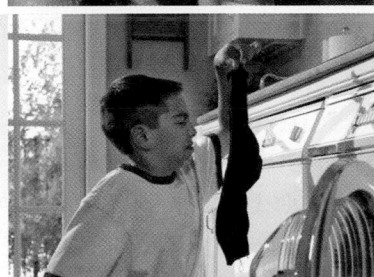

Taste

On the tongue are more than 10,000 tiny bumps, called taste buds. Each taste bud can sense four main tastes sweet, sour, salty, and bitter. The taste buds send information along a nerve to the brain.

Touch

Different nerve cells in the skin give the body its sense of touch. They help a person tell hot from cold, wet from dry, and hard from soft. Each cell sends information to the spinal cord. The spinal cord then sends the information to the brain.

Human Body Systems

Immune System

The immune system protects the body from germs. Germs cause disease and infection. Most of the time, the immune system is able to prevent germs from entering the body. Skin, tears, and saliva are parts of the immune system. They work to kill germs and keep them out of the body.

When germs do find a way into your body, white blood cells help find and kill them quickly before you become ill. White blood cells are part of the blood. They travel through blood vessels and lymph (LIMF) vessels. Lymph vessels are similar to blood vessels. However, instead of carrying blood, they carry a fluid called lymph. Many white blood cells are made and live in lymph nodes. Here, they filter out harmful materials from the body.

White blood cells might not always kill germs before the germs start to reproduce in your body. When germs reproduce, they cause illness. Even while you feel ill, the immune system works to kill and remove germs until you are well again.

lymph vessels

lymph nodes

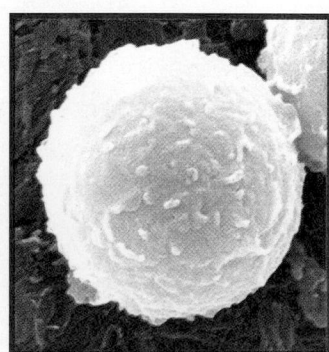

◄ This is how a white blood cell looks through a microscope.

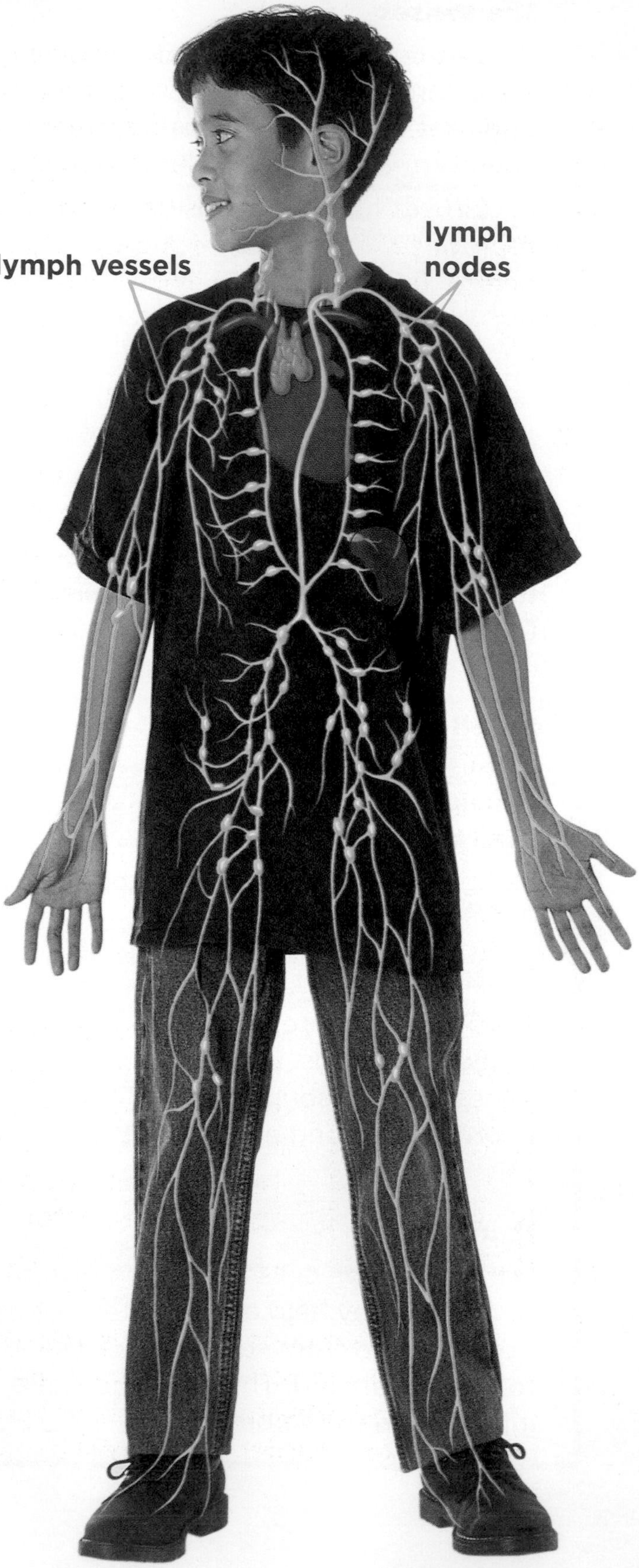

Viruses and Bacteria

One of the main types of germs that makes the body ill are viruses. Illness from a virus like a cold or flu can be a big deal. Yet viruses themselves are very small. In fact, you need a special microscope, an electron microscope, to look at a virus.

Viruses need to be inside living cells, called hosts, to reproduce. As they reproduce, viruses take nutrients and energy from the cell. They can even produce harmful materials that make the body itch or have dangerously high temperatures.

Bacteria are the other main type of germ that can make the body ill. Bacteria are tiny, one-celled organisms. They can live on most surfaces and are able to reproduce outside of cells. Some bacteria can have a harmful effect on the body. Other bacteria, however, are good for the body. Some bacteria in your body, for example, help you digest food.

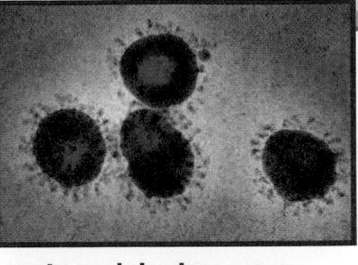

▲ A cold virus as seen through a microscope.

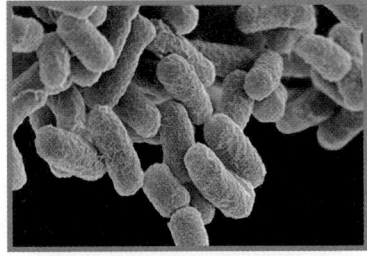

▲ *E. coli* bacteria as seen through a microscope.

You can help your body defend itself against germs. Here is what you can do.

▶ Eat healthful foods. This helps your body get all the nutrients it needs to stay healthy. A healthy body is better able to fight germs.

▶ Be active. Being active makes your body fit. A fit body is better able to fight germs.

▶ Get a yearly checkup. Make sure you get all of your immunizations. Follow directions when taking medicines given to you by a doctor.

▶ Get plenty of rest. You need about 10 hours of sleep each night. Sleeping helps repair your body. Get extra rest when you are ill.

▶ Do not share cups or utensils with other people. Germs can be on objects you touch. Wash your hands, especially before eating and drinking. By washing your hands, you kill germs and make it harder for harmful things to get into your body.

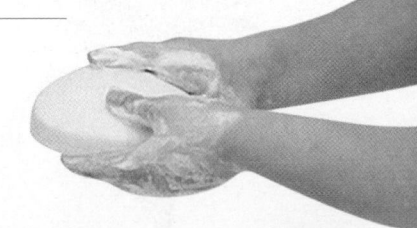

Healthy Living

Nutrients

Nutrients are materials in foods that help the body grow, get energy, and stay healthy. By eating a balance of healthful foods, your body gets the nutrients it needs to do all of these things.

There are six kinds of nutrients—carbohydrates, vitamins, minerals, proteins, water, and fats. Each nutrient helps the body in different ways.

Carbohydrates

Carbohydrates are the main source of energy for the body. Starches and sugars are two types of carbohydrates. Starches come from foods like bread, pasta, and cereal. They provide long-lasting energy. Sugars come from fruits and can be used immediately by the body for energy.

carbohydrates

Vitamins

Vitamins help keep the body healthy. They also help to build new cells in the body. The table below shows some vitamins and their sources.

Vitamin	Sources	Benefits
A	milk, fruit, carrots, green vegetables	keeps eyes, teeth, gums, skin, and hair healthy
C	citrus fruits, strawberries, tomatoes	helps heart, cells, and muscles function
D	milk, fish, eggs	helps keep teeth and bones strong

Minerals

Minerals help form new bone and blood cells. They also help your muscles and nervous system work properly. Here are some minerals and their sources.

Mineral	Sources	Benefits
calcium	yogurt, milk, cheese, green vegetables	builds strong teeth and bones
iron	meat, beans, fish, whole grains	helps red blood cells function properly
zinc	meat, fish, eggs	helps your body grow and helps to heal wounds

Fats

Fats help the body use other nutrients and store vitamins. Fats also help the cells of the body work properly. They even help keep the body warm. Fats can be found in foods such as meats, eggs, milk, butter, and nuts. Oils also contain fats. Though some fats help the body, some fats can cause health problems.

fats

Water

Water is one of the most important nutrients. About $\frac{2}{3}$ of the body is made of water! Water makes up most of the body's cells. It helps the body remove waste and protects joints. It also prevents the body from getting too hot.

Proteins

Proteins are a part of every living cell. Proteins help bones and muscles grow. They even help the immune system fight diseases. Foods high in protein are milk, eggs, meats, fish, nuts, and cheese.

proteins

Healthy Living

Stay Fit

MyPyramid

You can use MyPyramid as a guide to healthful eating. The pyramid will show you the amounts of foods you should eat from each of the five food groups. A food group is foods with the same kinds of nutrients. To find the correct amounts of foods that are right for you, visit www.MyPyramid.gov.

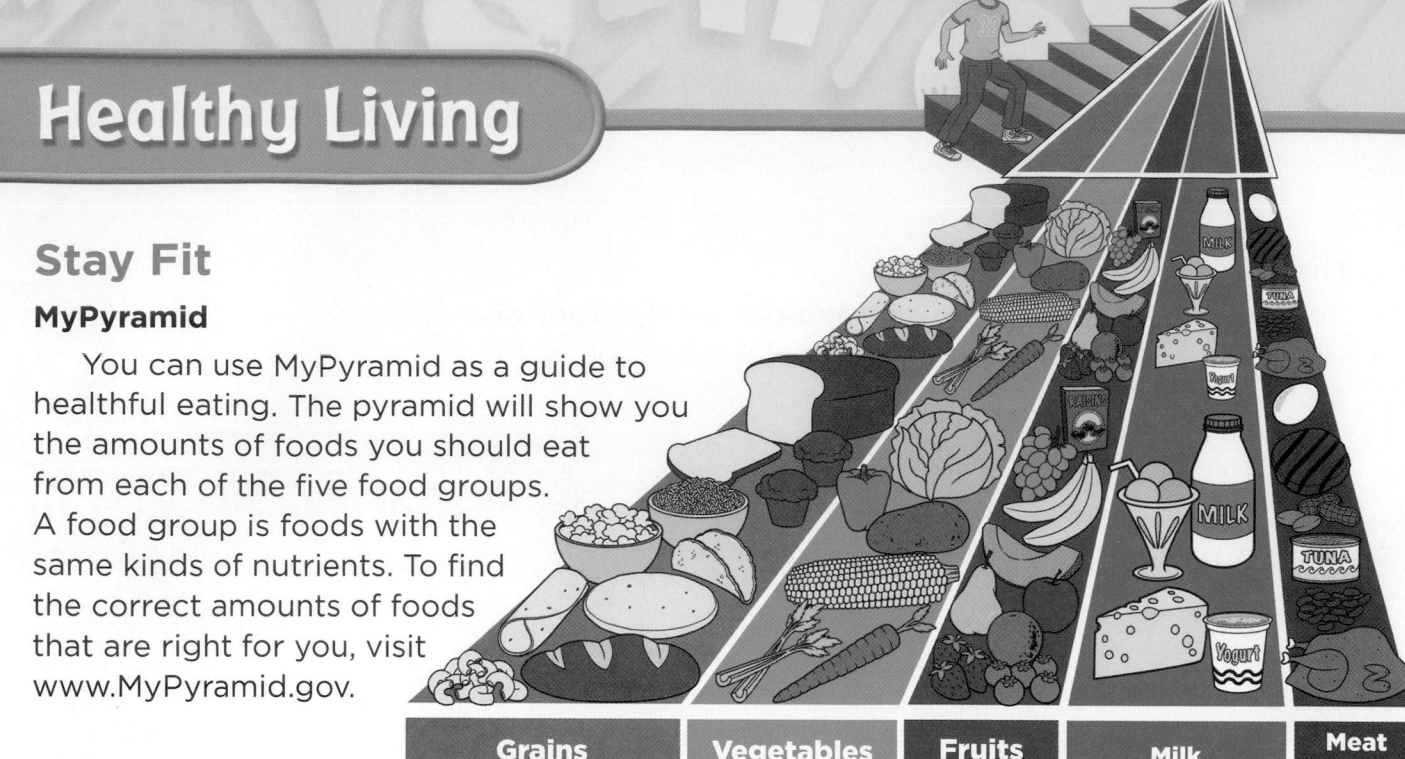

| Grains | Vegetables | Fruits | Milk | Meat & Beans |

Be Drug Free

Do not use cigarettes, illegal drugs, or alcohol. These things can harm your body. They can keep you from growing properly and becoming fit.

Be Physically Active

You need to be physically active for at least 60 minutes every day. When you are physically active, you become physically fit. When you are physically fit, your heart, lungs, bones, joints, and muscles stay strong. You keep a healthful weight and lower the risk of disease. You do not have to be on a sports team to be physically active. You just need to move your body. Running, biking, and swimming are just some ways to be physically active.

by Dinah Zike

Folding Instructions

The following pages offer step-by-step instructions to make the Foldables study guides.

Half-Book

1. Fold a sheet of paper ($8\frac{1}{2}''$ x 11") in half.
2. This book can be folded vertically like a hot dog or ...
3. ... it can be folded horizontally like a hamburger.

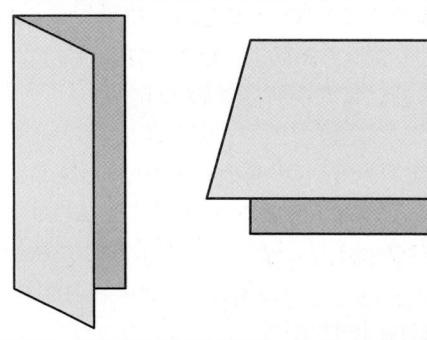

Folded Book

1. Make a half-book.
2. Fold in half again like a hamburger. This creates a ready-made cover and two small pages inside for recording information.

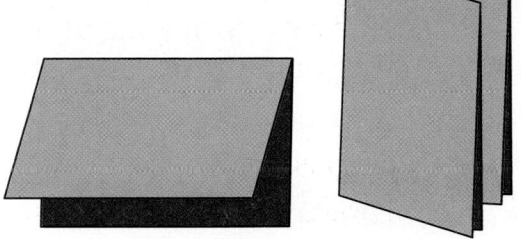

Pocket Book

1. Fold a sheet of paper ($8\frac{1}{2}''$ x 11") in half like a hamburger.
2. Open the folded paper. Fold one of the long sides up two inches to form a pocket. Refold along the hamburger fold so the newly formed pockets are on the inside.
3. Glue the outer edges of the two-inch fold with a small amount of glue.

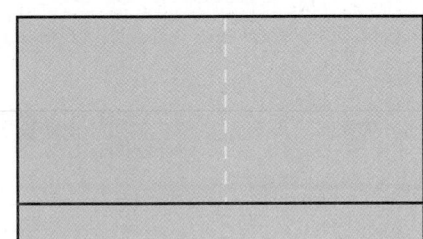

Shutter Fold

1. Begin as if you were going to make a hamburger, but instead of creasing the paper, pinch it to show the midpoint.
2. Fold the outer edges of the paper to meet at the pinch, or midpoint, forming a shutter fold.

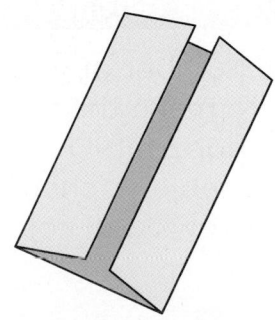

Trifold Book

1. Fold a sheet of paper ($8\frac{1}{2}$" x 11") into thirds.
2. Use this book as is, or cut it into shapes.

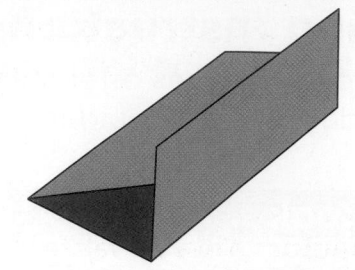

Three-Tab Book

1. Fold a sheet of paper like a hot dog.
2. With the paper horizontal and the fold of the hot dog up, fold the right side toward the center, trying to cover one half of the paper.
3. Fold the left side over the right side to make a book with three folds.
4. Open the folded book. Place one hand between the two thicknesses of paper and cut up the two valleys on one side only. This will create three tabs.

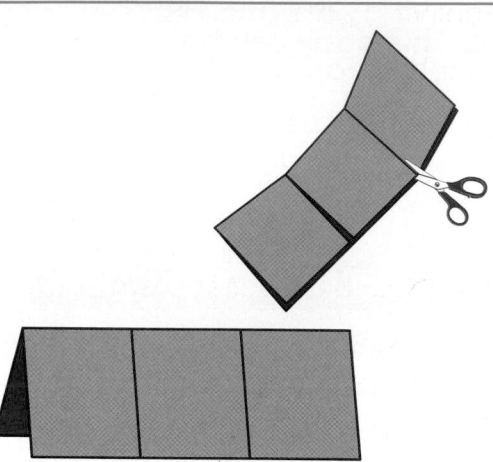

Layered-Look Book

1. Stack two sheets of paper ($8\frac{1}{2}$" x 11") so the back sheet is one inch higher than the front sheet.
2. Bring the bottoms of both sheets upward and align the edges so all the layers or tabs are the same distance apart.
3. When all the tabs are an equal distance apart, fold the papers and crease well.
4. Open the papers and glue them together along the valley, or inner center fold, or staple them along the mountain.

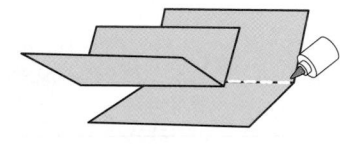

Folded Table or Chart

1. Fold the number of vertical columns needed to make the table or chart.
2. Fold the horizontal rows needed to make the table or chart.
3. Label the rows and columns.

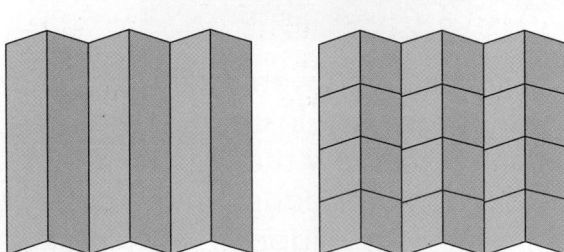

Glossary

Use this glossary to learn how to pronounce and understand the meanings of the science words used in this book. The page number at the end of each definition tells you where to find that word in the book.

A

abiotic factor (ā′bī·ā′tik fak′tər) a nonliving part of an ecosystem (p. 130)

acceleration (ak·se′lərā′shən) any change in the speed or direction of a moving object (p. 486)

accommodation (əkä′mədā′shən) an individual organism's response to changes in its ecosystem (p. 188)

acid (a′səd) a substance that tastes sour and turns blue litmus paper red (p. 470)

acid rain (a′səd rān) harmful rain caused by the burning of fossil fuels (p. 296)

adaptation (a′dap·tā′shən) a trait that helps a living thing survive in its environment (p. 166)

air mass (âr mas) a large region of the atmosphere where the air has similar properties throughout (p. 336)

air pressure (âr pre′shər) the force of air pushing down on an area (p. 317)

Pronunciation Key
The following symbols are used throughout this glossary.

a	at	e	end	o	soft	u	up	hw	white	ə	about
ā	ape	ē	me	ō	go	ū	use	ng	song		taken
ä	farther	i	tip	ôr	fork	ü	rule	th	thin		pencil
âr	care	ī	ice	oi	oil	u̇	pull	th	this		lemon
ô	law	îr	fear	ou	out	ûr	turn	yü	rule		circus
								zh	measure		

′ = primary accent; shows which syllable takes the main stress, such as **at** in **atmosphere** (at′məsfîr′)

′ = secondary accent; shows which syllables take lighter stresses, such as **sphere** in **atmosphere**

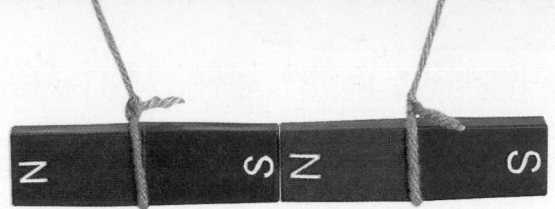

alloy (a′loi) a mixture of one metal with one or more metals or substances (p. 459)

amber (am′bər) hardened tree sap, yellow to brown in color, that can be a source of insect fossils (p. 274)

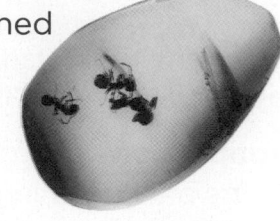

amphibian (am·fi′bē·ən) a cold-blooded vertebrate that spends part of its life in water and part of its life on land (p. 92)

amplitude (am′plətüd′) a measure that relates to the amount of energy in a sound wave (p. 545)

area (âr′ē·ə) the number of unit squares that fit inside a surface (p. 423)

arthropod (är′thrəpod′) an invertebrate with jointed legs, a body that is divided into sections, and an exoskeleton (p. 82)

asteroid (as′təroid′) a chunk of rock or metal that orbits the Sun (p. 388)

atmosphere (at′məsfîr′) the blanket of gases that surrounds Earth (p. 314)

atom (a′təm) the smallest particle of an element; all atoms of one element are alike but are different from those of any other element (p. 432)

attract (ətrakt′) to pull (p. 576)

avalanche (a′vəlanch′) a large, sudden movement of ice and snow down a hill or mountain (p. 242)

axis (ak′səs) a real or imaginary line that a spinning object turns around (p. 360)

B

balanced forces (ba′lənst fôrs′əz) forces that cancel each other out when acting together on a single object (p. 494)

barometer (bərä·mətər) a device used for measuring air pressure (p. 318)

base (bās) a substance that tastes bitter and turns red litmus paper blue (p. 470)

biome (bī′ōm) one of Earth's large ecosystems with its own kind of climate, soil, and living things (p. 138)

biotic factor (bī·ä′tik fak′tər) a living part of an ecosystem (p. 130)

bird (bərd) a vertebrate that has a beak, feathers, two wings, and two legs (p. 93)

buoyancy (boi′ən·sē) the upward force of a liquid or a gas on an object (p. 413)

C

camouflage (ka′məfläj′) an adaptation by which an animal can hide by blending in with its surroundings (p. 168)

cast (kast) a fossil formed or shaped within a mold (p. 275)

cell (sel) the smallest unit of living matter (p. 22)

change of state (chānj əv stāt) a physical change of matter from one state—solid, liquid, or gas—to another state because of a change in the energy of the matter (p. 448)

chemical change (kə′mi·kəl chānj) a change that produces new matter with different properties from the original matter (p. 450)

circuit (sûr′kət) a path through which electric current can flow (p. 566)

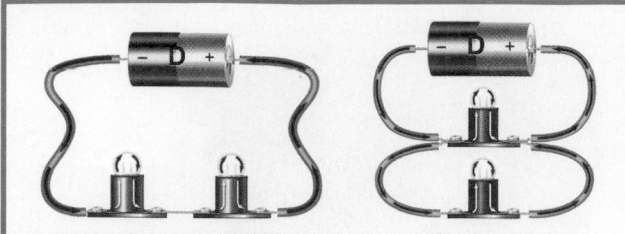

circulatory system (sûr′kyələtôr′ē sis′təm) the organ system that moves blood throughout the body (p. 103)

climate (klī′mət) the average weather pattern of a region over time (p. 346)

clone (klōn) an offspring that is an exact copy of its parent (p. 114)

cloud (kloud) a collection of tiny water droplets or ice crystals in the atmosphere (p. 325)

cnidarian (nī·dâr′ē·ən) an invertebrate with stinging cells on the ends of tentacles (p. 80)

cold-blooded (kōld′ blə′dəd) a type of animal that gets its heat from outside its body (p. 90)

cold front (kōld frunt) a boundary where a cold air mass slides under a warm air mass (p. 337)

comet (kä′mət) a chunk of ice, rock, and dust that orbits the Sun (p. 388)

community (kəmyü′nətē) all the populations in an ecosystem (p. 132)

competition (kom′pəti′shən) the struggle among living things for water, food, or other resources (p. 155)

compost (kom′pōst) a mixture of decaying matter that helps plants grow in soil (p. 298)

compound (kom′pound) a substance made when two or more elements join and lose their own properties (p. 468)

compound machine (kom′pound məshēn′) a combination of two or more simple machines (p. 520)

condensation (kon′den·sā′shən) the process of a gas changing to a liquid (p. 325)

conduction (kən·duk′shən) the transfer of thermal energy between two objects that are touching (p. 532)

conductor (kən·duk′tər) a material through which heat or electricity flows easily (p. 533)

conservation (kon′sər·vā′shən) the act of saving, protecting, or using resources wisely (p. 298)

constellation (kon′stə·lā′shən) a group of stars that appear to form a pattern in the night sky (p. 396)

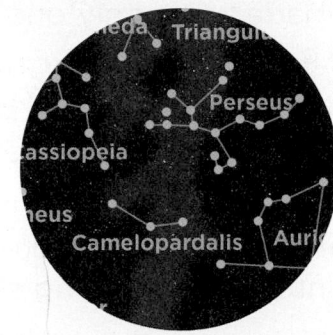

consumer (kən·sü′mər) an organism that cannot make its own food (p. 151)

convection (kən·vek′shən) the transfer of thermal energy by flowing gases or liquid, such as the rising of warm air from a heater (p. 532)

crater (krā′tər) a hollow area or pit in the ground (p. 371)

crust (krust) rock that makes up the Moon's and Earth's outermost layers (p. 208)

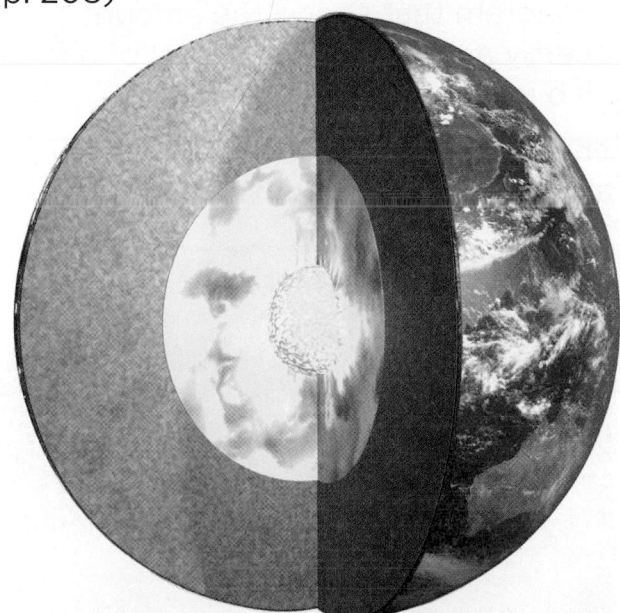

current (kûr′ənt) the directed flow of a gas or liquid (p. 348)

current electricity (kûr′ənt i·lek′tri′sə·tē) the flow of electrical charges (p. 566)

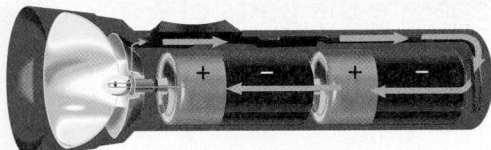

deciduous forest (di·si′jü·əs fôr′əst) a biome with many kinds of trees that lose their leaves each autumn (p. 140)

decomposer (dē′kəm·pō′zər) an organism that breaks down wastes and the remains of other organisms into simpler substances (p. 151)

density (den′sə·tē) the amount of matter in a given space; in scientific terms, density is the amount of mass in a unit of volume (p. 424)

deposition (de′pə·zi′shən) the dropping off of eroded soil and bits of rock (p. 229)

desert (de′zərt) a biome with very little rainfall (p. 142)

digestive system (dī·jes′tiv sis′təm) the organ system that breaks down food for fuel (p. 104)

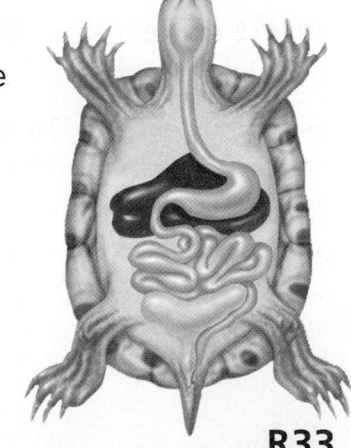

discharge (dis′chärj) the sudden movement of electric charges from one object to another object (p. 565)

distillation (dis′tələ′shən) the use of evaporation and condensation to separate the parts of a mixture (p. 462)

earthquake (ûrth′kwāk′) a sudden shaking of the rock that makes up Earth's crust (p. 216)

echinoderm (i·kī′nədûrm′) a spiny-skinned invertebrate (p. 81)

echo (e′kō) a repetition of a specific sound produced by reflection of sound waves from a surface (p. 542)

ecosystem (ē′kō·sis′təm) the living and nonliving things in an environment and all of their interactions (p. 131)

effort force (e′fərt fôrs) the force applied to move a load (p. 515)

electromagnet (i·lek′trō·mag′nət) a magnet formed when electric current flows through wire wrapped in coils around an iron bar (p. 581)

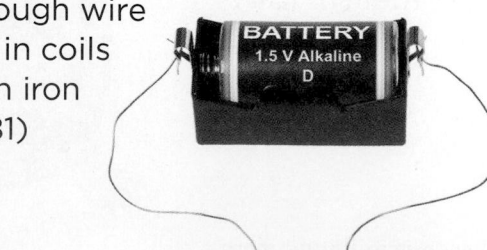

electromagnetic spectrum (i·lek′trō·mag·ne′tik spek′trəm) a range of all light waves of varying wavelengths, including the visible spectrum (p. 552)

element (e′ləmənt) a substance that is made up of only one type of matter (p. 432)

endangered (in·dān′jərd) close to becoming extinct; having very few of its kind left (p. 189)

endoskeleton (en′dō·ske′lətən) an internal supporting structure (p. 81)

energy (e′nər·jē) the ability to do work (p. 504)

energy pyramid (e′nər·jē pir′əmid′) a diagram that shows the amount of energy available at each level of a food web in an ecosystem (p. 156)

environment (in·vī′rən·mənt) all the living and nonliving things in an area (p. 296)

erosion (i·rō′zhən) the weathering and removal of rock or soil (p. 228)

evaporation (i·va′pərā′shən) when a liquid changes slowly into a gas (pp. 324, 449)

excretory system (ek′skrətôr′ē sis′təm) the organ system that removes wastes from the body (p. 103)

exoskeleton (ek′sō·ske′lətən) a hard covering that protects the bodies of some invertebrates (p. 82)

experiment (ik·sper′əment) a test designed to support or disprove a hypothesis; to perform such a test (p. 6)

extinct (ik·stingkt′) when the last of a species dies (p. 189)

fault (fôlt) a crack in Earth's crust along which movement has taken place (p. 215)

fertilization (fur′tələzā′shən) the joining of a female sex cell, the egg, and a male sex cell, the sperm, to produce a fertilized egg (p. 63)

filter (fil′tər) a tool that physically

separates matter by size; has mesh or a screen that retains the bigger pieces but allows smaller pieces to fall through the holes (p. 461)

filtration (fil·trā′shən) a method of separating the parts of a mixture using a filter (p. 461)

flood (flud) a great flow of water over land that is usually dry (p. 238)

fold (fōld) a bend in layers of rock (p. 215)

food chain (füd chān) the path that energy takes from one organism to another in the form of food (p. 152)

food web (füd web) the food chains that overlap in an ecosystem (p. 154)

force (fôrs) a push or a pull (p. 486)

forecast (fôr'kast') a prediction about the weather; the act of making such a prediction (p. 339)

fossil (fo'səl) any evidence of an organism that lived in the past (p. 274)

fossil fuel (fo'səl fū'əl) a source of energy made from the remains of ancient, once-living things (p. 278)

freeze (frēz) to change state from a liquid to a solid (p. 325)

frequency (frē'kwən·sē) the number of wavelengths that pass a reference point in a given amount of time (p. 544)

friction (frik'shən) a force between surfaces that slows objects or stops them from moving (p. 487)

front (frunt) a boundary between air masses with different temperatures (p. 337)

G

gas (gas) a state of matter that has no definite shape or volume (p. 415)

generator (je'nərā'tər) a device that produces alternating current by spinning an electric coil between the poles of a magnet; changes motion into electrical energy (p. 582)

turbine
shaft

N   S

Current flows in the opposite direction.

germination (jər'mənā'shən) when something begins to grow, as when a seed sprouts into a new plant (p. 64)

grassland (gras'land') a biome where grasses are the main plant life (p. 140)

gravity (gra'vətē) a force of attraction, or pull, between all objects (pp. 381, 426, 488)

groundwater (ground'wä'tər) water that flows in the cracks and spaces of underground rock (p. 287)

H

habitat (ha'bətat') the home of an organism (p. 131)

heat (hēt) the movement of thermal energy from a warmer object to a cooler object (p. 530)

heredity (hə·re'də·tē) the passing of traits from parent to offspring (p. 114)

hibernate (hī'bər·nāt) to rest or sleep through the cold winter (p. 168)

horizon (hə·rī'zən) a layer of soil that is different from the layers above it and below it (p. 265)

humidity (hū·mi'də·tē) a measurement of how much water vapor is in the air (p. 316)

humus (hū'məs) decayed plant and animal matter in soil (p. 264)

hurricane (hûr'ə·kān') a very large, swirling storm with strong winds and heavy rains (p. 240)

hypothesis (hī·po'thə·sis) a statement that can be tested using the scientific method (p. 5)

igneous rock (ig'nē·əs rok) a rock that forms when melted rock cools and hardens (p. 254)

imprint (im'print) a fossil made by an impression (p. 275)

inclined plane (in·klīnd' plān) a flat, slanted surface that is used as a simple machine (p. 518)

inertia (i·nûr'shə) the tendency of an object to remain in motion or to stay at rest unless acted upon by an outside force (p. 486)

inherited behavior (in·her'ət·əd bi·hā'vyûr) a set of actions that a living thing is born with and does not need to learn (p. 116)

inner core (i'nər kôr) a sphere of solid material at the center of Earth (p. 208)

instinct (in'stingkt') an inherited behavior that is automatic (p. 116)

insulator (in'səlā'tər) a material that slows or stops the flow of energy, such as thermal energy, electricity, and sound (p. 533)

invertebrate (in·vûr'təbrāt) an animal without a backbone (p. 79)

irrigation (ir'əgā'shən) the bringing of water to soil through pipes or ditches (p. 290)

 K

kinetic energy (kī·ne'tik e'nər·jē) the energy an object has because it is moving (p. 505)

kingdom (king'dəm) the largest group into which an organism can be classified (p. 35)

 L

landslide (land'slīd') a sudden movement of rock and soil down a slope (p. 242)

learned behavior (lûrnd bi·hā'vyûr) an action or set of actions that an animal changes through experience (p. 116)

length (lengkth) the number of units that fit along one edge of an object (p. 423)

lever (le'vər) a simple machine made of a bar and a fixed point, called a fulcrum, that allows the bar to pivot (p. 514)

lichen (lī'kən) algae and fungi that grow together; the algae provide food, while the fungi provide a place to live (p. 199)

life cycle (līf sī'kəl) the stages of growth and change that a particular kind of organism goes through (pp. 65, 110)

life span (līf span) how long an organism can be expected to live (p. 111)

liquid (li'kwəd) a state of matter that has a definite volume but no definite shape (p. 414)

load (lōd) the object being lifted or moved by a machine (p. 515)

lunar eclipse (lü'nər i·klips') a blocking of the Moon's light when the Moon passes into Earth's shadow; happens when Earth is directly between the Sun and the Moon (p. 374)

magnetic field (mag·ne'tik fēld) the region around a magnet where its force attracts or repels (p. 578)

mammal (ma'məl) a warm-blooded vertebrate with hair or fur; female mammals produce milk to feed their young (p. 94)

mantle (man'təl) the layer of rock below Earth's crust (p. 208)

mass (mas) the amount of matter making up an object (p. 412)

matter (ma'tər) anything that has mass and takes up space (p. 412)

melt (melt) to change state from a solid to a liquid (p. 330)

metal (me'təl) any of a group of elements that conducts heat and electricity, has a shiny luster, and can be hammered into a sheet (p. 433)

metamorphic rock (me'təmôr'fik rok) rock formed from another kind of rock under heat and pressure (p. 256)

metamorphosis (me'təmôr'fəsis) a series of separate body forms during an animal's development (p. 112)

meteor (mē'tē·ôr) a piece of rock, ice, or metal that burns up in Earth's atmosphere, causing a streak of light to appear in the sky (p. 388)

meteorite (mē′tē·ərīt′) a meteor that hits Earth's surface (p. 388)

metric system (me′trik sis′təm) a system of measurement based on units of ten; used in most countries and in all scientific work (p. 422)

mimicry (mi′mi·krē) an adaptation in which one kind of organism has similar traits to another (p. 169)

mineral (mi′nərəl) a natural, nonliving, solid material found in rock (p. 252)

mixture (miks′chər) two or more types of matter that are blended together and keep their own chemical properties (p. 458)

mold (mōld) an empty space in rock that once held the remains of a living thing (p. 275)

mollusk (mo′ləsk) an invertebrate that has a soft body and a shell (p. 81)

moraine (mərān′) a large body of weathered rock deposited at the edge of a glacier (p. 231)

motor (mō′tər) a device that changes electricity or other forms of energy into mechanical energy (p. 581)

mountain (moun′tən) a tall landform that rises to a peak (p. 215)

muscular system (mus′kyələr sis′təm) the organ system made up of muscles that help the body move (p. 100)

nervous system (nûr′vəs sis′təm) the set of organs that uses information from the senses to control all body systems (p. 101)

newton (nü′tən) a metric unit for weight measuring an amount of force (p. 495)

nonrenewable resource (non′ri·nü′əbəl rē′sôrs′) a natural material or source of energy that is useful to people and cannot be replaced easily (p. 278)

opaque (ō·pāk′) completely blocking light from passing through (p. 558)

orbit (ôr′bət) the path an object takes as it travels around another object (p. 362)

organ (ôr′gən) a group of tissues that work together to do a certain job (p. 27)

organism (ôr′gəni·zəm) a living thing that carries out basic life functions on its own (p. 22)

organ system (ôr′gən sis′təm) a group of organs that work together to carry out a life function (p. 27)

outer core (ou′tər kôr) the liquid layer below Earth's mantle (p. 208)

output force (out′put fôrs) the force applied to a load by a simple machine (p. 515)

ovary (ō′vərē) a structure containing egg cells (p. 62)

oxygen (äk′si·jən) a gas found in air and water that plants and animals need to live (p. 22)

parallel circuit (pâr′ələl′ sûr′kət) a circuit in which electric current flows through more than one path (p. 569)

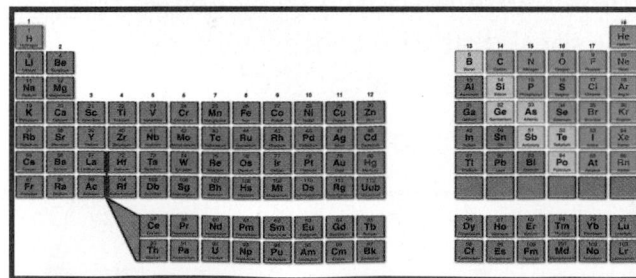

periodic table (pîr′ē·ä′dik tā′bəl) a chart that classifies all the known elements by their properties (p. 434)

permeability (pûr′mē·əbi′lətē) a measure of how fast water can pass through a porous material (p. 266)

phase (fāz) a temporary state of being, often used to describe a change in the appearance of the Moon (p. 373)

photosynthesis (fō′tō·sin′thəsis) the process in green plants and certain other organisms that uses energy from sunlight to make food from water and carbon dioxide (p. 50)

physical change (fi′zi·kəl chānj) a change that begins and ends with the same type of matter (p. 446)

pitch (pich) the highness or lowness of a sound as determined by its frequency (p. 545)

planet (pla′nət) a large sphere in space that orbits a star (p. 380)

plateau (pla·tō′) a high landform with a flat top (p. 215)

pole (pōl) one of two ends of a magnet, where the magnetic force is strongest (p. 577)

pollination (po′lənā′shən) the transfer of a flower's pollen from anther to pistil (p. 63)

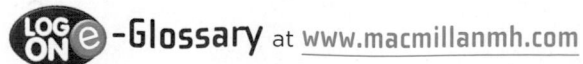

LOG ON **e-Glossary** at www.macmillanmh.com

pollution (pəlü'shən) harmful or unwanted material that has been added to the environment (p. 296)

population (po'pyələ'shən) all the members of a single type of organism in an ecosystem (p. 132)

pore space (pôr spās) the space between particles of soil (p. 266)

porous (pôr'əs) having pore spaces through which air and water can pass (p. 266)

potential energy (pəten'shəl e'nər·jē) the energy that is stored inside an object (p. 504)

precipitation (pri·si'pətā'shən) water in the atmosphere that falls to Earth as rain, snow, hail, or sleet (p. 325)

prism (pri'zəm) a thick piece of glass that bends light, separating white light into bands of colored light (p. 552)

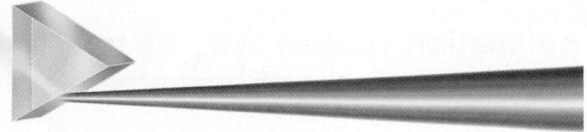

producer (prədü'sər) an organism, such as a plant, that makes its own food (p. 150)

property (pro'pər·tē) a characteristic of matter that can be observed or measured (p. 412)

radiation (rā'dē·ā'shən) the transfer of energy through space (p. 533)

rain gauge (rān gāj) a device that measures how much precipitation has fallen (p. 318)

recycle (rē·sī'kəl) to make new objects or materials from old objects or materials (p. 300)

reduce (ri·düs') to use less of something (p. 300)

reflection (ri·flek'shən) the bouncing of light waves off a surface (p. 556)

refraction (ri·frak'shən) the bending of light as it passes from one transparent material into another (p. 554)

relative age (re'lətiv āj) the age of one thing as compared to another (p. 255)

renewable resource (ri·nü'əbəl rē'sôrs') a useful material that is replaced quickly in nature (p. 280)

repel (ri·pel') to push away (p. 576)

reproduction (rē·prəduk'shən') the making of offspring (p. 62)

reptile (rep'tīl) a cold-blooded vertebrate that has scaly, waterproof skin, breathes air with lungs, and lays eggs (p. 92)

reservoir (re'zəvwär') a storage area for holding and managing freshwater (p. 288)

resource (rē'sôrs) a material or object that has useful properties (p. 258)

respiration (res'pərā'shən) the using and releasing of energy in a cell (p. 51)

respiratory system (res'pɑ·ətôr'ē sis'təm) the organ system that brings oxygen to body cells and removes waste gas (p. 102)

reuse (rē·ūz') to use something again (p. 300)

revolution (re'vəlü'shən) one complete trip around an object in a circular or nearly circular path (p. 362)

rock cycle (rok sī'kəl) the process in which rocks continuously change from one type to another (p. 256)

root (rüt) the part of a plant that takes in water and minerals (p. 49)

root hair (rüt hâr) one of the threadlike cells on a root that takes in water and minerals (p. 49)

rotation (rō·tā'shən) the act of spinning around an axis (p. 360)

runoff (run'ôf) water that flows over the surface of the land but does not evaporate or soak into the ground (p. 288)

rust (rust) a solid brown compound formed when iron combines chemically with oxygen (p. 450)

S

scientific method (sī'ən·ti'fik me'thəd) an organized process that scientists use to answer questions (p. 4)

sedimentary rock (se'dəmen'tərē rok) a rock that forms when small bits of matter are pressed together in layers (p. 255)

seed (sēd) an undeveloped plant with stored food sealed in a protective covering (p. 60)

seismic wave (sīz'mik wāv) a vibration caused by an earthquake (p. 218)

seismograph (sīz'məgraf') an instrument that detects and records earthquakes; shows seismic waves as jagged lines along a graph (p. 218)

series circuit (sîr'ēz sûr'kət) a circuit in which electric current flows in the same direction along a single path (p. 568)

simple machine (sim'pəl məshēn') anything that has few parts and makes it easier to do a task (p. 514)

skeletal system (ske'lətəl sis'təm) the organ system made up of bones that support the body (p. 100)

soil profile (soil prō'fīl) a view of the different horizons in a soil, from the surface down to the bedrock (p. 265)

soil water (soil wä'tər) water that soaks into and collects in soil (p. 287)

solar eclipse (sō'lər i·klips') a blocking of the Sun's light that happens when Earth passes through the Moon's shadow; at that time the Moon is between Earth and the Sun (p. 374)

solar system (sō'lər sis'təm) the Sun and all the objects that orbit around it (p. 380)

solid (so'ləd) a state of matter that has a definite shape and volume (p. 414)

solution (səlü'shən) a mixture in which one or more kinds of matter are mixed evenly in another kind of matter (p. 458)

sound wave (sound wāv) a wave that transfers energy through matter and spreads outward in all directions from a vibration (p. 541)

speed (spēd) the distance an object moves in an amount of time (p. 485)

sponge (spunj) a simple invertebrate that has a hollow body with a single opening and lives in water (p. 80)

R44
GLOSSARY

spore (spôr) a cell in a seedless plant that can grow into a new plant (p. 52)

star (stär) a hot sphere of gases in space that makes its own light (p. 394)

static electricity (sta′tik i·lek′tri′sətē) the buildup of electrical charge on an object or material (p. 565)

stationary front (stā′shəner′ē frunt) a boundary between air masses that are not moving (p. 337)

stem (stem) the part of a plant that holds the plant up and carries food, water, and other materials to and from the roots and leaves (p. 49)

stimulus (stim′yələs) *sing.*
stimuli (stim′yəlī) *pl. n.*, something in the environment that causes a living thing to respond (p. 176)

stomata (stō′mətə) *pl. n.*, **stoma** (stō′mə) *sing.* pores in the bottom of leaves that open and close to let in air or give off water vapor (p. 51)

subsoil (sub′soil′) a hard layer of clay and minerals beneath topsoil (p. 265)

taiga (tī′gə) a cool forest biome of conifers found in the upper northern hemisphere (p. 142)

tarnish (tär′nish) discoloration of metal by exposure to air (p. 450)

telescope (te′ləskōp′) a device that collects light and makes distant objects appear closer and larger (p. 382)

temperature (tem′pərəchur) a measurement of how hot or cold something is (p. 316)

terminus (tûr′mənəs) the downhill end of a glacier where glacial till and other debris are deposited (p. 231)

thermometer (thûr′mä′mətər) a tool used to measure temperature (p. 318)

tissue (ti′shü) a group of similar cells that work together to carry out a job (p. 27)

topsoil (top′soil′) the dark top layer of soil, rich in humus and minerals, in which many organisms live and grow (p. 265)

tornado (tôr·nā′dō) a swirling column of wind that moves across the ground in a narrow path (p. 240)

trait (trāt) a characteristic of a living thing (p. 34)

translucent (trans·lü′sənt) letting only some light through so objects on the other side appear blurry (p. 558)

transparent (trans·pâr′ənt) letting all the light through so objects on the other side can be seen clearly (p. 558)

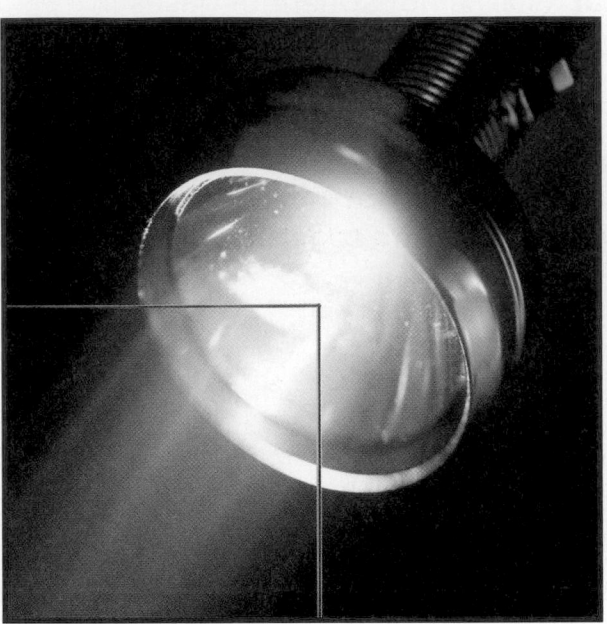

transpiration (trans′pərā′shən) the release of water vapor through the stomata of a plant (p. 51)

tropical rain forest (tro′pi·kəl rān fôr′əst) a hot, humid forest biome near the equator with much rainfall and a wide variety of organisms (p. 141)

tropism (trō′pi′zəm) the reaction of a plant to a stimulus (p. 177)

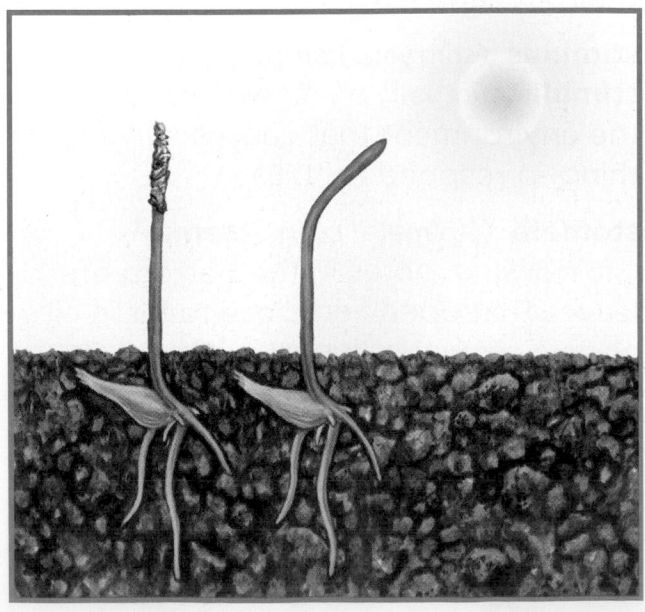

tundra (tun′drə) a large, treeless biome where the ground is frozen all year (p. 143)

unbalanced forces (un·ba′lənst fôrs′əz) forces that do not cancel each other out when acting together on a single object (p. 495)

variable (ver′ē·ə·bəl) something that can be changed or controlled in an experiment (p. 5)

velocity (vəlo′sətē) the speed and direction of a moving object (p. 485)

vertebrate (vûr′təbrāt′) an animal with a backbone (p. 90)

vibration (vī·brā′shən) a back-and-forth motion (p. 540)

volcano (vol·kā′nō) a mountain that builds up around an opening in Earth's crust (p. 220)

volume (vol′yūm) a measure of how much space matter takes up (p. 413); the loudness or softness of a sound (p. 545)

warm-blooded (wôrm′blə′dəd) a kind of animal whose body temperature does not change much (p. 90)

warm front (wôrm frunt) a boundary between air masses that allows a warm air mass to slide up and over a cold air mass (p. 337)

water cycle (wä′tər sī′kəl) the constant movement of water between Earth's surface and the atmosphere (p. 326)

watershed (wä′tər·shed′) an area of land where water flows downhill to a common stream, lake, or river (p. 287)

water vapor (wä′tər vā′pər) water in the form of a gas (p. 324)

wavelength (wāv′lengkth) the distance from the top of one wave to the top of the next (p. 544)

weathering (we′thəring) the breaking down of rocks into smaller pieces (p. 226)

weight (wāt) the measure of the pull of gravity between an object and Earth (p. 426)

well (wel) a hole drilled or dug below the ground to reach groundwater (p. 288)

wind vane (wind vān) a device that moves to show which way the wind is blowing (p. 318)

work (wûrk) the use of force to move an object a certain distance (p. 504)

Index

Note: Page references followed by an asterisk indicate activities.

Credits

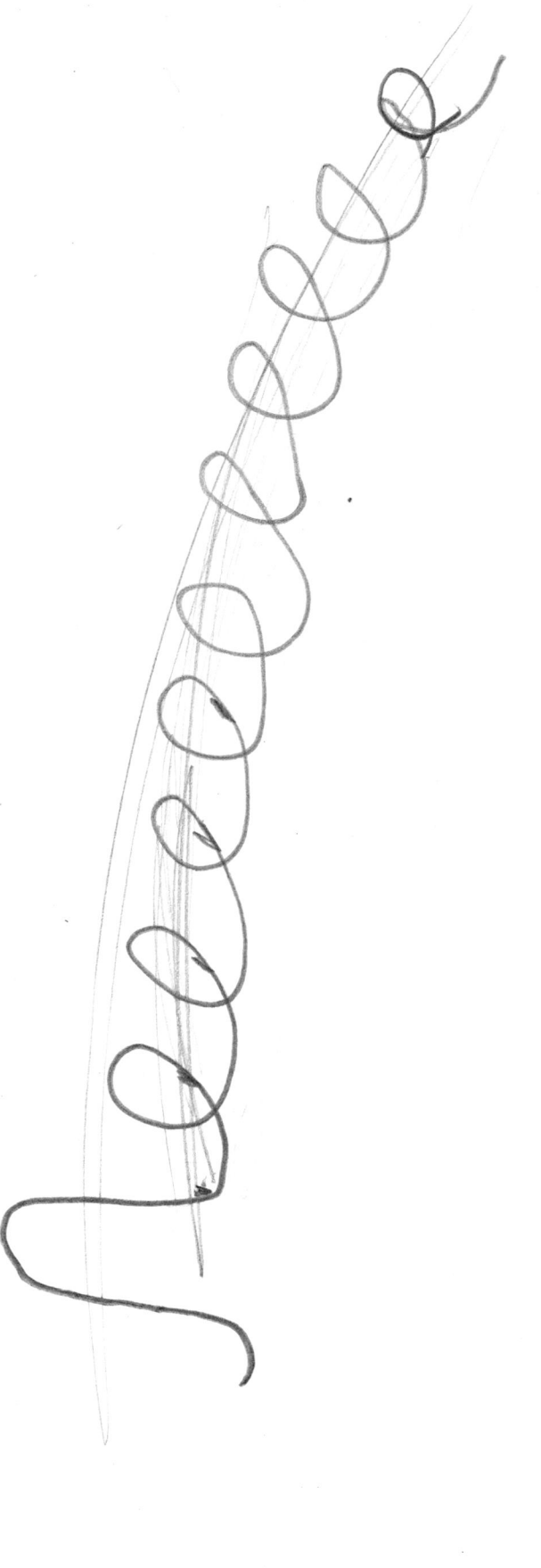